Logic in Games

Johan van Benthem

The MIT Press
Cambridge, Massachusetts
London, England

MIT Press books may be purchased at special quantity discounts for business or sales promotional use. For information, please email special_sales@mitpress.mit.edu.

This book was set in the LaTeX programming language by Fernando Velázquez Quesada and the author. Printed and bound in the United States of America.

Library of Congress Cataloging-in-Publication Data

Benthem, Johan van, 1949–
Logic in games / Johan van Benthem.
p. cm.
Includes bibliographical references and index.
ISBN 978-0-262-01990-3 (hardcover : alk. paper)
1. Game theory. 2. Logic. I. Title.
QA269.B36 2014
519.3—dc23

2013015059

10 9 8 7 6 5 4 3 2 1

Contents

Preface

This book is about encounters between logic and games. My interest in this interface started in my student days when I read the classic *Games and Decisions* (Luce & Raiffa 1957). It was reinforced by teaching my first course on philosophical logic in 1975, where the exciting, and exasperating, intricacies of Lorenzen dialogue games were a key theme. Later on, Hintikka's evaluation games entered my radar as a natural companion. Even so, my first systematic search was only in 1988, when I wrote a literature survey on all contacts I could find between logic and games for the Fraunhofer Foundation, commissioned for the then large amount of 2,000 German marks. I found an amazing number of interesting lines, even though there was nothing like a systematic field. Then in the 1990s, the TARK conferences on reasoning about knowledge brought real game theorists into my world who were transforming the area of epistemic logic. One person who convinced me that there was much to learn for logicians here was my Stanford colleague Yoav Shoham. As a result, my Spinoza project "Logic in Action" (1996–2001) had games as a main line, and we organized a number of meetings with game theorists and computer scientists working on the foundations of interaction. This theme has persisted at the ILLC in Amsterdam, with highlights such as the lively Marie Curie Center "Gloriclass" (2006–2010), that produced some 12 dissertations related to games in mathematics, computer science, linguistics, and the social sciences.

The origins for this book are lecture notes for the course *Logic in Games* taught in the years 1999–2002 in Amsterdam, Stanford, and elsewhere, for students coming from philosophy, mathematics, computer science, and economics. It was my way of exploring the area, with bits and pieces of established theory, and a lot of suggestions and hunches, many of them since taken up by students in papers and dissertations. Now, 10 years later, I still do not have a stable view of the subject: things keep shifting. But there is enough of substance to put up for public scrutiny.

This book has two entangled strands, connected by many bridges. First, it fits in my program of Logical Dynamics, putting information-driven agency at a center place in logic. Thus, it forms a natural sequel to the earlier books *Exploring Logical Dynamics* (van Benthem 1996) and *Logical Dynamics of Information and Interaction* (van Benthem 2011e). While this earlier work emphasized process structure and social informational events, this book adds the theme of multi-agent strategic interaction. This logical dynamics perspective is particularly clear with the first main strand in this book, the notion of a Theory of Play emerging from the combination of logic and game theory. It occupies roughly half of the book, and is prominent in Parts I and especially II, while Part III (and to some extent Part V) provide natural continuations to more global views of games.

The book also has a serious second strand that is not about the logical dynamics eschatology. The "in" of the title *Logic in Games* is meant to be ambiguous between two directions. The first is "logic *of* games," the use of logic to understand games, resulting in systems that are often called "game logics." But there is also a second direction of "logic *as* games," the use of games to understand basic notions of logic such as truth or proof. This is explained in Part IV on "logic games," of which there exists a wide variety, and it is also the spirit of various sorts of game semantics for logical systems. I find these two directions equally fundamental, and Part VI explores a number of ways in which they interact, even though the precise duality still escapes me. I believe that this interplay of the words "of" and "in" is not particular to logic and games, but that it is in fact a major feature of logical theory anywhere. Eventually, this may also throw new light on logical dynamics, as we will see in the Conclusion of this book.

Like some students and many colleagues, readers may find these perspective changes confusing. Therefore, the material has been arranged in a way that allows for several independent paths. The sequence of Parts I, II, III, and V forms an introduction to logics of games, addressing basic themes such as levels of game equivalence or strategic reasoning, with the Theory of Play as a major highlight, integrating game logics with the dynamic-epistemic logic of informational events. Part IV is a freestanding introduction to logic games, while Part V can be read as a natural continuation crossing over to general game logics. Part VI then extends this interface between our two main directions.

Read in whatever way, this book is meant to open up an area, not to close it. Its way of arranging the material brings to light quite a few open research problems, listed throughout, extending an earlier list in my survey paper *Open Problems in Logic and Games* (van Benthem 2005b).

While this book is not primarily technical, placing its main emphasis on exploring ideas, it is not a self-contained introduction to all the logics that will be linked with games. The reader is supposed to know the basics of logic, including modal logic and its computational interfaces. Many textbooks will serve this purpose. For instance, van Benthem (2010b) covers most of the basics that will be needed in what follows. Also, it will help if the reader has had prior exposure to game theory of the sort that can be achieved with many excellent available resources in that field.

What remains is the pleasant duty of mentioning some important names. As usual, I have learned a lot from supervising Ph.D. students working on dissertations in this area, in particular, Boudewijn de Bruin, Cédric Dégrémont, Amélie Gheerbrant, Nina Gierasimczuk, Lena Kurzen, Sieuwert van Otterloo, Marc Pauly, Merlijn Sevenster, and Jonathan Zvesper. I also thank my co-authors on several papers that went into the making of this book: Thomas Ågotnes, Cédric Dégremont, Hans van Ditmarsch, Amélie Gheerbrant, Sujata Ghosh, Fenrong Liu, Ştefan Minică, Sieuwert van Otterloo, Eric Pacuit, Olivier Roy, and Fernando Velázquez Quesada. And of course, many colleagues and students have been inspirational in several ways, of whom I would like to mention Krzysztof Apt, Sergei Artemov, Alexandru Baltag, Dietmar Berwanger, Giacomo Bonanno, Adam Brandenburger, Robin Clark, Jianying Cui, Paul Dekker, Nic Dimitri, Jan van Eijck, Peter van Emde Boas, Valentin Goranko, Erich Grädel, Davide Grossi, Paul Harrenstein, Jaakko Hintikka, Wilfrid Hodges, Wiebe van der Hoek, Guifei Jiang, Benedikt Löwe, Rohit Parikh, Ramaswamy Ramanujam, Robert van Rooij, Ariel Rubinstein, Tomasz Sadzik, Gabriel Sandu, Jeremy Seligman, Sonja Smets, Wolfgang Thomas, Paolo Turrini, Yde Venema, Rineke Verbrugge, Mike Wooldridge, and especially, Samson Abramsky. Of course, this is just a register of debts, not a list of endorsements. I also thank the readers who sent detailed comments on this text, the three anonymous reviewers for the MIT Press, and, especially, Giacomo Bonanno. In addition, Fernando Velázquez Quesada provided indispensable help with the physical production of this book. Finally, many of the acknowledgments that were stated in my preceding book *Logical Dynamics of Information and Interaction* (van Benthem 2011e) remain just as valid here, since there are no airtight seals between the compartments in my research. Thanks to all.

Johan van Benthem
Bloemendaal, The Netherlands
December 2012

Introduction
Exploring the Realm of Logic in Games

There are many valid points of entry to the interface zone between logic and games. This Introduction explains briefly why the interface is natural, and then takes the reader on a leisurely, somewhat rambling walk along different sites linking logic and games in a number of ways. In the course of this excursion, many general themes of this book will emerge that will be taken up more systematically later on. Readers who have no time for leisurely strolls can skip straight ahead to Chapter 1.

1 Encounters between logic and games

The appeal of games Games are a long-standing and ubiquitous practice, forming a characteristic ingredient of human culture (Huizinga 1938, Hesse 1943). Further, to the theorist of human interaction, games provide a rich model for cognitive processes, carrying vivid intuitions. The two perspectives merge naturally: a stream of ever new games offers a free cognitive laboratory where theory meets practice. Not surprisingly then, games occur in many disciplines: economics, philosophy, linguistics, computer science, cognitive science, and the social sciences. In this book, we will focus on connections between games and the field of logic. In this Introduction, we will show by means of a number of examples how this is a very natural contact. Many themes lightly touched upon here will then return in the more technical chapters of the book.

Logic of games and logic as games In what follows, we will encounter two aspects of the title of this book *Logic in Games*. Its connective 'in' is deliberately ambiguous. First, there is logic *of* games, the study of general game structure, which will lead us to contacts between logic, game theory, and also computer science and philosophy. This study employs standard techniques of the field: "game logics" capture essential aspects of reasoning about, or inside, games. But next, there is also logic *as* games, the study of logic by means of games, with "logic games" capturing basic reasoning activities and suggesting new ways of understanding what logic is. Thus, we have a cycle

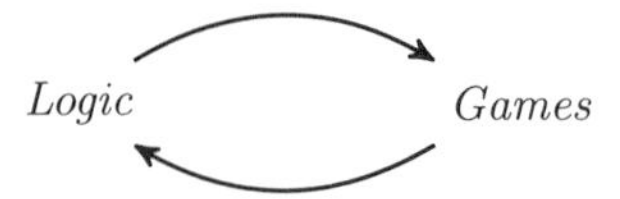

Moreover, cycles invite spinning round in a spiral, or a carousel, and one can look at game logics via associated logic games, or at logic games in terms of matching game logics. Some students find this dual view confusing, preferring one direction while ignoring the other. But in this book, both sides will be present, even though we are far from understanding fully how they are intertwined. Our main focus will be on logic of games throughout, but logic as games remains an essential complementary viewpoint.

This Introduction is an informal tour of this interface. First, we introduce some simple logic games, showing how these naturally give rise to more general questions about games. This brings us to the topic of defining games in general, their analogies with processes in computer science, and their analysis by means of standard process logics. Next, we consider game theory, an area with its own motivations and concerns. We discuss a few basic themes, note their logical import, and suggest some contours of an interface between logic and game theory. What typically emerges in this mix is an analysis of players, leading to what may be called a "Theory of Play" involving many standard topics from philosophical logic. Finally, we explain what this book contains.

2 Logical games

Argumentation games While the origins of logic in antiquity are not well understood, reflection on legal, political, or philosophical debate seems a key factor in its emergence in the Greek, Indian, and Chinese traditions. Consider that debate clearly resembles a game. A person may win an argument, upholding a claim against an opponent, but there is also the bitter taste of defeat. In this process, well-timed responses are crucial. This discourse aspect has persisted in the history of logic, although descriptive aspects have also shaped its course. After the mathematical turn of the field, exact models of dialogue emerged (Lorenzen 1955). As an illustration, consider the well-known inference

from premises $\neg A$, $A \vee B$ to conclusion B

In the descriptive view of logic, this inference spells out what the world is like given the data at our disposal. However, we can also view it in discourse mode, as a basic subroutine in an argumentation game. There is a proponent $\boldsymbol{P}$ defending the claim B against an opponent $\boldsymbol{O}$ who has already committed to the premises $\neg A$, $A \vee B$. The procedure lets each player speak in turn. We record some possible moves:

1 $\boldsymbol{O}$ starts by challenging $\boldsymbol{P}$ to produce a defense of B.
2 $\boldsymbol{P}$ now presses $\boldsymbol{O}$ on the commitment $A \vee B$, demanding a choice.
3 $\boldsymbol{O}$ must respond to this, having nothing else to say.

There are two options here that we list separately:

$3'$ $\boldsymbol{O}$ commits to A.
$4'$ $\boldsymbol{P}$ now points at $\boldsymbol{O}$'s commitment to $\neg A$,

and $\boldsymbol{P}$ wins because of $\boldsymbol{O}$'s self-contradiction.

$3''$ $\boldsymbol{O}$ commits to B.
$4''$ Now $\boldsymbol{P}$ uses this concession to make a defense to 1.

$\boldsymbol{O}$ has nothing further to say, and loses.

One crucial feature emerges right here. Player $\boldsymbol{P}$ has a *winning strategy*: whatever $\boldsymbol{O}$ does, $\boldsymbol{P}$ can counter to win the game. This reflects an idea of logical validity as safety in debate. An inference is valid if the proponent has a winning strategy for the conclusion against an opponent granting the premises. This pragmatic view is on a par with semantic validity as the preservation of truth or syntactic validity as derivability. Valid arguments are those that can always be won in debate, as long as the moves are chosen well.

Note to the game-theoretic reader The above scenario can easily be cast as an extensive game in the standard sense of game theory. Likewise, the notion of a strategy employed here is the standard one of a function that prescribes an action for a player at each turn, without constraining what other players do at their turns. Extensive games and strategies will return throughout this book, and many precise connections between logical and game-theoretic views will unfold as we proceed.

Consistency games Argumentative dialogue is one way in which logic involves games for different actors. The other side of the coin is maintaining consistency. Player $\boldsymbol{O}$ has a positive purpose, claiming that the set $\{\neg A, A \vee B, \neg B\}$ is consistent. Indeed, maintaining consistency is an important feature of ordinary communication. Medieval logic had an "Obligatio Game" testing debating skills by requiring a student to maintain consistency while responding to challenges issued by a teacher:

> A number of rounds n is chosen, the severity of the exam. The teacher gives abstract assertions $P_1, \ldots, P_n$ that the student has to accept or reject as they are put forward. In the former case, P_i is added to the student's current commitments, and otherwise, the negation $\neg P_i$ is added. The student passes by maintaining consistency throughout.

This presentation is from Hamblin (1970). For more historical detail and accuracy, see Dutilh-Novaes (2007), Uckelman (2009). In principle, the student always has a winning strategy, by choosing some model $\boldsymbol{M}$ for the complete language of the teacher's assertions and then committing according to whether a statement is true or false in $\boldsymbol{M}$. But the realities of the game are of course much richer than this simple procedure.

Evaluation games Another famous logic game arises with understanding assertions. Consider two people discussing a quantifier statement $\forall x \exists y \varphi(x, y)$ about numbers. One player, $\boldsymbol{A}$, chooses a number x, and the other, $\boldsymbol{E}$, must come up with some number y making $\varphi(x, y)$ true. Intuitively, $\boldsymbol{A}$ challenges the initial assertion, while $\boldsymbol{E}$ defends it. To make this more concrete, consider a simple model with two objects s and t, and a relation $R = \{\langle s, t\rangle, \langle t, s\rangle\}$ (a so-called 2-cycle). An evaluation game for the assertion $\forall x \exists y Rxy$ can be pictured as a tree whose leaves are wins for the player who is right about the atomic statement reached there:

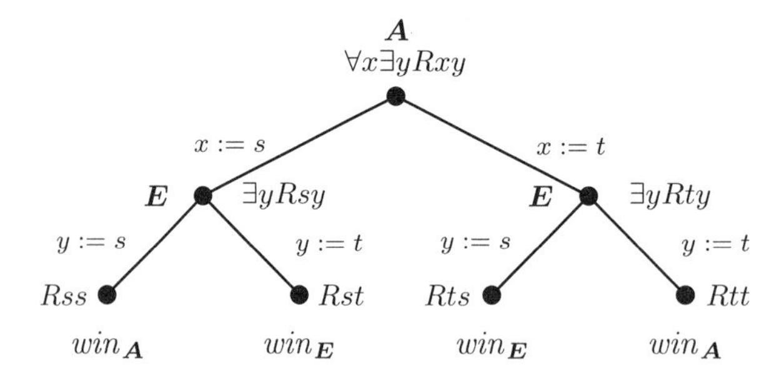

Game theorists will recognize a simple extensive form game here with four histories.

The obvious fact that $\forall x \exists y Rxy$ is true in the 2-cycle model is again reflected in a game-theoretical feature. Player $\boldsymbol{E}$ has a winning strategy in this evaluation game, which may be stated as: "choose the object different from that mentioned by $\boldsymbol{A}$." Thus, as in the preceding examples, a logical notion (this time, truth) corresponds to the existence of a winning strategy in a suitable game. This fact can be made precise by providing a general definition of evaluation games $\boldsymbol{game}(\varphi, \boldsymbol{M}, s)$ for arbitrary first-order formulas φ, models $\boldsymbol{M}$, and variable assignments s (cf. Chapter 14). Here a player called verifier $\boldsymbol{V}$ claims that the formula is true, while a falsifier $\boldsymbol{F}$ claims that it is false.

FACT The following two assertions are equivalent for all models, assignments, and formulas: (a) $\boldsymbol{M}, s \models \varphi$, (b) $\boldsymbol{V}$ has a winning strategy in $\boldsymbol{game}(\varphi, \boldsymbol{M}, s)$.

Logic games By now there are logic games for a wide variety of tasks (Hodges 2001). Much of modern logic can be usefully cast in the form of games for model checking (Hintikka 1973), argumentation and dialogue (Lorenzen 1955), comparing models for similarity (Fraïssé 1954, Ehrenfeucht 1961), or constructing models for given formulas (Hodges 2006, Hirsch & Hodkinson 2002). We will introduce the major varieties in Part IV of this book, suggesting that any significant logical task can be "gamified" once we find a natural way of pulling apart roles and purposes. Whatever the technical benefits of this shift in perspective, it is an intriguing step in reconceptualizing logic, away from lonesome thinkers and provers, to interactive tasks for several actors.

3 From logic games to general game structure

Now we take a step toward the other direction of this book. As a subspecies of games, logic games are very specialized activities. Nevertheless, they involve various broader game-theoretical issues. Nice examples can be found with the above evaluation games. We now present short previews of three fundamental issues, determinacy, game equivalence, and game operations.

Determinacy The above evaluation games have a simple but striking feature: either verifier or falsifier has a winning strategy. The reason is the logical law of the excluded middle. In any semantic model, either the given formula φ is true or its negation is. Thus, either $\boldsymbol{V}$ has a winning strategy in the game for φ, or $\boldsymbol{V}$ has a winning strategy in the game for the negation $\neg\varphi$, an operation that triggers a role switch between $\boldsymbol{V}$ and $\boldsymbol{F}$ in the game for φ. Equivalently, in the latter case, there is a winning strategy for $\boldsymbol{F}$ in the game for φ. Two-player games in which some player has a winning strategy are called *determined.*

The general game-theoretic background of our observation is a result in Zermelo (1913); see also Euwe (1929). We state this background here for two-person games whose players $\boldsymbol{A}$ and $\boldsymbol{E}$ can only win or lose, and where there is a fixed finite bound on the length of all runs.

THEOREM All zero-sum two-player games of fixed finite depth are determined.

Proof The proof is a style of solving games that will return at many places in this book. We provide a simple bottom-up algorithm determining the player having the winning strategy at any given node of a game tree of this finite sort. First, color those end nodes black that are wins for player $\boldsymbol{A}$, and color the other end nodes white, being the wins for $\boldsymbol{E}$. Then extend this coloring stepwise as follows. If all children of node n have been colored already, do one of the following:

(a) If player $\boldsymbol{A}$ is to move, and at least one child is black: color n black; if all children are white, color n white,

(b) If player $\boldsymbol{E}$ is to move, and at least one child is white: color n white; if all children are black, color n black.

This procedure colors all nodes black where player $\boldsymbol{A}$ has a winning strategy, while coloring those where $\boldsymbol{E}$ has a winning strategy white. The key to the adequacy of the coloring can be proved by induction: a player has a winning strategy at a turn iff this player can make a move to at least one daughter node where there is again a winning strategy. ■

This algorithm stands at a watershed of game theory and computer science. It points to the game solution method of *Backward Induction* that we will discuss at many places in this book. Used as a computational device, sophisticated modern versions have solved real games such as Checkers, as well as central tasks in Artificial Intelligence (Schaeffer & van den Herik 2002). Zermelo and Euwe were concerned with Chess, an old interest in computer science and cognitive science, which also allows draws. Here the theorem implies that one of the players has a non-losing strategy. Today, it is still unknown which one, as the game tree is so huge.

REMARK Infinite games
Infinite two-player games of winning and losing need not be determined. Never-ending infinite games are of independent interest, and they raise logical issues of their own that we will study in Chapters 5 and 20.

The main point of interest for this book is how close a basic game-theoretic fact can be to a standard logical law. In fact, one way of proving Zermelo's Theorem is merely by unpacking the two cases of the excluded middle for finite iterated quantified assertions "For every move of ***A***, there is a move for ***E*** (and so forth) such that ***E*** wins" (cf. Chapter 1).

Game equivalence Determinacy is important, but it is just a special property of some simple games. Logic also raises basic issues concerning arbitrary games.

EXAMPLE Propositional distribution
Consider the propositional law of distribution for conjunction over disjunction:

$$p \wedge (q \vee r) \leftrightarrow (p \wedge q) \vee (p \wedge r)$$

The two finite trees in the following figure correspond to evaluation games for the two propositional formulas involved, letting ***A*** stand for falsifier and ***E*** for verifier.

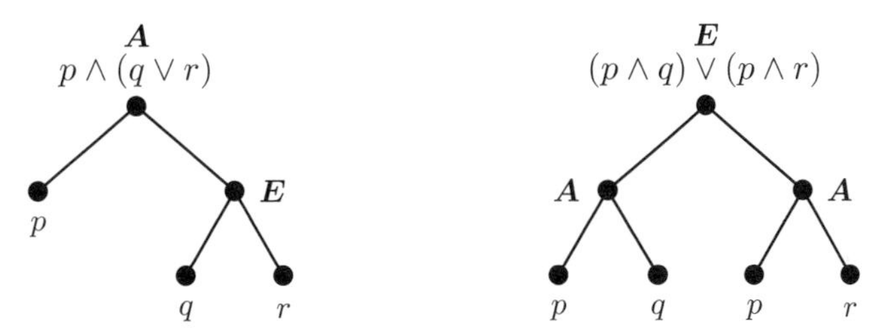

This picture raises the following intuitive question: "When are two games the same?" In particular, does the logical validity of distribution mean that the pictured games are the same in some natural sense? Game equivalence is a fundamental issue, which has been studied in game theory (Thompson 1952 is a famous early source), and it will be investigated in more detail in Chapter 1. For now, intuitively, if we focus on *turns and moves*, the two games are not equivalent: they differ in etiquette (who gets to play first) and in choice structure.

This is one natural level for looking at games, involving details of the fun of playing. But if one's focus is on *achievable outcomes* only, the verdict changes.

Both players have the same powers of achieving outcomes in both games: ***A*** can force the outcome to fall in the sets $\{p\}$, $\{q, r\}$, ***E*** can force the outcome to fall in the sets $\{p, q\}$, $\{p, r\}$. Here, a player's "powers" are those sets U of outcomes for which the player has a strategy making sure the game will end inside U, no matter what all the others do. On the left, ***A*** has two strategies, *left* and *right*, yielding powers $\{p\}$ and $\{q, r\}$, and ***E*** has strategies yielding powers $\{p, q\}$, $\{p, r\}$. On the right, player ***E*** has strategies *left* and *right* giving ***E*** the same powers as on the left. By contrast, player ***A*** now has four strategies:

left: L, *right*: L, *left*: L, right: R, *left*: R, *right*: L, *left*: R, *right*: R

The first and fourth give the same powers for ***A*** as on the left, while the second and third strategy produce merely weaker powers subsumed by the former. ■

We will see later what game equivalences make sense for what purpose. For now, we note that distribution is an attractive principle about safe scheduling shifts that leave players' powers intact. Thus, once more, familiar logical laws encode significant game-theoretic content.

Game operations It is not just logical laws that have game-theoretic content. The same holds for the logical constants that make predicate logic tick. Evaluation games give a new game-theoretical take on the basic logical operations:

(a) conjunction and disjunction are choices $G \wedge H$, $G \vee H$

(b) negation is role switch, also called dual $\neg G$, or G^d

Clearly, choice and switch are completely general operations forming new games out of old. Here is another such operation that operates inside evaluation games. Consider the earlier rule for an existentially quantified formula $\exists x \psi(x)$:

V picks an object d in ***M***, and play then continues with $\psi(d)$

Perhaps surprisingly, the existential quantifier $\exists x$ does not serve as a game operation here: it clearly denotes an atomic game of object picking by verifier. The general operation in this clause hides behind the phrase "continues," which signals

(c) sequential composition of games $G\,;H$

Still, these are just a few of the natural operations that form new games out of old. Here is one more. So far we have two forms of game conjunction. The Boolean $G \wedge H$ forces a choice at the start, and the game not chosen is not played at all. Sequential composition $G\,;H$ may lead to play of both games, but only if the first ever gets completed. Now consider two basic games, "family" and "career." Neither Boolean choice nor sequential composition seems the right conjunction, and most of us try to cope with the following operation:

(d) parallel composition of games $G\,||H$

We play a stretch in one game, then switch to the other, and so on.

We will study game operations much more systematically in Part V, including connections with several systems of logic.

4 Games as interactive processes

Toward real games We have now seen how games for logical tasks have a general structure that makes sense for all games. Let us now go all the way toward real games, in economic or social behavior, sports, or war. All of these involve rule-governed action by intelligent players. We switch perspectives here, using logic as a general tool for analyzing these games. In this broader realm, logic clarifies process structure, but also the mechanics of deliberation and action by players as the game proceeds. For a start, we consider the first strand on its own, using a perspective from computer science.

Extensive games as processes Games are an enriched form of computational process, having participants with possibly different goals. Thus, games have started replacing single machines as a realistic model of distributed computation, complex computer systems, and the Internet today. An *extensive game* is a tree consisting of possible histories of moves by players taking turns indicated at nodes, while outcomes are either marked by numerical utility values for all players, or ordered by qualitative preference relations for all players. Without the preference relations, one has an extensive game form. Chapter 1 has more formal definitions of these notions, and the further chapters in Parts I and II will add more as matters are discussed in greater depth.

Trees with admissible runs such as this are very familiar from a logical point of view. They occur as "labeled transition systems" in computer science, and also as standard models for modal or temporal logics, being of the general form

$$\boldsymbol{M} = (S, \{R_a\}_{a \in A}, V)$$

where S is a universe of states or worlds, the R_a are binary transition relations on S for atomic state-changing actions a in some given set A, while the valuation V marks, for each atomic property p in some given base vocabulary, at which states p holds. There is a large logical literature on these process graphs (cf. Blackburn et al. 2001, or Chapter 1), that may be viewed either as abstract machines, or when unraveled to tree structures, as spaces of all possible executions of a process.

Specialized to extensive games, the states are a domain of action stages related by transitions for available moves for players, and decorated with special predicates:

$$\boldsymbol{M} = (NODES, MOVES, \boldsymbol{turn}, \boldsymbol{end}, VAL)$$

Non-final nodes are turns for players with outgoing transitions for available moves. Special proposition letters $\boldsymbol{turn}_i$ mark turns for player i, and $\boldsymbol{end}$ marks final points. The valuation VAL may also interpret other predicates at nodes, such as utility values for players. In this book, we will mainly use extensive game trees, although much of what we say applies to process graphs in general.

Process equivalences Now the earlier topic of structural equivalence returns. A basic concern in computer science is the level at which one wants to model processes.

EXAMPLE Levels of equivalence
A well-known case of comparison in the computational literature involves the following two machines (or one-player games):

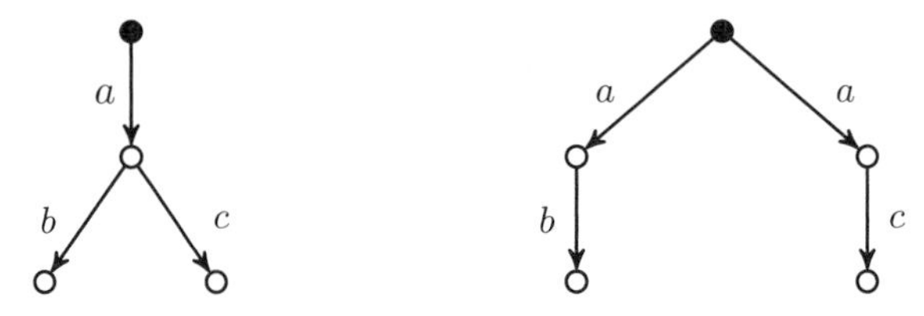

Do these diagrams represent the same process? Both produce the same finite sequences of observable actions $\{ab, ac\}$, although the first machine starts deterministically, and the other with an internal choice. In terms of external input-output behavior then, the machines are the same, given their "finite trace equivalence." But if one is also interested in internal control, in particular, available choices, then a better measure is a well-known finer structural comparison tracking choices, called "bisimulation." Indeed, the two machines have no bisimulation, as we will see in Chapter 1, which will present precise definitions for the relevant notions. ■

In the field of computation, there is a hierarchy of process equivalences, from coarser finite trace equivalence to finer ones such as bisimulation. No best level exists: it depends on the purpose. The same is true for games. Extensive games go well with bisimulation, but the earlier power level is natural, too, being an input-output view closer to trace equivalence.

Games and process logics The ladder of simulations also has a syntactic counterpart. The finer the process equivalence, the more expressive a matching *language* defining the relevant process properties. In particular, bisimulation is correlated with the use of *modal logic*, which will be one of the main working languages for games in this book.

EXAMPLE Games as modal process graphs
Consider a simple two-step game between two players $\boldsymbol{A}$ and $\boldsymbol{E}$, with end nodes indicated by numbers, and one distinguished proposition letter p:

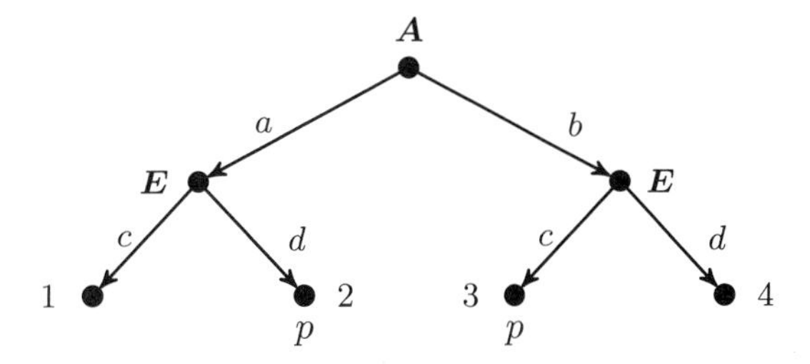

Clearly, $\boldsymbol{E}$ has a strategy making sure that a state is reached where p holds. This power is expressed by the following typical modal formula that is true at the root:

$$[a]\langle d\rangle p \wedge [b]\langle c\rangle p$$

The left-hand conjunct of this formula says that after every execution of a (marked by the universal modality $[\ ]$) there exists an execution of d (marked by the existential modality $\langle\ \rangle$) that results in a state satisfying p. The right-hand conjunct is similar. Since all actions are unique, the difference between every and some is slight here, but its intent becomes clearer in a related $[\ ]\langle\ \rangle$-type claim about the game with actions involving choice. The following response pattern will return at many places in our logical analysis of players' strategic powers:

$$[a \cup b]\langle c \cup d\rangle p$$

where $a \cup b$ stands for the choice program of executing either a or b at the agent's discretion, a notion from modal logics of actions (see Chapter 1 for details). ∎

This is just a start, and the structure of games supports many other logical languages, as we will see later in this book.

From algorithms to games The link between processes and games is not just theory. Computational tasks turn into games quite easily. Consider the key search problem of graph reachability: "Given two nodes s and t in a graph, is there a sequence of successive arrows leading from s to t?" There are fast *Ptime* algorithms finding such a path if one exists (Papadimitriou 1994). But what if there is a disturbance, a reality in travel?

EXAMPLE Sabotage games
The following network links two European centers of logic and computation:

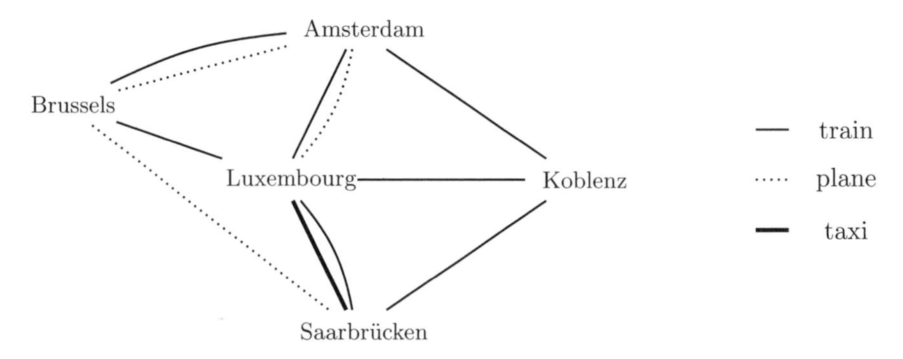

Let us focus on two nodes s and t, namely, Amsterdam and Saarbrücken. It is easy to plan trips either way. But what if the usual transportation system breaks down, and a malevolent demon starts canceling connections, anywhere in the network? Let us say that, at every stage of our trip, the demon first takes out one connection. Now we have a genuine two-player game, and the question is who can win where.

From Saarbrücken to Amsterdam, a German colleague has a winning strategy. The demon's opening move may block Brussels or Koblenz, but the player gets to Luxembourg in the first round, and to Amsterdam in the next. The demon may also cut a link between Amsterdam and a city in the middle, but the player can then go to at least one place with two intact roads. But from the Dutch side, the demon has the winning strategy. It first cuts a link between Saarbrücken and Luxembourg. If the Dutch player goes to any city in the middle, the demon has time in the next rounds to cut the last link to Saarbrücken. ■

One can gamify any algorithmic task into a "sabotage game" with obstructing players. In general, the solution complexity will go up, as we will see in Chapter 23. By now, sabotage games have been used for quite different tasks as well, such as scenarios for learning.

Interactive computation as games Sabotage games exemplify a more general phenomenon, related to our earlier dichotomy between logic of games and logic as games. In addition to providing notions and tools for analyzing games, modern computer science has also started using games themselves as models for interactive computation where systems react to each other and their environment. Some recent paradigms exemplifying this perspective will be discussed in Chapters 18 and 20.

Process logics and game theory Many process calculi coexist in computer science, including modal, dynamic, and temporal logics. These will return in Parts I and II of this book, in describing fine structure of general games. As for coarser views, we will study logics of strategic powers in Part III, and matching global game operations in Part V, presenting two relevant calculi, dynamic game logic and linear game logic.

It is important to note that fine or coarse are not cultural qualifications here. This diversity of perspectives reflects two legitimate uses of logical methods in any area, namely, providing different levels of *zoom*. Sometimes, logic is used to zoom in on a topic, providing finer details of formulation and reasoning that were left implicit before. But sometimes also, logical calculi provide a higher-level abstraction, zooming out from details of a given reasoning practice to make general patterns visible. This book will provide recurrent instances of both of these zoom functions.

5 Logic meets game theory

Real games are not just about actions and information. They also crucially involve players' evaluation of outcomes, encoded in utility values or qualitative preferences. It is the balance between information, action, and evaluation that drives rational behavior. We now explore how this affects the earlier style of thinking, using logic to analyze basic assumptions about how players align their information and evaluation.

Preference, Backward Induction, and rationality How can we find, not just any, but a best course of action in a game? Assuming players to be rational, how can theorists predict behavior, or make sense of play once observed? Game theorists are after equilibria that show a stability making deviation unprofitable, although off-equilibrium behavior can sometimes be important, too: see Schelling (1978). In finite extensive games, the basic procedure of Backward Induction extends the Zermelo coloring to find such equilibria.

EXAMPLE Predicting behavior in the presence of preferences
Consider an earlier game, with players' view of outcomes displayed in ordered pairs (***A***-value, ***E***-value):

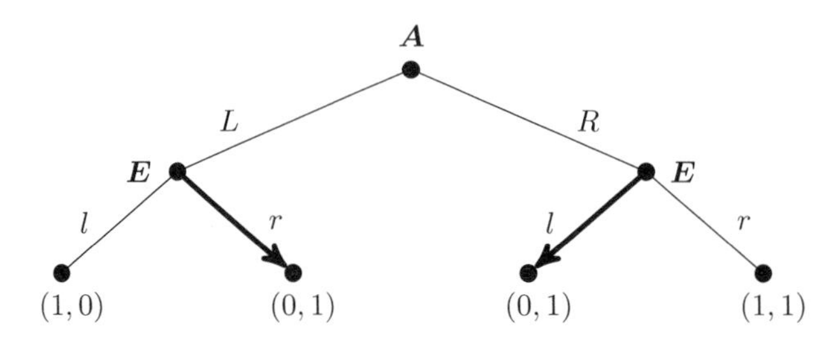

In the earlier game of just winning and losing, ***E*** had a winning strategy marked by the black arrows, and it did not matter what ***A*** did. But now suppose that ***A*** has a slight preference between the two sites for ***A***'s defeat, being the end nodes with values $(0, 1)$. The defeat to the left takes place on an undistinguished beach, but that to the right on a picturesque mountaintop, and bards might well sings ballads about ***A***'s last stand for centuries. The new utility values for the outcomes might then be as follows, with a tiny positive number ε for the mountaintop:

$$(1, 0) \qquad (0, 1) \qquad (\varepsilon, 1) \qquad (1, 0)$$

With these preferences, ***A*** goes right at the start, and then ***E*** goes left. ■

The algorithm computing this course of action finds values for each node in the game tree for each player, representing the best outcome value that the player can guarantee through best possible further play (as far as within the player's power).

DEFINITION Backward Induction algorithm
Here is a more precise description of the preceding numerical calculation:

> Suppose ***E*** is to move, and all values for daughter nodes are known. The ***E***-value is the maximum of all the ***E***-values on the daughters, and the ***A***-value is the minimum of the ***A***-values at all ***E***-best daughters. The dual case for ***A***'s turns is completely analogous.

Different assumptions yield modified algorithms. For instance, with a benevolent opponent, in case of ties, one might reasonably expect to get the maximal outcome from among the opponent's best moves. ■

Nash equilibrium The general game-theoretic notion here is this. We state it for two players, but it works for any number (cf. Osborne & Rubinstein 1994). Any two strategies σ and τ for two players *1* and *2*, respectively, determine a unique outcome $[\sigma, \tau]$ of the game, obtained by playing σ and τ against each other. This outcome can be evaluated by both players. Now we say that

> a pair of strategies σ, τ is a *Nash equilibrium* if, for no $\sigma' \neq \sigma$, $[\sigma', \tau] >_1 [\sigma, \tau]$, and similarly for player *2* with τ.

That is, neither player can improve its own outcome by changing strategies while the other sticks to the one given. Backward Induction yields strategies that are in equilibrium, even "subgame perfect equilibrium": best strategies at nodes remain best when restricted to lower nodes heading sublimes underneath.

Backward Induction is an attractive style of analysis that often makes sense. Even so, it also has instances that may give one pause.

EXAMPLE A debatable equilibrium
In the following game, the algorithm tells player ***E*** to turn left at the relevant turn, which then gives player ***A*** a belief that this will happen, and so, based on this belief about the other player, ***A*** should turn left at the start:

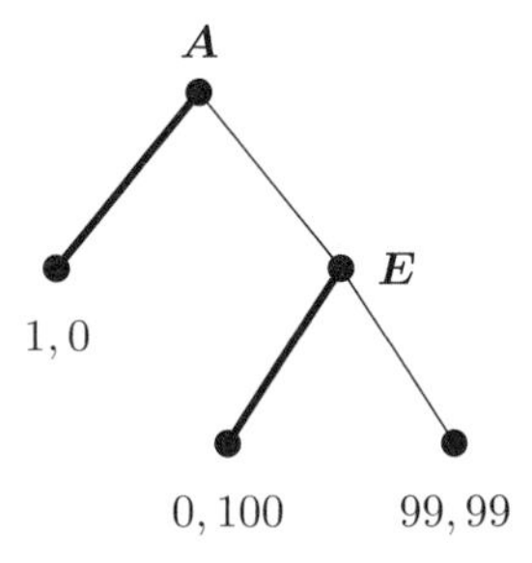

As a result, both players are worse off than in the outcome (99, 99). ■

This perhaps surprising outcome raises the question of why players should act this way, and whether a more cooperative behavior could be stable. Indeed, the example is reminiscent of game-theoretic discussions of non-self-enforcing equilibria (Aumann 1990) or of so-called assurance games (Skyrms 2004). Our aim here is not to either criticize or improve game theory. The example is intended as a reminder of the subtlety of even the simplest social scenarios and the choice points that we have in understanding them. In particular, the reasoning leading to the above outcome is a mixture of many long-standing interests of logicians, including action, preference, belief, and counterfactuals. We will return to it at several places in Part I of this book, probing the logical structure of rationality. In Part II, we will go further, and analyze backward induction as a dynamic deliberation procedure that transforms a given game by successive announcements of rationality of players that gradually create a pattern of expectations. Here is an illustration in our particular case.

EXAMPLE Building up expectations in stages
Think of expectations as ordering the histories of a game by relative plausibility. In the picture below, this is marked by symbols >. Order appears as we keep announcing that players are "rational-in-beliefs," never playing a dominated move whose outcomes they believe to be worse than those of some other available move:

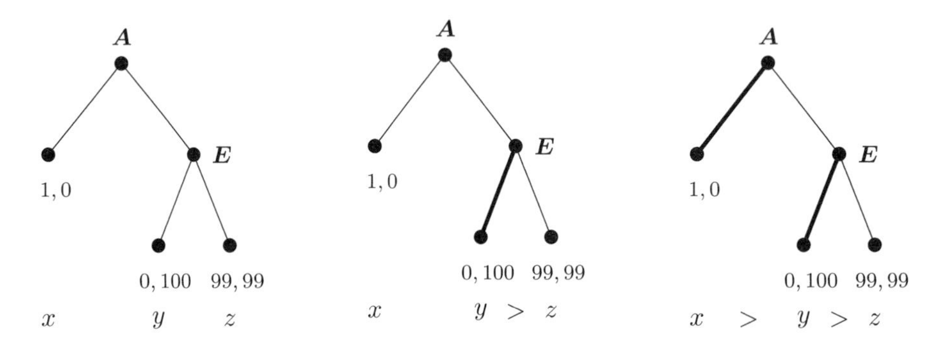

If all this is still too cryptic now, things will become fully clear in Chapter 8. ■

Imperfect information So far, we have looked at fully transparent games, where players' only uncertainty is about the future. But many social scenarios have further kinds of uncertainty.

EXAMPLE A card mini-game
Simple card games are an excellent setting for studying communication. Three cards red, white, and blue are given to three players: *1* gets red, *2* white, and *3* blue. All players see their own cards, but not the others. Now *2* asks *1* "Do you have the blue card?" and the truthful answer is forthcoming: "No." Who knows what?

If the question is genuine, player *1* will know the cards after it was asked. After the answer, player *2* knows, too, while *3* still does not. But there is also knowledge about others involved. At the end, all three players know that *1* and *2*, but not *3*, have learned the cards, and this fact is common knowledge between them. ■

This way of understanding the scenario presupposes that questions are sincere, as seems reasonable with children. The reader will find it interesting to analyze the case with a possibly insincere question in similar terms (our later methods in this book will cover both).

Iterated knowledge about others or common knowledge in groups are crucial to games, as they help keep social behavior stable. We will study these notions in Part II of this book using epistemic logics with update mechanisms explaining the information flow in our example. As a preview, the following sequence of diagrams (it helps to play it in one's mind as a video) shows how successive information updates shrink an initial model with six possible deals of the cards. The question rules out the two worlds where ***2*** has the blue card, the answer rules out the two worlds where ***1*** has the blue card:

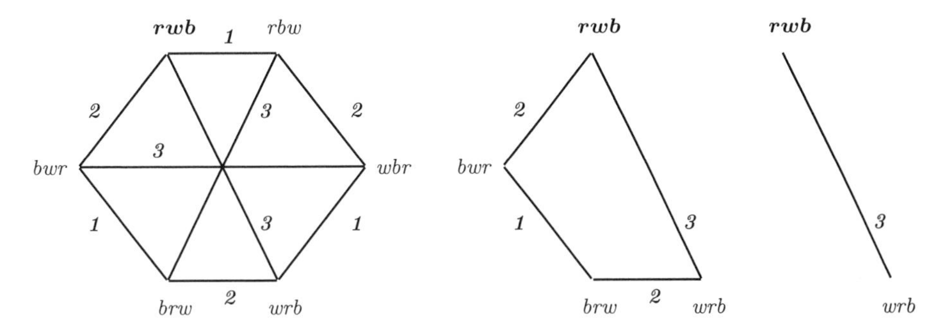

Let us now look at such games of "imperfect information" more abstractly, as general processes. Here is one more illustration of the sort of information structure that then emerges.

EXAMPLE Evaluating formulas under imperfect information

Consider the earlier evaluation game for the formula $\forall x \exists y Rxy$, but now assuming that verifier is ignorant of the object chosen by falsifier in the opening move. The new game tree looks as follows, with a dotted line indicating ***E***'s uncertainty:

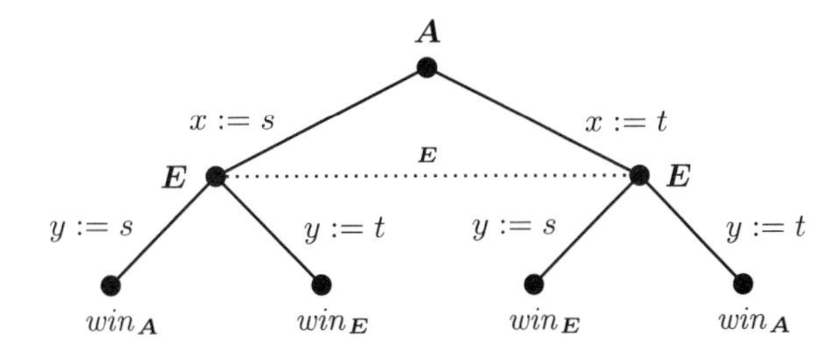

This game is quite different from the earlier one. In particular, allowing only strategies that can be played without resolving the uncertainty, as seems reasonable, player $\boldsymbol{E}$ has only two of the original four strategies left in this game: *left* and *right*. We easily see that determinacy is lost: neither player has a winning strategy! ■

Logics of imperfect information games The richer structure in these games invites a modal process language with an epistemic modality $K\varphi$ for knowledge, defined as truth of φ in all states one cannot tell apart from the current one.

Let us explore how this formalism works in the above setting. The following logical formulas describe player $\boldsymbol{E}$'s plight in the central nodes of the above game:

(a) $K_{\boldsymbol{E}}(\langle y := t\rangle win_{\boldsymbol{E}} \vee \langle y := s\rangle win_{\boldsymbol{E}})$
Here $\boldsymbol{E}$ knows that some move will force a win for $\boldsymbol{E}$, picking either s or t.

(b) $\neg K_{\boldsymbol{E}}\langle y := t\rangle win_{\boldsymbol{E}} \;\wedge\; \neg K_{\boldsymbol{E}}\langle y := s\rangle win_{\boldsymbol{E}}$
Now, there is no particular move of which $\boldsymbol{E}$ knows that it will force a win.

This is the well-known "de re, de dicto" distinction from philosophical logic. A person may know that the ideal partner is out there (de dicto), without ever finding out who is the one (de re). The epistemic logic of imperfect information games encodes interesting properties of players, such as perfect recall or bounded memory. We will take this up in Chapter 3, and again in the study of types of players and styles of play in Part II of this book.

We have now seen at least two ways in which knowledge enters games: "forward ignorance" of the future course of play, and "sideways ignorance" about where we are in the game. We will discuss such different forms of information in much greater detail later on, especially in Chapters 5 and 6 on various models for games.

REMARK Logic about logic
If you recall that the preceding example was itself a logical game, the modal logic introduced here is a logic about logic. If these sudden shifts in perspective are bothersome, please realize that dizzying self-reflective flights occur quite often in logic, and if that does not help, just take a gulp of fresh air before reading on.

Strategic form games While extensive games are appealing, game theory also has a quite different, and more widely used format, depicted in familiar matrix pictures. A *strategic game* consists of (a) a finite set N of players, (b) for each player $i \in N$ a non-empty set A_i of actions available to the player, and (c) for each player $i \in N$ a preference relation $\geq_i$ on $A = \Pi_{j\in N} A_j$.

This encodes global strategies as atomic actions, with the tuples in A standing for the total outcomes of the game that can be evaluated by the players. This level of structure is like the earlier power view, although the precise analogy is somewhat delicate (cf. Chapter 12). Strategic games are often depicted as matrices that encode basic social scenarios.

EXAMPLE Matrix games

A simple example of a strategic game is Hawk versus Dove. In many settings, agents can choose between two behaviors: aggressive or meek. Here are some matching preferences, annotating outcomes in the order (***A***-value, ***E***-value):

		E	
		dove	*hawk*
A	*dove*	3, 3	1, 4
	hawk	4, 1	0, 0

The understanding here is different from the earlier extensive game trees, in that players now choose their actions simultaneously, independently from each other. What is optimal behavior in this scenario? The two straightforward Nash equilibria of this game are (*hawk, dove*) and (*dove, hawk*).

Another evergreen of this format is the famous Prisoner's Dilemma. Consider two countries caught in the following situation:

		E	
		arm	*disarm*
A	*arm*	1, 1	4, 0
	disarm	0, 4	3, 3

Here the only Nash equilibrium (*arm, arm*) is suboptimal in that disarming would benefit both players. The Prisoner's Dilemma raises deep issues about social cooperation (Axelrod 1984), but we have nothing to add to these in this book. ■

Not all matrix games have Nash equilibria with just the pure strategies as stated. A counterexample is the common game Matching Pennies, discussed in Chapters 3 and 21. However, one can increase the space of behaviors by adding "mixed strategies," probabilistic mixtures of pure strategies. Strategic games then bring their own theory beyond the Zermelo tradition, starting from results going back to the work of Borel, von Neumann, and Nash (cf. Osborne & Rubinstein 1994).

THEOREM All finite strategic games have Nash equilibria in mixed strategies.

Logics of strategic games Like extensive games, strategic games can be seen as models for logical languages of action, preference, and knowledge (van Benthem et al. 2011). In particular, defining Nash equilibria in strategic games has long been a benchmark problem for logics in game theory (cf. van der Hoek & Pauly 2006). In this book, strategic games will be mostly a side topic, since rational action and reasoning are often better studied at the level of detail offered by extensive games. Still, Chapters 12 and 13 will take a closer look at the modal logic of matrix games, while including some well-known conceptual issues in modeling simultaneous action.

Solving games by logical announcements Strategic games have their own solution procedures, but again these invite connections with logic, viewing matrices themselves as models for logical languages of knowledge and action. A classical method is *iterated removal of strictly dominated strategies* (SD^{ω}), where one strategy dominates another for a player i if its outcomes are always better for i. The following example explores how this works.

EXAMPLE Pruning games by removing strictly dominated strategies
Consider the following matrix, with the legend (***A***-value, ***E***-value) for pairs:

		E		
		a	b	c
A	d	$2, 3$	$2, 2$	$1, 1$
	e	$0, 2$	$4, 0$	$1, 0$
	f	$0, 1$	$1, 4$	$2, 0$

First remove the dominated right-hand column: ***E***'s action c. After that, the bottom row for ***A***'s action f is strictly dominated, and after its removal, ***E***'s action b becomes strictly dominated, and then ***A***'s action e. The successive removals leave just the Nash equilibrium (d, a). ∎

There is an extensive game-theoretic literature analyzing game solution in terms of players' knowledge and epistemic logic (cf. de Bruin 2010), defining optimal profiles in terms of common knowledge or belief of rationality. But there is a simple dynamic take as before, using an analogy to the well-known puzzle of the Muddy Children (Fagin et al. 1995), where iterated announcement of the children's ignorance leads to a stable solution in the limit. The above matrices may be viewed as epistemic models where players have decided on their own action, but are yet in ignorance of what others have chosen. Now, these models may change as further information comes in, say through deliberation, and as with our card game, models will then get smaller. For SD^{ω}, a statement of rationality that drives a matching deliberation procedure is this: "No one plays a strategy that one knows to be worse than another option."

As this gets repeated, information flows, and the game matrix shrinks in successive steps until a first fixed point is reached, much in the style of information flow that we saw already with the earlier three cards example:

1	2	3
4	5	6
7	8	9

1	2
4	5
7	8

1	2
4	5

1
4

1

Each box acts as an epistemic model as described above: for instance, ***E***'s ranges of ignorance are the vertical columns. Each successive announcement increases players' knowledge, until the first fixed point is reached, where rationality has become common knowledge.

Thus, as with Backward Induction, there is a natural logic to solving strategic games, and we will investigate its details in Chapter 13.

Probability and mixed strategies A crucial aspect of real game theory is its use of probabilities. As we noted, all finite strategic games have equilibria in probabilistic mixed strategies. The notion of equilibrium is the same in this larger strategy space, with outcomes of profiles computed as expected values in the obvious sense. The following illustration is from the newspaper column Savant (2002), although there may also be official game-theoretical versions.

EXAMPLE The Library Puzzle

"A stranger offers you a game. You both show heads or tails. If both show heads, she pays you \$1, if both tails, she pays \$3, while you must pay her \$2 in case you show different things. Is this game fair, with expected value $1/4\times(+1)+1/4\times(+3)+1/2\times(-2) = 0$?" In her commentary, Vos Savant said the game was unfair to you with repeated play. The stranger can play heads two-thirds of the time, which gives you an average payoff of $2/3\times(1/2\times(+1)+1/2\times(-2))+1/3\times(1/2\times(+3)+1/2\times(-2) = -1/6)$. But what if you choose to play a different strategy, namely, "heads all the time"? Then the expected value is $2/3 \times (+1) + 1/3 \times (-2) = 0$. So, what is the fair value of this game?

It is easy to see that, if a strategy pair (σ, τ) is in equilibrium, each pure strategy occurring in the mixed strategy σ is also a best response for player ***1*** to τ. Then one can analyze the library game as follows. In equilibrium, let the stranger play heads with probability p and tails with $1-p$. You play heads with probability q

and tails with probability $1-q$. Now, your expected outcome against the p-strategy must be the same whether you play heads all the time, or tails all the time. That is: $p \times 1 + (1-p) \times -2 = p \times -2 + (1-p) \times 3$, which works out to $p = 5/8$. By a similar computation, q equals $5/8$ as well. The expected value for you is $-1/8$. Thus, the library game is indeed unfavorable to you, although not for exactly the reason given by the author. ■

There are several ways of interpreting what it means to play a mixed strategy (Osborne & Rubinstein 1994). For instance, besides its equilibria in pure strategies, the earlier Hawk versus Dove game has an equilibrium with each player choosing *hawk* and *dove* 50% of the time. This can be interpreted biologically in terms of stable populations having this mixture of types of individual. But it can also be interpreted in terms of degrees of belief for players produced by learning methods for patterns of behavior in evolutionary games (cf. Leyton-Brown & Shoham 2008, Hutegger & Skyrms 2012).

Logic and probability will not be a major theme in this book, although many of the logical systems that we will study have natural probabilistic extensions.

Infinite games and evolutionary game theory Here is the last topic in our tour of relevant topics in game theory. While all games so far were finite, infinite games arise naturally as well. Infinite computational processes are as basic as finite ones. Programs for standard tasks aim for termination, but equally important programs such as operating systems are meant to run forever, facilitating the running of finite tasks. Likewise, while specific conversations aim for termination, the overarching game of discourse is in principle unbounded. These are metaphors, but there is substance behind them (see Lewis 1969 and Benz et al. 2005 on the use of signaling games in understanding the conventions of natural language).

Infinite games have been used to model the emergence of cooperation (Axelrod 1984) by infinite sequences of Prisoner's Dilemma games. In such games, Backward Induction fails, but new strategies emerge exploiting the temporal structure. A key example is the Tit for Tat strategy: "copy your opponent's last choice in the next game," with immediate rewards and punishments. Tit for Tat is in Nash equilibrium with itself, making cooperation a stable option in the long run. It will find logical uses in Chapters 4 and 20.

Many new notions arise in this setting, such as "evolutionary stability" for strategies against mutant invaders (Maynard-Smith 1982). Evolutionary game theory is a branch of dynamical systems theory (Hofbauer & Sigmund 1998). It has not been studied very much by logicians so far (but see Kooistra 2012 on defining evolutionary stable equilibria), and this book will have little to say about it, except in Chapters 5 and 12. In the realm of infinite evolutionary games, much of our later analysis in this book may need rethinking.

6 From logic and game theory to Theory of Play

The preceding discussion might suggest that all is well in the foundations of games, and that logic just serves to celebrate that. But things are more complex, as there are serious issues concerning the interpretation and logical structure of games.

What justifies Backward Induction? A good point of entry is Backward Induction. Its solutions have been criticized for not making coherent sense in some cases, as the following example shows.

EXAMPLE The paradox of Backward Induction
Consider the following game (longer versions are sometimes called "centipedes"):

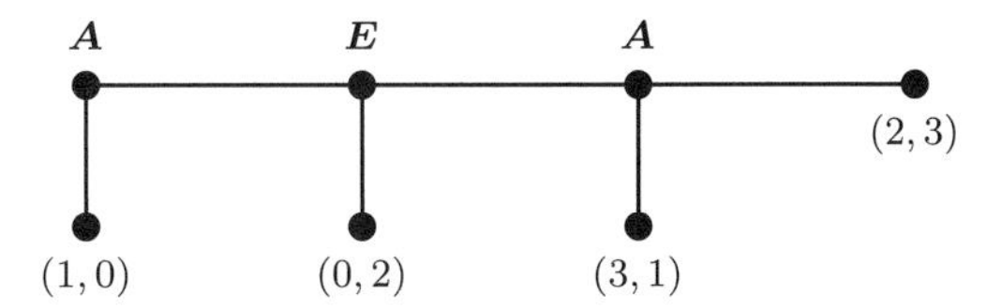

Backward Induction computes the value $(1, 0)$ for the initial node on the left, i.e., $\boldsymbol{A}$ plays down. Now this is strange, as both players, even when competitive, can see that they would be better off going across to the end, where $\boldsymbol{A}$ is sure to get more than 1, and $\boldsymbol{E}$ gets more than 0. ■

Motivating this solution leads to epistemic-logic based foundations of game theory, with insights such as the following result from Aumann (1995).

THEOREM The Backward Induction solution must obtain in extensive games whose players act while having common knowledge of rationality.

Even so, questions remain about the intuitive interpretation of the off-equilibrium path, which is crucial to the counterfactual reasoning keeping the behavior in place. Is the standard intuitive story behind Backward Induction coherent, when one really thinks it through?

Suppose that ***A*** moves right, deviating from Backward Induction, would ***E*** really stick to the old reasoning? ***E*** may think that ***A*** is now shown to be a different type of player whose behavior at the end might differ from initial expectations.

Making the players explicit The general issue here is the transition from a priori deliberation to analyzing actual real-time play of a game, where crucial information may become available about other participants. Backward Induction looks toward the future of the tree only, and hence deviations are not informative: they can be seen as mistakes without further effects for the rest of the game. This is one extreme line to take, maybe most appropriate to pregame analysis, and not so much to actual behavior during play. In the latter context, other lines of reasoning exist, where the past is informative. Given that players have come to the current node, what are the most plausible expectations about their future behavior? In an appealing picture, game trees will now mark a distinguished point s that indicates how far play has progressed, and players' accessibility relations for knowledge or belief can depend on that stage s:

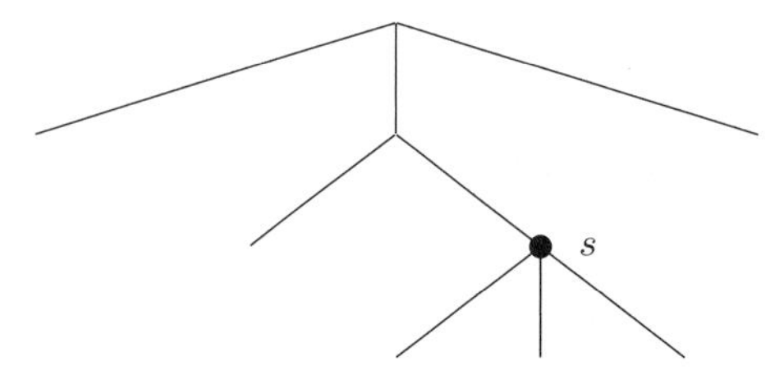

Thus, players know what has happened, and let their behavior depend on two things: what the remaining game looks like, and what happened so far in the larger game. Now the space of reasoning gets much larger, and in making recommendations or predictions, we need to know more about the agents, their belief revision policies, and their styles of play (cf. Stalnaker 1999). For algorithmic alternatives to Backward Induction in this line, see the game-theoretic literature on Forward Induction: for instance, Perea (2012).

We will explore this broader program in Parts I and II of this book, expressed in the equation

$$\text{Logic} + \text{Game Theory} = \text{Theory of Play}$$

Theory of Play involves input from various areas of logic. Process structure provides the playground for games, and as we have seen, tools here come from computational logic. But play also involves agency, and hence we move to philosophical logic, and current multi-agent systems. Reasoning about social scenarios involves philosophical themes such as knowledge, belief, strategies, preferences, and goals of agents. All this is crucial to games, where players deliberate, plan, and interact over time. This calls for enriched process logics of various sorts: epistemic, counterfactual, and preference-based. We will encounter many such combinations in the course of this book. To a logician, this is gratifying, since it shows how games provide unity to a fast-expanding field.

7 Conclusion

The examples in this Introduction offer a first look at the interface of logic and games. They raise a perhaps bewildering variety of issues at first glance. Let us emphasize once more that their purpose at this stage is merely to prove that there is an interface. We hope to have shown how almost every aspect of logic has some game behind it, while vice versa, about every feature of games has some logical ramifications. The subsequent chapters of this book will present a more systematic theory behind all this.

The main duality The examples offered here came in two flavors: "logic as games," and "logic of games." The reader will have noticed that the latter line gained the upper hand as we proceeded. This book is indeed largely about the logical study of games, using notions and results from computational logic and philosophical logic. Still, as we said at the outset, this strand is not our exclusive focus. We will also devote serious attention to logical core tasks cast in the form of games in Part IV, and as our finale, the interplay of game logics and logic games will be at center stage in Part VI. This reflects our view that the duality shown here is also crucial to logic in general.

A meeting of disciplines Our examples have shown that many disciplines meet in their interest in games, often with a role for logic as an analytical tool. First of all there is game theory itself, with topics ranging from abstract mathematical analysis to empirical studies of social behavior. Entangled with this, we encountered basic themes from computer science where new flavors of game theory are emerging that incorporate ideas from process theory, automata theory, and studies of agency. Our introduction also touched on uses of games in mathematics, philosophy, and linguistics. In this book, we will develop an integral logical perspective on games, while freely using relevant ideas from all of these areas. This is not to say that our treatment contains or supersedes these other approaches. They have their own achievements and style that we cannot do full justice to. For a better perspective on what we do, and do not do, in this book, the reader can consult several excellent recent texts written with different aims, and in different styles. We will list a few at the end of this Introduction.

Do games change logic as we know it? Now let us return to what we do cover in this book. Our final point for this Introduction is about the historical thrust of the topics raised here. Logic in games, in both its senses, proposes a significant extension of classical agendas for logic, and also in some parts, a radical departure from established ways of viewing the basic logical notions. Still, it will be clear to any reader of this book that the methodology employed throughout is the standard mathematical modus operandi of the field, running from formal modeling to the design of formal systems and their meta-theoretical properties. Mathematics is neutral as to our view of logic. This is historically interesting. The famous criticism in Toulmin (1958) was that over the centuries, logic had developed an obsession with form, and accordingly, with abstract mathematical concepts that miss the essence of reasoning. In contrast, a competing alternative was offered of "formalities," the procedure that forms the heart of specialized skills such as legal argumentation, or of debate in general. The perspective on games pursued in this book shows that this is an entirely false opposition. Formalities have form. There is a surprising amount of logical structure to activities, procedures, and intelligent interaction, and it is brought out by the tools that logicians have known and loved ever since the mathematical turn of the field in the 19th century.

8 Contents of this book

The chapters of this book develop the interface of logic and games using both existing and new material. Each chapter has a clearly indicated purpose, but its nature may differ. Some are about established lines of research, others propose a new perspective and raise new questions, while still others are mainly meant to make a connection between different fields. Likewise, quite a few chapters contain original results by the author, but some merely give a didactic presentation of known techniques.

In Part I of this book, we will look at games as rich interactive processes, pursuing logics of games. Chapter 1 is about the bare process structure of extensive games, and its analysis by means of modal logics of action, systematically related to thinking in terms of bisimulations for game equivalence. Chapter 2 adds preference structure, and shows how the crucial notions of rationality operative in Backward

Induction and other game solution methods can be defined in modal logics of action and preference, or, zooming in on details of their computation, in more expressive fixed point logics. Chapter 3 adds considerations of information flow in games of imperfect information, showing how these support epistemic extensions of the preceding logics. Combining action and knowledge allows us to analyze well-behaved types of players, including those with perfect memory, in terms of special axioms. Chapter 4 then turns to the topic of strategies that players use for achieving global goals in a game, often neglected as objects of study in their own right, showing how these can be defined in propositional dynamic logic and related formalisms. Chapter 5 is a brief discussion of infinite games and temporal logics, extending the earlier modal analysis of mostly finite games. Finally, Chapter 6 is a systematic discussion of successively richer "models of games" that are needed when we want to encode more information about players.

Part II of the book then moves toward the logical dynamics of the many kinds of action that occur in and around games. Chapter 7 is a brief introduction to relevant ideas and techniques from dynamic-epistemic logic, although we will assume that the reader has access to more substantial sources beyond this book. Chapter 8 analyzes pre-play processes of deliberation, connecting Backward Induction with dynamic procedures of iterated hard or soft update with rationality, and through these, to the field of belief revision and learning scenarios. Chapter 9 then considers the dynamic logic of the many processes that may occur in actual real-time game play: observing moves, receiving other kinds of relevant information, or even changes in players' preferences. The chapter also shows how these same techniques can analyze post-play rationalizations of a game, and even what happens under actual changes in a current game. These technical results suggest the emergence of something more general, a Theory of Play whose contours and prospects are discussed in Chapter 10.

Part III of the book investigates more global perspectives on games, still with a view to the earlier concerns in Parts I and II. Chapter 11 shows how our earlier modal techniques generalize to players' powers for influencing outcomes. Chapter 12 investigates strategic form games and shows how modal logics still make good sense for analyzing information, action, and freedom. Chapter 13 then analyzes solution algorithms for strategic games in dynamic-epistemic logics of deliberation scenarios, suggesting a dynamic-epistemic foundation for game theory.

Next, Part IV makes a turn to logic games, introducing the major varieties of logic as games. Chapter 14 has evaluation games for several logical languages (first-order, modal, fixed point logics). Chapter 15 deals with comparison games between models that provide fine structure for earlier notions of structural equivalence. Chapter 16 has games for model construction in tasks of maintaining consistency, and Chapter 17 has related games for dialogue and proof, the other side of this coin. Finally, Chapter 18 draws some general lines running through and connecting different logic games, and shows how similar ideas play in the foundations of computation.

Part V mixes ideas from game logics and logic games in two major calculi of game-forming operations. Chapter 19 discusses dynamic game logic, a generalization of dynamic logic of programs to neighborhood models for players' powers. Chapter 20 has linear logics for parallel game constructions, linking up with infinite games viewed as models for interactive computation involving different systems.

Part VI of the book then draws together logic of games and logic as games, not in one grand unification, but in the form of a number of productive interfaces and merges. Chapter 21 is mainly about logical evaluation games with imperfect information, Chapter 22 discusses recent knowledge games played over epistemic models, and Chapter 23 presents sabotage games as a model of generalized computation that has features of both game logics and logic games. The final chapters raise more theoretical issues. Chapter 24 shows how logic games can serve as a complete representation for certain kinds of general game algebra, and Chapter 25 presents further general themes, including hybrid, but natural systems merging features of logic games and game logics.

The book ends with a Conclusion drawing together the main strands, identifying the major lacunae in what we have so far, and pointing at prospects for the way ahead. In particular, what the reader can learn from our material, besides the grand view as such, are many techniques of general interest. These include game equivalences, logics for internal and external analysis of game structure, algebras of game operations, and dynamic logics for information flow. Moreover, there are conceptual repercussions beyond techniques, including changes in our way of viewing games, and logic.

Further sources Our book is just one pass through the field, without any pretense at achieving completeness. As we have said earlier, the subject of games can be approached from many angles, starting in many disciplines. Here are a few sources. Hodges (2001) and van der Hoek & Pauly (2006) are compact surveys of logic and games; Osborne & Rubinstein (1994), Binmore (2008), Gintis (2000), Perea (2012), and Brandenburger et al. (2013) are lively presentations of modern game theory; the parallel volumes Leyton-Brown & Shoham (2008) and Shoham & Leyton-Brown (2008) present game theory with many links to agency in computer science; Grädel et al. (2001) presents the powerful theory of games as a model for reactive systems that is emerging in contacts with computational logic and automata theory, while Apt & Grädel (2011) adds contacts with standard game theory; Abramsky (2008) gives game semantics for programming languages and interactive computation; Mann et al. (2011) presents game-theoretic logics in philosophy and linguistics; Gintis (2008) connects game theory to the big issues in social epistemology; Pacuit & Roy (2013) present epistemic game theory for logicians and philosophers; de Bruin (2010) is a critical philosophical study of this same interface; Clark (2011) gives a broad canvas of classical and evolutionary games applied to logic, linguistics, and empirical cognitive science; Hodges (2006) and Väänänen (2011) are elegant mathematical treatises showing games at work in mathematical logic; and Kechris (1994) has sophisticated game methods in the set-theoretical foundations of mathematics.

Two major current interface conferences between logic and game theory are TARK `http://www.tark.org/` on reasoning about rationality and knowledge, and LOFT `http://www.econ.ucdavis.edu/faculty/bonanno/loft.html` on foundations of games and decision theory. Many computer science conferences on logical foundations of computation and agency also have important games-related material. A website listing relevant publications and events is `http://www.loriweb.org`.

I Game Logics and Process Structure

Introduction to Part I

Our interest in this opening part is reasoning about interaction with standard logics of computation and information. To see what we are after, consider the following simple scenario, where players have preferences encoded in pairs (***A***-value, ***E***-value). The standard solution method of Backward Induction for extensive games tells player ***E*** to choose left at ***E***'s turn, expecting which gives player ***A*** a belief that this will happen, and hence, based on this belief about ***E***, player ***A*** should turn left at the start:

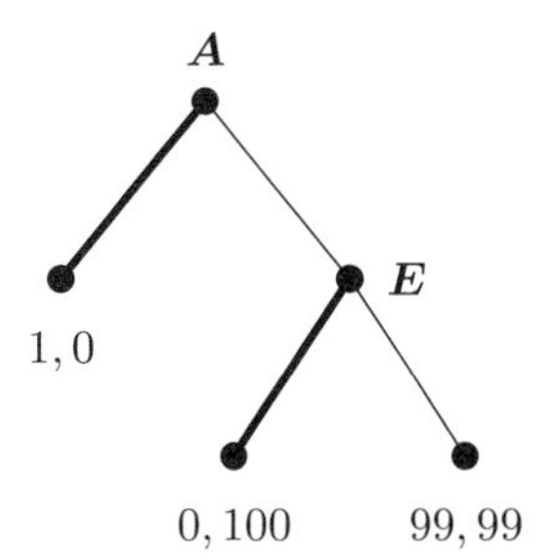

This piece of reasoning about rational behavior may be considered surprising, since the outcome $(99, 99)$ would appear to be better for both.

So, why should players act this way? The logical approach is about understanding the reasoning underlying Backward Induction, rather than proposing new game solutions. Interestingly, that reasoning is a mix of many notions often studied separately. It is about actions, players' knowledge of the game, their preferences, but also their beliefs about what will happen, plans, and counterfactual reasoning about situations that will not even be reached with the plan decided on. One simple social scenario involves just about the entire agenda of philosophical logic in a coherent manner. The mechanics of human minds intertwines action, belief, and preference, just as the mechanics of physical bodies intertwines their mass, position, and forces. The bridge law for these notions driving Backward Induction is usually taken to be *rationality*: "players never choose an action whose outcomes they believe to be worse than those of some other available action." Evidently, this law is packed with assumptions, and logic wants to clarify these, rather than endorsing any specific game-theoretic recommendation.

All this will happen in the following chapters, moving from bare computational models to eventually richer views of agency. We first look at the bare process structure of actions, then add preferences, then imperfect information, and after that, we consider two longer-term notions: strategies and infinite games in branching time. Finally, we discuss the methodology of small models for games used in this book. The result is a rich survey of game logics, although by no means an exhaustive one. Further topics in the literature will be mentioned as we proceed.

1
Games as Processes: Definability and Invariance

We start in a very simple fashion. A game tree with nodes and moves is a model for a process with states and transitions, of a kind that has been studied in logic, computer science, and philosophy under names such as Kripke models or labeled transition systems. Such models invite the use of logical languages that define basic process structure in a perspicuous way, while representing intuitive patterns of reasoning about social scenarios. A further current use of logic in games viewed in this manner is the analysis of computational complexity for model checking desirable properties, but such computational concerns will be less prominent in this book. In this chapter, we will see that, in particular, modal logics do many useful jobs, making them a first candidate for a *language of games*. However, a further fundamental perspective from logic makes sense. A choice of language is at the same time a choice for a level of structural invariance between semantic structures. Accordingly, our second main theme is *structural invariance between games*, looking for natural levels of viewing interaction. In brief: "When are two games the same?" There is no consensus on this important issue, but we put forward some proposals. Finally, we state a brief conclusion, mention some key literature that went into the making of this chapter, and list some further directions that continue themes raised in the main text.

NOTE The games in this chapter and most of this book are normally taken to be *finite*. When it is important to include infinite games, we will say so explicitly (cf. Chapters 5, 14, 20, and others). Many notions and results in this book also apply to infinite games, but we will not pursue this generalization systematically.

CAVEAT We repeat an earlier warning. This chapter is not an introduction to modal logic, and the same is true for other logics in this book. The textbook van Benthem (2010b) covers most of the basics that will be needed.

1.1 Games as process graphs in modal logic

Extensive games as process graphs We start by defining our basic models, using a slight variation on the presentation in Osborne & Rubinstein (1994).

DEFINITION 1.1 Extensive games
An *extensive game* $\boldsymbol{G}$ is a tuple $(I, A, H, t, \{\leq_i\})$ consisting of (a) a set I of players, (b) a set of actions or moves A, (c) a set H of finite or infinite sequences of successive actions from A (the histories), closed under taking finite prefixes and infinite limits of histories, (d) a turn function t mapping each non-terminal history having a proper continuation in H to a unique player whose turn it is, and finally, (e) for each player i, a connected preference relation $h \leq_i h'$ (player i weakly prefers h' to h) on terminal histories. Dropping preference relations from an extensive game yields an *extensive game form* being just the underlying pure action structure. ■

Several things can be generalized here, say, allowing partial preference orders with incomparable outcomes, allowing simultaneous moves, or dropping limit closure to model a wider class of temporal processes. However, the basic format is the most widely used, and understanding it will facilitate logical generalizations later.

To readers from other fields, extensive game forms will exhibit a familiar structure: they are just multi-agent "process graphs" in logic and computer science with a universe of states and transitions. In this chapter, we start at the level of game forms, looking at action structure only. Preference will return in Chapter 2.

DEFINITION 1.2 Extensive game models
An *extensive game model* is a tree $\boldsymbol{M} = (S, M, \boldsymbol{turn}, \boldsymbol{end}, V)$ with a set of nodes S and a family M of binary transition relations for the available moves, pointing from parent to daughter nodes. There are two special properties of nodes: non-final nodes have unary proposition letters $\boldsymbol{turn}_i$ indicating the player whose turn it is, while $\boldsymbol{end}$ marks end nodes. On top of this, the valuation V serves to interpret other relevant properties of nodes, such as utility values for players, or more ad hoc features of game states. ■

Extensive game models, also called labeled transition systems or Kripke models, are natural models for a logical language. In this chapter, we will concentrate on the use of modal logics for extensive games, although we will also look at extensive games as models for temporal logics in Chapters 5 and 6.

Basic modal logic Extensive game trees support a standard modal language.

DEFINITION 1.3 Modal game language and semantics
Modal formulas are defined using the following inductive syntax rules for Boolean operations and modalities, where the p come from a set *Prop* of atomic proposition letters, and the a from a set *Act* of atomic action symbols:

$$p \mid \neg\varphi \mid \varphi \vee \psi \mid \langle a \rangle \varphi$$

Boolean conjunction $\wedge$ and implication $\rightarrow$, as well as the universal modality $[a]$ are defined in terms of these as usual. Modal formulas are interpreted at nodes s as local properties of game stages in the earlier models $\boldsymbol{M}$, in the format

$$\boldsymbol{M}, s \models \varphi \qquad \text{formula } \varphi \text{ is true at node } s \text{ in model } \boldsymbol{M}$$

The inductive clauses of this truth definition are as usual for the Booleans, while there is the following key truth condition for labeled modalities:

$$\boldsymbol{M}, s \models \langle a \rangle \varphi \quad \text{iff} \quad \text{there exists some } t \text{ with } sR_a t \text{ and } \boldsymbol{M}, t \models \varphi$$

This says that some particular instance of the move a is available at the current node leading to a next node in the game tree satisfying φ. The universal modality $[a]\varphi$ says that φ is true at all successor nodes reachable by an a-transition. ■

Modal operator combinations now describe possible interactions in games. We briefly repeat an example discussed at greater length in the Introduction.

EXAMPLE 1.1 Modal operators and strategic powers
Consider a simple two-step game like the following, between two players $\boldsymbol{A}$ and $\boldsymbol{E}$:

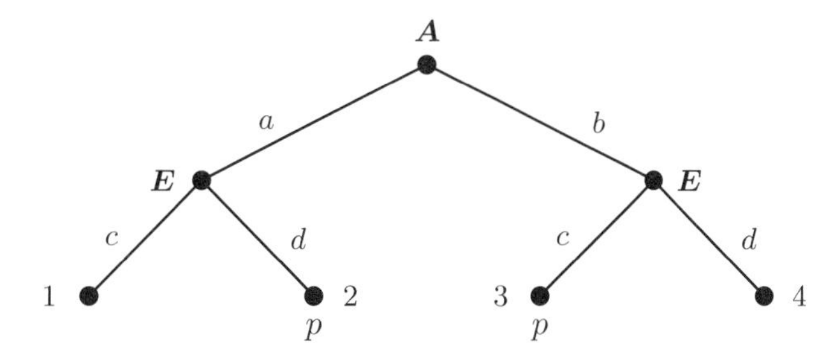

Player $\boldsymbol{E}$ clearly has a strategy for making sure that a state is reached where p holds, that is, a rule for responding to the opponent's moves producing some desired effect.

This feature of the game is directly expressed by a modal formula like this:

$$[a]\langle d\rangle p \wedge [b]\langle c\rangle p$$

Other modal formulas express other interesting properties of nodes in this game. More generally, we will see modal notations at work throughout this book. ■

As noted in the Introduction, notions like strategy and other items in game forms match their definition from game theory, unless explicitly noted otherwise.

Dynamic programs We can also introduce explicit notation for basic and complex moves in a game. For instance, letting the symbol *move* stand for the union of all relations available to players, in the preceding game, the modal combination

$$[\mathit{move}\text{-}\boldsymbol{A}]\langle \mathit{move}\text{-}\boldsymbol{E}\rangle\varphi$$

says that, at the current node, player $\boldsymbol{E}$ has a strategy responding to $\boldsymbol{A}$'s initial move which ensures that the property defined by φ results after two steps of play. Union is one of the program operations of an extension of the basic modal language called *propositional dynamic logic*, PDL (Harel et al. 2000).

DEFINITION 1.4 PDL programs
Starting from atomic action symbols in a set *Act* as before, the *PDL programs* are defined by the following inductive syntax

$$a \mid \pi_1 \cup \pi_2 \mid \pi_1 ; \pi_2 \mid \pi^* \mid ?\varphi$$

of atomic actions, union, composition, finite iteration, and tests. In the above models, programs π are interpreted as binary transition relations R_π between nodes, according to the following inductive clauses:

$R_{\pi_1 \cup \pi_2} = R_{\pi_1} \cup R_{\pi_2}$	Choice as union of transition relations
$R_{\pi_1 ; \pi_2} = R_{\pi_1} \circ R_{\pi_2}$	Sequential composition of relations
$R_{\pi^*} = (R_\pi)^*$	Reflexive-transitive closure of relations
$R_{?\varphi} = \{(s,t) \mid s = t \;\&\; \boldsymbol{M}, s \models \varphi\}$	Successful test of a property ■

All of these program operations make sense in the action structure of games, e.g., in defining strategies, and we will encounter them in many chapters to follow.

Excluded middle and determinacy Extending our observations to extensive games up to depth k, and using alternations $\Box\Diamond\Box\Diamond\cdots$ of modal operators up to length k, we can express the existence of winning strategies in any given finite game.

In this manner, standard logical laws acquire immediate game-theoretic import. In particular, consider the valid law of excluded middle

$$\Box\Diamond\Box\Diamond\cdots\varphi \lor \neg\Box\Diamond\Box\Diamond\cdots\varphi$$

or after some logical equivalences $\neg\Box = \Diamond\neg$, $\neg\Diamond = \Box\neg$ pushing negations inside:

$$\Box\Diamond\Box\Diamond\cdots\varphi \lor \Diamond\Box\Diamond\Box\cdots\neg\varphi$$

where the dots indicate the depth of the tree.

FACT 1.1 Modal excluded middle expresses the determinacy of finite games.

Here, determinacy is the basic property of win-lose games that one of the two players has a winning strategy. Since winning can be any condition φ of histories, in the above formula, the left-hand disjunct says that the responding player ***E*** has a winning strategy, and the right-hand disjunct that the starting player ***A*** has a winning strategy. (This may fail in infinite games: players cannot both have a winning strategy, but perhaps neither one has. Determined infinite games have been studied extensively in Descriptive Set Theory; Moschovakis 1980, Kechris 1994.)

Zermelo's theorem This brings us to an early game-theoretic result predating Backward Induction, proved by Ernst Zermelo in 1913 for zero-sum games, where what one player wins is lost by the other (win vs. lose is the typical example).

THEOREM 1.1 Every finite zero-sum two-player game is determined.

Proof Recall the algorithm in the Introduction for determining the player having the winning strategy at any given node of a game tree of this finite sort. The method worked bottom-up through the game tree. First, color end nodes *black* that are wins for player ***A***, and color the other end nodes *white*, being wins for ***E***. Then, if all children of node s have been colored already, do one of the following:

(a) If player ***A*** is to move, and at least one child is black: color s *black*; if all children are white, color s *white*.
(b) If player ***E*** is to move, and at least one child is white: color s *white*; if all children are black, color s *black*.

This algorithm eventually colors all nodes black where player ***A*** has a winning strategy, while coloring those where ***E*** has a winning strategy white. Once the root has been colored, we see the determinacy, and even who has the winning strategy.

The reason for the correctness of this procedure is that a player i whose turn it is has a winning strategy iff i can make a move to at least one daughter node where i has a winning strategy. ■

Application: The teaching game Zermelo's Theorem is widely applicable. A variant of the sabotage game in our Introduction shows another flavor of multi-move agent interaction.

EXAMPLE 1.2 Teaching, the grim realities
A student located at S in the following diagram wants to reach the escape E, whereas the teacher wants to prevent the student from getting there. Each line segment is a path that can be traveled. In each round of the game, the teacher first cuts one line, anywhere, and the student must then travel one link still open at the current position:

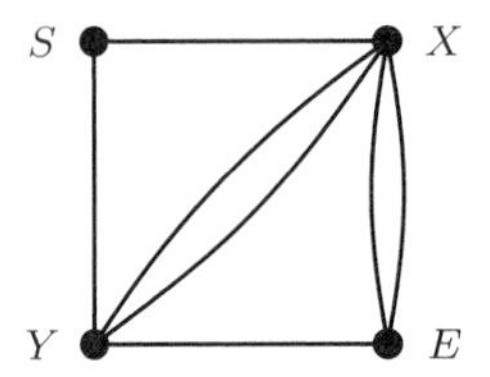

In this particular game, the teacher has a winning strategy, by first cutting a line to the right between X and E, and then waiting for the student's moves, cutting lines appropriately. General games like this arise on any graph with single or multiple lines. Gierasimczuk et al. (2009) provide links with real teaching scenarios. ■

Zermelo's Theorem explains why the student or the teacher has a winning strategy: cutting lines must end. Many games fall under this result, like most logic games in Part IV. For applications to computer games, see van den Herik et al. (2011).

Modal μ-calculus for equilibrium notions A good test for expressive power of logics is their capacity for representing basic arguments in the field under study. Our basic modal language cannot yet express the proof of Zermelo's Theorem. Starting from atomic predicates $\boldsymbol{win}_i$ at end nodes marking which player has won, that proof inductively defined new predicates WIN_i (i.e., player i has a winning strategy at the current node, where we use i and j to indicate the opposing players) through a recursion of the form

$$WIN_i \leftrightarrow (\boldsymbol{end} \wedge \boldsymbol{win}_i) \vee (\boldsymbol{turn}_i \wedge \langle move\text{-}i \rangle\, WIN_i) \vee (\boldsymbol{turn}_j \wedge [move\text{-}j]\, WIN_i)$$

Here *move-x* is the union of all moves for player x. Note how the defined expression WIN_i occurs recursively in the body of its own definition.

Despite its modal shape, this inductive definition has no explicit version in our modal base language. We need a logic that supports inductive definitions and complex recursions in computation. The *modal μ-calculus* extends modal logic with operators for "smallest fixed points"

$$\mu p \bullet \varphi(p)$$

where formulas $\varphi(p)$ must have a special syntax. The propositional variable p may occur only positively in $\varphi(p)$, that is, in the scope of an even number of negations.[1] This ensures that the following approximation function on sets of states:

$$F_\varphi^M(X) = \{s \in \boldsymbol{M} \mid \boldsymbol{M}, [p := X], s \models \varphi\}$$

is monotonic in the inclusion order:

$$\text{whenever } X \subseteq Y, \text{ then } F_\varphi^M(X) \subseteq F_\varphi^M(Y)$$

On complete lattices (such as power sets of models), the *Tarski-Knaster Theorem* says that monotonic maps F always have a "smallest fixed point," a smallest set of states X where $F(X) = X$. One can reach this smallest fixed point F_* through a sequence of approximations indexed by ordinals until there is no more increase:

$$\varnothing, F(\varnothing), F^2(\varnothing), \ldots, F^\alpha(\varnothing), \ldots F_*$$

The formula $\mu p \bullet \varphi(p)$ holds in a model $\boldsymbol{M}$ at just those states that belong to the smallest fixed point for the map $F_\varphi^M(X)$. Completely dually, there are also "greatest fixed points" for monotonic maps, and these are denoted by formulas

$$\nu p \bullet \varphi(p), \qquad \text{with } p \text{ occurring only positively in } \varphi(p).$$

Greatest fixed points are definable from smallest ones (and vice versa), as shown in the valid formula $\nu p \bullet \varphi(p) \leftrightarrow \neg \mu p \bullet \neg \varphi(\neg p)$, where $\neg \varphi(\neg p)$ still has p positive.

---

1 Alternately, $\varphi(p)$ has to be a formula constructed from arbitrary p-free formulas of the language plus occurrences of p, using only $\vee$, $\wedge$, $\Box$, $\Diamond$, and μ-operators.

These are just some of the basic properties of this system, which embodies the decidable modal core theory of induction and recursion. There is a fast-growing literature on the μ-calculus (cf. Bradfield & Stirling 2006, Venema 2007). This system will return in several parts of this book, including Chapters 14 and 18.

FACT 1.2 The Zermelo solution is definable as follows in the modal μ-calculus:

$$WIN_i = \mu p \bullet (\boldsymbol{end} \wedge \boldsymbol{win}_i) \vee (\boldsymbol{turn}_i \wedge \langle move\text{-}i \rangle p) \vee (\boldsymbol{turn}_j \wedge [move\text{-}j]p)$$

Proof The key points are the modal format of the inductive response clauses, plus the positive occurrences of the atom p standing for the predicate defined. ■

Fixed point definitions fit well with the equilibrium notions of game theory that often have some iterative intuition attached of successive approximation. Hence, the μ-calculus has many uses in games, as discussed further in Chapters 7 and 13 where we will use extended fixed point logics to analyze other equilibrium notions.[2]

Control over outcomes and forcing Winning is just one aspect of control in a game. Games are all about powers over outcomes that players can exercise via their strategies. This suggests further logical notions with a modal flavor.

DEFINITION 1.5 Forcing modalities $\{i\}\varphi$

$\boldsymbol{M}, s \models \{i\}\varphi$ iff player i has a strategy for the subgame starting at s that guarantees only end nodes will be reached where φ holds, whatever the other player does. ■

FACT 1.3 The modal μ-calculus can define forcing modalities for games.

Proof The modal fixed point formula

$$\{i\}\varphi = \mu p \bullet (\boldsymbol{end} \wedge \varphi) \vee (\boldsymbol{turn}_i \wedge \langle move\text{-}i \rangle p) \vee (\boldsymbol{turn}_j \wedge [move\text{-}j]p)$$

defines the existence of a strategy for player i making proposition φ hold at the end of the game, whatever the other player does.[3] ■

---

2 In infinite games, it seems better to use greatest fixed points defining a largest predicate satisfying the recursion. This does not build strategies inductively from below, but views them as rules that can be called, and always remain at our service. This is the perspective of co-induction in *co-algebra* (Venema 2006). Chapters 4, 5, and 18 discuss such strategies.

3 One can easily modify the definition of $\{i\}\varphi$ to enforce truth of φ at all intermediate nodes, a variant that will be used in Chapter 5.

Analogously, shifting modalities slightly, the formula

$$COOP\varphi = \mu p \bullet (\boldsymbol{end} \wedge \varphi) \vee (\boldsymbol{turn}_i \wedge \langle move\text{-}i \rangle p) \vee (\boldsymbol{turn}_j \wedge \langle move\text{-}j \rangle p)$$

defines the existence of a cooperative outcome φ. However, this notion is already definable in PDL as well, using the program modality

$$\langle ((?\boldsymbol{turn}_i; move\text{-}i) \cup (?\boldsymbol{turn}_j; move\text{-}j))^* \rangle (\boldsymbol{end} \wedge \varphi)$$

The program is an explicit strategy that makes the forcing statement true. We will say more on the issue of explicit strategies versus forcing modalities in Chapter 4.

1.2 Process and game equivalences

We have seen how many properties of game forms can be defined in modal formulas, making modal logic a good medium for reasoning about the basics of interaction. But next to language design, there is another basic perspective on games. To see this, recall the view of games as processes as explored in our Introduction.

Process equivalence Process graphs represent processes, and a natural question arises of when two graphs stand for the same process. As elsewhere in mathematics, we want an appealing invariance relation. When are two processes the same?

EXAMPLE 1.3 The same process, or not?
We repeat two pictures of processes, or machines, from our Introduction:

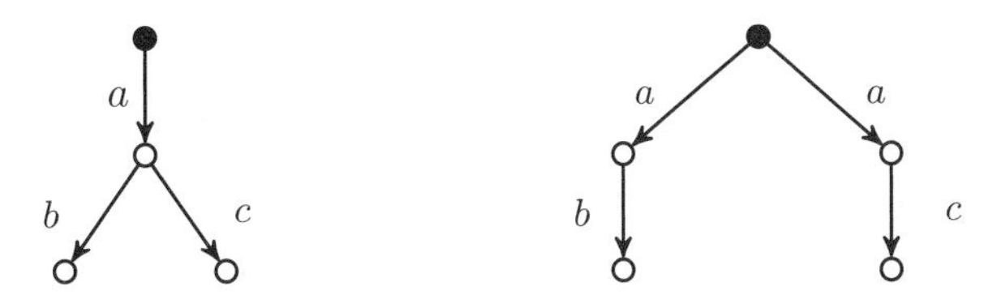

Both diagrams produce the same finite traces of observable actions $\{ab, ac\}$, so qua input-output behavior the machines are the same under "finite trace equivalence." But in doing so, the first machine starts deterministically, and the other with an internal choice. In order to explore such internal workings of a process, the measure must have a finer structural comparison distinguishing these models. ■

Bisimulation As will be clear from the example, there is no unique answer: different structural invariances may set natural levels of process structure. Finite trace equivalence is one extreme, another extreme is the standard notion of isomorphism, preserving every detail of size and relational structure of a process that is definable in first- or higher-order languages. For present purposes, however, we want to be in between these two levels. The structural invariance matching the expressive power of modal logic is a notion that has been proposed independently in many settings, including computer science and set theory. It works on any graph model of actions and states, providing an account of "simulating" (the key notion in relating different processes) both observable actions and internal choices that led to these.

DEFINITION 1.6 Bisimulation

A *bisimulation* is a binary relation E between states of two graphs $\boldsymbol{M}$ and $\boldsymbol{N}$ such that, if xEy, then we have (1) atomic harmony, and (2) back-and-forth clauses,

(1) x, y verify the same proposition letters.

(2a) If xRz, then there exists u in $\boldsymbol{N}$ with yRu and zEu.

(2b) If yRu, then there exists z in $\boldsymbol{M}$ with xRz and zEu.

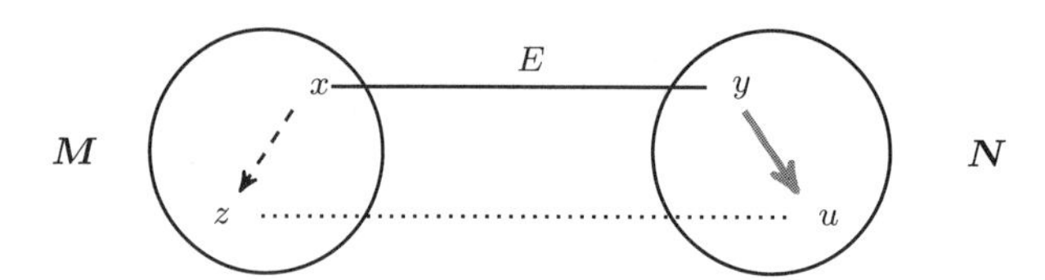

■

Bisimulation respects local properties plus options available to the process at any stage. Our earlier two finite-trace equivalent graphs are not bisimilar in their roots. Before we continue with exploring its properties, let us point out some major uses.

EXAMPLE 1.4 Uses of bisimulation

Bisimulation contracts process graphs to a simplest equivalent graph, as in:

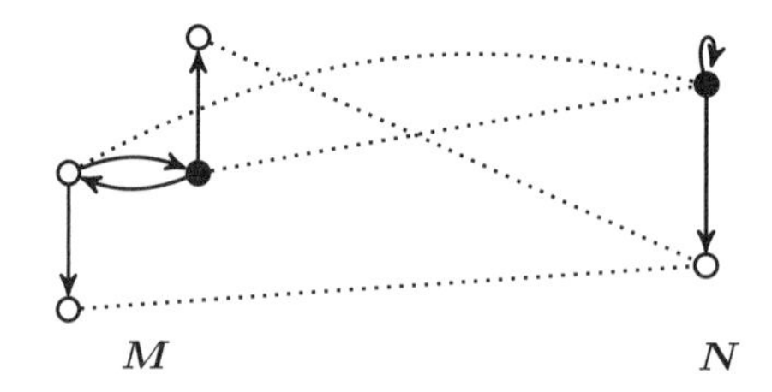

A reverse use of bisimulation unravels any process graph $\boldsymbol{M}$, s to a rooted tree. The states are all finite paths through $\boldsymbol{M}$ starting with s and passing on to R-successors at each step. One path has another path accessible if the second is one step longer. The valuation on paths is copied from the value on their last nodes.

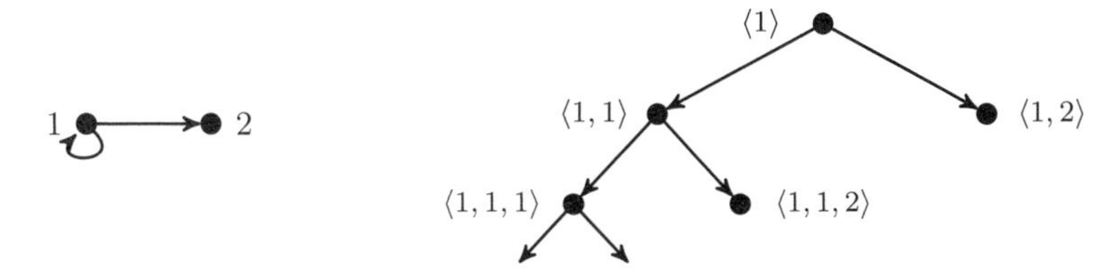

An unraveling is like a game tree, with the branches encoding all possible runs. This format is very convenient for theoretical purposes in logic and computation.

Invariance of modal formulas Bisimulation preserves the truth of modal and dynamic formulas across models, and there are converses, too. The following are the basic model-theoretic facts in this area.

THEOREM 1.2 For all graphs $\boldsymbol{M}$ and $\boldsymbol{N}$ with nodes s, t, condition (a) implies (b):

(a) There is a bisimulation E between $\boldsymbol{M}$ and $\boldsymbol{N}$ with sEt.

(b) $\boldsymbol{M}, s$ and $\boldsymbol{N}, t$ satisfy the same formulas of the μ-calculus.

This implies invariance of modal and dynamic formulas under bisimulation. The following partial converse says that the basic modal language and the similarity relation match on finite models.

THEOREM 1.3 For any two finite models $\boldsymbol{M}$ and $\boldsymbol{N}$ with given nodes s and t, the following statements are equivalent:

(a) $\boldsymbol{M}, s$ and $\boldsymbol{N}, t$ satisfy the same modal formulas.

(b) There is a bisimulation E between $\boldsymbol{M}$ and $\boldsymbol{N}$ with sEt.

Our third result says that, at this same level of description, the dynamic language even provides complete descriptions for any finite graph.

THEOREM 1.4 For each finite graph $\boldsymbol{M}$ and node s, there exists a dynamic logic formula $\delta(\boldsymbol{M}, s)$ such that the following are equivalent for all graphs $\boldsymbol{N}, t$:

(a) $\boldsymbol{N}, t \models \delta(\boldsymbol{M}, s)$.

(b) $\boldsymbol{N}, t$ is bisimilar to $\boldsymbol{M}, s$.

Proofs of all this can be found in van Benthem (2010b). The last two facts even hold for arbitrary graphs when we increase the expressive power of basic modal logic with arbitrary infinite conjunctions and disjunctions in its construction rules.[4]

Hierarchy of languages and process equivalences There is a hierarchy of natural process equivalences, from coarser ones like finite trace equivalence to finer ones like bisimulation. No best level of sameness exists in process theory: it depends on the purpose. This latitude is also known from studies of space, from fine-grained in geometry to coarse-grained in topology: it all depends on what you mean by "is." Crucially, the same seems true for games. Extensive games match well with bisimulation. However, our earlier strategic power perspective was closer to an input-output process view such as trace equivalence.

Hierarchy of languages The ladder of structural simulations has a syntactic counterpart. The finer the process equivalence, the more expressive a matching language for the relevant process properties. In this setting, the previously discussed modal results generalize. Similar invariance and definability results hold for many kinds of process equivalence (van Benthem 1996). Further, we submit that the same is true for games.

Game equivalence and invariances The same style of invariance thinking applies to the question when two games are the same (van Benthem 2002a).

EXAMPLE 1.5 Local versus global game equivalence
Recall an earlier example from the Introduction. Consider the following two games:

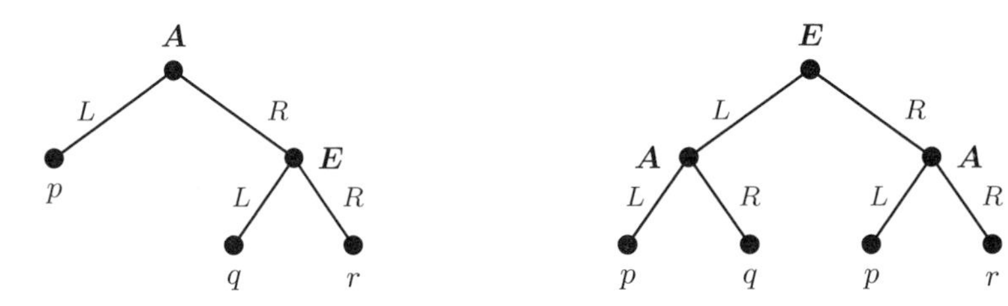

Are these two games the same? The answer depends on our level of interest. In terms of turns and local moves, the games are not equivalent, and some kind of bisimulation seems the right invariance, whose absence is matched by a difference

4 Bisimulation also applies to infinite games with infinite histories as outcomes. In this case, matching languages are the branching temporal logics of Chapters 5, 6, and 9.

that is expressible in a modal language. For instance, the game on the right, but not that on the left, satisfies the modal formula $\langle L\rangle\langle R\rangle \mathit{true}$ at its root. ■

More generally, in line with our definability results for process graphs, game forms can be described either via structural equivalences or in terms of game languages with logical formulas (Bonanno 1993, van Benthem 2001b).

Power equivalence and forcing modalities Let us explore the variety. If our focus is on achievable outcomes of the two games only, then the verdict changed.

EXAMPLE 1.5, CONTINUED
Both players have the same powers of achieving outcomes in both games:

(a) ***A*** can force the outcome to fall in the sets $\{p\}$, $\{q, r\}$.
(b) ***E*** can force the outcome to fall in the sets $\{p, q\}$, $\{p, r\}$.

As in the Introduction, a player's powers are those sets U of outcomes for which the player has a strategy making sure the game will end inside U, no matter what others do. In the game on the left, player ***A*** has strategies *left* and *right*, yielding powers $\{p\}$ and $\{q, r\}$, and player ***E*** has two strategies yielding powers $\{p, q\}$ and $\{p, r\}$. On the right, ***E*** has *left* and *right*, giving the same powers as on the left. But ***A*** now has four strategies:

left: L, *right*: L, *left*: L, *right*: R, *left*: R, *right*: L, *left*: R, *right*: R

The first and fourth strategies give the same powers for ***A*** as on the left, while the second and third strategies produce weaker powers subsumed by the former. ■

We will return to these simple but important examples at many places in this book. Again, there is a matching notion of power bisimulation plus a modal language for this level of description, involving the strategic forcing modalities $\{i\}\varphi$ of Section 2.1 of Chapter 2. This forcing perspective will be explored at greater length in Chapters 11 and 19. It is also the intuitive equivalence view behind the logic games that we will study in Part IV.

REMARK Levels and transformations in game theory
Issues of grain size have long been studied in game theory. In particular, extensive games induce a notion of extracted information that can be captured by *transformations* between games having the same normal or reduced form (cf. Thompson

1952, Kohlberg & Mertens 1986).[5] On a different tack, Bonanno (1992a) identifies extracted information with a set-theoretic form close to forcing, and gives a matching transformation of interchange of contiguous simultaneous moves.

1.3 Further notions of game equivalence

Having introduced our two main topics of definability in logical languages for games and matching notions of structural simulation, we now explore a few variations, some getting closer to the game-theoretic literature.

Alternating bisimulation Are there other natural game equivalences beyond the present two? Finite trace equivalence seems too coarse, but bisimulation sometimes seems too fine. The following two non-bisimilar single-player games seem equivalent:

One does not normally distinguish players' internal actions this finely, and the switch is in fact one of the "Thompson transformations" of Thompson (1952). But things matter with switches between different players, moving into another zone of control. We might not call game shapes equivalent if the turn patterns were very different, as with the earlier game, but now for formulas $(A \vee B) \wedge C$ and $A \vee (B \wedge C)$. Formulating a matching notion calls for a mixture of our earlier ideas.

Definition 1.7 Alternating bisimulation

An *alternating bisimulation* only requires the zigzag conditions of bisimulation with respect to "extended moves" that consist of a finite action sequence by one player ending in a state that is a move for the other player, or an endpoint. Also, it disregards the exact nature of these moves, viewing them solely as transitions. ■

---

5 Transformations are close to mathematical invariances in geometry and other fields. The closest analogue in our setting are bisimulations, although these are not functions. One associated transformation on models is the earlier bisimulation contraction.

EXAMPLE 1.6 Compacting games
Alternating bisimulation will identify the following two game trees:

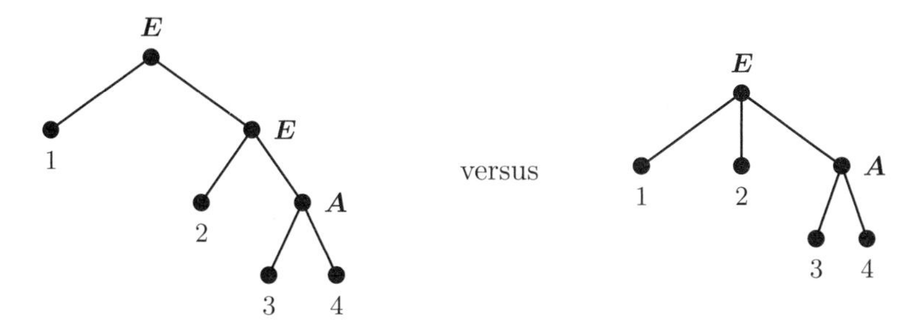

This allows for normalizing games so that each move leads to a turn change. ■

Alternating bisimulation can again be tied up with matching modal game languages, but no complete characterization seems to be known. It resembles the intermediate forcing bisimulations of Chapter 11 that record players' powers for getting to any position of a game.[6]

Simulating strategies A different intuition about game equivalence that many people share might be phrased as follows:

players should have matching strategies in both games.

Ordinary bisimulation implies this property of matching.

FACT 1.4 Games that admit a bisimulation between their roots allow all players to copy strategies for forcing sets of end nodes.

The reason is that the zigzag conditions allow both players to copy moves across. But bisimulation seems too strong, and weaker invariances suffice, depending on how the matching is spelled out. So far, there seems to be no best formulation of game equivalence as having the same strategies modulo effective simulation.

DIGRESSION Perhaps we must think differently here, using two simulation relations, one for player ***E*** and one for ***A***, comparing their separate powers across nodes in two games. When turns are the same, ordinary zigzag clauses for available moves seem acceptable. But the situation gets harder when the turns are not the same:

6 It is also related to stuttering bisimulations in computer science (Gheerbrant 2010).

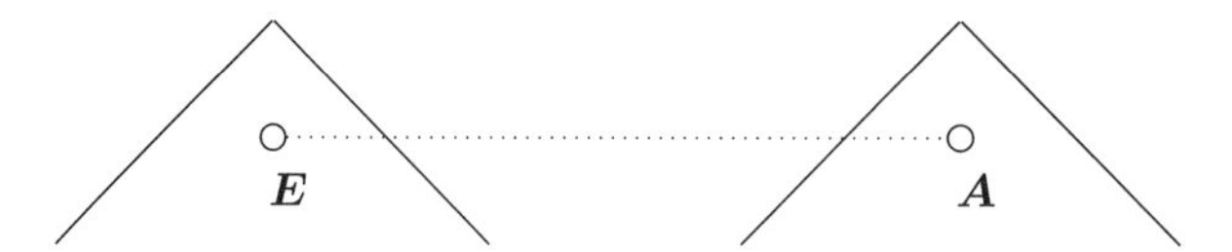

With such a turn mismatch, single moves by player $\boldsymbol{E}$ must correspond to responses to all of $\boldsymbol{A}$'s moves on the other side. But $\boldsymbol{E}$'s powers on the left need not match her powers at each $\boldsymbol{A}$-successor on the right: a counterexample is easily found in our propositional distribution games. We may need inclusion of players' powers via "directed simulations" (Kurtonina & de Rijke 1997).

Links with logic games Game comparison also makes sense for the special realm of logic games in Part IV. But logic games also add to the current style of analysis. In Chapter 15, we will study Ehrenfeucht-Fraïssé games that test two given models for similarity over some specified number of rounds. These games add fine structure to notions of simulation, measuring differences that can be detected by concrete modal or first-order formulas. In this finer perspective, a bisimulation is a global winning strategy for establishing similarity in a game with an infinite number of rounds. Thus, logic games will provide a way of refining our earlier question of structural equivalence to the question of:

How similar are two given games?

Different levels once more Different game equivalences need not stand for deep competing modeling choices; instead, they may just reflect a frequent phenomenon in practice. The same game $\boldsymbol{G}$ can often be described at different levels, as one can individuate moves more or less finely. In the limit, we can even form a kind of extreme coarsening $coarse(\boldsymbol{G})$ by keeping the old nodes, but taking just two relations for the players as follows:

R_E is the union of all moves for player $\boldsymbol{E}$, and R_A is defined likewise.

Since $coarse(\boldsymbol{G})$ has less information than $\boldsymbol{G}$ itself, it does not support a reduction of properties of $\boldsymbol{G}$: things go the other way around. To see this, take any modal formula φ in the language of $coarse(\boldsymbol{G})$, and translate it into a formula $fine(\varphi)$ replacing the new relation symbols by their definitions. This results in the equivalence

$$coarse(\boldsymbol{G}), s \models \varphi \quad \text{iff} \quad \boldsymbol{G}, s \models fine(\varphi)$$

A related view in Chapter 19 will distinguish between games themselves and "game boards" that games are played on. One level is an external setting for observable effects of the game. The other is a richer game structure with internal predicates, such as turns, preferences, wins, and losses. A typical connection between the two ways of viewing one and the same game is the "adequacy lemma" for evaluation games in the Introduction that stated an equivalence between

$$\text{(a) } \boldsymbol{M}, s \models \varphi, \quad \text{(b) } \boldsymbol{game}(\boldsymbol{M}, \varphi), \langle s, \varphi \rangle \models \mathit{WIN}_{\boldsymbol{V}}$$

This relates a statement about powers of players in a game with properties of its game board. Two-level views are often illuminating. In particular, the above links may be used to analyze the complexity of strategies in games. We postpone a further analysis of multiple views on games until Parts III and V of this book.

1.4 Conclusion

The main points We have found our first serious connection between logic and games. Extensive games are processes of a kind well-known in computer science, and for a start, we have shown how modal languages fit well for the purpose of defining game-theoretic notions and capturing basic game-theoretic reasoning. We gave a systematic family of languages (modal logic, dynamic logic, μ-calculus) that link game theory with computational logic, allowing for traffic of ideas. In particular, modal languages correlate with natural notions of structural invariance, reflected in different notions of process equivalence. Our second main point was taking the same invariance thinking to games, with levels ranging from finer to coarser, a common mathematical perspective for defining a family of structures.

Benefits By paying systematic attention to links between structure and language, logic brings to light key patterns in definitions and inferences, such as the modal quantifier interchange underlying strategic behavior. Moreover, one can use this framework for model-checking given games for specific properties, proving completeness for calculi of reasoning about interaction, determining computational complexity of game-theoretic tasks, or investigating model-theoretic behavior such as transfer of properties from one game to another. In this book, we will not pursue such applications in any technical detail, but they are there, and we will continue to develop many further relevant interfaces.

Open problems One benefit of a logical stance is new problems. We conclude with some open problems raised by the analysis in this chapter. What are the best logical languages for formalizing basic game-theoretic proofs and defining major structures in games? We have suggested that a modal language, perhaps with fixed point operators, is a suitable vehicle, but is this borne out with further game-theoretic phenomena? This question of fit to basic reasoning about interaction will be addressed in the chapters to follow in Parts I and II. Related to this issue of language design is another: What are the most natural notions of structural equivalence between games? As we have seen, there are several natural levels here, and none of the invariances that we have mentioned seems to exhaust the topic.

1.5 Literature

This chapter is based on the process view of games from van Benthem (2002a).

There is a broad literature on logics for basic game structure. Pioneering papers on logical languages for games and their extracted information content are Bonanno (1992a, 1993), and a good survey of many varieties of game logics is the chapter van der Hoek & Pauly (2006) in the *Handbook of Modal Logic*. Structural equivalences, view levels, and transformations in game theory were discussed in famous papers such as Thompson (1952) and Kohlberg & Mertens (1986). Further interfaces between logic and game theory will occur throughout Parts I and II of this book, but also in Chapters 12 and 13 on logics for strategic games.

2
Preference, Game Solution, and Best Action

Real games arise over bare game forms when we take into account how players *evaluate* possible outcomes. Available actions give the kinematics of what can happen in a game, but it is only their interplay with evaluation that provides a more explanatory dynamics of well-considered intelligent behavior. How players evaluate outcomes is encoded in utility values, or in preference orderings.

Logic has long been applied to preference structure. In this chapter, we first review some basic preference logic from philosophy and computer science, showing how it applies to games. Then we analyze the paradigmatic solution procedure of Backward Induction as a pilot case, focusing on two features: (1) the role of *rationality* as a bridge law between information, action, and preference, and (2) the role of *recursive approximation* to the optimal outcome. There need not be a unique best level of syntactic detail for such a conceptual analysis, and we will present two levels of zoom. We first define the Backward Induction solution in a first-order fixed point logic of action and preference, making a junction with the area of computational logic. Next, we hide the recursive machinery, and use a modal logic to study the basic properties of "best actions" as supplied by the Backward Induction procedure. As usual, the chapter ends with a statement of further directions, conclusions, and open problems, as well as some selected literature behind the results presented here.

2.1 Basic preference logic

Models To model preferences, we start with a simple setting that lies behind many decision problems, games, and other scenarios (cf. Hanson 2001, Liu 2011).

Definition 2.1 Preference models

Preference models $\boldsymbol{M} = (W, \leq, V)$ are standard modal structures with a set of worlds W standing for any sort of objects that are subject to evaluation and comparison, a binary betterness relation $s \leq t$ between worlds (i.e., t is at least as good as s), and a valuation V for proposition letters encoding unary properties of worlds or other relevant objects. ■

The comparison relation $\leq$ will usually be different for different agents, but in defining the basic logic, we will suppress agent subscripts $\leq_i$ for greater readability. We use the artificial term betterness to stress that this is an abstract comparison relation, making no claim yet concerning the intuitive term preference. Note that we are comparing individual worlds here, not properties of worlds. In actual parlance, preference often runs between generic properties of worlds or events, as in preferring tea over coffee. We will see in a moment how the latter view can be dealt with, too.

Very similar models with plausibility orderings are used for modeling belief (Girard 2008, van Benthem 2007c), and there are also connections with conditional logic and non-monotonic logics. These links will return in Part II of this book.

Constraints on betterness orders Which properties should a genuine betterness relation have? Total orders satisfying reflexivity $\forall x : x \leq x$, transitivity $\forall xyz : ((x \leq y \wedge y \leq z) \rightarrow x \leq z)$, and connectedness $\forall xy : (x \leq y \vee y \leq x)$, are common in decision theory and game theory, as these properties are induced by agents' numerical utilities for outcomes. But in the logical and philosophical literature on preference, a more general medium has been proposed, too, of *pre-orders*, satisfying just reflexivity and transitivity. Then there are four intuitively irreducible basic relations between worlds:

$w \leq v$, $\neg v \leq w$	$(w < v)$	w strictly precedes v
$v \leq w$, $\neg w \leq v$	$(v < w)$	v strictly precedes w
$w \leq v$, $v \leq w$	$(w \sim v)$	w, v are indifferent
$\neg w \leq v$, $\neg v \leq w$	$(w \# v)$	w, v are incomparable

The latter two clauses, although often confused, describe different situations.

One can also put two relations in betterness models: a "weak order" $w \leq v$ for at least as good, and a "strict order" $w < v$ for 'better', defined as $w \leq v$ and $\neg v \leq w$, respectively. This setup fits belief revision (Baltag & Smets 2008) and preference merge (Andréka et al. 2002). The logic of this extended language is axiomatized in van Benthem et al. (2009c), and related to the philosophical literature.

Modal logics Our base models interpret a standard modal language. In particular, a modal formula $\langle\leq\rangle\varphi$ will make the following local assertion at a world w:

$$\boldsymbol{M}, w \models \langle\leq\rangle\varphi \quad \text{iff} \quad \text{there exists a } v \geq w \text{ with } \boldsymbol{M}, v \models \varphi$$

that is, there is a world v at least as good as w that satisfies φ. In combination with other standard modal operators, in particular, a *universal modality* $U\varphi$ saying that φ holds at all worlds, this formalism can express quite a few further notions.

EXAMPLE 2.1 Defining conditionals from preference
Consider the bimodal formula

$$U\langle\leq\rangle[\leq]\varphi$$

This says that everywhere, there is some better world upward from which φ holds. In finite pre-orders, this says that all maximal elements in the ordering (having no properly better worlds) satisfy φ. But then we are close to other basic notions turning on maximality. Boutilier (1994) showed how a preference modality can define conditionals $\psi \Rightarrow \varphi$ in the style of Lewis (1973) with the following combination:

$$U(\psi \to \langle\leq\rangle(\psi \wedge [\leq](\psi \to \varphi)))$$

This is simply the standard ψ-relativized form of $U\langle\leq\rangle[\leq]\varphi$, saying, at least in finite models, that φ is true in all maximal worlds satisfying ψ. ■

Later on, we will use this modal language to define the Backward Induction solution for extensive games. By doing things this way, we can use the standard machinery of modal deduction to analyze the behavior of conditionals, or game-theoretic strategies.

The modal base logic of preference pre-orders is the system $S4$, while connectedness validates $S4.3$. In general, assumptions on orderings induce modal axioms by the standard technique of frame correspondences (cf. Blackburn et al. 2001).

Propositional preference The modal language describes local properties of betterness at worlds. But a betterness relation need not yet determine what we mean by agents' preferences in a more colloquial sense. Many authors consider preference to be a relation between propositions calling for comparison of sets of worlds. For a given relation $\leq$ among worlds, this may be achieved by "set lifting" of relations given on points. One ubiquitous proposal in such lifting is the $\forall\exists$ stipulation that

$\forall\exists$-*rule* a set Y is preferred to a set X if $\forall x \in X\, \exists y \in Y : x \leq y$

However, alternatives are possible. Von Wright's pioneering view of propositional preference is analyzed in van Benthem et al. (2009c) as the $\forall\forall$ stipulation that

$\forall\forall$-*rule* a set Y is preferred to a set X if $\forall x \in X\, \forall y \in Y : x \leq y$

Liu (2011) reviews proposals for set lifted relations in various fields, but concludes that no consensus on one canonical notion of preference seems to have ever emerged. Preference as a comparison between propositions may depend on the scenario. For instance, in a game, when comparing sets of outcomes that can be reached by available moves, players have options. They might prefer a set whose minimum utility value exceeds the maximum of another: this is like the $\forall\forall$ reading, which can produce incomparable nodes. But it would also be quite reasonable to require that the maximum of one set exceeds the maximum of the other, producing a connected order that would be rather like the $\forall\exists$ reading.

EXAMPLE 2.2 Options in set lifting
Consider the choice between move L and move R in the following:

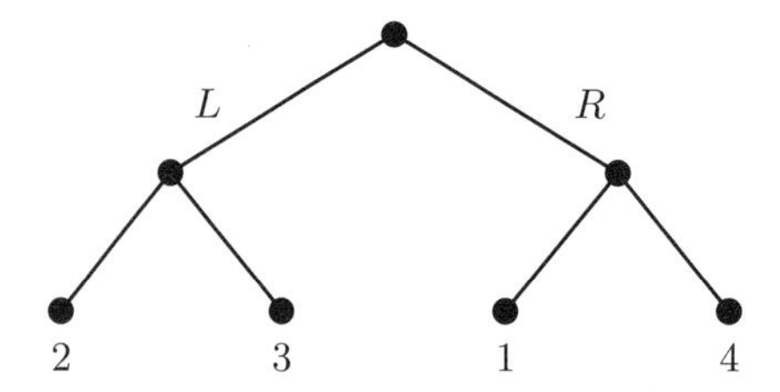

The $\forall\exists$ reading prefers R over L, the $\forall\forall$ reading makes them incomparable, while maximizing minimal outcomes would lead to a preference for L over R. ■

Modal logics again Many lifts are definable in a modal base logic, with a few extra gadgets. For instance, using the universal modality $U\varphi$ (saying that φ is true in all worlds), with formulas standing for sets of worlds, $\forall\exists$ preference is expressed by $U(\varphi \to \langle\leq\rangle\psi)$. For $\forall\forall$ preference, things are more complex: see van Benthem et al. (2009c) for a proposed solution.[7]

7 An alternative view in Jiang (2012) defines set preferences by a process of comparing samples, and presents a dynamic logic of object picking.

2.2 Relational strategies and options for preference

Strategies as subrelations of the move relation To make our models of preference fit with games, we make a few changes from the usual setup. A strategy is usually taken to be a function on nodes in a game tree, yielding a unique recommendation for play. But in many settings, it makes sense to think of strategies as nondeterministic binary subrelations of the total relation *move* (the union of all actions in the game) that merely constrain further moves by marking one or more as admissible. This reflects a common sense view of strategies as plans for action, and it facilitates defining strategies in dynamic logic (cf. Chapter 4). Our numerical version of Backward Induction already had this relational flavor. Its computed relation *bi* linked nodes to all daughters with maximal values for the active player, of which there may be more than one.[8]

Solution algorithms and notions of preference Here is another important point about what look like obvious rules of computation: they embody *assumptions about players*. Recall that our Backward Induction clause for the non-active player took a minimal value. This is a worst-case assumption that the active player does not care at all about the other player's interests. But we might also assume some minimal cooperation, choosing maximal values for the other player among the maximal nodes. This variety of versions highlights an important feature: solution methods are not neutral, they tend to encode significant views of a game. Here is another way of phrasing this: things depend on what we mean by rationality.

Rationality: Avoiding stupid moves An active player must compare different moves, each of which, given the relational nature of the procedure, allows for many leaves that can be reached via further *bi*-play. A minimal notion of rational choice says that

> I do not play a move when I have another move whose outcomes I prefer.

---

8 Different views of strategies have been discussed in game theory in terms of plans of action, recommendations for action, or predictions of actions. Different choices may favor relational or functional views (cf. Greenberg 1990, Bonanno 2001). In this chapter, we will mainly follow the recommendation view, although Chapter 8 also casts strategies as beliefs or expectations about future behavior.

This seems plausible, but what notion of preference is meant here? In our first Backward Induction algorithm, a player i preferred a set Y to X if the minimum of its values for i is higher. This is the earlier $\forall\exists$ pattern for a set preference

$$\forall y \in Y\, \exists x \in X : x \leq_i y$$

However, the same notion of rationality allows for alternatives. A common notion of preference for Y over X that we saw already is the $\forall\forall$ view that

$$\forall y \in Y\, \forall x \in X : x \leq_i y$$

Relational Backward Induction The latter view suggests a minimal version of game solution where players merely avoid "strictly dominated" moves that are worse no matter what (Osborne & Rubinstein 1994). This will be our running example.

EXAMPLE 2.3 Relational Backward Induction
Call a move *a dominated* if it has a sibling move all of whose reachable endpoints are preferred by the current player to all reachable endpoints via *a* itself.

Now, first, mark all moves as active. The algorithm works in stages. At each stage, mark dominated moves in the $\forall\forall$ sense of set preference as passive, leaving all others active. In this comparison, reachable endpoints by an active move are all those that can be reached via moves that are still active at this stage. ■

This is a cautious notion of game solution making weaker assumptions about the behavior of other players than our earlier version.[9] We write *bi* for the subrelation of the total *move* relation produced at the end.

EXAMPLE 2.4 Some comparisons
Consider the following games, where the values indicated are utilities for player ***A***. For simplicity, we assume that player ***E*** has no preference between ***E***'s moves:

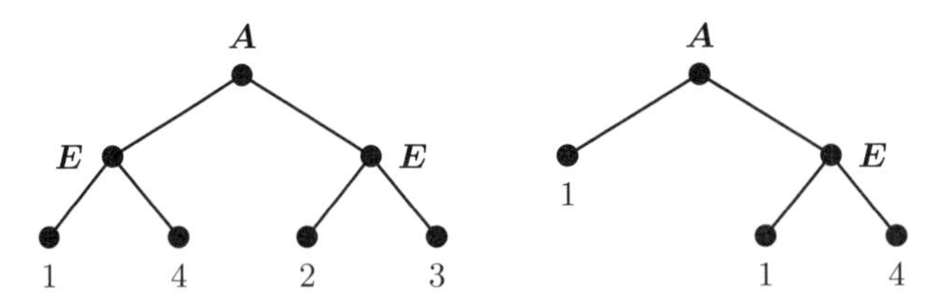

9 Many versions agree on so-called "distinguished games" that separate all histories.

In the game on the left, our original Backward Induction algorithm makes $\boldsymbol{A}$ go right, as the minimum 2 is greater than 1. But cautious *BI* accepts both moves for $\boldsymbol{A}$, as none strictly dominates the other. This may be viewed as set-lifted preference for very risk-averse players.

Interestingly, both versions pass all moves in the game to the right. This seems strange, as $\boldsymbol{A}$ might go right at the start, having nothing to lose, and a lot to gain. But analyzing all variants for preference comparisons between moves is not our goal here, and we move on to other topics. ■

EXAMPLE 2.5 More comparisons
Different views of Backward Induction also emerge when we think of Nash equilibria for *functional strategies*. Consider the following game:

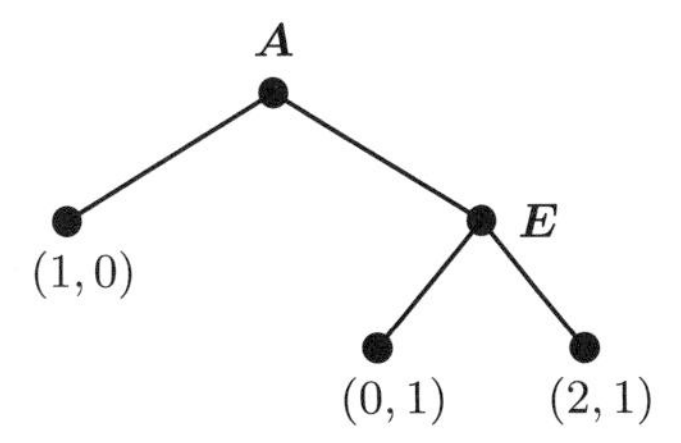

Our cautious Backward Induction computation allows all moves in this game, since no move is dominated. However, there are two Nash equilibria in functional strategies: (L, L) and (R, R), corresponding to pessimistic or optimistic views on $\boldsymbol{A}$'s part as to what $\boldsymbol{E}$ would do at $\boldsymbol{E}$'s turn. Instead of focusing on relations, it would also be possible to analyze Backward Induction in terms of strategy profiles in equilibrium, using the models of games in Chapters 6 and 12. We will not explore this alternative logical route in this book. ■

2.3 Defining Backward Induction in fixed point logics

Defining the Backward Induction solution is a benchmark for logics of games.[10] We start by citing a result from van Benthem et al. (2006b).

10 The game-theoretic literature distinguishes the Backward Induction path: the actual history produced, and the Backward Induction strategy that includes off-path behavior whose interpretation may be questioned. The characterizations of Aumann (1995, 1999) are for the Backward Induction path only. Our approach covers the whole strategy.

THEOREM 2.1 On finite extensive games, the Backward Induction strategy is the largest subrelation σ of the total *move* relation that has at least one successor at each node, while satisfying the following property for all players i:

> *Rationality* (RAT) No alternative move for the current player i yields outcomes via further play with σ that are all strictly better for i than all outcomes resulting from starting at the current move and then playing σ all the way down the tree.

The following picture illustrates more concretely what this says:

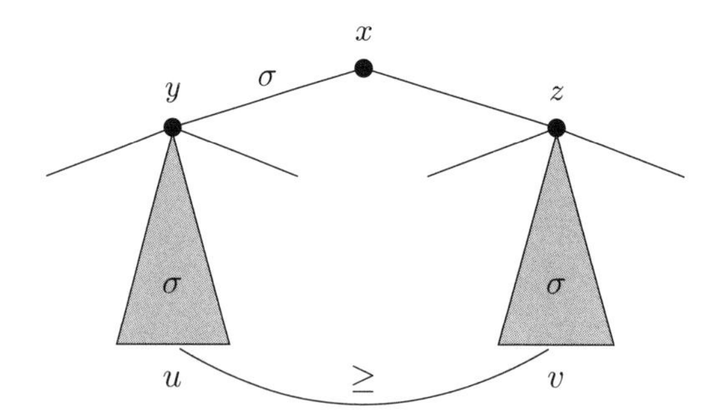

The shaded area in this diagram is the part that can be reached via further play with our strategy σ. The proof that the property RAT is necessary and sufficient for capturing the Backward Induction solution is by a straightforward induction on the depth of finite game trees.

Stated in more syntactic terms, RAT expresses a "confluence property" for action and preference, where steps in different subtrees get linked by preference:

$$CF \quad \forall x \forall y \Big(\big(Turn_i(x) \wedge x\sigma y \big) \rightarrow \forall z \big(x \; move \; z \rightarrow \exists u \exists v (end(u) \wedge end(v) \wedge y\sigma^* v \wedge z\sigma^* u \wedge u \leq_i v) \big) \Big)$$

This is the basis for definability of the Backward Induction solution in a well-known system from computational logic, *first-order fixed point logic* LFP(FO) (Ebbinghaus & Flum 1999). The system LFP(FO) extends first-order logic with smallest and greatest fixed point operators producing inductively defined new predicates of any arity, a highly expressive device going far beyond the μ-calculus of Chapter 1 that only defined new unary predicates.[11]

THEOREM 2.2 The Backward Induction relation is definable in LFP(FO).

Proof In the consequent of the above $\forall\forall\exists\exists$ format for CF (confluence), all occurrences of the symbol σ are syntactically positive. This positive syntax allows for a greatest fixed point operator defining *bi* in LFP(FO) as the following relation S:

$$\nu S, xy \bullet \forall x \forall y \Big(\big(Turn_i(x) \land Sxy \big) \to \forall z \big(x \, move \, z \to \exists u \exists v (end(u) \land end(v) \land yS^*v \land zS^*u \land u \leq_i v) \big) \Big)$$

It is shown in detail in van Benthem & Gheerbrant (2010) how the successive steps of the cautious Backward Induction algorithm match the approximation stages for the greatest fixed point denoted by this formula, starting from the total *move* relation, and refining it downward. ■

LFP(FO) is a fundamental system in its field, and hence our analysis has made a significant junction between game solution methods and logics of computation.

Variants Variants can be defined, too, with, say, an $\forall\forall\forall\exists$ format $\forall x \forall y((Turn_i(x) \land x\sigma y) \to \forall z(x \, move \, z \to \forall u((end(u) \land y\sigma^* u) \to \exists v(end(v) \land z\sigma^* v \land v \leq_i u))))$. This syntax is no longer positive for σ, and existence and uniqueness results now require an appeal to the well-foundedness of game trees, as well as the use of special logics for trees. These matters are studied extensively in Gheerbrant (2010).

2.4 Zooming out to modal logics of best action

Fixed point logics make solution procedures for games fully explicit in their notation. But logical description of behavior can take place at various levels, either zooming in on formal details that lie below a natural reasoning practice, or doing precisely the opposite, zooming out to useful abstractions that lie above the surface level. In the latter vein, often, we want to hide the details of a computation, and

---

11 The price for this expressive power is that validity in LFP(FO) is non-axiomatizable, and indeed of very high complexity. Still, this system has many uses, for instance, in finite model theory (Ebbinghaus & Flum 1999, Libkin 2004).

merely record properties of the notion of *best action*. An agent may just want to know what to do, without being bothered with all the details behind the relevant recommendation.

For this purpose, it would be good to extract a simple surface logic for reasoning with the ideas in this chapter, while hiding most of the earlier machinery. At this level of grain, modal preference logics become a good alternative again, on top of a standard logic of action as in Chapter 1. In particular, a new modality $\langle best \rangle$ now talks about the best actions that agents have available, encoding some particular style of recommendation. For concreteness, we will interpret it as follows:

> $\langle best \rangle \varphi$ says that is φ true in some successor of the current node that can be reached in one step via the *bi* relation.

THEOREM 2.3 The Backward Induction strategy is the unique relation σ satisfying the following modal axiom for all players i, and for all propositions p, viewed as sets of nodes:

$$(turn_i \wedge \langle best \rangle [best^*](\mathbf{end} \to p)) \;\to\; [move\text{-}i]\langle best^* \rangle (\mathbf{end} \wedge \langle pref_i \rangle p).$$

Proof The proof is a modal frame correspondence, applied to the earlier confluence property CF, which can be computed by standard modal methods.[12] ■

It is easy to find further valid principles in this language, expressing, for instance, that best actions are actions. The following natural issue in this setting goes back to van Otterloo (2005):

OPEN PROBLEM Axiomatize the modal logic of finite game trees with a *move* relation and its transitive closure, turn predicates and preference relations for players, plus a new relation *best* as computed by Backward Induction.[13]

This is just one instance of a global logic for practical reasoning that can be extracted from more detailed game structure. Section 2.7 will mention others.

---

12 The overall form here is a Geach-style convergence axiom (see van Benthem et al. 2012 for the latest results in frame correspondence for modal fixed point logics).

13 We get at least the basic modal logics for moves and for preference as discussed earlier, while the above bridge axiom fixes relevant connections between them. However, looking at details of this calculus, in Chapter 9, we also find a need for relativized predicates $best^P$ referring to moves that are best with a Backward Induction computation restricted to the submodel of all worlds satisfying P.

Pitfalls of complexity Global logics of best action may also be interesting from an unexpected computational perspective. You may think that surface logics should be easy, but there is a snag. The rationality in the above confluence picture entangles two binary relations in tree models for games: one for actions, and another for preference. The resulting *grid structure* for two relations on game trees can encode complex geometric "tiling problems," making the bimodal logics undecidable and nonaxiomatizable (cf. Harel 1985, van Benthem 2010b). We will elaborate on this phenomenon in Chapter 3 when discussing information processing agents with perfect memory (cf. Halpern & Vardi 1989), and once more in Chapter 12 on logics for strategic games with grid-like matrix structures. There is an interesting tension here between two views of simplicity. On the one hand, rationality is an appealing property guaranteeing uniform predictable behavior of agents, but on the other hand, rationality may have a high computational cost in the complexity of its induced logic of agency.[14]

2.5 Conclusion

The main points In this chapter, we have seen how games support a natural combination of existing logics for action and preference. We showed how the resulting game logics with action and preference can deal with the preference structure that is characteristic for real games, up to the point of defining the standard benchmark of the Backward Induction solution procedure. These logics came in two natural varieties of detail. Zooming out on basic global patterns, we found a modal logic of best action that seems of general interest in practical reasoning. Zooming in on details of solution procedures, we showed how the Backward Induction strategy can be defined in richer fixed point languages for games, in particular, the logic LFP(FO) for inductive definitions. This reinforces the general point made in Chapter 1 that game-theoretic equilibrium notions match up well with fixed point logics. In this way, our analysis creates a junction between game theory, computational logic, and philosophical logic. This combination of strands will continue as this book proceeds.

Open problems The themes of this chapter also suggest a number of open problems. These include axiomatizing modal logics of best action, and exploring the

---

14 On the computational complexity of game solution procedures as analyzed in this chapter, cf. Szymanik (2013).

computational complexity of logics of action and preference, especially the effects of varying bridge principles between the two components, of which rationality was just one. Of fundamental importance also would be finding the right structural equivalence for games with preferences, continuing a major theme from Chapter 1. For instance, should we now identify processes when only their best actions can be simulated? Further issues include defining strategies explicitly: in Chapter 1, programs from dynamic logic served this purpose; how can these be extended to deal with preference? Finally, there is the challenge of an extension to infinite games, where Backward Induction has a problematic status (Löwe 2003). Fixed point logics fit well with infinite models, and greatest fixed points like the one we found denote co-inductive objects such as never-ending strategies (cf. Chapter 18). However, the precise relation is still unclear in our setting.

Some of the issues mentioned here will be discussed in our final Section 2.7 on further directions, which can be skipped without loss of continuity. Also, a number of relevant considerations will return later on in this book, in Chapter 4 on strategies and in Chapter 8 on dynamic logics for game solution procedures.

2.6 Literature

This chapter is based on van Benthem (2002a), van Benthem et al. (2006b), and especially, van Benthem & Gheerbrant (2010).

Further texts with many additional insights on preference, games, and logic are Dégremont (2010), Gheerbrant (2010), Liu (2011), and Zvesper (2010).

2.7 Further directions

For the reader who wants more food for thought, we list a few further directions.

Entangling preference and belief Our first topic is an important reinterpretation of what we have done in this chapter. Looking more closely at the earlier confluence property, we see that it makes comparisons between current moves based on an assumption about future play, viz. that Backward Induction will be played henceforth. This reveals another aspect of game solution: it entangles preference with belief. Backward Induction, and indeed also other game solution methods, are really procedures for *creating expectations* of players about optimal behavior.

In line with this, our earlier version of rationality can be amended to a perhaps more sophisticated notion of "rationality-in-beliefs": Players play no move whose results they believe to be worse than those of some other available move. Beliefs will be the topic of later chapters, especially in Part II where we analyze the dynamics of game solution procedures as a process of successive belief changes. As will be shown in Chapter 8, strategies are very much like beliefs, in a precise formal sense.

Other social scenarios Backward Induction is just one scenario, and many other scenarios in games, or social settings generally, can be studied in the above spirit.

Variety of logics for games It is not written in stone that a language of games has to be a basic modal one. Stronger formalisms are also illuminating, such as the use of temporal Until operators in van Benthem (2002a) to define preferences between nodes in game trees. Further extensions occur in defining Nash equilibria (cf. Chapter 12): these require intersections of relations, a program operation that goes beyond PDL. Going back to Chapter 1, such alternative logics may also suggest new notions of process equivalence for games.

One more rough level: Deontic logic Our line of zooming out to best action has counterparts in recent top-level logics of complex social scenarios in terms of common sense notions such as "may" and "ought" that pervade ordinary discourse. Tamminga & Kooi (2008), Roy (2011), and Roy et al. (2012) provide further illustrations of this more global zoom level, relating games to deontic logics for reasoning about permissions and obligations.

Priority models for preference Our models for preference were sets of abstract worlds with a primitive betterness relation. This does not record why agents would have this preference, based on what considerations. Reasons for preference are explicitly present in the "priority graphs" of Liu (2011) that list the relevant properties of worlds with their relative importance. Betterness order on worlds is then derived from priority graphs in a natural lexicographical manner.

This richer style of analysis makes sense for games, too, since we often judge outcomes in terms of ordered goals that we want to see satisfied, rather than an immediate preference ordering. See Chapter 22 on some uses of players' goals in knowledge games over epistemic models, and Grossi & Turrini (2012) and Liu (2012) for an application of priority graphs to short-horizon games. Osherson & Weinstein (2012) present another take on reason-based preference, with various modal connectives reflecting reasons to desire that a sentence be true.

Game equivalence with preferences As in Chapter 1, a greater variety of languages also suggests variety in structural notions of game equivalence. Our earlier notions of game invariance become more delicate in the presence of preferences.

As a simple illustration, we can intuitively identify the following two non-bisimilar one-player games as the same:

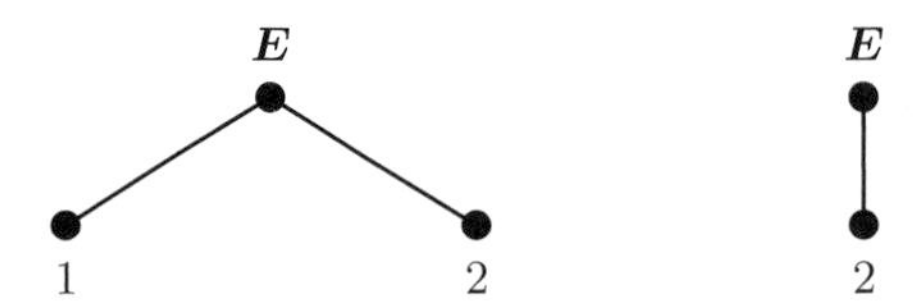

Three approaches to equivalence have been explored in van Benthem (2002a): (i) direct simulation on preference links, (ii) pruning games to their best actions, and then using the standard action invariances of Chapter 1, and (iii) preference-equivalence of games as supporting the same Nash equilibria.

What complicates intuition here is an issue of equivalence for whom? The above games only seem equivalent for rational players ***E***, not for erratic or stupid ones. But then, a more general point arises. Should our recurrent question of game equivalence perhaps be relativized to "agent types" for players? Such an agent orientation would be a significant shift in perspective compared to standard process theories in computational logic, of a kind discussed further in Part II of this book.

3
Games with Imperfect Information

Games do not just consist of moves and outcomes that can be evaluated; they are also *played*, and focusing on players naturally involves their knowledge and information. In particular, in games of imperfect information such as card games, or many natural social scenarios, players need not know exactly where they are in the game tree. Reasons for this may be diverse: limited observation (say, hidden information in card games), processing limitations (say limited memory), or yet other factors. Thus, in addition to the actions and preferences studied so far, we need knowledge and eventually also belief, drawing once more on ideas from philosophical logic. In this chapter, we will show how games of imperfect information fit with standard epistemic logic, and then resume some of our earlier themes of Chapters 1 and 2 in this richer setting.

A conspicuous trend in all this is the emergence of players and play as an object of study in its own right. Already in Chapter 1, while game forms were just a static playground of all possible actions, they were navigated by actual players as they chose their strategies. Chapter 2 then added more information about how players viewed this playground in terms of their preferences. In this chapter, we emphasize one more aspect of players, namely their information processing as a game proceeds. All of this will eventually lead to the dynamic-epistemic logics of Part II providing a Theory of Play.

3.1 Varieties of knowledge in games

In this chapter, we will mainly use knowledge as modeled in epistemic logic (cf. Fagin et al. 1995, van Benthem 2010a). This is essentially the fundamental notion of an agent having semantic information that something is the case, referring to the current range of possibilities for the actual situation.[15] Foundations of game theory also involve the beliefs of players, and so do more general logics of agency, but doxastic logic will only be a side theme in this chapter, coming into play mostly in Part II of this book, by enriching the epistemic models presented here.

Perfect information Knowledge arises in games in different ways. Most scenarios discussed so far in this book are games of perfect information, where, intuitively, players know exactly which node of the game tree they are at as play proceeds. This corresponds to several informational assumptions: players know the game they are playing, and also, their powers of observation allow them to see exactly what is going on as play unfolds. But even then, in a branching tree, there is uncertainty as to the future: the players do not know which history will become actual as play proceeds. In this sense, the modal logic of branching actions in Chapter 1 is already an epistemic logic talking about possible future continuations of the current stage, and one can think of Backward Induction as a way of predicting the future in this setting of uncertainty. But for now, we discuss another important sense in which knowledge enters games.

Imperfect information In games of imperfect information, ignorance gets worse, and players need not know their exact position in a game tree as play proceeds. As before, this can be analyzed at two levels: that of local actions, and that of powers over outcomes, each with their own notions of structural invariance. We will focus on the former, deferring the latter view mainly to Chapter 11. Imperfect information games support an epistemic language, and with actions added, a combined modal-epistemic logic.

Varieties of knowledge in games Yet other kinds of knowledge are relevant to players, less tied to the structure of the game, such as knowledge about the strategies of others. We will discuss all varieties mentioned here: procedural, observational, and multi-agent, in Chapter 6, and once again in Chapter 10 when exploring our Theory of Play.

---

15 Other views of knowledge, based on fine-grained information closer to the syntax of propositions, are also relevant to understanding games, for instance, in studies of play that involve awareness (cf. Halpern & Rêgo 2006). However, they will remain marginal in this book, except for a brief appearance in Chapter 7.

3.2 Imperfect information games at a glance

Games of imperfect information are process graphs as in Chapter 1, but with a new feature of uncertainty marking between certain nodes.

EXAMPLE 3.1 Game trees with and without uncertainty
First consider a typical picture of an extensive game as before:

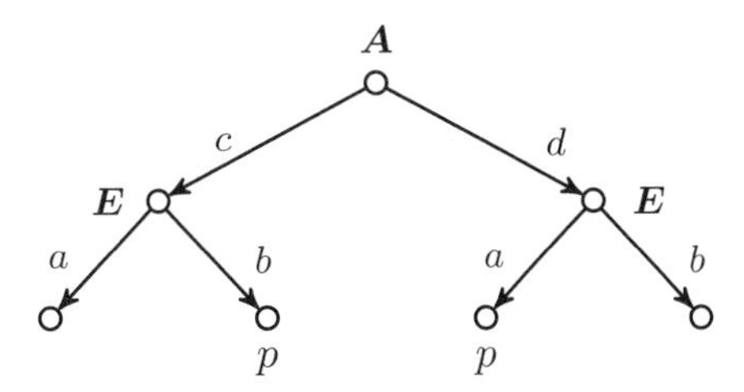

Here, at the root, a modal formula $[c \cup d]\langle a \cup \rangle p$ expressed player $\boldsymbol{E}$'s having a strategy for achieving outcomes satisfying p. To deal with imperfect information, game theorists draw dotted lines, whose equivalence classes form "information sets." Consider the above game with a new feature, an uncertainty for player $\boldsymbol{E}$ about the initial move played by $\boldsymbol{A}$. Perhaps $\boldsymbol{A}$ put the initial move in an envelope, or $\boldsymbol{E}$ did not pay attention ...

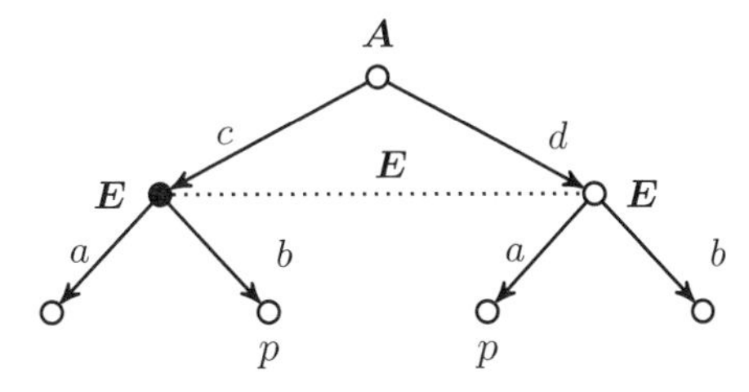

Intuitively, $\boldsymbol{E}$'s situation has changed considerably here. While the old strategy for achieving p still exists, and $\boldsymbol{E}$ knows this, it is unclear whether to choose *left* or *right*, since the exact location is unknown. $\boldsymbol{E}$'s powers have diminished. ■

Books on game theory are replete with more sophisticated examples of imperfect information games (cf. Osborne & Rubinstein 1994), and further illustrations will occur in the course of this chapter, and at various places later on in this book.

3.3 Modal-epistemic logic

Process graphs with uncertainty Games of imperfect information are process graphs as in Chapter 1, but with a new structure among states, as brought out in the following notion.

DEFINITION 3.1 Epistemic process graphs
Epistemic process graphs are of the form $\boldsymbol{M} = (S, \{R_a \mid a \in A\}, \{\sim_i \mid i \in I\}, V)$ consisting of game models as in Chapter 1, with added binary equivalence relations $\sim_i$ for players i that represent their uncertainties as to which state is the actual one. When relevant, we also mark one particular state s in S as the actual one. ■

Thus, the relations $\sim_i$ encode when player i cannot tell one node from the other over the course of the game. Epistemic process graphs can be generalized to arbitrary binary epistemic accessibility relations in a standard fashion, but we will have no need of this generality for the points to be made in this chapter. We can also add preference relations as in Chapter 2, where, of course, players may have quite different preferences between epistemically possible states (cf. Liu 2011).

In principle, any sort of uncertainty pattern might occur in epistemic process graphs. At various stages, players need not know what the opponent has played, what they have played themselves, whether it is their turn, or even whether the game has ended. This soon takes us beyond the usual game-theoretic setting where uncertainty links only run between players' own turns. (Uncertainty about what others will do is explored in Battigalli & Bonanno 1999a.)

EXAMPLE 3.2 Further imperfect information games
Plausible scenarios for the following pictures are left as an exercise for the readers.

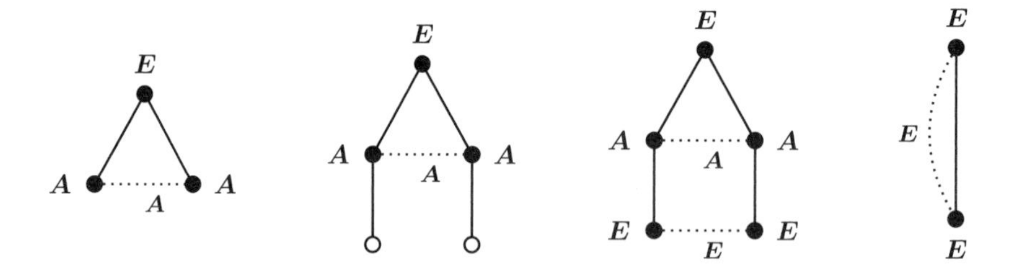

We will take these diagrams in great generality, since epistemic logic over arbitrary structures fits well with the literature on imperfect information games (cf. Bonanno 1992a, Battigalli & Bonanno 1999a, and Battigalli & Bonanno 1999b). ■

REMARK General models
It is important to note that epistemic process graphs are just one way of associating epistemic models with games. As we will see in Chapters 5 and 6, and in Part II, it is equally feasible to take other objects as states, say complete histories of a game or more abstract possible worlds, to model other forms of information that players may have as they navigate a game. Indeed, general models $\boldsymbol{M} = (S, \{\sim_i \mid i \in I\}, V)$ of epistemic logic leave the choice of the set of states S completely free, subsuming all of these special cases. The epistemic definitions to follow in this chapter work in this full generality, even where stated only for epistemic process graphs.

Modal-epistemic language Epistemic process graphs are models for a joint modal-epistemic language. All the modal formulas of Chapter 1 still make sense for the underlying action structure, but now we can also make more subtle assertions about players' adventures en route, using the dotted lines as a standard accessibility relation for epistemic knowledge operators interpreted in the usual way.

DEFINITION 3.2 Truth definition for knowledge
At stage s in a model, a player *knows* exactly those propositions φ that are true throughout the information set of s, i.e., at all points connected to s by an epistemic uncertainty link:

$$\boldsymbol{M}, s \models K_i\varphi \quad \text{iff} \quad \boldsymbol{M}, t \models \varphi \text{ for all } t \text{ with } s \sim_i t.$$

This says that formula φ is true all across agent i's current semantic range. ■

This epistemic logic view applies at once to imperfect information games.

EXAMPLE 3.3 Ignorance about the initial move
In the imperfect information game depicted earlier, after $\boldsymbol{A}$ has played move c in the root, in the state marked by the black dot (in fact, in both states in the middle), $\boldsymbol{E}$ knows that playing move a or b will give p, as the disjunction $\langle a \rangle p \vee \langle b \rangle p$ is true at both middle states. This may be expressed by the epistemic formula

$$K_{\boldsymbol{E}}(\langle a \rangle p \vee \langle b \rangle p)$$

On the other hand, there is no specific move of which $\boldsymbol{E}$ knows that it guarantees a p-outcome, which shows in the black node in the truth of the formula

$$\neg K_{\boldsymbol{E}}\langle a\rangle p \wedge \neg K_{\boldsymbol{E}}\langle b\rangle p$$

This is the famous "de dicto" versus "de re" distinction from philosophical logic.

Such finer distinctions are typical for a language with both actions and knowledge for agents. They also occur in fields like philosophy, AI, and computer science. We will see similar patterns in the epistemic temporal logics of Chapters 5 and 9.

Iterations and group knowledge An important and characteristic feature of epistemic languages is iteration. Players can have knowledge about each other's knowledge and ignorance via formulas such as $K_{\boldsymbol{E}}K_{\boldsymbol{A}}\varphi$ or $K_{\boldsymbol{E}}\neg K_{\boldsymbol{A}}\varphi$, and this may be crucial to understanding a game. Indeed, very basic informational episodes such as asking a question and giving an answer crucially involve knowledge and ignorance about the information of others.

EXAMPLE 3.4 A model for a question/answer scenario
A question answer episode might start as follows. Agent $\boldsymbol{E}$ does not know whether p is the case, but $\boldsymbol{A}$ is fully informed about it ($\sim_{\boldsymbol{A}}$ is just the identity relation). The black dot indicates the actual world, the way things really are:

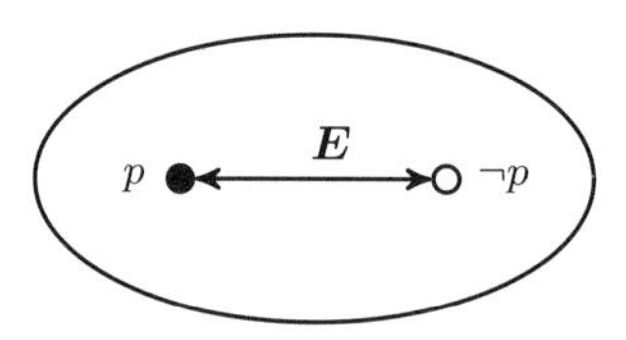

In the black dot to the left, the following epistemic formulas are true

$$p,\ K_{\boldsymbol{A}}p,\ \neg K_{\boldsymbol{E}}p \wedge \neg K_{\boldsymbol{E}}\neg p,\ K_{\boldsymbol{E}}(K_{\boldsymbol{A}}p \vee K_{\boldsymbol{A}}\neg p)$$
$$C_{\{\boldsymbol{E},\boldsymbol{A}\}}(\neg K_{\boldsymbol{E}}p \wedge \neg K_{\boldsymbol{E}}\neg p),\ C_{\{\boldsymbol{E},\boldsymbol{A}\}}(K_{\boldsymbol{E}}(K_{\boldsymbol{A}}p \vee K_{\boldsymbol{A}}\neg p))$$

Now $\boldsymbol{E}$ can ask $\boldsymbol{A}$ whether p is the case: $\boldsymbol{E}$ knows that $\boldsymbol{A}$ knows the answer. ■

The new notation $C_{\{\boldsymbol{E},\boldsymbol{A}\}}$ is significant here. Communication and playing games are forms of shared agency that create groups of agents with knowledge of their own. A ubiquitous instance is the following notion.

DEFINITION 3.3 Common knowledge
A group G has *common knowledge* of a proposition φ in an epistemic model $\boldsymbol{M}$ at state s if φ holds throughout the reachable part of the relevant epistemic model.

Formally, $\boldsymbol{M}, s \models C_G\varphi$ iff φ is true in all those states that can be reached from s in a finite number of $\sim_i$ steps, where the successive indices i can be any members of the group G. ■

In the above imperfect information game, $\boldsymbol{E}$'s plight was common knowledge between the players. Here is a scenario showing how common knowledge may arise:

EXAMPLE 3.4, CONTINUED Epistemic dynamics
There is a dynamics to the earlier question that can be modeled separately. Intuitively, the truthful answer "Yes" is an event that changes the initial model, taking it to the following one-point model where maximal information has been achieved:

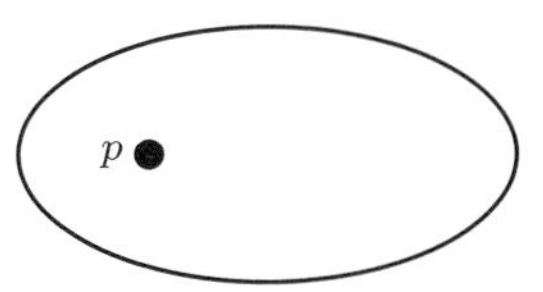

The common knowledge formula $C_{\{\boldsymbol{E},\boldsymbol{A}\}}p$ now holds at the black dot. ■

We will not make much use of iterations in this chapter, but they will occur at many places in this book, as they are crucial to understanding social interaction. The latter theme is of course much broader than what we have just shown (see van Ditmarsch et al. 2007, van Benthem 2011e for logical studies of communicative acts), and Chapter 7 will take up this theme in much greater generality.

Uniform strategies and nondeterminacy Let us now return to the topic of actions. A striking aspect of the preceding imperfect information game is its nondeterminacy. Uncertainty links have effects on playable strategies. Telling player $\boldsymbol{E}$ to do the opposite of player $\boldsymbol{A}$ was a strategy guaranteeing outcome p in the original game, but, although still present, it is unusable now. For, $\boldsymbol{E}$ cannot tell what $\boldsymbol{A}$ did. We can formalize this notion using a special kind of strategies.

DEFINITION 3.4 Uniform strategies
A strategy in an imperfect information game is *uniform* if it prescribes the same move for a player at epistemically indistinguishable turns for that player. ■

This restriction has an epistemic flavor, as players must know what the strategy tells them to do. In Example 3.1, neither player has a uniform winning strategy, interpreting p as the statement that player $\boldsymbol{E}$ wins. Player $\boldsymbol{A}$ did not have one to begin with, and $\boldsymbol{E}$ has lost the one that used to work.

We will discuss the logical form of uniform strategies in more detail in Chapter 4, including the epistemic issue of what players know about their effects.

Calculus of reasoning In addition to defining basic notions, a logic of games should provide the means for analyzing systematic reasoning about players' available actions, knowledge, and ignorance in arbitrary models of the above kind.

FACT 3.1 The complete set of axioms for validity in modal-epistemic logic is:

(a) The minimal modal logic for each operator $[A]$.

(b) Epistemic $S5$ for each knowledge operator K_i.

With a common knowledge operator added to the language, we get a minimal logic incorporating principles for program iteration from propositional dynamic logic (cf. Fagin et al. 1995).[16] In Chapter 5, we will see it at work when analyzing the behavior of programs for uniform strategies.

3.4 Correspondence for logical axioms

Our minimal epistemic action logic is weak. In particular, it lacks striking axioms linking knowledge and action. Still, some commutation principles look attractive, as we shall see now.

EXAMPLE 3.5 Interchanging knowledge and action
Consider the modal-epistemic interchange law

$$K_i[a]\varphi \to [a]K_i\varphi$$

This seems valid for many scenarios. A person knows a priori that after dropping a full teacup, there will be tea on the floor. And after having dropped this cup, the person knows that there is tea on the floor. Even so, this was not in our minimal logic, since it can be refuted. A person may know that drinking too much makes people boring, but after drinking too much, the person is not aware of being boring.

16 One might extract a few further special-purpose axioms from Osborne & Rubinstein (1994). The fact of who is to move is taken to be common knowledge between players: $turn_i \to C_{\{1,2\}}turn_i$. Also, all nodes in the same information set have the same possible actions: $\langle a\rangle\top \to C_{\{1,2\}}\langle a\rangle\top$. Both properties will reappear in later chapters.

This actually tells us something interesting. The axiom is valid for an action without "epistemic side-effects," but it may fail for actions with epistemic import. ∎

This is not just an amusing example. In games, players often know at the start what particular strategies may achieve, and of course, they also want to have that knowledge available as they play successive moves of these strategies.

Similar observations hold for the converse $[a]K_i\varphi \to K_i[a]\varphi$. It fails in the imperfect information game of Example 3.1. In the black intermediate point, $\boldsymbol{E}$'s going right will reveal afterward that p holds, even though $\boldsymbol{E}$ does not know right now that going right leads to a p-state.

But then, this game is a strange scenario, in that the uncertainty in the middle has suddenly evaporated at the end. Indeed, the law $[a]K_i\varphi \to K_i[a]\varphi$ is valid in games where agents learn only by observing new events and not suddenly by a miracle. To remove the miracle in the above scenario, one could add an explicit information-producing action at the end of seeing where we are, a theme that will return at the end of this chapter.

Correspondence analysis of special axioms We can be more precise here, using a well-known technique. What the above logical axioms say about the special games where they are valid can be stated exactly in terms of modal frame correspondences. Typically, what we find then are special assumptions about epistemic effects of actions in games, and players' abilities.

In what follows, we say that a formula holds in a graph if it is true at all points in that graph under all valuations for its proposition letters.

FACT 3.2 $K_i[a]\varphi \to [a]K_i\varphi$ holds in an epistemic process graph $\boldsymbol{G}$ iff $\boldsymbol{G}$ satisfies the following property: $\forall xyz : ((xR_a y \wedge y \sim_i z) \to \exists u : (x \sim_i u \wedge uR_a z))$.

Proof (a) If the property holds, then so does the axiom. For any valuation V on $\boldsymbol{G}$, suppose that $\boldsymbol{M} = (\boldsymbol{G}, V), s \models K_i[a]\varphi$. Now consider any point t with $sR_a t$: we show that $\boldsymbol{M}, t \models K_i\varphi$. So, let v be any world epistemically related to t (i.e., $t \sim_i v$). By the given property, there is also a point u with $s \sim_i u \wedge uR_a v$. But then, by the assumption about s, we have $\boldsymbol{M}, u \models [a]\varphi$, and hence $\boldsymbol{M}, v \models \varphi$. (b) Assume that the interchange axiom holds at point s in $\boldsymbol{G}$. Now, we choose a special valuation V making φ true only in those worlds w that satisfy the condition $\exists u : s \sim_i u \wedge uR_a w$. Clearly, we have $\boldsymbol{M} = (\boldsymbol{G}, V), s \models K_i[a]\varphi$. Thus, it follows from our assumption about the axiom that also $\boldsymbol{M}, s \models [a]K_i\varphi$. This means that, for any t and v with $sR_a t \wedge t \sim_i v$, $\boldsymbol{M}, v \models \varphi$. But by the definition of V, this says that $\exists u : s \sim_i u \wedge uR_a v$, proving the stated property. ∎

More graphically, this relation between an action a and the information of an agent i expresses a well-known property of "confluence" for two binary relations:

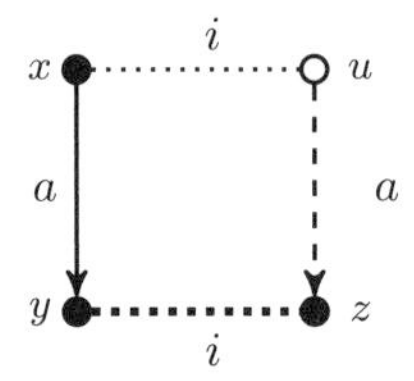

The commuting diagram says that agent i's performing action a can never create new uncertainties for i: all dotted lines after a must have been inherited from before.

This property has a clear import for games. It says that the player has *perfect recall*, in the following strong sense: one remembers what one knew before. Indeed, new uncertainty can only arise when comparing two different events a and b, which can only happen if a player makes a partial observation of a current action, and does not know just which one. This could be because the source of the action is another player, as in our earlier running example, or for more general reasons of privacy in communication. We will analyze scenarios of the latter sort in much greater detail in Part II of this book.[17]

REMARK Learning by observation only
The converse axiom $[a]K_i\varphi \to K_i[a]\varphi$ can be analyzed in the same style, leading to a corresponding property $\forall xyz : ((x \sim_i y \wedge yR_az) \to \exists u : (xR_au \wedge u \sim_i z))$. Its matching commuting diagram says that old uncertainties are inherited under publicly observed actions a. A generalized form of this property is sometimes called "no miracles" (cf. Chapter 9): old uncertainties of an agent i only disappear when new events make distinctions.

General logical methods The above proof exhibits a simple pattern that can be generalized. Many similar equivalences follow by standard modal techniques (cf. Blackburn et al. 2001). Indeed, frame correspondences between modal-epistemic axioms and constraints on imperfect information games need not be found ad hoc.

17 There are also other versions of perfect recall in the game-theoretic literature: see Bonanno (2004a) and Bonanno (2004b) for some alternatives to the current version, expressible in epistemic-temporal logics.

They can be computed using effective algorithms, because many natural axioms concerning games have so-called "Sahlqvist forms."

Coda: Diversity of players Highlighting the role of players as information-processing agents, as done in this chapter, opens up a natural dimension of diversity for players themselves. We have emphasized perfect recall, but many other kinds of players are possible, such as automata with limited memory: by now, these are perhaps the majority of the companions we interact with in our lives. The logics of this chapter are not all biased toward agents having flawless memory. Indeed, it is also easy to write axioms for memory-free agents that can only remember the last move that they observed.

FACT 3.3 Memory-free players satisfy $K\varphi \leftrightarrow \bigvee_e \big(\langle e^{\cup}\rangle\top \wedge U(\langle e^{\cup}\rangle\top \to \varphi)\big)$.

Using a universal modality U, and a backward-looking converse modality $\langle e^{\cup}\rangle$ over past actions or events, this formula says that an agent knows that φ is the case if φ is true in all worlds of the model that arise from the same preceding action.[18]

3.5 Complexity of rich game logics

Logics of games may look simple, but they do combine many different notions, such as action, preference, and knowledge. Now it is known that logics with combinations of modalities can have surprising complexity for their sets of valid principles. We first raised this theme in Section 2.4 of Chapter 2, but now elaborate on this topic (see also Chapter 12). What determines this behavior is the mode of combination.[19] In particular, what may look like natural commutation properties can actually drive up complexity immensely, away from decidable logics to undecidable ones.

THEOREM 3.1 The minimal logic of two modalities [1], [2] satisfying the axiom $[1][2]\varphi \to [2][1]\varphi$, and with an added universal modality U, is undecidable.

In other words, simple decidable component logics may combine into undecidable systems, or sometimes even unaxiomatizable or worse, depending on the mode of

18 Liu (2008) studies agent variety in terms of memory and powers of observation and inference, a perspective that will become important in the Theory of Play of Chapter 10.

19 This high complexity was first shown in Halpern & Vardi (1989) for epistemic-temporal logics of agents with perfect recall. The paper also cataloged complexity effects of other relevant epistemic properties for a wide array of logical languages.

combination. For instance, the preceding bimodal logic is Π^1_1-complete (i.e., the validities are as complex as the full second-order theory of arithmetic).

The technical reason for this explosion is that satisfiability in such logics may encode complex geometrical "tiling problems" on the structure $\mathbb{N} \times \mathbb{N}$ (see Harel 1985, Marx 2006, and van Benthem 2010b for mathematical details). Such an encoding works if the models for the logic have a structure with grid cells that looks enough like $\mathbb{N} \times \mathbb{N}$. The earlier commuting diagram imposed by our notion of perfect recall is precisely of this kind, and it is easy to show that epistemic action logics over game trees in a language with a universal modality, or common knowledge plus a future modality, can be non-axiomatizable. Details and further examples can be found in van Benthem & Pacuit (2006).

This complexity danger is sometimes made concrete in a simple picture. Modal logics of trees are harmless, while modal logics of grids are dangerous:

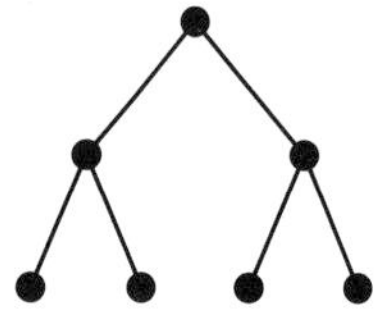
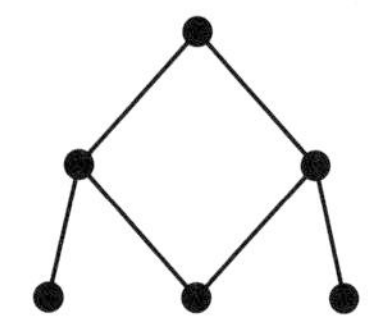

It is the looseness of trees that keeps their logics simple, and it is the close-knit structure of grids that makes logics complex. While extensive games are trees, adding further epistemic relations may induce grid structure, and the same may be true for adding preference order, as we saw in Chapter 2 with the complexity of rationality. Thus, while logic of games is a natural enterprise, it may actually lead to complex systems.

Discussion It is a moot point what these high complexity results mean. An optimistic line is that they are good news, since they show that logic of games is rich mathematically. Next, complexity need not always hurt: it may go down for important areas of special interest, such as finite games. Also, it has been suggested that the type of statement used in tiling reductions is not likely to occur in natural reasoning about interaction, so the part of the logic that is of actual use might escape the clutches of the complexity results.

Logic complexity versus task complexity But the discussion continues. Intuitively, there is a tension in high complexity results for logics of agents with perfect recall, since this property makes agents simple and well behaved. Likewise, commuting diagrams are devices that make reasoning smoother. While this is not a paradox (a theory of regular objects may be richer than one for arbitrary objects), there is an issue whether complexity of a logical system tells us what we really want to know about the complexity of significant tasks faced by agents of the sort we are describing in that system.[20]

3.6 Uncovering the mechanics of the dotted lines

We conclude with a desideratum that will be taken up fully only later in this book. Our discussion of imperfect information games has served its purpose of linking up logics of games with epistemic logic, and bringing the players explicitly into the picture. But even so, it is not yet completely satisfactory. An imperfect information game gives us the traces of some *process* that left players with the uncertainties marked as dotted lines in the game tree. However, one is left guessing what that process may have been, and therefore, we have briefly discussed adding informational events of observation that remove "miracles." Sometimes, it is hard to think of any convincing scenario at all that would leave the given traces. It would be far more attractive to analyze the sort of player and the sort of play that produces imperfect information games directly as a process in its own right. One expects a good deal of variety here, since information in games can come from so many different sources, as we have indicated at the beginning of this chapter.

Logical dynamics To do this, we need a shift in focus, to a logical dynamics (van Benthem 1996) making informational actions and other relevant events explicit. Games involve a wide variety of acts by players: prior deliberation steps, playing actual moves, observing moves played by others, perhaps acts of forgetting or remembering, and even acts of post-game analysis. Beyond these, the relevant dynamic repertoire may include acts of belief revision, preference change, or even of changing the current game. Making the logic of such events explicit is precisely what we will do in Part II of this book, using techniques from dynamic-epistemic logic (van Benthem 2011e).

Here is one instance of immediate relevance to this chapter. By analyzing the dynamics of information flow in games for concrete kinds of players, we see precisely what sorts of imperfect information games arise in natural scenarios. The

20 In the setting of epistemic logic, this issue of internal task complexity versus external logical system complexity was taken up in Gierasimczuk & Szymanik (2011).

representation theorems to be proved in Chapter 9 provide concrete explanations for the characteristic traces left in a game by players with perfect recall endowed with powers of observation and policies of belief revision.

3.7 Conclusion

The main points Imperfect information games record scenarios where players have only limited knowledge of where they are in the game tree. We have shown how the resulting structures are models for a standard epistemic extension of the action logic of games that serves as their minimal logic. How agents traverse such games depends on special assumptions on their memory or powers of observation, and we have shown how perfect recall and other important properties of players (including bounded memory) are reflected precisely in natural additional modal-epistemic axioms via modal frame correspondences. We also noted that there can be hidden complexities in the system of validities, depending on the manner of putting component logics together.

Open problems A number of open problems come to light in our analysis. We already noted the need for uncovering the underlying procedural mechanics that produces the uncertainty lines in our game trees. We have also seen the need for a deeper analysis of the different kinds of knowledge that occur in games: either forward toward the future, or sideways or backward toward the past. Next, there is the obvious challenge of merging imperfect information and players' preferences that drives the true dynamics of imperfect information games. Finally, our brief discussion of uniform strategies in imperfect information games highlighted the important distinction between "knowing that" and "knowing how": a relatively neglected topic in logical studies of knowledge. Many of these topics will return in later chapters of Parts I and II of this book.

3.8 Literature

This chapter is based on van Benthem (2001b) and van Benthem & Liu (1994).

There is a vast literature on imperfect information in games and related social settings such as agency and planning, of which we mention Moore (1985), Fagin et al. (1995), Battigalli & Bonanno (1999a), van der Hoek & Wooldridge (2003), and Bolander & Andersen (2011).

3.9 Further directions

As always, our chapter points at many further directions. We list a few.

Bisimulation and characteristic formulas Modal-epistemic logic is a game-internal language of action and knowledge. As in Chapter 1, then, one can line up two perspectives on its expressive power: (a) the roots of two games satisfy the same modal-epistemic formulas, (b) there is a bisimulation between the games linking the roots. Our earlier theory then generalizes. With imperfect information games, bisimulations need two back-and-forth conditions, one for actions and one for uncertainty links. Further, formulas of dynamic-epistemic logic characterize finite games up to bisimulation equivalence.

Uniform strategies We have seen how the notion of strategy changes in imperfect information games to having the same action at epistemically indistinguishable points. In Chapter 1, strategies were defined as programs in propositional dynamic logic. This approach is generalizable, but we now need the "knowledge programs" of Fagin et al. (1995), whose only test conditions for actions are knowledge statements. Knowing one's strategy is related to the distinction between knowing that and knowing how. A strategy represents know-how for achieving certain goals, and this notion will be studied on its own in Chapter 4.

Powers and game equivalence As in earlier chapters, imperfect information games can be studied at different levels. In this chapter we have taken the fine-grained level of actions and local knowledge. One can also focus on strategic powers of players for influencing the outcomes of the game. To enter into the latter spirit, consider the following game. ***E***'s two uniform strategies give powers $\{1, 3\}$ and $\{2, 4\}$, while ***A***'s two uniform strategies LL and RR give powers $\{1, 2\}$ and $\{3, 4\}$:

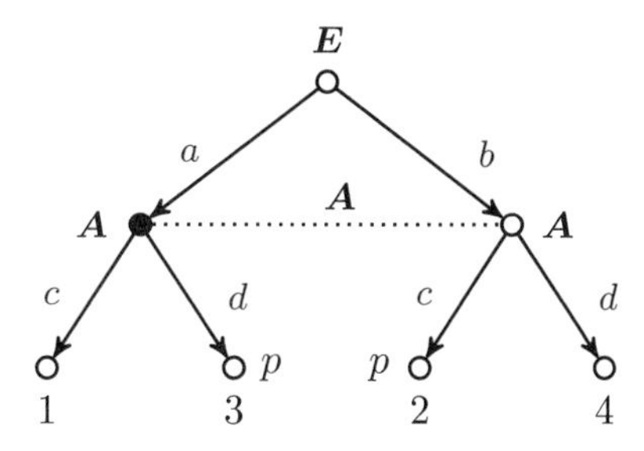

In terms of pure powers given by uniform strategies, this game tree is equivalent to one that interchanges the roles of the two players plus some outcomes. We refer to Chapter 21 for logical studies of imperfect information games like this, and once more to Thompson (1952), Kohlberg & Mertens (1986), and Bonanno (1992b) for game-theoretic studies of such settings. Within our logical framework, we will pursue this more global strategic level of game analysis in Chapter 11.

Fixed point logics for game solution In Chapter 2, the mechanics of game solution came out particularly clearly in fixed point logics. We have not given similar systems here, and indeed the fixed point theory of imperfect information games seems virtually nonexistent. One reason is the proliferation of equilibrium concepts in this area, and hence less of a grip on the logical essentials than what we found with Backward Induction.

Adding preferences An important feature left out altogether in this chapter is players' preferences. While it is not hard to combine the logics of action and knowledge in the above with the logics of preference of Chapter 2, there are many subtleties in trying to understand the precise reasoning that is needed in concrete instances. Just to illustrate the entanglement of imperfect information and preference, recall a distinction made at the start of this chapter. We had knowledge linked to observation in imperfect information games, but also knowledge linked to uncertainty about how any game will proceed. We did not bring the two views together, but both arise in concrete settings.

Here are two scenarios, with our usual outcome order ($\boldsymbol{A}$-value, $\boldsymbol{E}$-value).[21]

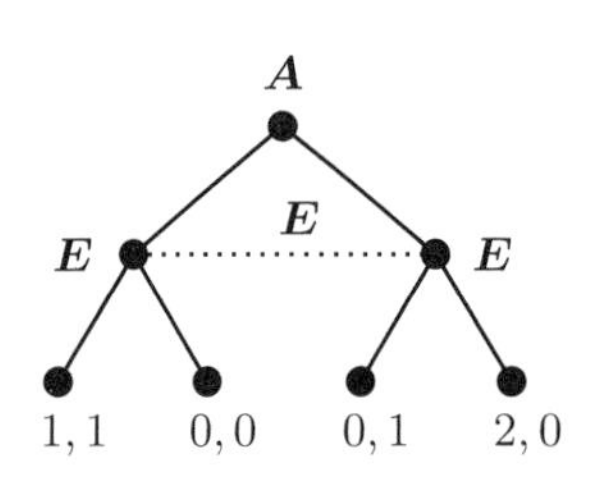

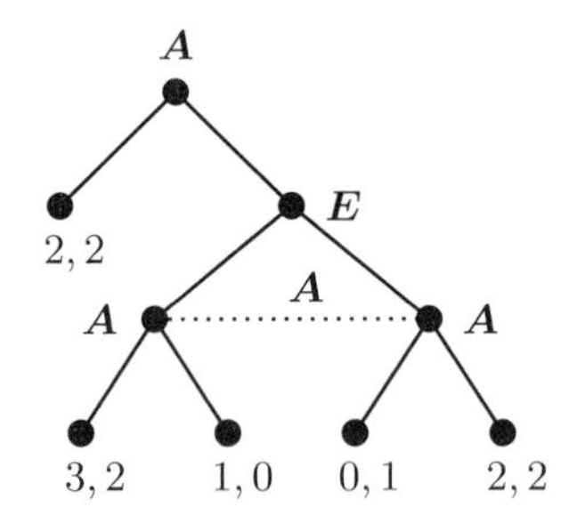

21 The tree to the right is slightly adapted from an example in an invited lecture by Robert Stalnaker at the *Gloriclass Farewell Event*, ILLC Amsterdam, January 2010.

The game on the left suggests an extended use of Backward Induction, but the one on the right raises tricky issues of what $\boldsymbol{A}$ is telling $\boldsymbol{E}$ by moving right. We leave the question of what should happen in both games to the reader. Dégremont (2010) and Zvesper (2010) have more extensive discussion of what logic might say here.

Trading kinds of knowledge in games The issue of relating the different kinds of knowledge in games at the beginning of this chapter has not been settled. Many ideas are circulating about trading one kind for another: say, replacing uncertainty about the future by imperfect information-style uncertainty right now between different histories.[22] We will return to combining and transforming notions of knowledge in Chapter 6, although eventually, we may just have to accept the diversity. Also, we may need other forms of knowledge beyond the purely semantic, such as syntactic awareness-based views (Fagin et al. 1995, van Benthem & Martínez 2008).

Adding beliefs As we have noted several times, games are not just driven by knowledge, but also by beliefs. A common representation for this richer setting uses epistemic models, adding a relation of comparative plausibility on epistemically accessible worlds (cf. Chapter 7). Belief is then what is true in the most plausible accessible worlds. This adds fine structure to all earlier topics, but we leave its exploration to Part II of this book.

Adding probability More expressive quantitative logics for numerical degrees of belief or frequency information about process behavior arise when we add probability functions to our models. Logic and probability mix well in the study of agency, as can be seen in the systems of Halpern (2003b) and van Benthem et al. (2009b), but in this book, we will ignore probabilistic perspectives, regarding them as an extra for later.

Game logics in other areas While the logics in this book are tailored to games, their scope is more general. Many points in the preceding apply to a much broader class of social scenarios. In particular, our modal-epistemic logic also occurs in planning, when agents act under uncertainty (Moore 1985). There are also analogies between what we do in this book and recent work on dynamic-epistemic logics for multi-agent planning (cf. Bolander & Andersen 2011, Andersen et al. 2012).

---

22 Compare this to the Harsanyi Doctrine in game theory (Osborne & Rubinstein 1994).

4
Making Strategies Explicit

4.1 Strategies as first-class citizens

As we saw in our Introduction, much of game theory is about the existence of strategic equilibria. In the same spirit, many game logics have existential quantifiers saying that players have a strategy for achieving some purpose (cf. Parikh 1985, Alur et al. 2002, and Chapter 19), much like the power modalities discussed in earlier chapters. But the strategies themselves are not part of the formal language, and this leaves out a protagonist in the story of interaction. Strategies are what drives rational agency over time, unfolding via successive interactive moves. The widespread tendency in logic toward hiding information under existential quantifiers has been called ∃-sickness in van Benthem (1999).[23] To cure this, it makes sense to move strategies explicitly into our logics, and reason about their behavior, as one would do with plans. There are several ways of doing this. One is to use logics of programs to deal with strategies, and we will show how this works with propositional dynamic logic PDL. Another approach is by inspecting the elementary reasoning about strategies underlying basic game-theoretic results, and designing an abstract calculus of strategies that can represent these naturally. In particular, this emancipation makes sense for the logic games of Part IV of this book, where hiding strategies under existential power quantifiers is common (cf. Chapter 25).

23 Other instances of ∃-sickness are doing logic of provability rather than of proofs, or of knowability rather than of knowing. In general, suffixes such as -ility should be red flags!

The logic of strategies is a recent area, and we will not present any definitive version, but the theme will return in Chapters 5 and 18 and in Part V.

4.2 Defining strategies in dynamic logic

Our first approach to strategies treats them as programs in a well-known logic.

Dynamic logic of programs Let us recall the basics of propositional dynamic logic, PDL, introduced in Chapter 1. PDL was designed originally to study imperative computer programs, or complex actions, that use the standard operations of sequential composition (;), guarded choice (IF... THEN... ELSE...), and guarded iteration (WHILE... DO...).

DEFINITION 4.1 Propositional dynamic logic
The language of PDL defines formulas and programs in a mutual recursion, with *formulas* denoting sets of states (i.e., they are local conditions on states of a process), while *programs* denote binary transition relations between states, consisting of ordered pairs (input state, output state) for the successful executions. Programs are built from atomic actions (moves) $a, b, \ldots$, and tests $?\varphi$ for all formulas φ, using the three "regular operations" of ; (sequential composition), $\cup$ (nondeterministic choice) and * (nondeterministic finite iteration). Formulas are constructed from atoms and Booleans as in a basic modal language, but now with dynamic modalities $[\pi]\varphi$ interpreted as follows in the process models $\boldsymbol{M}$ of Chapter 1

$$\boldsymbol{M}, s \models [\pi]\varphi \quad \text{iff} \quad \varphi \text{ is true after every successful execution of } \pi \text{ starting at } s.$$

This is the standard way of describing effects of programs or actions. ■

This system applies directly to games. In Chapter 1, the total *move* relation was a union of atomic relations, and the modal pattern for the existence of a winning strategy was $[a \cup b]\langle c \cup d\rangle p$. PDL focuses on finitely terminating programs, a restriction that one can question in general (see Chapter 5 on strategies in infinite games). However, it makes good sense in finite games, or with infinite strategies whose local steps are finite terminating programs. For safety's sake, we will work with finite games in this chapter, unless explicitly stated otherwise.

As for a calculus of reasoning about complex actions, the valid laws of PDL are decidable, and it has a complete set of axioms analyzing the regular program constructions in a perspicuous way (Harel et al. 2000, van Benthem 2010b).

Strategies defined by programs Strategies in game theory are partial functions on players' turns, given by instructions of the form "if my opponent plays this, then I play that." But as we have seen in Chapters 1 and 2, more general strategies are transition relations with more than one best move. They are like plans that can be useful by just constraining moves, without fixing a unique course of action. Thus, on top of the hard-wired *move* relation in a game, we now get new defined relations, corresponding to players' strategies, and these can often be defined explicitly in a PDL program format.[24]

As an example, the earlier "forcing modality" can be made explicit as follows.

FACT 4.1 For any game program expression σ, PDL can define an explicit forcing modality $\{\sigma, i\}\varphi$ stating that σ is a strategy for player i forcing the game, against any play of the others, to pass only through states satisfying φ.

Proof The formula $[((?\boldsymbol{turn_E} ; \sigma) \cup (?\boldsymbol{turn_A} ; move\text{-}\boldsymbol{A}))^*]\varphi$ defines the forcing. This says that following the program σ at $\boldsymbol{E}$'s turns and any move at $\boldsymbol{A}$'s turns always yields φ-states. ■

A related observation is that, given definable relational strategies for players $\boldsymbol{A}$ and $\boldsymbol{E}$, we get to an outcome for the game that can be defined as well.

FACT 4.2 Outcomes of running joint strategies σ, τ can be described in PDL.

Proof The formula $[((?\boldsymbol{turn_E} ; \sigma) \cup (?\boldsymbol{turn_A} ; \tau))^*](\boldsymbol{end} \to \varphi)$ does the job. ■

Flat strategy programs Our program format has some overkill. A strategy prescribes one move at a time, subject to local conditions. Then, local iterations * make little sense, and one can restrict attention to PDL programs using only atomic actions, tests, ; , and $\cup$. These can easily be brought into a normal form consisting of a union of "guarded actions" of the form

$$?\varphi_1 ; a_1 ; \cdots ; ?\varphi_n ; a_n ; ?\psi$$

This makes a strategy a set of conditional rules that are applicable only under local conditions and with specified postconditions.[25]

---

24 It seems reasonable to require that relations for strategies must be non-empty on turns of the relevant player. All that we have to say is compatible with this.

25 This format for strategies of belief revision and model change was proposed in van Benthem & Liu (2007), and more general versions occur elsewhere (van Eijck 2008, Girard et al. 2012, Ramanujam & Simon 2008).

Even this may still be too much, since prescribing sequences of actions $a_1 ; \cdots ; a_n$ only makes sense when a player has several consecutive turns, while most games have alternating scheduling. “Flat programs” often suffice, being unions of guarded actions of the form $?\varphi ;\ a ;\ ?\psi$.

Expressive completeness On a model-by-model basis, the expressive power of PDL is high (Rodenhäuser 2001). Consider any finite game $\boldsymbol{M}$ with strategy σ for player i. As a relation, σ is a finite set of ordered pairs (s, t). In case we have an “expressive model” $\boldsymbol{M}$ whose states s are definable in our modal language by formulas def_s,[26] we can define pairs (s, t) by formulas $def_s ; a ; def_t$, where a is the relevant move, and take the relevant union (note that this is indeed a flat program in the earlier syntactic sense).

FACT 4.3 In expressive finite extensive games, all strategies are PDL-definable.

PDL and strategy combination PDL also describes combinations of strategies in terms of operations on relations (van Benthem 2002a). A basic operation is union, allowing all possible moves according to both strategies. Union merges two plans, constraining players' moves into a common weakening. The laws of PDL describe how this operation behaves in single steps:

$$\langle \sigma \cup \tau, i \rangle \varphi \leftrightarrow \langle \sigma, i \rangle \varphi \vee \langle \tau, i \rangle \varphi$$

It may be of more interest to look at strategy modalities $\{\sigma, \boldsymbol{E}\}\varphi$ defined as before with repeated steps. In that case, it is easy to see that distribution fails (see Chapters 11 and 19 for more on this). There are obvious counterexamples to:

$$\{\sigma \cup \tau, i\}\varphi \leftrightarrow \{\sigma, i\}\varphi \vee \{\tau, i\}\varphi$$

and we only have monotonicity laws such as:

$$\{\sigma, i\}\varphi \rightarrow \{\sigma \cup \tau, i\}\varphi$$

Perhaps a more important operation on relational strategies is the *intersection* $\sigma \cap \tau$, which combines two separate recommendations. In some cases, this may not leave any moves at all. But in general, intersection mimics an intuitive composition of strategies. Think of different players, with a strategy for one player allowing any

---

26 This can be achieved using temporal past modalities to describe the history up to s.

move for the other. Intersection then produces joint outcomes when both players play their given strategy. But we can also think of intersection as composing different partial strategies for the same player. Suppose that a strategy σ works up to a certain level in the tree, imposing no restrictions afterward, while strategy τ imposes no restrictions first, but kicks in after the domain of σ has come to an end:

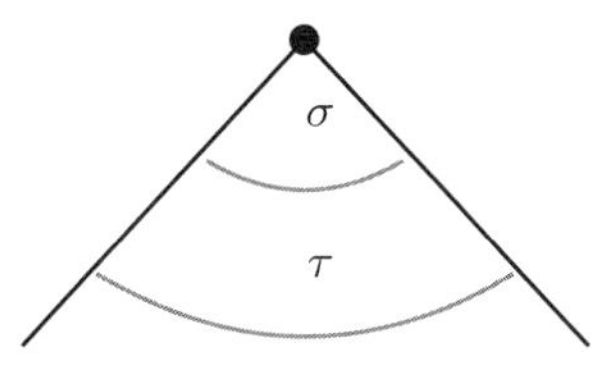

In this case, the intersection has the effect of the intuitive composition $\sigma\,;\tau$.

PDL has no reduction axiom for $\sigma \cap \tau$, not even for single steps, since intersection is not a regular program operation. But we do keep some general laws of strategy calculus, such as the validity of

$$(\{\sigma, i\}\varphi \wedge \{\tau, i\}\psi) \to \{\sigma \cap \tau, i\}(\varphi \wedge \psi)$$

Finding a complete axiomatization for forcing statements with union and intersection of strategies seems an open problem, but given the earlier PDL definition of forcing, an implicit calculus lies embedded inside PDL with intersection (cf. Harel et al. 2000). A recent, more in-depth study following up on our theme of strategy combination is van Eijck (2012).

Zooming out: Powers and explicit strategies The forcing modalities $\{\sigma, i\}\varphi$ used in this chapter so far have explicit strategies inside them. They may be viewed as explicit counterparts to the more implicit forcing modalities $\{i\}\varphi$ of Chapter 1 that quantified existentially over strategies for player i, zooming out on mere powers without the methods for achieving them. As we noted earlier, different zoom levels may have their own uses, and in particular, it makes sense to have both here.

Mere power modalities for players have an important use that is not subsumed by our explicit approach. If we want to say that a player *lacks a strategy* for achieving a certain goal, the following formula will do:

$$\neg\{i\}\varphi$$

No explicit version can express the same, since, in general, we cannot list all possible strategies σ in order to deny that they force φ. Thus, it makes sense to have logics that combine both implicit and explicit forcing statements.

A further and perhaps surprising use of programs plus existential forcing modalities occurs when defining strategies that make powers explicit. Given any player i and formula φ, define the PDL program

$$\sigma_{\varphi,i} = ?\boldsymbol{turn}_i\,;\ move\,;\ ?\{i\}\varphi$$

For instance, this is one way to define the success strategy in the Zermelo coloring algorithm of Chapter 1: "make sure that you keep going to winning positions." Now we can state a simple equivalence that follows easily from our earlier definitions.

FACT 4.4 $\{i\}\varphi \leftrightarrow \{\sigma_{\varphi,i}, i\}\varphi$ is valid.

Proof From right to left, if we have the specific strategy, then, a fortiori, the existential forcing modality holds. From left to right, we check that current truth of the forcing statement will persist along the stated strategy. This follows a recursion as in Chapter 1. (a) If it is i's turn, then there exists at least one successor where $\{i\}\varphi$ is true, (b) If it is j's turn, then all successors have this property, (c) If we are in an endpoint, then φ holds, and we are done. On finite games, this says that we can force total histories whose endpoints satisfy φ.[27] ■

On infinite games, the preceding equivalence still holds with the following understanding of forcing modalities: "the player can force a set of total histories at all of whose stages φ is true." The logic of this temporal forcing modality has some delicate aspects that we postpone until Chapter 5.

Details of definability What borderline separates the above trivializing rule "be successful" from strategies with real content? One crucial factor is *definability* of strategies in suitably restricted formalisms. For instance, while the above forcing modality of success looks at the whole future of the game, in practice, strategies often lack such forward-looking tests. Rather, they check for what happened in the past of the game tree, or test even just local assertions at the current node that require neither futurology nor history. Examples of strategies that are definable in such restricted ways, including memory-free ones that can be surprisingly powerful,

---

27 Variant conditions arise here by structuring φ. For instance, using suitable dynamic modalities, one can express that φ is true everywhere on the branches forced.

will be found throughout this book (cf. Chapters 18 and 20). Still, the picture is diverse. Some natural strategies seem to be intrinsically forward-looking, such as the Backward Induction strategy of Chapter 2. Programs in dynamic logic can define and calibrate some of these kinds of strategies, although the more general format for inspecting past and future would be the temporal languages to be discussed in the next chapter.

In summary, dynamic logic does a good job at defining strategies in simple extensive games. While we have also seen a need for richer computational logics serving similar purposes,[28] our point of existence for useful logical calculi of strategies has been made. Beyond analyzing strategic reasoning, such formalisms could also be used for calibrating strategies in hierarchies of definability, from simpler to more complex rules of behavior.

4.3 General calculus of strategies

While PDL programs can define strategies in specific games, this is done on a local game-by-game basis. And also when viewed as a generic proof calculus, we proposed this system mainly because of its track record for process graphs, not as a result of analyzing specific pieces of strategic reasoning. In other words, we started from systems at hand, not from an initial reasoning practice.

Another approach to designing strategy logics proceeds by inspecting standard game-theoretic arguments. We now present two examples of this approach to analyzing strategic reasoning, although we will not propose a final new calculus. Further examples of the same "quasi-empirical" approach will be found in Chapter 5 and in Part V of this book on logics of strategic powers in complex games.

Looking ahead, consider the logic of players' powers presented in Chapter 11.

EXAMPLE 4.1 Strategizing dynamic power logic
Consider a Boolean choice game $G \cup H$ where player $\boldsymbol{E}$ starts by choosing to play either game G or game H. In terms of the forcing notation of Chapter 1, now with games explicitly marked, the following principle is obviously valid:

$$\{G \cup H, \boldsymbol{E}\}\varphi \leftrightarrow \{G, \boldsymbol{E}\}\varphi \vee \{H, \boldsymbol{E}\}\varphi$$

---

28 A simpler view of strategies makes them finite sequences of basic actions by automata (cf. Ramanujam & Simon 2008). Such strategies have also been added to temporal logics of games (cf. Broersen 2009, Herzig & Lorini 2010).

The intuitive reasoning is as follows. From left to right, if $\boldsymbol{E}$ has a strategy forcing outcomes satisfying φ in $G \cup H$, then the first step in that strategy describes $\boldsymbol{E}$'s choice, left or right, and the rest of the strategy gives outcomes with φ in the chosen game. And vice versa, say, if $\boldsymbol{E}$ has a strategy forcing φ in game G, then prefixing that strategy with a move of going left gives a strategy forcing φ in $G \cup H$. ■

Right under the surface of this example lies a calculus of arbitrary strategies σ. Our first argument introduced two operations: $head(\sigma)$ takes the first move of the strategy, and $tail(\sigma)$ gives the remainder. Clearly, there are natural laws governing these operations, in particular:

$$\sigma = \big(head(\sigma), tail(\sigma)\big)$$

Next, the converse argument talks about prefixing an action a to a strategy σ, yielding a concatenated strategy $a\,;\sigma$ equally satisfying natural laws such as:

$$head(a\,;\sigma) = a \qquad\qquad tail(a\,;\sigma) = \sigma$$

This suggests that there is a basic strategy calculus underneath our common reasoning about games.[29] The following example takes its exploration a bit further.

EXAMPLE 4.2 Exploring basic strategy calculus
Consider the following simple sequent derivation for a propositional validity:

$$\begin{array}{ccc}
A \Rightarrow A & B \Rightarrow B & \\
A, B \Rightarrow A & A, B \Rightarrow B & C \Rightarrow C \\
\multicolumn{2}{c}{A, B \Rightarrow A \wedge B} & A, C \Rightarrow C \\
\multicolumn{2}{c}{A, B \Rightarrow (A \wedge B) \vee C} & A, C \Rightarrow (A \wedge B) \vee C \\
\multicolumn{3}{c}{A, B \vee C \Rightarrow (A \wedge B) \vee C} \\
\multicolumn{3}{c}{A \wedge (B \vee C) \Rightarrow (A \wedge B) \vee C}
\end{array}$$

Here is a corresponding richer form indicating strategies:

29 Interestingly, the head and tail operations suggest a co-inductive view (cf. Venema 2006) where strategies are observed and then return to serving mode, rather than the inductive construction view of PDL programs. Chapters 5 and 18 have more on this.

$$
\begin{array}{c}
x : A \Rightarrow x : A \qquad\qquad y : B \Rightarrow y : B \\
x : A, y : B \Rightarrow x : A \qquad x : A, y : B \Rightarrow y : B \qquad\qquad\qquad z : C \Rightarrow z : C \\
x : A, y : B \Rightarrow (x, y) : A \wedge B \qquad\qquad x : A, z : C \Rightarrow z : C \\
x : A, y : B \Rightarrow \langle l, (x, y) \rangle : (A \wedge B) \vee C \qquad x : A, z : C \Rightarrow \langle r, z \rangle : (A \wedge B) \vee C \\
x : A, u : (B \vee C) \Rightarrow \mathit{IF}\ \mathit{head}(u) = l\ \mathit{THEN}\ \langle x, \mathit{tail}(u) \rangle\ \mathit{ELSE}\ \mathit{tail}(u) : (A \wedge B) \vee C \\
v : A \wedge (B \vee C) \Rightarrow \mathit{IF}\ \mathit{head}((v)_2) = l\ \mathit{THEN}\ \langle (v)_1, \mathit{tail}((v)_2) \rangle\ \mathit{ELSE}\ \mathit{tail}((v)_2) : (A \wedge B) \vee C
\end{array}
$$

The latter format can be read as proof construction, but we can also read what we have produced as a construction of strategies.[30] Its key operations are

storing strategies for a player who is not to move	$\langle\ ,\ \rangle$
using a strategy from a list	$(\ )_i$
executing the first action of a strategy	*head*( )
executing the remaining strategy	*tail*( )
making a choice dependent on some information	IF THEN ELSE

Clearly, this repertoire is different from PDL, and it may also have a quite different logical rationale, with a calculus more like those found in type theories.[31] A matching general strategy calculus might manipulate statements of the form "σ is a strategy forcing outcomes satisfying φ for player i in game G." ■

We will return to the logical structure of concrete instances of strategic reasoning at various places in this book, starting in Chapter 5 on infinite games.

4.4 Strategies in the presence of knowledge

Strategies also work in more complex settings, such as the imperfect information games of Chapter 3. Interesting new issues arise in this setting. For a start, in this case, the important strategies were the uniform ones, prescribing the same move at positions that a player cannot distinguish, or equivalently, making the action

---

30 For instance, in addition to being about proofs, taking a leaf from Chapter 14 of this book, the given proof is also a recipe for turning any winning strategy of the verifier in an evaluation game for $A \wedge (B \vee C)$ into a winning strategy for the game $(A \wedge B) \vee C$.

31 One analogy are type-theoretic statements $t : P$ saying that t is a proof of proposition P, or an object with property P. Type theories have rules encoding constructions of complex types (Barendregt 2001). See Abramsky & Jagadeesan (1994) for systematic links with games and strategies.

chosen dependent only on what a player knows. As in Section 4.2, we can add PDL-style programs to this setting. But there is a twist.

Knowledge programs It makes sense to impose a restriction now to the "knowledge programs" of Fagin et al. (1995), whose only test conditions for actions are knowledge formulas of the form $K\varphi$. More precisely, in our setting, we want knowledge programs that can serve as executable plans for an agent i, whose test conditions φ must have the property that i always knows whether φ is the case. Without loss of generality, this means that we can restrict to test conditions of the form $K_i\varphi$, since the following equivalence is valid:

$$U(K_i\varphi \lor K_i\neg\varphi) \to U(\varphi \leftrightarrow K_i\varphi)$$

As a special case, we take *flat knowledge programs*, with the special format defined earlier for flat PDL programs, with only one-step moves. These seem close to uniform strategies, but how close, precisely? First, we need a generalization. Uniform strategies as defined in Chapter 4 were functional, but our PDL programs are relational. Let us say that a relational strategy σ for player i is *uniform* if, whenever $\boldsymbol{turn}_i x$, $\boldsymbol{turn}_i y$, and $x \sim_i y$, then σ allows the same actions at x and y.

This is related to a general epistemic requirement on game models that occurs in the literature (cf. Osborne & Rubinstein 1994), namely, that players should know all actions available to them.[32] In modal terms, this imposes the condition:

$$(\boldsymbol{turn}_i \land \langle a\rangle\top) \to K_i\langle a\rangle\top, \text{ or even } (\boldsymbol{turn}_i \land \langle a\rangle\top) \to C_{\{i,j\}}\langle a\rangle\top$$

Making this assumption, the following implication becomes valid.

FACT 4.5 Flat knowledge programs define uniform strategies.

Proof Knowledge conditions have the same truth value at epistemically indistinguishable nodes for a player, and by the assumption, the available moves are the same. Hence the transitions defined by the program are the same as well. ■

The following result states a converse, given some conditions on the power of models for defining nodes such as those in Section 4.2 (van Benthem 2001b).[33]

32 In other words, players get no new knowledge from inspecting available moves.

33 Here we assume that models are bisimulation-contracted for action and uncertainty links. This ensures that each epistemic equivalence class will have a unique modal definition in our language, by the results of Chapter 1.

FACT 4.6 On expressive finite games of imperfect information, the uniform strategies are definable by knowledge programs in epistemic PDL.

Proof One enumerates the action of the given uniform strategy as we did for PDL programs in the non-epistemic case. We only need to say what the strategy allows on each epistemic equivalence class for the relevant player. This can be stated using the modal definitions of the equivalence classes, while noting that these definitions are known, making them suitable as test conditions for a knowledge program. ■

Further entanglements of knowledge and action The merge of epistemic logic with logics of strategies operates at two levels (see van Benthem 2012e for more on what follows). One level is propositional: preconditions and postconditions of actions can now be epistemic. We saw epistemic preconditions in knowledge programs, and Chapter 22 will give examples of epistemic postconditions in knowledge games, such as being the first to know a certain secret. The second level is dynamic, since actions can now be "epistemized" in different ways. Knowledge programs showed how test conditions can be epistemic. Further, actions can be epistemic by themselves, such as making an observation or asking a question. This special sort of epistemic action will be a main focus in Chapter 7 and beyond.

The two levels, propositional and dynamic, interact. For instance, one natural issue that arises is to which extent agents know the effects of their actions. For instance, should there be introspection given the epistemic nature of knowledge programs? Suppose that a knowledge program π for player i guarantees effect φ in a model, that is, the forcing statement $\{\pi, i\}\varphi$ holds everywhere, does the player know this? Actually, it is easy to see that this can fail.

EXAMPLE 4.3 Not knowing what you are doing
Let a single player have two indistinguishable worlds s and t, and only one action a at each, leading to worlds u and v where u has the property p, while v does not.

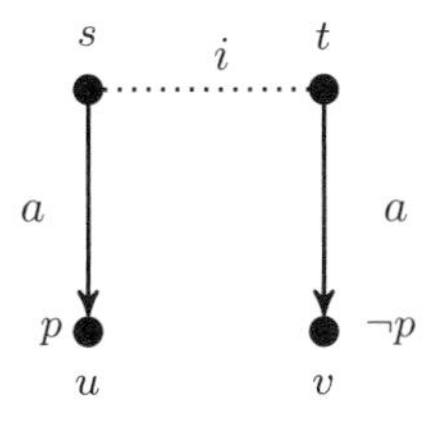

In this model, introspection fails: $?\top \,;\, a$ is a knowledge program for which $[?\top \,;\, a]p$ is true at s, but not at t, whence $K[?\top \,;\, a]p$ is false at s. ■

However, there is another issue of entanglement here. Suppose that we do know the effects of our basic actions, then what about complex actions defined by knowledge programs? Here the axioms of PDL have interesting things to say. For instance, suppose that we know all the effects of some program π_1. Since those effects may include statements $[\pi_2]\varphi$ for other programs π_2, it follows that we know $[\pi_1][\pi_2]\varphi$, and hence all the effects of longer programs $\pi_1 ; \pi_2$.

A final issue of entanglement has to do with our notion of perfect recall from Chapter 3. It might be called "epistemic grip": even if players do know the effect of a program, what will they know at intermediate stages of following a strategy? Consider the problem of following a guide in a bog, but having forgotten why we trusted the guide in the first place. In imperfect information games, if a player knows at the start that $\{\pi, i\}\varphi$ is the case, does it follow that, at each later stage, the player knows that the remaining strategy played yields outcomes with φ? Here recall the commutation of knowledge and action for players with perfect memory:

$$K_i[a]\varphi \to [a]K_i\varphi$$

FACT 4.7 If players have perfect recall for atomic actions, then they have it for all complex knowledge programs.

Proof To see this, it suffices to look at the following strings of implications, with the relevant inductive hypothesis presupposed at various steps:

(a) $K[\pi_1 ; \pi_2]\varphi \to K[\pi_1][\pi_2]\varphi \to [\pi_1]K[\pi_2]\varphi \to [\pi_1][\pi_2]K\varphi \to [\pi_1 ; \pi_2]K\varphi$

(b) $K[\pi_1 \cup \pi_2]\varphi \to K([\pi_1]\varphi \wedge [\pi_2]\varphi) \to (K[\pi_1]\varphi \wedge K[\pi_2]\varphi) \to$
$([\pi_1]K\varphi \wedge [\pi_2]K\varphi) \to [\pi_1 \cup \pi_2]K\varphi$

These show how intermediate knowledge proceeds. The crucial step now is that of tests. For arbitrary tests, $K[?\alpha]\varphi$ implies $K(\alpha \to \varphi)$, but there is no guarantee at all that this implies $\alpha \to K\varphi$. However, this is different for knowledge tests:

$$K[?K\alpha]\varphi \to K(K\alpha \to \varphi) \to (K\alpha \to K\varphi) \text{ is valid in epistemic } S4. \qquad \blacksquare$$

These are just a few of the issues connecting knowledge and action in games.[34]

---

34 In imperfect information games for players with perfect recall, all uniform strategies lead to epistemic grip (van Benthem 2001b). The converse question seems open.

Know-how and understanding strategies Programs are also interesting as epistemic objects in their own right, since they represent the intuitive notion of knowing how as opposed to merely knowing that. There is an extensive literature on what know-how is, but the game setting makes the issue quite concrete.

Again, the entanglement of the two notions is of particular interest. In addition to knowing how involving knowing that, there is also the fundamental notion of "knowing a plan" that seems crucial to rational agency. There is no generally accepted explication of what this means. One line is to ask, as we did in the above, for propositional knowledge about the effects that a plan will achieve.[35] However, intuitively, more is involved in knowing a plan. Consider what we want genuine learning to achieve: not just correct propositional knowledge, but also the ability to engage in a practice based on the plan. In education, we teach know-how at least as much as know-that, but what is it, really?

The contrast may be highlighted in terms of understanding a strategy or a plan versus merely knowing it. In addition to propositional knowledge of effects of a plan, or parts of it, there are other key features such as "robustness": that is, counterfactually knowing the effects of a plan under changed circumstances, or the ability to modify it as needed. And there are further tests, such as a talent for zoom: that is, being able to describe a plan at different levels of detail, moving up or down between grain levels as needed.[36]

4.5 Conclusion

The main points Strategies are so important to games and to social agency in general that they deserve explicit attention as objects of study. We have shown how to do this, by adding explicit strategies to existing logics of games. First we showed how the programs of propositional dynamic logic can be used for defining strategies and reasoning about them. Next, changing tacks, we showed how a more general

35 This issue also plays in the area of epistemic planning (Bolander & Andersen 2011, Andersen et al. 2012), where different kinds of knowledge or beliefs are important: about where we are in following some current plan, but also about how we expect the process to develop over time.

36 Similar issues arise in analyzing what it means for someone to understand a formal proof, and useful intuitions might be drawn from mathematical practice.

core calculus of strategic reasoning can be extracted from basic game-theoretic arguments. Finally, we showed how strategy logics mix well with epistemic structure, proving a number of results about imperfect information games, and raising some issues of better understanding know-how as opposed to know-that.

Open problems No established strategy logic exists comparable to game logic. Thus, finding general core languages for defining strategies and matching calculi for strategic reasoning seems a natural goal. As we have suggested, this may require fieldwork in analyzing good benchmarks. As we will see with various examples in Chapters 5 and 25, there are at least two sources for this in the present book. One is the logic games of Part IV that provide many instances of basic strategic reasoning. Another source are key strategies in game theory, including simple widespread rules such as Tit for Tat, discussed in our Introduction. We will continue with this exploration in Chapter 5.

Additional open problems arise with understanding the interplay of strategies with knowledge and information change, as discussed above, including a better understanding of knowing how. Beyond that, we need to understand the entanglement with other epistemic attitudes that are crucial to action, such as belief, and with acts of belief revision.

The next obvious desideratum in the context of this book is extending our analysis, local or generic, to games with preference structure. Defining the Backward Induction strategy of Chapter 2 in a basic logic of strategies would be an obvious benchmark. Finally, making strategies, which are complex actions, into first-class citizens, fits well with the logical dynamics program of Part II of this book, but an optimal integration has not happened yet.

4.6 Literature

This chapter is based on van Benthem (2012a). Some parts have also been taken from van Benthem (2012e) and van Benthem (2013). This program for putting strategies in the limelight was presented originally at a half-year project at the Netherlands Institute for Advanced Studies (NIAS) in 2007.

A useful anthology on logics of strategies from the 1990s is Bicchieri et al. (1999). In the meantime, many relevant publications have appeared, including work by the Chennai Group working with automata theory (cf. Ramanujam & Simon 2009, Ghosh & Ramanujam 2011), by the CWI group on strategies in PDL and related formalisms (cf. Dechesne et al. 2009, Wang 2010), the STRATMAS

Project (`http://www.ai.rug.nl/~sujata/documents.html`) on many different approaches to strategic reasoning, and on the borderline with the dynamic-epistemic logic of Part II of this book (Pacuit & Simon 2011).

4.7 Further directions

We have mentioned a number of open problems already in the main text of this chapter. Some further detail now follows.

Systems thinking versus a quasi-empirical approach As we have suggested at several places, there are at least two approaches to designing logics of strategies. One starts from general system considerations and analogies with notions such as programs or automata, and it is the dominant approach in this book. But one could also start by independently compiling a repertoire of basic strategies of wide sweep (as has been done for algorithms), plus the kind of basic reasoning that establishes their properties. Striking phenomena here are the ubiquity and power of simple identity strategies such as Tit for Tat in game theory, or copy-cat in computation, or indeed variable identification in logic itself. In particular, this suggests looking at definability of strategies in simple logical languages, although we are not aware of any established calibration hierarchy. Hand in hand with this would be logical analysis of proofs establishing key results about strategies in the literature, a working style that will return at a number of places in this book.

Extending the dynamic logic approach We have only shown PDL at work for strategies in games with sequential turns. But parallel action is a common feature in games. This requires endowing atomic actions with more structure, giving them components for each player, and perhaps also the environment, as in Fagin et al. (1995). The logics of Chapters 12 and 20 provide examples of how this may be done, adding fine structure to basic events in terms of control by the players. Another type of extension concerns effects of strategies. So far, we looked at either endpoints, or at all future stages. But there are other intuitive success notions, in terms of intermediate effects. Let $\{\sigma\}^*\varphi$ say that strategy σ guarantees reaching a "barrier" of φ-positions in the game (a set that intersects each maximal chain). This may be the optimal setting for composition of strategies. What happens to game logic when we add barrier modalities?

Modal fixed point logics Richer formalisms than PDL also make sense for strategies. In Chapter 1, we saw a move from PDL to the modal μ-calculus that defines a much richer set of recursive notions, including talk about infinite computations. But there is a problem: the μ-calculus has no explicit programs. Still, many formulas suggest a match. For instance, "keep playing a" (the non-terminating program WHILE $\top$ DO a) is a witness to the infinite a-branches claimed to exist by a greatest fixed point formulas $\nu p \bullet \langle a \rangle p$. An explicit program version of the μ-calculus might be a useful calculus of strategies. Attempts so far have only focused on terminating programs (Hollenberg 1998), and perhaps a better paradigm are the μ-automata of Bradfield & Stirling (2006). In Chapter 18, we will return to this theme, including strategic reasoning in graph games (cf. Venema 2006). In Chapter 2, we even used the much stronger fixed point logic LFP(FO), where similar points apply. Even so, one might say that fixed point logics do give explicit dynamic information about strategies, since the fixed point operators themselves refer to a fixed computational approximation procedure close to strategy construction.

Strategies and invariants Our discussion of strategies in PDL, and in particular, that of the trivializing power strategy "nothing succeeds like success," also suggests a further perspective. Many good strategies consist in maintaining a suitable *invariant* throughout the course of a game. This is true for parlor games such as Nim, but also for many of the logic games to be discussed in Part IV of this book. Indeed, our modal forcing statements about future success can themselves be viewed as abstract invariants, and the same is true for logical formulas in many game languages. But invariants can also be other structures: in some sense, the epistemic-doxastic models to be discussed in later chapters serve as invariants recording some (but not all) memory of past behavior. This book has no systematic theory of invariants to offer, but several topics have some bearing on this, such as the discussion in Chapters 18 and 25 of games viewed simultaneously at different levels.

Temporal logics Another broad paradigm that supports strategy calculus is that of temporal logics. Recent proposals include "strategizing" alternating temporal logic ATL (Alur et al. 2002, Ågotnes et al. 2007) or epistemic ATEL (van der Hoek & Wooldridge 2003, van Otterloo 2005), and analyses of games in interpreted systems (Halpern 2003a) or situation calculus (Reiter 2001). Temporal logic might be a better focus for the study of strategies in infinite games than the systems in our chapter, and we will give a few illustrations in Chapter 5.

From concrete to generic strategies The strategies in the systems of this chapter were concrete objects defined inside specific games. But there is also another level of generic strategies that achieve their effects across all games of some appropriate type. Examples are simple but powerful copying strategies such as Tit for

Tat or copy-cat, that work across a wide range of games. Generic strategies lie behind the logics of the game-constructing operations in Part V of this book, based on dynamic logic and linear logic, and they will appear in Chapters 19, 20, and 21.

Yet other approaches There are yet other approaches for making strategies explicit, such as automata theory (Ramanujam 2008) or type theory (Jacobs 1999). It remains to be seen how these fit with the logics in this chapter, although the Appendix to Chapter 18 has some relevant discussion. The study of strategies may also profit from concrete instances in other fields, such as the area of planning mentioned already in Chapter 3 (cf. Moore 1985, Bolander & Andersen 2011). Another mathematical paradigm for concrete strategies linked to logical tasks is formal learning theory (Kelly 1996).

Adding knowledge and belief There is more to adding informational attitudes to PDL than we have shown so far. First, one can perform a more drastic epistemization than we did, not just on propositions, but also on programs, making transitions themselves objects that can have epistemic structure, in the line of event models in dynamic-epistemic logic (cf. Chapter 7).[37] It is possible to go further in other ways. Our view of understanding a strategy as knowing its effects under changed circumstances mirrors counterfactual views in philosophy that make knowledge of φ a true belief that tracks truth in the following sense: if the world had been slightly different, we would still have a correct belief or disbelief about φ (Nozick 1981). This suggests incorporating belief into the structure of strategies, and indeed, Chapter 8 will view strategies as encoding beliefs. This twist implies a radical perspective, since beliefs come with belief revision, an ability to correct ourselves when contradicted by the facts. We will return to strategy revision in one of the points below (see also Chapter 9).

Adding preferences An intuitive sense of strategic behavior is based on motives and goals. Indeed, action and preference were deeply entangled in our logic for Backward Induction in Chapter 2. What about explicit strategies in this richer realm? We need extensions of our earlier modal preference logics that can define benchmarks like the Backward Induction strategy, perhaps in the style of the deontic dynamic logics of van der Meyden (1996).

---

37 An epistemic variant of "arrow logic" is used for this purpose in van Benthem (2011d).

Strategies across changing situations So far, we have studied strategies in fixed situations, which can lead to local solutions. What happens when a strategy has a proven effect, but we change the game to a new one? We will address in detail the topic of strategies in changing games in Chapter 9, but briefly explore some points presently. Recall our earlier point about understanding a strategy. A good plan should still work when circumstances change; it will be robust under at least small changes. But many strategies fall apart under change. For instance, the Backward Induction strategy can shift wildly with addition or deletion of moves.[38] Two options arise here, "recomputation" and "repair." Should we compute a new plan in a changed game, or repair the old plan? We often start with repair, and only recompute when forced. Could there be a serious theory of plan revision? Can we say more precisely when gradual changes are sufficient, and when they fail?[39] The issues of strategy structure, definability and preservation behavior arising here are still to be addressed systematically. Even for PDL, we know of no model-theoretic preservation theorems that relate program behavior across different models.

Conceptual clarification A final difficulty is the proliferation of undefined terms in the field, such as plan, strategy, agent type, or protocol. All point to similar things, and discussions become confusing. It would be good to fix terminology (cf. van Benthem et al. 2013). Protocols might be general constraints on a process; strategies might be ways of using one's freedom within these constraints; and agent types could be reserved for repeatable styles of action that agents have available.

---

38 One might say that the flexibility we seek is already given in the standard notion of a game, where a strategy has to work under any eventuality. One could collect all relevant cases of change into one "supergame," asking for one strategy working there. But such a pre-encoding seems far removed from our ordinary understanding of plans. In Chapter 6 and also in Part II, we will opt for working with small models instead.

39 For a nice example of repairing programs, see Huth & Ryan (2004).

5
Infinite Games and Temporal Evolution

All games studied so far were finite. But as we saw in our Introduction, infinite games are equally important. Never-ending processes model a wide range of phenomena, from operating systems to social norms. Indeed, infinite games are the paradigm in evolutionary game theory and much of computer science. A transition to the infinite occurs at various places in this book, starting with the infinite evolutionary games in our Introduction, and continuing with the limit scenarios with iterated announcements of Part II, the infinite evaluation games for fixed point logics in Chapter 14, and the models for branching time that will occur repeatedly. Infinite games can be studied with the tools we have introduced so far, since the modal μ-calculus of Chapters 1 and 4 can describe infinite histories. Still, the major paradigm in this realm is *temporal logic*, including the systems for action and knowledge evolving over time in Chapter 8 (see also Chapter 25). While this book places emphasis on modal and dynamic logics, this is not a rigid choice. In this chapter, we provide some basic background on temporal logics for studying games, while illustrating how these apply to basic issues in games such as information flow and strategic reasoning.

REMARK Names of players
Like most of this book, this chapter has two-player games for its paradigm, and as always, we appreciate diversity in life forms of notation. The two players, or roles, in the games to be discussed will be indicated by either $\boldsymbol{E}$ and $\boldsymbol{A}$, or by i and j, largely depending on customs in the literature of origin.

5.1 Determinacy generalized

We start with a special area close to our original starting point for game logic in Chapter 1. One striking result for finite games was Zermelo's Theorem. Much stronger results exist for infinite games, where players produce infinite runs, marked as winning or losing. Winning conditions for histories in infinite games vary a lot.

Gale-Stewart Theorem An important special case occurs when winning a run for a player depends only on what happens in some finite initial segment of play.

DEFINITION 5.1 Open sets of runs
A set of runs O is *open* if every infinite sequence that belongs to O has a finite initial segment X such that all runs of the game sharing X are also in O. An infinite game is called open if one of the players has an open winning condition. ■

In what follows, the Gale-Stewart Theorem from the 1950s generalizing Zermelo's Theorem is explored.

THEOREM 5.1 All open infinite games are determined.

Proof First we state the following completely general auxiliary fact. It goes under the name of *Weak Determinacy*, for reasons that will be clear from its formulation.

FACT 5.1 If player $\boldsymbol{E}$ has no winning strategy at stage s of an infinite game, then $\boldsymbol{A}$ has a strategy forcing a set of runs from s at all of whose stages $\boldsymbol{E}$ has no winning strategy for the remaining game from then on.

Weak Determinacy is proved as follows. $\boldsymbol{A}$'s strategy arises by the same reasoning as for Zermelo's Theorem. If $\boldsymbol{E}$ is to move, then no successor node can guarantee a win, since $\boldsymbol{E}$ has no winning strategy now, and so $\boldsymbol{A}$ can just wait and see. If $\boldsymbol{A}$ is to move, then there must be at least one possible move leading to a state where $\boldsymbol{E}$ has no winning strategy; otherwise, $\boldsymbol{E}$ has a winning strategy right now after all. By choosing his moves this way, $\boldsymbol{A}$ is bound to produce runs of the kind described.

Now, assume without loss of generality that $\boldsymbol{E}$ is the player who has the open set of winning runs. Then $\boldsymbol{A}$'s nuisance strategy just described is a winning strategy. Consider any run r produced. If r were winning for $\boldsymbol{E}$, then some finite initial segment $r(n)$ would have all its continuations winning, by openness. But then, $\boldsymbol{E}$ would have had a winning strategy by stage $r(n)$: namely, play whatever move. Quod non. ■

Open winning conditions occur, for instance, with our logic games of comparison and construction in Chapters 15 and 16. The Gale-Stewart Theorem is itself a special case of Martin's Theorem stating that all Borel games are determined. In this case, winning conditions lie in the Borel Hierarchy of sets of sequences (cf. Moschovakis 1980, Vervoort 2000). Natural non-open Borel conditions are fairness of operating systems (all requests eventually get answered), and the parity winning conditions for fixed point games in Chapter 14.

Nondeterminacy With non-Borel winning conventions, infinite games can be nondetermined. To define an example we need a basic set-theoretic notion.

DEFINITION 5.2 Ultrafilters
An *ultrafilter* on the natural numbers $\mathbb{N}$ (the only case we will need here) is a family U of non-empty sets of natural numbers with the following three properties:

(a) The family U is closed under taking supersets.

(b) The family U is closed under taking intersections.

(c) $X \in U$ iff $\mathbb{N} - X \notin U$ for all sets $X \subseteq \mathbb{N}$.

A *free* ultrafilter is a family U of this sort that contains no finite sets. ■

Free ultrafilters exist by the Axiom of Choice. Let U^* be one. Here is the game.

EXAMPLE 5.1 A nondetermined game
Two players, $\boldsymbol{A}$ and $\boldsymbol{E}$, pick successive neighboring closed initial segments of the natural numbers, of arbitrary finite sizes, producing a succession:

$$\boldsymbol{A}: [0, n_1], \text{ with } n_1 > 0, \quad \boldsymbol{E}: [n_1 + 1, n_2], \text{ with } n_2 > n_1 + 1, \quad \text{etc.}$$

We stipulate that player $\boldsymbol{E}$ wins if the union of all intervals chosen by $\boldsymbol{E}$ is in $\boldsymbol{U}^*$, otherwise, $\boldsymbol{A}$ wins. Winning sets in this game are not open for either player, as sets in U^* are not determined through finite initial segments. ■

FACT 5.2 The interval selection game is not determined.

Proof The proof is a "strategy stealing" argument. Player $\boldsymbol{A}$ has no winning strategy. For, if $\boldsymbol{A}$ had one, $\boldsymbol{E}$ could use that strategy with a delay of one step to copy $\boldsymbol{A}$'s responses disguised as $\boldsymbol{E}$'s moves. Both resulting sets of intervals (disjoint up to some finite initial segment) have their unions in $\boldsymbol{U}^*$; however, this cannot be, as $\boldsymbol{U}^*$ is free. But neither does $\boldsymbol{E}$ have a winning strategy, for similar reasons. ■

Through copying, modulo initial segments, the players' powers are the same in this game. We will encounter this symmetric situation once more in Chapters 11 and 20, when studying players' powers in possibly nondetermined infinite games. We will analyze the preceding arguments more deeply in Sections 5.2 and 5.5.

The dependence on the Axiom of Choice is significant. Set theorists have studied mathematical universes based on an alternative Axiom of Determinacy, stating that all infinite games are determined.[40]

5.2 Branching time and temporal logic of powers

Now that we have explored infinite games, we make a transition to temporal logic, starting with a simple system. Infinite games remind us of the larger temporal playground of possible histories that processes live in. What sort of logics are appropriate here? A simple point of entry is the proof of the Gale-Stewart Theorem. As we noted several times in Chapter 4, one good design method for logics is to look at the minimal expressive power needed to formalize basic game-theoretic arguments. The proof of the Gale-Stewart Theorem revolved around Weak Determinacy. What is the logical structure of the key argument establishing this principle?

The straightforward formalization is in the world where all games live: that of a *branching time* with forking histories, as in the following familiar picture:

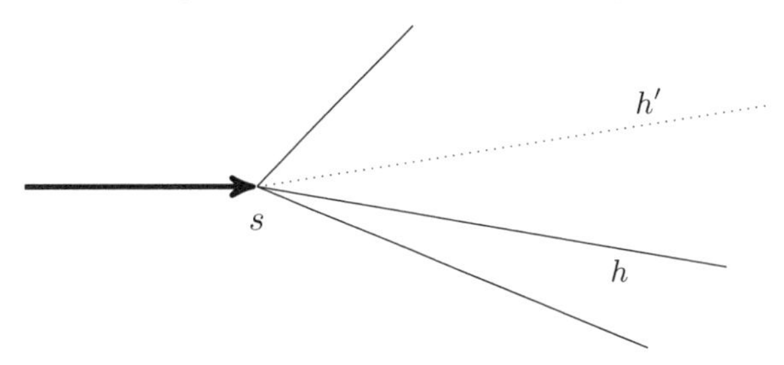

The bold line entering from the left indicates the actual history, only known up to stage s so far. In such models, points can have local properties encoded, while whole branches can have relevant global properties that are stage-independent. Examples of the latter are winning conditions on branches in some infinite logic games (see Chapters 14 and 15), or discounted payoffs in evolutionary games (Osborne &

40 There is a rich literature on determinacy results in descriptive set theory (Moschovakis 1980, Kechris 1994, Vervoort 2000), but it is tangential to our study of games.

Rubinstein 1994). Moreover, in these models, it is possible to distinguish between external properties and game-internal properties, such as the marking of nodes as a turn for one or the other player.

Such structures typically suggest a branching temporal logic, evaluating its formulas at ordered pairs (h, t) of a current branch h and a current point t on it. In this formalism, we have a simple perspicuous formula expressing weak determinacy. Let the two different players be i and j.

FACT 5.3 Weak Determinacy is the validity of the formula $\{i\}\varphi \vee \{j\}A\neg\{i\}\varphi$.

Here $\{i\}\varphi$ is a temporal extension of the forcing modalities of Chapters 1 and 4 stating that there exists a strategy for player i ensuring that only runs result, having the current history h up to point t as an initial segment, that satisfy the temporal logic formula φ. The temporal logic shows in the operator $A\psi$ saying that ψ is always true on the current branch.

REMARK An alternative notation
We can also suppress the temporal operator to obtain a pure forcing notation that may make our main point more clear. The two relevant principles in our discussion so far are then as follows:

Strong Determinacy	$\{i\}\varphi \vee \{j\}\neg\varphi$	SD
Weak Determinacy	$\{i\}\varphi \vee \{j\}\neg\{i\}\varphi$	WD

Strong Determinacy does not always hold in reasoning about games, but Weak Determinacy does. Deriving SD depends on further assumptions that enable the transition from WD, and these, too, often have a simple logical form (see Section 5.3 for more on this, including the relevant forcing notion).

We have turned a game-theoretic lemma into a law of temporal logic with game modalities. Call the set of valid principles on branching models *temporal forcing logic*. This system codifies a temporal base theory of action in infinite games as a companion to earlier modal logics.[41] Its complexity can be analyzed as follows.

FACT 5.4 Temporal forcing logic is decidable.

Proof All of the above modalities, including that for forcing, can be defined in the well-known system MSOL of *monadic second-order logic*, extending first-order logic with quantification over subsets. In particular, on trees with successor relations,

41 For natural merges of modal and temporal logics on trees, see Stirling (1995).

MSOL can quantify over histories, since they are maximal linearly ordered subsets. Crucially, MSOL can also define strategies, since binary subrelations of the *move* relation can be coded uniquely as subsets, too, as shown in the analysis of Backward Induction in Chapter 8. By Rabin's Theorem (Rabin 1968, Walukiewicz 2002), MSOL is decidable on our tree models, and hence so is temporal forcing logic.[42] ■

The preceding result still leaves the following challenge of a complete description.

Open Problem Provide a complete axiomatization for temporal forcing logic.

In the next section, we will look closely at the more detailed level of analysis of the preceding chapter, and give further fine structure by scrutinizing the details of the strategic reasoning underlying the earlier results.

5.3 Strategizing temporal logic

To see in more detail what can be done with temporal logics, consider representation of strategies, the theme of Chapter 4. The traditional business of logic is analyzing a given reasoning practice. We apply this perspective to a few set pieces of strategic reasoning (see van Benthem 2013 for further details and motivation). As usual, for convenience, we restrict attention to games with only two players, i and j.

Temporal logic of powers Our starting point is the temporal setting of Section 5.2. Zooming in on the nodes of these branching trees, assuming time to be discrete, we can interpret a branching temporal language of a sort found in many flavors in the literature, in the format

$$\boldsymbol{M}, h, s \models \varphi \qquad \text{formula } \varphi \text{ is true at stage } s \text{ on history } h$$

with formulas constructed using proposition letters, Boolean connectives, temporal operators F (sometimes in the future), G (always in the future), H (always in the past), P (sometimes in the past), and O (at the next moment) referring to the current branch, and modalities $\Diamond$ and $\Box$ over all branches at the current stage. Some typical clauses are as follows.

---

42 Indeed, many of the MSOL-definable strategy-related modalities on trees that are used in this book are also bisimulation-invariant. Hence, by the Janin-Walukiewicz Theorem (Janin & Walukiewicz 1996), they are also definable in the modal μ-calculus, reflecting our analysis in Chapters 1 and 18.

$\boldsymbol{M}, h, s \models F\varphi$ iff $\boldsymbol{M}, h, t \models \varphi$ for some point $t \geq s$,
$\boldsymbol{M}, h, s \models O\varphi$ iff $\boldsymbol{M}, h, s+1 \models \varphi$ with $s+1$ the direct successor of s on h,
$\boldsymbol{M}, h, s \models \Diamond\varphi$ iff $\boldsymbol{M}, h', s \models \varphi$ for some h' equal to h up to stage s.

We will say more about such logics in Section 5.4, but for now, given our topic, we add a new strategic modality $\{i\}\varphi$ describing the powers of player i at the current stage of the game, a temporal version of the forcing modalities of Chapter 1:

$\boldsymbol{M}, h, s \models \{i\}\varphi$ Player i has a strategy from s onward which ensures that only histories h' result for which, at each stage $t \geq s$, $\boldsymbol{M}, h', t \models \varphi$.

This looks local to stages s, but φ can also be a global stage-independent property of the histories h'. It is common to speak of "winning conditions" φ in this setting. Sometimes, these are external features of histories, such as having built a partial isomorphism in the model comparison games of Chapter 15, but sometimes also, they refer essentially to game-internal features such as what happened at turns of specific players.

It is also important to note that, although our forcing modality $\{i\}\varphi$ is strong, it does not say that φ must be true on the current branch. We can have a strategy for salvation, and yet be on the road to perdition.[43]

Valid laws Principles of reasoning for this language are a combination of components that occur at several places in this book.

THEOREM 5.2 The following principles are valid in temporal forcing logic:

(a) The standard laws of branching temporal logic.

(b) The standard minimal logic of a monotonic neighborhood modality for $\{i\}\varphi$, plus one axiom stating its modal character: $\{i\}\varphi \rightarrow \Box\{i\}\varphi$.

(c) Three game-based principles in the spirit of Chapters 1 and 4:

(i) $\{i\}\varphi \leftrightarrow (((\boldsymbol{end} \wedge \varphi) \vee (\boldsymbol{turn}_i \wedge \Diamond O\{i\}\varphi) \vee (\boldsymbol{turn}_j \wedge \Box O\{i\}\varphi)))$

(ii) $\big(\alpha \wedge \Box G(((\boldsymbol{turn}_i \wedge \alpha) \rightarrow \Diamond O\alpha) \wedge ((\boldsymbol{turn}_j \wedge \alpha) \rightarrow \Box O\alpha)\big) \rightarrow \{i\}\alpha$

(iii) $(\{i\}\varphi \wedge \{j\}\psi) \rightarrow \Diamond(\varphi \wedge \psi)$

43 Our strategy operator is modal, referring to all possible branches. A natural local variation $\{i\}^+\varphi$ would only require the current history h to be played according to the strategy. We have not yet found a need for this further operator.

Proof For (a), references can be found in Section 5.4. For (b), see the modal logic of forcing presented in Chapter 11. As for (c), the first principle (i) is the fixed point recursion that we explored in Chapter 1, while (ii) is a strong introduction law reminiscent of the axiom for the universal iteration modality in PDL, although the antecedent is stronger.[44] The third principle (iii) is distinctly different from the first two. It is a simple form of independence of strategy choices for the two players. This independence is much like the modal logic of forcing, or the STIT-type logics of players' abilities in Chapter 12. ■

It is easy to derive further valid principles from these laws.

FACT 5.5 The *S4* law $\{i\}\alpha \to \{i\}\{i\}\alpha$ is valid in temporal forcing logic.

This tells us something interesting about strategic powers. If we follow a strategy, we have a guarantee that, in principle, we never have to leave the safe area of having that strategy available. In epistemic logic, this formula would be a principle of introspection. Here, however, it expresses a sort of safety of strategies.

In addition to valid principles, non-validities can be informative, too. A striking example follows from our explanation of the forcing modality: $\{i\}\alpha \to \alpha$ is not valid. More remarkable non-validities only appear on infinite histories.

EXAMPLE 5.2 Informative non-validities
The principle $G\{i\}\alpha \to (\alpha \vee F\alpha)$ is valid on finite games, since on endpoints of the current history, α will be forced. But now consider the following infinite model (an evergreen in many areas), viewed as a one-player game:

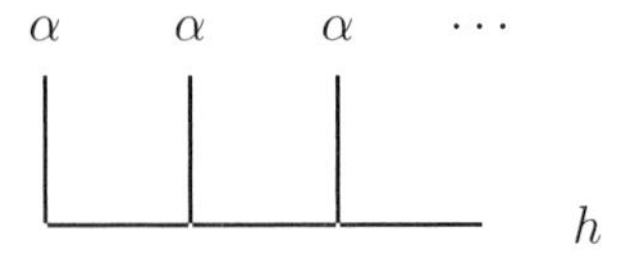

On each point of the unique infinite history h, a player has a strategy for ending in an endpoint. But staying on the infinite branch forever will produce a history that fails to satisfy the condition of being finite. ■

---

44 These principles have the most power when applied to genuinely local properties of stages on histories. The reader will find it instructive to see how little the second law says when the predicate α is global winning of infinite branches. This has to do with the difference between finite-stage properties and the potentially more complex winning conditions discussed in connection with the Gale-Stewart Theorem.

Presumably, our logic encodes many further valid and non-valid principles demonstrating the interplay of finite and infinite winning conditions. However, right here, we end with a small illustration going back to an earlier section.

Proving Weak Determinacy Weak Determinacy looks as follows in our system:

$$\{i\}\varphi \vee \{j\}\neg\{i\}\varphi$$

As an illustration of our logic, we now derive this formally, using two clauses from item (c) in Theorem 5.2.

$$\begin{array}{ll} (\boldsymbol{turn}_i \wedge \neg\{i\}\varphi) \rightarrow \Box O\neg\{i\}\varphi & \text{from (i)} \\ (\boldsymbol{turn}_j \wedge \neg\{i\}\varphi) \rightarrow \Diamond O\neg\{i\}\varphi & \text{from (i)} \\ \neg\{i\}\varphi \rightarrow \{j\}\neg\{i\}\varphi & \text{from (ii)} \end{array}$$

Now we can also derive the Gale-Stewart Theorem. Let φ be an open condition:

$$\varphi \rightarrow F\Box G\varphi$$

Using this, it is easy to show formally that $\{j\}\neg\{i\}\varphi \rightarrow \{j\}\neg\varphi$, and in combination with Weak Determinacy, it follows that the game is determined:

$$\{i\}\varphi \vee \{j\}\neg\varphi$$

Finally, Zermelo's Theorem is a simple corollary, since the fact of having an endpoint is an open property of branches that can be represented as:

$$F\boldsymbol{end} \rightarrow F\Box GF\boldsymbol{end}$$

As for invalid principles, Weak Determinacy does not imply Strong Determinacy, and it turns out to be of interest to analyze our earlier counterexample.

The logic of strategy stealing Within the above general logic of strategies, further interesting properties come to light in special models. To see this, going beyond the Gale-Stewart Theorem, consider the nondetermined interval game of Example 5.1. Analyzing the strategy stealing argument in more detail reveals an interesting logical structure.

Suppose that $\boldsymbol{E}$ starts; the case with $\boldsymbol{A}$ starting is similar. The strategy σ then gives a first move $\sigma(-)$. Now let $\boldsymbol{A}$ play any move e. $\boldsymbol{E}$'s response will be $\sigma(\sigma(-), e)$, after which it is $\boldsymbol{A}$'s turn again. Crucially, this same sequence of events can also be viewed differently, as a move $\sigma(-)$ played by $\boldsymbol{E}$, followed by a move $e : \sigma(\sigma(-), e)$ played by $\boldsymbol{A}$, after which it is $\boldsymbol{E}$'s turn. This shift in perspective is what generated the contradiction in the earlier proof.

This argument presupposes the following special property of games.[45]

> *Composition Closure* Any player can play any concatenation of available successive moves.

The tree for the interval game has an interesting feature that is of independent interest. The two stages described here start the same subgames in terms of available moves, but with the difference that all turn markings are interchanged. Thus, in terms of a notion from general game logic (cf. Chapters 19 and 20), one subgame is a *dual* of the other in that all turns have been interchanged. $\boldsymbol{A}$'s strategy now uses $\boldsymbol{E}$'s strategy in the other game to produce two identical runs in both subgames, except for the inverted turn marking. But then, the winning conditions are in conflict: the union of $\boldsymbol{E}$'s choices should be in the ultrafilter, but so should the union of $\boldsymbol{A}$'s choices on that same history.

More important than this single scenario may be the following positive principle of game logic underneath, that we can state as a "copy law."

FACT 5.6 In games with composition closure, the following principle holds:

$$\{i\}\varphi \to \Diamond OO\{j\}\varphi^d$$

where φ^d is φ with all occurrences of $turn_i$, $turn_j$ interchanged.

Essentially, in games of the sort described here, players only have powers whose duals in the preceding sense are compatible.[46]

---

45 One can define this property in a suitably extended modal-temporal language.

46 More can be said here. The two subgames are not full duals in the sense of our other chapters, since that would also require inverting the sets of winning conditions for the two players. But if we do that, we do not get the above contradiction. In fact, games have two options for dualizing: turns and winning conditions. Both can be done independently. Sometimes, one only wants to dualize winning conditions, as in the games for smallest versus greatest fixed points in Venema (2007). For a logical analogy, compare negating a formula: dualizing all connectives, and then reversing polarity of atoms, versus just doing one or the other. Perhaps the game algebra of both forms of dualization is worth studying.

REMARK Copying, borrowing, and knowledge
One may compare our strategy stealing argument with the use of the copy-cat strategy of Chapter 20, although that involves full dualization. A striking further difference is that here, $\boldsymbol{A}$ needs to know $\boldsymbol{E}$'s whole strategy to run the simulation, whereas in copy-cat, it is only necessary to see one move at a time. This reflects a difference between actual parallel play versus playing a "shadow match" in the sense of Venema (2006). What also emerges here is a difference between the rhetoric of our arguments and their formalization. Intuitively, being able to steal or borrow a strategy involves knowledge, but our formalizations in this chapter do not make this feature explicit. This issue will come up again in Chapter 18.

Our conclusion is that temporal logic helps in analyzing strategic reasoning, while also extracting general principles that may not have met the eye before.[47]

5.4 Epistemic and doxastic temporal logics

Having studied specific actions and strategies, we now turn to a broader canvas. Infinite playgrounds are also the domain of temporal logics of information-based agency, and we now consider these on their own. We will present a few systems that add knowledge and belief (and in principle also preference). These form a natural continuation of the process logics for games in earlier chapters.

The grand stage of branching time The appealing structure we have been exploring in this chapter occurs in many fields, from logic, mathematics, and computer science, to philosophy and game theory. Tree models for branching time consist of legal histories h, say, the possible evolutions of a game. At each stage, players are in a node s on some actual history whose past they know completely or partially, but whose future is yet to be revealed. This is a grand stage view of agency, with histories as complete runs of some information-driven process, which is ubiquitous in the literature on interpreted systems (Fagin et al. 1995), epistemic-temporal logic (Parikh & Ramanujam 2003), STIT (Belnap et al. 2001), or game semantics (Abramsky 2008b). Such structures invite languages with temporal operators and other notions important to agency, such as knowledge. These languages describe game structure, but also important computational system properties such

47 There is a growing literature on strategic reasoning in temporal logics of agency, of which we mention Ågotnes et al. (2007) and Pacuit & Simon (2011).

as safety and liveness that underlie the infinite logic games of Part IV of this book. This is also the global playground for the local informational steps that will be studied in Part II of this book, and the connection will return in Chapter 8.

Branching temporal logic When thinking of agency, these structures suggest an action language with knowledge, belief, and added temporal operators. Its models come in two flavors (cf. Hodkinson & Reynolds 2006, van Benthem & Pacuit 2006) that will both be used in this book.

The first kind of model evaluates on pairs of complete histories and stages, which we have already seen in our case study of strategic powers. For instance, with games of perfect information, we evaluated formulas at points on histories. Looking at an extended repertoire, we use the following notations: sa denotes the unique point following s after event a has taken place (we assume that events are unique), and $s < t$ says that point t comes after s.

DEFINITION 5.3 Basic temporal operators
The most important temporal operators are given below, including those in Section 5.3 plus some counterparts to modalities used in earlier chapters:

$\boldsymbol{M}, h, s \models F_a\varphi$ iff sa lies on h, and $\boldsymbol{M}, h, sa \models \varphi$ (after a next event)
$\boldsymbol{M}, h, s \models O\varphi$ iff t directly follows s and $\boldsymbol{M}, h, t \models \varphi$ (at the next stage)
$\boldsymbol{M}, h, s \models F\varphi$ iff for some point $t > s$, $\boldsymbol{M}, h, t \models \varphi$ (in the future)
$\boldsymbol{M}, h, s \models P_a\varphi$ iff $s = s'a$ for some s', and $\boldsymbol{M}, h, s' \models \varphi$ (at the previous stage)
$\boldsymbol{M}, h, s \models P\varphi$ iff for some point $t < s$, $\boldsymbol{M}, h, t \models \varphi$ (in the past)

These are purely temporal operators staying on the same history. The next operator is modal, looking at all available histories fanning out from the present stage:

$\boldsymbol{M}, h, s \models \Diamond_i\varphi$ iff $\boldsymbol{M}, h', s \models \varphi$ for some h' equal to h for i up to stage s.[48]

This modality ranges over histories that may yet become realized in the future. ■

Combining operators yields further modalities, such as $\Diamond F\varphi$: φ may become true later on some possible history, but not necessarily on the actual history.

---

48 A universal version $\Box$ can be defined as usual as the dual operator $\neg\Diamond\neg$. The modality $\Diamond$ may also be viewed as an agent's knowledge about how the process may still unfold. In general, $\Diamond$ can then be agent indexed, since not all agents need to find the same future histories possible. Such modeling options will be discussed in Chapter 6.

We will return to this double-index view of histories in Chapter 6 when discussing various representations of games. But there is also an alternative flavor of temporal logic available to us. We now take a different tack from another broad strand in the literature, interpreting formulas only at finite stages of histories.

A temporal logic for epistemic forests A second widely used perspective focuses on a modal universe of temporal stages. In this approach, we use only finite histories as points of evaluation living in a modalized future of possible histories extending the current one. Take sets $\mathbb{A}$ of agents and $\mathbb{E}$ of events. A *history* will now be just a finite sequence of events, and $\mathbb{E}^*$ denotes the set of all such histories. Looking at the future as events occur, we write he for the unique history after e has happened in h. Also, we will write $h \leq h'$ if h is a prefix of h', and $h \leq_e h'$ if $h' = he$. Models are now as follows, starting from the basic notion of a protocol.

DEFINITION 5.4 Epistemic forest models
A *protocol* is a set of histories $\mathbb{H} \subseteq \mathbb{E}^*$ closed under prefixes. An *ETL frame* is a tuple $(\mathbb{E}, \mathbb{H}, \{\sim_i\}_{i \in \mathbb{A}})$ with a protocol $\mathbb{H}$, and accessibility relations $\sim_i$. An *epistemic forest model* (also called ETL model) is an ETL frame plus a valuation V sending proposition letters to sets of histories in $\mathbb{H}$. ■

In what follows, when we write h, he, etc., we always assume that these histories occur in the relevant protocol.

These models describe how knowledge evolves over time in an informational process. The epistemic relations $\sim_i$ represent uncertainty of agents about how the current history has gone, due to their limited powers of observation or memory. Thus, $h \sim_i h'$ means that from i's point of view, the history h' looks the same as the history h. Importantly, not all of these models are trees. Depending on the scenario, there may be multiple initial points, whence the term "epistemic forest" that we will use occasionally as a more poetic name for these models.

The *epistemic temporal language* L_{ETL} extends the basic epistemic logic of Chapter 3 with event modalities for forest models. It is generated from atomic propositions At by the following syntax:

$$p \mid \neg\varphi \mid \varphi \wedge \psi \mid [i]\varphi \mid \langle e \rangle \varphi$$

where $i \in \mathbb{A}$, $e \in \mathbb{E}$, and $p \in At$. Here the modality $[i]\varphi$ stands for epistemic $K_i\varphi$. Booleans and dual modalities $\langle i \rangle$, $[e]$ are defined in the usual way.

DEFINITION 5.5 Truth of L_{ETL} formulas
Let $\boldsymbol{M} = (\mathbb{E}, \mathbb{H}, \{\sim_i\}_{i \in \mathbb{A}}, V)$ be an epistemic forest model. The truth of a formula φ at a history $h \in \mathbb{H}$ ($\boldsymbol{M}, h \models \varphi$) is defined with the following key clauses:

$$\boldsymbol{M}, h \models [i]\varphi \quad \text{iff} \quad \text{for each } h' \in \mathbb{H}, \text{ if } h \sim_i h', \text{ then } \boldsymbol{M}, h' \models \varphi.$$

$$\boldsymbol{M}, h \models \langle e \rangle \varphi \quad \text{iff} \quad \text{there exists } h' = he \in \mathbb{H} \text{ with } \boldsymbol{M}, h' \models \varphi.$$

Further epistemic and temporal operators are easily defined in the same style. ■

We now review a few features of this second epistemic-temporal system that are reminiscent of the modal-epistemic concerns discussed earlier in Chapter 3.

Types of agents Further constraints on models reflect special features of agents, or of the current informational process. These intertwine epistemic and action accessibility, with matching epistemic-temporal axioms in the same style as the axioms in Chapter 3. The following correspondence result shows the analogy.

FACT 5.7 The axiom $K[e]\varphi \to [e]K\varphi$ corresponds to ETL-style perfect recall:
If $he \sim k$, then there is a history h' with $k = h'e$ and $h \sim h'$.

This says that agents' current uncertainties can only come from previous uncertainties.[49] The axiom presupposes perfect observation of the current event e. In the dynamic-epistemic logics of Part II that deal with imperfect information games, this will change to allow for uncertainty about the current event.[50]

Again, as in Chapter 3, epistemic-temporal languages also describe other agents. Take a "memory-free" automaton that only remembers the last-observed event, making any two histories he and ke ending in the same event epistemically accessible. Then, with finitely many events, knowledge of the automaton can be defined in the temporal part of the language. Using the above backward modalities P_e plus a universal modality U over all histories, we have the valid equivalence

$$K\varphi \leftrightarrow \bigvee_e \left(\langle e^{\cup} \rangle \top \wedge U(\langle e^{\cup} \rangle \top \to \varphi) \right).[51]$$

49 An induction on distance from the root then derives "synchronicity": uncertainties $h \sim k$ only occur between h and k at the same tree level. Weaker forms of perfect recall in games also allow uncertainty links across levels (Bonanno 2004b, Dégremont et al. 2011).

50 The dual $[e]K\varphi \to K[e]\varphi$ corresponds to the no miracles principle: for all ke with $h \sim k$, we have $he \sim ke$. Thus, learning can only take place by observing different signals.

51 Similar ideas work for bounded memory in general (Halpern & Vardi 1989, Liu 2008).

Expressive power and complexity One can vary the expressive power of these languages in the temporal part (one common addition are the temporal expressions Since and Until), but also in the epistemic part, adding operators of common or distributed knowledge (cf. Chapter 3 and 7). But then we meet the balance discussed in Chapters 2 and 3. Increases in expressive power may lead to upward jumps in computational complexity of combined logics of knowledge and time. The first investigation of these phenomena was made in Halpern & Vardi (1989). The table below collects a few observations from this pioneering paper, showing where dangerous thresholds lie for the complexity of validity.

	K, P, F	K, C_G, F_e	K, C_G, F_e, P_e	K, C_G, F
All ETL models	Decidable	Decidable	Decidable	RE
Perfect Recall	RE	RE	RE	Π_1^1-complete
No Miracles	RE	RE	RE	Π_1^1-complete

In this table, complexities run from decidable through axiomatizable (RE) to Π_1^1-complete, which is the complexity of truth for universal second-order statements in arithmetic.[52] The survey of expressive power and complexity for epistemic temporal logics in van Benthem & Pacuit (2006) also cites relevant background on tree logics and modal product logics. As in Chapter 3, complexity comes from defining grid cells and encoding tiling problems, whose feasibility depends on a delicate balance in language design.

Beliefs over time Epistemic temporal models generalize to other attitudes that drive rational agency, in particular, to the notion of belief. Again, this can be done in two ways, either in the earlier (h, s) style, or on finite stages h of histories. In the latter line, *epistemic-doxastic-temporal* models are branching event forests as before, with nodes in epistemic equivalence classes now also ordered by *plausibility relations* for agents. Forest models then interpret belief modalities $B\varphi$ as saying that φ is true in the most plausible accessible histories. A plausibility ordering carries much more information, however, and it can also interpret conditional beliefs in a natural way, as we will see later on. This way of modeling beliefs will be discussed in Chapter 6, and again in Chapter 7. For applications of temporal plausibility models

52 There is a gap between RE and Π_1^1-complete: few epistemic-temporal logics fall in between these classes. This also occurs with extensions of first-order logic, where being able to define a copy of the natural numbers $\mathbb{N}$ is a watershed. If you cannot, as is the case for first-order logic, complexity stays low: if you can, as is the case for first-order fixed point logics or second-order logic, complexity soars.

to game theory, see Bonanno (2001, 2007). Further issues in temporal modeling of games will come up in Chapter 6. In addition, temporal frameworks will play a role in Part II on the dynamics of actions in games.

Representation by underlying mechanisms All temporal models discussed here come with hardwired epistemic accessibility relations or doxastic plausibility orderings. But one can ask, as in Chapter 3, what underlying mechanism or process would produce this pattern of ignorance or preference. Answers will come in the representation theorems of Chapter 8. These show precisely what traces are left by agents updating their knowledge according to general dynamic-epistemic product update (Baltag et al. 1998) and revising their beliefs using a master rule of "priority update" (Baltag & Smets 2008).

5.5 Conclusion

The main points Passing from finite to infinite structures is natural in logic and in computation, and forms no barrier to a study of games. We have shown how temporal logics in two flavors (i.e., history-based or merely stage-based) can analyze basic strategic features of infinite games in an illuminating way. These logics were a natural extension of the modal logics used in earlier chapters for finite games, whose themes returned with new flavors. We also showed how epistemic and doxastic temporal logics are available for a further study of agency in games.

Open problems Still, the move from finite to infinite raises important open problems that will not be addressed in this book. One is a shift in focus for game logics from classical game theory to evolutionary game theory (Hofbauer & Sigmund 1998). Logical systems will then meet the theory of dynamical systems. An interesting program in this direction is the dynamic topological logic of Kremer & Mints (2007). On a much smaller scale, another relevant line are the logics for game matrices in Chapter 12 that can define notions such as evolutionary stability.

The move from finite to infinite may also affect our very understanding of strategies. In much of the preceding, these are finite rules with an inductive character, as in the bottom-up procedure of Backward Induction. However, repeating a point made earlier, a strategy may also be viewed "co-inductively," as a never-ending resource that we invoke as needed. We maintain our health by consulting our doctor, closing the episode, and expecting the doctor to be available as fresh as ever afterward. This sort of intuition fits better with using greatest rather than smallest fixed points in the μ-calculus or LFP(FO), and at a deeper level, with the non-wellfounded structures of co-algebra (Venema 2006). As noted in Chapter 18 on games in computational logic, transposing the theory of this book to the co-algebra setting is an interesting challenge.

5.6 Literature

Our discussion of forcing and strategies is taken from van Benthem (2013), and our general picture of epistemic-temporal frameworks comes from the survey van Benthem & Pacuit (2006).

Descriptive set theory is explained in many modern sources, such as Moschovakis (1980) and Kechris (1994). An interesting connection between solving infinite games and key methods in the foundations of mathematics (realizability, bar induction) is explored in Escardo & Oliva (2010). Epistemic temporal logics exist in great variety, with Fagin et al. (1995) and Parikh & Ramanujam (2003) as major flavors. An important computational source is the game semantics of Abramsky (2008b), a philosophical source is the STIT framework of Belnap et al. (2001). Bonanno (2001, 2007) applies temporal logics to game solution including policies for belief revision. Also relevant to the themes of this chapter are many topics in the logic games of Part IV and systems emanating from these in Part V, in particular, the linear logics of infinite games in Chapter 20.

6
From Games to Models for Games

Reasoning about games reflects social settings we are all familiar with, as we constantly make sense of what others do in interaction with ourselves. But upon closer inspection, this practice is subtle. Do we really stick to a given game tree, or should further aspects be represented, such as our view of other players we are dealing with? We will start with some simple pieces of reasoning that logics of social scenarios should be able to deal with. These will be our cues for a discussion of natural levels of modeling games, identifying thresholds where new structure has to come in. Our preference will be for keeping models small, or thin, while putting additional aspects into the dynamics of play, a topic that will be continued in Part II of this book. This bottom-up methodology of logical modeling may be contrasted with the use of sometimes huge "type spaces" in game theory,[53] where every relevant consideration about players has been packed into the initial model.

This chapter will repeat many points from earlier chapters, providing a common thread among the often bewildering variety of logical models for games. At the same time, it will highlight the need for a finer information dynamics of games, and logics in that vein will be the main topic of Part II on the Theory of Play.

---

53 For information on type spaces, cf. Osborne & Rubinstein (1994), Brandenburger & Keisler (2006), Perea (2012), Brandenburger et al. (2014), and Pacuit & Roy (2013).

6.1 Some simple scenarios

EXAMPLE 6.1 Interpreting a choice
Let agent $\boldsymbol{E}$ have actions *left* and *right*, while the second stage to the right is a fork that is not under $\boldsymbol{E}$'s control. The numbers at the leaves record utilities:

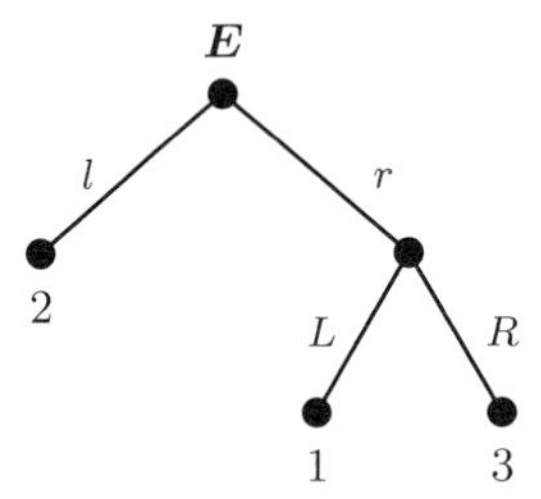

Assume that $\boldsymbol{E}$ does not know what will happen when r is chosen, although $\boldsymbol{E}$ may have expectations, for simplicity, in the form of unique continuations considered most plausible. We can make sense of different behavior here. If $\boldsymbol{E}$ in fact chooses action l with outcome 2, we would think that $\boldsymbol{E}$ considers R less plausible at the second stage, since outcome 2 is better than 1. But if r were to be chosen, we would probably think that $\boldsymbol{E}$ considers R, rather than L, most plausible, and that her actions are aiming for outcome 3.

But we can also complicate matters. Suppose that in fact $\boldsymbol{E}$ considers R less plausible than L for the second stage. Could a choice for r still make sense, trying to exploit the fact that some other player with influence over outcomes will now be misled, like us, into thinking $\boldsymbol{E}$ considers R a more plausible outcome?[54] ■

Looking more specifically at solution procedures for games, similar options arise.

EXAMPLE 6.2 Off-path steps violating Backward Induction
Let two players $\boldsymbol{A}$ and $\boldsymbol{E}$ play the following simple extensive game, with their payoffs at the endpoints marked in the order ($\boldsymbol{A}$-value, $\boldsymbol{E}$-value).

54 All this makes sense only when we think of the choice point at the right as leading to a choice between longer scenarios where information about $\boldsymbol{E}$ may matter to $\boldsymbol{A}$.

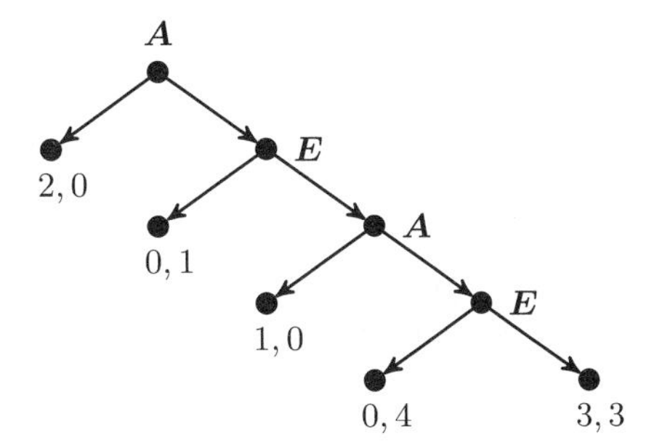

When computed as in Chapter 2, Backward Induction recommends moving left at each stage of the game. Thus, strikingly, player ***A*** should already opt out right at the start. But what if ***A*** deviates and goes right? Sticking with Backward Induction, players would have to view this as a mistake. However, there is another plausible line of reasoning. If ***A*** goes right, then player ***E*** can make sense of this surprising move by assuming that ***A*** expects to get the outcome $(3, 3)$. This new belief will change ***E***'s behavior to playing right at ***E***'s first turn, and if ***A*** in fact expected $(3, 3)$ right will be chosen once more. ■

Of course, what happens at the end of this game is up for grabs. You may think that ***E*** will go left at the end, callously ignoring ***A***, if this is a competitive game. But labels such as competitive or cooperative are not very useful in social scenarios. Many real settings are a bit of both (as in the case of academic life). The issue is how to act in a game while not knowing exactly which label applies. Maybe ***A*** has signaled something that influences ***E***'s behavior deeply. We want logics for games that can analyze scenarios such as this without preconceptions.

6.2 Different styles of reasoning and modeling choices

From deliberation to real-time dynamics of play Backward Induction is a powerful but very special way of deliberating about social scenarios, which only considers the *future* after the current node; the history so far does not count.[55]

---

55 Technically, the Backward Induction strategy is "invariant for generated submodels." Another special feature is its uniform node-independent plausibility ordering of histories, which is the same for both players.

In line with this, our logical analysis in Chapters 1 and 2 worked on game trees themselves. This simple setting has its limitations, and more structure may have to be represented. When actually playing a game, we may observe moves that seem significant to us, and there can be other changes in expectations. In such a richer setting, Backward Induction has rivals, as we just saw. Surprise moves may inform us about the type of player we are up against. Thus, Stalnaker (1999) identifies an additional parameter in analyzing games: we need to represent players' policies for *belief revision*. This seems just right, and what we know or believe about other players should indeed be crucial information. We will study this dynamic perspective on players in detail in Part II of this book, but here, we just mention a few themes that are relevant to designing models.

In principle, there is a danger to enriching our setting as indicated: the new structure might create a large and unstructured diversity of policies. Still, some general procedures exist, and we will now consider what these involve in terms of belief formation and choices of models.

Rationalization and belief formation If the past can be informative, then we should factor it in, as we did in the preceding example, where we interpreted an observed move in light of what players might have chosen earlier, but did not. A method staying close to our earlier analyses is "rationalizing" observed behavior in terms of ascribing beliefs to players that make them rational-in-beliefs. In game theory this is known as Forward Induction (cf. Battigalli & Siniscalchi 1999, and Chapters 9 and 10).

Different hypotheses to make sense of play Rationalization is not the only game in town. In other natural scenarios, the space of hypotheses about players can be very different. We may know that we are playing against a finite machine, or against an overworked colleague, and to optimize our own strategy, we need to find out how much memory our opponent has available. Chapter 10 discusses taxonomies of hypotheses about players, in order to tame the explosion of options in general agency.

Small models versus large models What does all this mean for the simple game models that we have studied so far? Here is a basic choice point in logical modeling. Game-theoretic analyses have tied more sophisticated reasoning styles such as Forward Induction to solution algorithms on strategic games such as iterated removal of weakly dominated strategies (Brandenburger 2007, Perea 2012). The usual semantic justification for these procedures involves type spaces encoding large sets of possible hypotheses about players' strategies, (Osborne & Rubinstein

1994), often in a category-theoretic setting.[56] But while this may work well as a theorist's perspective, the use of large, or thick, models is hard to relate to actual practice of agents. A typical trend in logical analysis has been the opposite idea of "small is beautiful," using small or thin models that stay close to an image that an agent might form of the situation. If needed, other relevant aspects can then be relocated to dynamic actions over such small models. In what follows, we take this stance, fitting small structures to reasoning scenarios of interest. We will also discuss matching logics and natural jumps in complexity.

6.3 Triggers for jumps: Diversity of knowledge in games

Intuitively, reasoning about games contains thresholds of complexity that may force us upward in what a model for a game should represent. We have seen one source for this pressure by now: information about revision policies of players when confronted with surprise events. Here is one more reason for choosing a richer set of modeling stages. We want to keep track of some natural intuitive distinctions that arise in games concerning the crucial notion of knowledge that permeates many of the chapters in this book.

As we have noted in Chapters 3 and 5, knowledge and beliefs of players come in at least three intuitively different kinds (van Benthem 1999). At any point in an extensive game of perfect information, players know the future in the sense of all possible continuations, and by the same token, they also know the past. This future knowledge is based on understanding the game as a process. However, in games of imperfect information, as in Chapter 3, players also had another kind of knowledge (and ignorance) about precisely where they are in the game tree. This time, the source of the information is different, depending on players' powers of observation and memory. And there is even a third intuitive kind of information, as we have seen in the preceding sections. Players may also have relevant knowledge *about other players*, their general types, or their current strategies. This third kind of knowledge may be based on prior experience, or on having been informed from outside of the game as defined officially. Our ambition in designing logical models

56 There may be a tradeoff between a simple solution algorithm and a complex semantic model for showing correctness. If this balance holds, then our simpler semantic models may have an algorithmic price.

is to make such distinctions visible, rather than wash them away in very powerful models for some uniform sort of knowledge.

Given these motivations, in the following sections, we will walk slowly along different levels for modeling games, taking our starting point in models used in earlier chapters, and often repeating points made there. Our discussion will put these models in one framework, and toward the end, we will pass on the torch to the logical dynamics of constructing and modifying models of games to be found in Part II of this book.

6.4 Modal logic of game trees

We start at the simple level of Chapter 1. Game trees are models of possible actions, and at the same time, they model future knowledge. As we have seen, much relevant structure can be brought out then by means of a propositional dynamic logic of simple and complex programs. This language can express actions, strategies, counterfactuals, and with one small addition, expectations about the future.

Pointed game trees as local models Consider game trees with atomic predicates at nodes, with a distinguished point s indicating how far play has progressed:

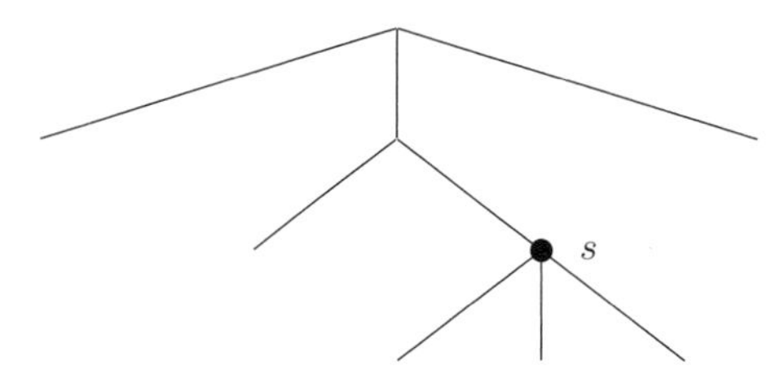

At the actual stage, players know what has happened, and let their behavior depend on two factors: (a) what the remaining game looks like, and (b) what happened so far in the larger game.

Modal language of action and future knowledge Pointed game trees support many languages, depending on the reasoning we want to represent. A good formalism was the modal action logic of Chapter 1, or its extension to a propositional dynamic logic, PDL, with propositions and programs. There are modalities talking about moves:

$\langle a \rangle \varphi$	after performing an a move, φ holds
$\langle a^* \rangle \varphi$	after performing some finite sequence of a moves, φ holds

This language can define strategic powers for players in games, as well as strategies themselves (cf. Chapter 4). But in addition to this action view of the language, it also encodes knowledge about the future. Adding a general term, *move*, for the union of all specific moves, $\langle move^* \rangle \varphi$ is a description of the epistemic future, acting as an existential tense running over all positions that can still be reached by further play. Unlike most epistemic logics in game theory, this is *S4* knowledge with reflexive and transitive accessibility, fitting the temporal character of information about the future.[57]

Looking backward While the modal logic PDL looks forward in time, there is also a case for looking backward. Consider the Backward Induction strategy as analyzed in Chapter 2 by the statement RAT that said a node is not the result of passing through a dominated move. This looks toward the past and we would add a converse program modality

$\langle a^{\cup} \rangle \varphi$ there is a node with φ reaching the current node by an a move

Given the structure of trees, we can then also define an operator Y for "yesterday," plus propositional constants marking the last move played (if we are not in the root). Another important definable type of expression is *counterfactuals*, since PDL with converse can talk about what would be the case if we had chosen different moves in the past.[58]

We conclude that a well-chosen propositional dynamic logic of extensive games does a creditable job in defining action structure, strategies, and even a form of future knowledge for players.

---

57 A similar procedural temporal aspect occurs in intuitionistic logic. For a study of analogies between dynamic-epistemic logics and intuitionistic logic as a model of information, see van Benthem (2009).

58 To fully define the rationality statement RAT, one can use a "hybrid extension" of the modal language (Areces & ten Cate 2006) referring to specific nodes using special devices. However, in terms of our zoom perspective, there is no duty in logical analysis to formalize everything down to the rock-bottom level.

6.5 From nodes to histories: Dynamic and temporal logics

Now we move on. Modal languages express local properties of nodes in a game tree. But reasoning about long-term social action refers to global histories that unfold over time (cf. Chapter 5). Now PDL can still do part of this job. The modality $\langle move^* \rangle \varphi$ already acted like a temporal assertion that at some history at some stage, φ will be the case. But we also want to say things such as $\exists G\varphi$: "along at least one future history, φ will always be the case." This modality is definable in PDL on finite game trees, since programs can express the statement that, until some endpoint, φ is always the case. On infinite games, this temporal modality becomes stronger, and we need extensions of PDL that can talk about infinite branches, such as the μ-calculus of Chapter 1. But there are attractive alternatives.

As we saw in Chapter 5, game trees also model a branching temporal logic. At each stage of the game, players are in a node s on some actual history whose past they know, either completely or partially, but whose future has yet to be fully revealed. This structure was accessed by a language with temporal and modal operators over branches whose intuitive meanings were as follows:

$\boldsymbol{M}, h, s \models F_a\varphi$	iff	sa lies on h, and $\boldsymbol{M}, h, sa \models \varphi$
$\boldsymbol{M}, h, s \models P_a\varphi$	iff	$s = s'a$ for some s', and $\boldsymbol{M}, h, s' \models \varphi$
$\boldsymbol{M}, h, s \models \Diamond_i\varphi$	iff	$\boldsymbol{M}, h', s \models \varphi$ for some h' equal for i to h up to stage s

The indices of evaluation here are history-point pairs (h, s), rather than local stages in the tree; but we will switch to a local modal approach of finite histories only in the rest of this chapter.

The temporal logic now encodes basic facts about play of a game. For instance, as moves are played publicly, players' knowledge adapts as follows.

FACT 6.1 The following principle is valid: $F_a\Diamond\varphi \leftrightarrow (F_a\top \wedge \Diamond F_a\varphi)$.

Proof This principle will be clarified in Chapter 9 in terms of a dynamic analysis of update actions. But it is useful to picture what it says. From left to right, if we can make an a move on the current history h to get a possible branch h' satisfying φ, then we can make an a move on h, and also, there is a possible branch right now (namely, h') on which an a move will bring us to a stage satisfying φ. From right to left, if a is the next move on h, and some (other) branch h' starts with a and then arrives at a stage satisfying the formula φ, then, given the unicity of events in our trees, h' and h coincide until after the a move, and so h itself satisfies $F_a\Diamond\varphi$. ■

Thus, game trees are also a natural model for social action where modal and temporal views combine well (Stirling 1995, van Benthem & Pacuit 2006).

6.6 Procedural information and epistemic forests

Now that we know what game trees can do for us, we make a next step. Depending on the scenarios of interest, we may need more complex models.

Lifting to forest models Suppose we do not know the strategies played by others, but we have some prior idea. Tree models may not suffice (van Benthem 2011d).

EXAMPLE 6.3 Strategic uncertainty
In the following game, let $\boldsymbol{A}$ know that $\boldsymbol{E}$ will play the same move throughout:

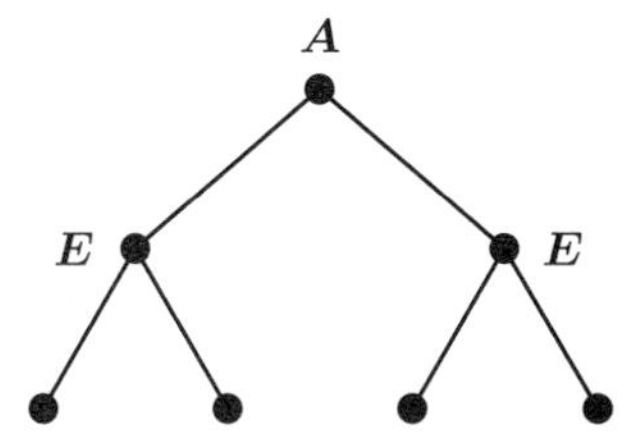

Then all four histories are still possible. But $\boldsymbol{A}$ only considers two future trees:

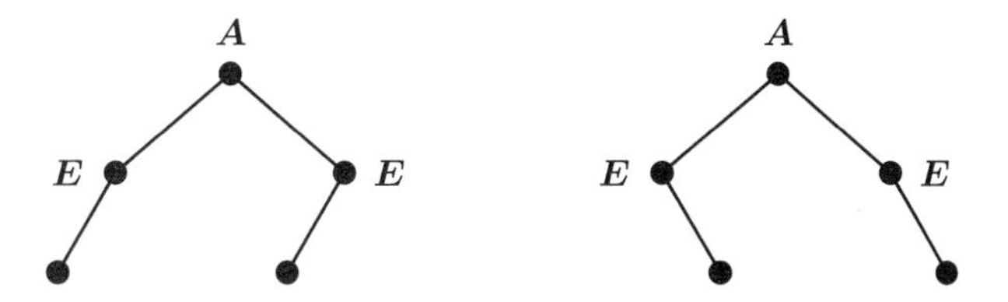

In longer games, this difference in modeling can be very important. Observing only one move by $\boldsymbol{E}$ tells $\boldsymbol{A}$ exactly what $\boldsymbol{E}$'s strategy will be in the whole game.[59] ■

---

59 The example is delicate. We could still flatten the distinction between the two strategies by allowing only histories where $\boldsymbol{E}$ always plays the same move (*left* or *right*), and then, one observation of a move by $\boldsymbol{E}$ tells $\boldsymbol{A}$ what $\boldsymbol{E}$ is going to do henceforth. Our forest model becomes essential with counterfactual assertions by $\boldsymbol{A}$ such as "$\boldsymbol{E}$ plays *left* now, and if I had played differently at the root, $\boldsymbol{E}$ would still have played *left*."

There are more reasons why trees may not suffice. We just saw one: uncertainty about actions or beliefs of other players. Another reason may be lack of information about which game is being played, even just for one agent. We will discuss further reasons in Section 6.8. To model such variation in scenarios, we need semantic structures containing different trees, linked by epistemic uncertainties between various views of the game. These models were introduced in Chapter 5 under the name of epistemic forests, but we repeat their definition here for convenience.

DEFINITION 6.1 Epistemic forest models
An *epistemic forest model* $\boldsymbol{M}$ is a family of finite sequences h of events from some given set E, related by extension with further events (the possible moves), and also by epistemic accessibility relations $\sim_i$ for agents i. Moreover, there is a valuation map V from proposition letters to those finite sequences h where they hold. ■

The finite histories encode possible stages of a process that goes from available sequences to longer ones. This need not be a single tree, as there may be different views of the process taking place. The epistemic uncertainty links make agents uncertain where they are and in which process, which can be due to many reasons.

A special case We may have gone too fast. We are still talking about perfect information games, although we may not know exactly which one, or how it is played. This imposes the following special restriction on forest models.

FACT 6.2 Epistemic forests for perfect information games satisfy the following condition: if $h \sim h'$ (for any player i) and $h(move \cup move^{\cup})^* h'$, then $h = h'$.

Proof This says precisely that players can only have uncertainties between different actual games, not between different positions in the same game.[60] ■

Logics of action and a new kind of knowledge Epistemic forest models support modal languages such as PDL with epistemic operators. Suppressing agent indices for convenience, we add the knowledge modality $K\varphi$ of Chapter 3:

$$\boldsymbol{M}, h \models K\varphi \quad \text{iff} \quad \boldsymbol{M}, h' \models \varphi \text{ for all histories } h' \text{ with } h \sim h'$$

This represents two kinds of knowledge in games: future knowledge expressed by the earlier $\langle move^* \rangle$ and standard K-type knowledge from other sources. The logic

60 The condition of Fact 6.2 can be defined syntactically in a matching modal logic of action and uncertainty, using hybrid "nominals" to denote unique finite histories h.

will show us how these can interact in epistemic forest models, and we will see an example below. Incidentally, as in Chapter 5 , we will be switching freely between temporal (h, s)-based approaches and purely modal h-based ones. The two seem very close, although there may not currently be any definitive mutual reduction.

Digression: Epistemic linear time One may wonder whether the branching epistemic structure of epistemic forests is really needed. Can we trade off forward uncertainty in a game tree against sideways epistemic uncertainty between linear histories, as in the interpreted systems of Fagin et al. (1995)? This seems to be the sense of the Harsanyi Doctrine in game theory (Osborne & Rubinstein 1994). We will not pursue this issue here, but briefly mention a very simple example.

EXAMPLE 6.4 Tradeoff
Intuitively, the following two models seem close indeed:

Assertions in our language for epistemic forests can be translated from any of these models to the other (cf. van Benthem 2009). ∎

There are appealing folklore ideas for converting one kind of structure into another.[61] Pacuit (2007) studies the related issue of epistemic forests versus the interpreted systems of Fagin et al. (1995). Still, intuitive differences in the two styles of thinking remain, as will be discussed later in Section 6.8.

We have shown how epistemic forest models can represent varieties of global knowledge inside and about games. As we have seen, they also support different

---

61 One idea for converting global uncertainty about the future into local uncertainty about the present works as follows. Given any tree G, assign epistemic models $\boldsymbol{M}_s$ to each node s, whose domain is the set of histories sharing the same past up to s, letting the agent be uncertain about all of them. Worlds in these models may be seen as pairs (h, s) with h any history passing through s. This cuts down current histories in just the right manner. The epistemic-temporal language of Chapter 5 matches this construction.

logical frameworks, including the epistemic-temporal logics of Chapter 5. This variety seems a strength rather than a weakness, in terms of offering flavors of modeling. See van Benthem et al. (2013) for an even broader landscape of logical frameworks.

6.7 Observation uncertainty and imperfect information games

We have dealt with perfect information games, and uncertainty about future play. The next step are games with imperfect information, where players may not know exactly where they are in the game tree, because of limitations on observation or memory. We have explored this realm in Chapter 3, using a combined modal-epistemic language. We also found correspondences between special modal-epistemic axioms and properties of players such as perfect recall. More generally, axioms in our languages reflected styles of play and types of players.

Imperfect information games fit well with epistemic forest models. Their notion of information from observation is just another way of producing uncertainty lines. Again they are a special case, but a virtual opposite of an earlier one. Uncertainties can only run between positions in some game that can actually be played.

FACT 6.3 Epistemic forest models for imperfect information games satisfy this condition: if $h \sim h'$ (for any player i), then $h(move \cup move^{\cup})^* h'$.

Proof This says exactly that the only uncertainties in the forest arise from inside a game tree, which is the hallmark of imperfect information games.[62] ■

Epistemic forests in their generality handle key sources of knowledge or ignorance in games: branching future, procedural information, observation, and memory. They also support modal epistemic temporal languages that express elaborate scenarios of social action. At this point, let us reflect on the models used so far.

6.8 Taking stock: Models of extensive games

Modal process models To help visualize where we stand, it may be useful to summarize the various models that we have seen in this book for the logical analysis of social scenarios. Our starting point was extensive game trees, viewed as models

62 Again, modal epistemic axioms can capture this special branch of forest logic.

for modal languages, with nodes as worlds, and accessibility relations for actions, preferences, and later on, also uncertainty.

This is the level of analysis in Chapters 1 through 4, and it served us well in analyzing the logical basics of Backward Induction, strategies, and imperfect information. Even so, the models used in these chapters were not always pure game trees, since we decorated them with new relations representing the information or the intentions of the players. For instance, imperfect information was an epistemic annotation on action trees, and as another sort of annotation, the Backward Induction procedure created a plausibility ordering between nodes.

Temporal trees and forests The modal approach, although well-known and attractive, has its expressive limitations. That is why we added temporal structure in Chapter 5, with possibly infinite histories in branching trees. This is close to the definition of extensive games (Osborne & Rubinstein 1994) and at the same time, trees are a widely used model of action and agency in the computational and philosophical literature (cf. Stirling 1995, Belnap et al. 2001). Very often, as we have seen, all that is important in social interaction can be stated as properties of histories or their local stages.[63]

But we also found that, once information and knowledge come into play, pure trees do not suffice, and we needed epistemic forest models, like the ones used in the preceding sections, with uncertainty links between nodes in possibly different views of the actual process. Epistemic forests are very rich structures, with both perfect information games and imperfect information games as special cases. And there may be more still than meets the eye. We now consider a few further scenarios and interpretations.

Example 6.5 Different uses of epistemic forest models

(a) *Information dynamics on epistemic models.* Consider any card game, say, the Three Card example in our Introduction. There are several possible deals of the cards, forming different possible words, not all of which players can tell apart initially. Next, events take place that may allow the players to rule out some deals, finding out more. The resulting structure is an epistemic forest, where the root has a number of different initial events corresponding to the deals, that are followed up by histories of subsequent events, where some actions may tell worlds apart (say,

63 The difference with modal models is slight in finite games, where histories match endpoints. Still, a forest is not a tree, as we saw in modeling procedural information about strategies, or differences in what players believe about other players' beliefs.

public discarding), while others may leave uncertainty (like someone else drawing a card from the stack). Similar forest models have been used in logical studies of learning theory (cf. Gierasimczuk 2010), and more complex versions will return in Chapters 7 and 9 on information dynamics in imperfect information games.

(b) *Uncertainty about options or about choice.* As another illustration of what forest models can say, recall our possible tradeoff example of Section 6.6:

What is modeled intuitively by these two diagrams? The situation to the left is a standard game tree, usually interpreted as the availability of two actions, and hence a choice for the agent whose turn it is. The situation to the right might be interpreted as the same uncertainty about what the agent is going to do, but it makes more sense to think of it as the agent's uncertainty about the actual process and its available moves, since the dotted link does not tell us whether the two options come from a choice. This view of choice comes under some pressure when we compare our modal logics here with other logics of action such as STIT, and in Chapter 12 we will make a more careful comparison. Incidentally, this uncertainty should make us weary of very simple 'linear reductions' for tree patterns of the sort that we briefly mentioned earlier.

(c) *Uncertainty about what others think.* In the next model, in the left root, agent $\boldsymbol{A}$ knows he expects h rather than h' to occur, but does not know whether $\boldsymbol{E}$ agrees:

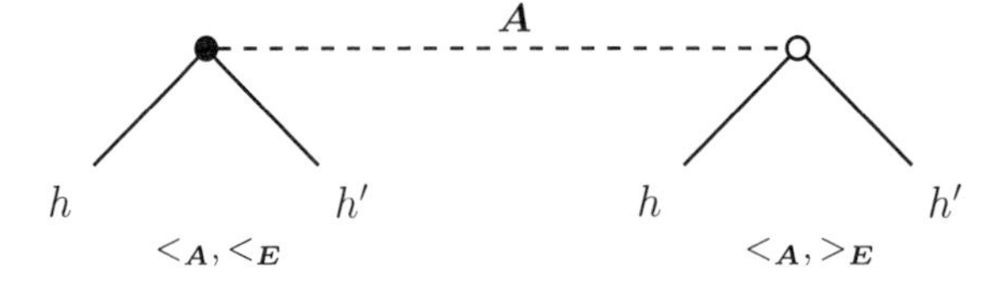

Note that we only placed links at the top level of epistemic forest models. The preceding interpretative options are also tied up, however, with how we place matching links further down, as described in earlier conditions on agents. ■

EXAMPLE 6.6 Inherited links and agent properties
Suppose event L takes place in case (c) of the preceding example, where we disregard beliefs. Intuitively, we end up in the following model (see Chapter 9 for details):

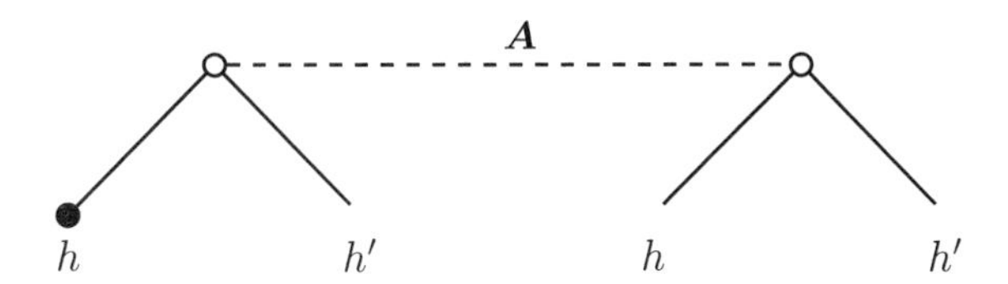

Here, a miracle has happened. By the structure of the model, $\boldsymbol{A}$ now knows his location, even though event L could also have happened in the other subtree. ■

In this way, epistemic forest models will model miracles of new knowledge without information flow, or in the other direction, cases of forgetfulness where uncertainty links pop up without any action causing them. But, as we have seen in Chapters 3 and 5, things change with standard kinds of agent, whose epistemic traces in forest models may become highly constrained. Thus, agents with perfect recall making only public observations of events e obeyed the condition

$$h \sim h' \quad \text{iff} \quad he \sim h'e$$

that fails in the preceding model. Such constraints will return in Chapter 9 on information dynamics in epistemic forest models.

Strategy models for games Now we go one step further. The literature also has modal models with abstract worlds (no longer nodes in extensive game trees) that contain whole strategy profiles for players and can also carry other structure. Strategy profiles can be represented in forest models, as subforests representing different views of the game (see Section 6.6), or as additional plausibility relations on trees in the style of Chapter 2, in a way that suffices for many purposes. But sometimes we want the full generality of the following kind of structures.

DEFINITION 6.2 Strategy models for extensive games
Strategy models for a game $\boldsymbol{G}$ are tuples $\boldsymbol{M} = (W, \{\sim_i\}_{i \in I}, \{\leq_i\}_{i \in I}, \{\propto_i\}_{i \in I}, V)$ whose worlds carry local information about all nodes in $\boldsymbol{G}$, plus strategy profiles: total specifications of each player's behavior throughout the game. Players' information about structure and procedure is encoded by uncertainty relations $\sim_i$ between the worlds. There are also preference orders $\leq_i$ over worlds, and plausibility orders $\propto_i$ encoding player's beliefs about which worlds are more likely than others. ■

Strategy models for games are a natural top level in our sequence, and they are the standard in epistemic game theory (Geanakoplos 1992, Battigalli & Siniscalchi 1999, Stalnaker 1999, Halpern 2003a). They are also congenial to modal logicians for their familiar flavor (cf. van der Hoek & Pauly 2006, Fitting 2011). We will use these structures in Chapters 12 and 13 to analyze solution procedures on games in strategic form.

Is there a simple knockdown example showing the need for moving to full-fledged strategy models? One interesting benefit is that their worlds provide a higher level for defining earlier notions. For instance, preference may now hold, not between specific outcomes of a game, but between whole strategies, the way one may prefer a strategy of governing by example over ruling with the rod. However, the most concrete use of these models in our setting may be complex cases of *counterfactual reasoning*. Epistemic forest models (or the epistemic-doxastic forest models of Section 6.9) can handle some counterfactual reasoning about off-path behavior, as we saw when discussing information about strategies. But still greater generality may be needed when thinking more globally about other kinds of strategies that players might have chosen, or other types that they might have had. Models for games support all of this thinking.[64]

In all, our preference in this book remains for using models that are as thin as possible, putting as much further structure as we can into the dynamics of Part II, which provides simply definable ways of changing models.

6.9 Adding beliefs and expectations

Our styles of modeling knowledge also make sense for beliefs, the more common fuel driving games and social action. There are several ways of introducing beliefs into our models so far. All of them present some interesting features of their own. Even so, we will discuss this next aspect of information-driven agency only lightly, since logics of belief will be more of a focus for our concerns in Part II.

64 Stalnaker (1999) discusses how counterfactuals arise naturally in reasoning about strategies. Bonanno (2012) is an up-to-date discussion of counterfactual reasoning in game theory, including careful distinctions between exploratory reasoning and on-line reasoning in games, and between counterfactual beliefs and beliefs in counterfactuals (see also Leitgeb 2007 on this distinction). On the other hand, Bonanno (2013) shows how counterfactual reasoning may not be needed after all for Backward Induction.

Belief among moves In Chapter 2, the Backward Induction procedure constructed the plausibility orderings that are normally primitives in doxastic-temporal models. These orderings matched strategies viewed as subrelations of best action inside the *move* relation. With such a local encoding of plausibility, we can have a language for belief adding just a modality $\langle best \rangle$ to our modal logic of game trees. While this looks attractive, a tricky side effect is the uniformity imposed by the Backward Induction algorithm.

EXAMPLE 6.7 Crossing expectations and ternary plausibility
Recall the following picture from our first example, but this time, with the numbers at the outcomes read, not as utility values, but as degrees of plausibility:

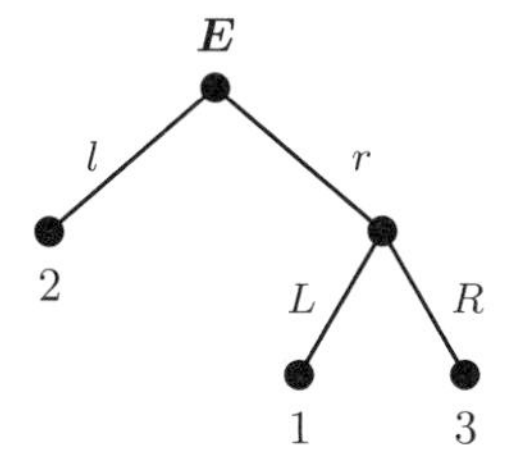

This pattern goes against the "node-compatible" plausibility order for Backward Induction that makes one local move more plausible than another, while all their outcomes then follow this decision. Yet it is easy to think of scenarios where the plausibility order depicted here is natural, and hence models with *ternary node-dependent plausibility relations* $x \leq_s y$ seem the more appropriate level of generality. The uniform plausibility relations of the Backward Induction algorithm induce a special subclass whose definition we will not spell out here. ■

This ternary approach will be discussed in greater depth in Chapter 10. In particular, Example 6.7 raises issues about how to reason counterfactually about parts of the game that lie off the current history: as mistakes, or as intentional deviations.

From trees to epistemic-doxastic forests The same reasons as for knowledge motivate a step from extensive game trees to epistemic doxastic forest models. Agents may have beliefs rather than knowledge about the structure of the game. They may also have beliefs about other agents' different beliefs or knowledge. And finally, extending the focus on knowledge in imperfect information games, they may have restricted powers of observation that generate fine-grained beliefs about where they are in the game. The resulting richer forest models for knowledge plus belief over time and their temporal logics have been investigated in Dégremont (2010).

Believing in forest models Let us be a bit more precise about the semantics of beliefs in forest models. As before, there are two broad ways of thinking in a temporal setting: one is history-based, evaluating assertions at pairs (h, s) of a history h and a current stage s on it, and one is stage-based, working with finite initial histories h only. This time, it may be easier to start with history-based models with indices (h, s), rather than the modalized language ETL of Chapter 5 that worked over stages h. The reason is this: expectations about the future seem largely about the future history, not about single nodes.

Plausibility and histories Belief can now be modeled by adding binary relations $\leq_i$ of relative plausibility between histories and interpreting a doxastic modality.

DEFINITION 6.3 Belief in history-based models
Here is how one extends the earlier setting of Chapter 5, thinking of a cone of most plausible histories inside the total fan of all possible histories at stage (h, s):

> $\boldsymbol{M}, h, s \models \langle B, i\rangle\varphi$ iff $\boldsymbol{M}, h', s \models \varphi$ for some history h' coinciding with h up to stage s and most plausible for i according to the given relation $\leq_i$.

Moving along a history, the plausibility relations will then encode belief changes for players such as the ones we saw with Backward Induction in Chapter 2. If one wants to model significant beliefs of players about beliefs of others in this manner, then single trees will not suffice, and we need the full generality of forest models.
This richer structure shows in the logic with an operator for conditional beliefs:

> $\boldsymbol{M}, h, s \models \langle B, i\rangle^{\psi}\varphi$ iff $\boldsymbol{M}, h', s \models \varphi$ for some history h' equal to h up to stage s and most plausible for i among all future histories satisfying ψ.

The more common universal variants of belief follow directly from these clauses. ■

We state one result to show how this temporal logic of belief works. It uses a modal-temporal operator F_a saying that the current branch starts with an a move.

FACT 6.4 The following principle is valid for the temporal logic of belief change:

$$F_a\langle B, i\rangle\varphi \leftrightarrow (F_a\top \wedge \langle B, i\rangle^{F_a\top} F_a\varphi).$$

Proof From left to right, suppose that, after move a at stage s on history h, we have $\boldsymbol{M}, h, sa \models \langle B, i\rangle\varphi$. This implies that $F_a\top$ holds at (h, s). Also, there is a most plausible history h' for i through sa with $\boldsymbol{M}, h', sa \models \varphi$. This h' is also most plausible for i at node s, but then only among the histories extending s that start with move a. If there were a more plausible one in that set, h' would not have been most plausible overall among the extensions of sa. Hence $\langle B, i\rangle^{F_a\top} F_a\varphi$ holds at (h, s).[65] Conversely, let $F_a\top \wedge \langle B, i\rangle^{F_a\top} F_a\varphi$ hold at (h, s). By a similar reasoning, using the inheritance of the plausibility order from extensions of s to extensions of sa (a node that exists by the truth of $F_a\top$), we see that $F_a\langle B, i\rangle\varphi$ holds. ■

A similar law can be stated for future truth of conditional beliefs. These principles may be compared to those for belief change in the dynamic logics of Chapter 7, or to the ones stated for AGM-style belief revision theory in Bonanno (2007).

Belief between nodes We can also analyze belief with a plausibility order between finite histories only, extending the ETL models of Chapter 5 to doxastic DETL models. This view will be used in Chapter 9 to deal with generalized imperfect information games with enriched information sets for players who develop beliefs about where they are in a game. As before, the history-based and stage-based approaches are technically close. The following statement illustrates this.

FACT 6.5 The preceding equivalence $F_a\langle B, i\rangle\varphi \leftrightarrow (F_a\top \wedge \langle B, i\rangle^{F_a\top} F_a\varphi)$ for new beliefs after observed events also holds in DETL models.

Proof This follows from the analysis in Chapter 9, or from the close analogy with the laws for new beliefs after "hard information" in Section 7.3 of Chapter 7. ■

Belief versus expectations Our discussion is not a routine extension of earlier themes. We conclude with a delicate point. Our treatment of belief really models two different notions. The first notion, inside game trees, may be called *expectation* (about the future course of the game), the other, between trees, is *belief* proper (about the process, or other agents). Even though the two notions are close, and their semantics involves plausibility comparisons, this distinction seems natural and illuminating. The usual single category of belief may not be coherent enough.

6.10 From statics of games to dynamics of play

Having looked at different kinds of models for games, let us now return to a theme raised on many occasions in earlier chapters.

---

65 This direction assumes that the plausibility order is not node-dependent: the order at nodes sa is inherited from the order at the preceding node s. In case we make a move to ternary plausibility orderings, as discussed earlier, our axiom would have to be generalized.

A game tree, forest model, or model for a game records what can happen, while adding traces of players' activities such as their knowledge or beliefs. But these traces have a reason: they are produced in real play, as players observe moves publicly, or receive other sorts of information about the game.

In the same spirit, we just saw how a stage-based view of belief might correspond to imperfect information games where players form beliefs during play.[66]

How can we make this dynamics explicit in a logic that does not just represent reasoning about a static situation, but also the effects of events and information flow? This explicit dynamics will be the theme of Part II of this book, and in particular, Chapter 9 will consider update mechanisms producing the temporal forests that we have defined here, while also looking into processes of game change.

These update actions can be quite sophisticated in social scenarios. For instance, as we have seen in our earlier discussion of counterfactuals, the proper way of observing moves is usually not by taking them at face value, but in more intentional ways. Thus, our repertoire for players will include the various actions:

(a) public observations $!e$ of events e, but also "uptake acts" such as
(b) "e was played intentionally, on the basis of rationality in beliefs," or
(c) "e was played by mistake, by deviating from Backward Induction," and so on.

These actions all arise with our initial scenarios in this chapter, but there are more.[67] All of these actions can be studied using the dynamic logics of Part II.

6.11 Conclusion

We have surveyed some of the many current logical models for games, emphasizing themes that link them.[68] Our main point has been that this variety is an asset. Instead of having huge worst-case models, much like wearing winter clothes in

66 Examples of more extraneous additional events during play, to be discussed in Chapters 9 and 10, are promises (van Benthem 2007e), announcements of players' plans (van Otterloo et al. 2004), and communicating intentions (Roy 2008).

67 Given that the extra features may just be hypotheses on the part of players, we may not want the hard information of public announcements ! here, but the soft information of the upgrades $\Uparrow$ defined in Chapter 7. Soft events are studied in Chapter 9, which also considers other events in games, such as changes in players' preferences (Liu 2011).

68 Our survey is by no means complete, and in particular, we have ignored the more fine-grained syntactic models of information that will be discussed in Chapter 7. These

summer, it seems better to let relevant scenarios dictate what one wants to capture, at the right level of lightness. We have shown how this goes in natural stages in game modeling, from lighter to heavier. Having done so, our next desideratum will be clear: getting a better grip on how these different models work in the dynamics of actually playing a game, the topic of Part II.

6.12 Literature

This chapter is based on van Benthem (2011d, 2012b), which contain more material, as well as the programmatic lecture "Keep It Simple" (van Benthem 2011f).

In addition, there are excellent sources on models for games, many mentioned in the text. We recommend Stalnaker (1999), Bonanno (2007), Fitting (2011), Brandenburger et al. (2014), and for a more systematic discussion of logic connections, the forthcoming textbook by Pacuit & Roy (2013).

6.13 Further directions

The following list of things to do is another virtue of the slow steps analysis in modeling for games that we have proposed. Doing so draws together ideas from sometimes competing logical frameworks, and it makes us see new issues all around.

Dynamics over thin models versus pre-encoding Making the dynamics of play explicit seems to be a natural complement to looking for small models (cf. Baltag et al. 2009). One relocates part of the relevant information about the game in the update process. How far does this go? Does it work both ways? Can we think of epistemic-temporal grand stage models as doing the reverse: simplify update to observing moves on a much larger static structure? In this light, how reasonable are epistemic forest models as a balance between the two views?

Transforming knowledge models There is a persistent intuition that the different kinds of knowledge we have distinguished can be turned into each other through some sort of representation. For instance, why not turn the earlier future ignorance based on branching in a game tree into uncertainty between histories in the style

---

structures can model additional phenomena such as players' awareness of the game they are in, or the detailed effects of inferences that players make.

of epistemic logic? We have briefly discussed folklore results to this effect (cf. van Benthem 2009) that support translations of logical languages. But we are not aware of a systematic study clarifying just how far these shifts in perspective go. Indeed, our discussion has thrown some doubt on these identifications, in terms of choice between future moves versus plain uncertainty about what will happen.[69]

Histories versus stages The preceding issue is linked to the distinction between history-based and stage-based models that we have seen throughout. Both make sense, and we saw how they support different intuitive notions in games: from expectations about which history will unfold to beliefs about were we are in a doxastic-epistemic forest. Even so, local stages made rich enough can encode whole histories, and hence the two perspectives seem close at an abstract level.

Beliefs versus expectations once more We have suggested two takes on belief in doxastic-epistemic forest models, one based on plausibility ordering for future histories, and one on plausibility ordering between stages in different components of the models. But this remains to be understood better, since there may be natural consistency constraints relating the two kinds of ordering. For instance, suppose one considers event a more likely to happen than event b from the current point. This seems to rule out a link of greater plausibility to some other subtree in the forest where the only action is b, although it would allow a case with a only. We are not aware of any precise analysis of such links.

Other constraints may involve choice points and entanglements between agents' expectations. When thinking about one of our opponent's turns, and knowing what the opponent expects to do, it would be wise to adopt the same expectation. The expectations of an active agent at a turn should dominate there, but just how? Backward Induction provided a strong instance, creating a uniform plausibility order that was the same for all players. But this is often too strong. For instance, at another extreme, an agent may think that, while another person expects to do the right thing, the actual chances of that happening are slim.

Long-term behavior The temporal setting offers a setting for phenomena in games beyond our emphasis in this chapter on what agents know or believe locally. Many

---

69 What may also be relevant here is the idea in van Benthem & Pacuit (2006) of a "Church Thesis" about agency saying that all serious systems of informational temporal logic have the same brute modeling power, if not the same intensional spirit.

dynamic analyses of game-theoretic solution procedures involve infinite limit behavior over time (cf. Dégremont & Roy 2009, Gierasimczuk 2010, Pacuit & Roy 2011) that goes beyond single update steps, eventually finding interesting stable models. This theme of long-term behavior will return in Chapters 7, 8, and 13.

Conclusion to Part I

In this first part of our book, we have surveyed a sequence of basic logics of games. Our survey covered many approaches, although some topics, such as representing beliefs, may have received too little attention, and will get a second chance in Part II. We have raised more issues than we solved, but still, by pursuing various levels of looking at games, our six chapters have brought some major logical themes to bear on games.

(a) Structural game equivalences and definability in matching modal languages.
(b) Combined logics of action and preference, systematic links with fixed point logics, various levels of zoom, and complexity of logic combinations.
(c) Knowledge in games, correspondence analysis for special agents.
(d) Explicit logics of strategies, local and generic strategy calculi, and the art of strategizing given logics.
(e) Lifting to infinite games, temporal logics for global playgrounds and strategies.
(f) Modeling stages for games: trees, forests, and models of games.

As for broader benefits, through these interconnected topics, we have shown how the traditional diverse agenda of philosophical logic becomes coherent by focusing attention on games. Also, we showed how philosophical and computational logic fit harmoniously in our setting. However, techniques from mathematical logic remain crucial; witness the open problems that we brought up along the way on topics such as structural game equivalences, definability in fragments of fixed point logics, computational complexity of game logics, or preservation behavior across changing models of games.

The analysis offered here does not cover every aspect of games. One clear omission is the study of groups of agents as actors, collective actions, and group information.

In Part III of this book, at least, we will take our approach to games in other formats, looking at players' powers and strategic forms.

But before we do, the most urgent target to come out of these chapters is the issue of *dynamics* versus *statics*. The models so far have been collections of snapshots in time, a frozen record of what has happened. The next main topic of this book in Part II is the dynamics of actual events that produce the game models that have been largely taken for granted here.

II Logical Dynamics and Theory of Play

Introduction to Part II

In Part I, a sequence of logical systems was introduced that tied extensive games to many issues in philosophical and computational logic. Despite the appeal of that perspective, it still leaves something to be desired. A game is a record of a process that can be played, or that has already been played, by agents who mingle action, information, and desires. So, what about that *play itself*, and its unfolding over time? In this second part of our book we make a switch from logical statics to logical dynamics, looking at the actions and events that make information (and other things) flow as players engage in a game. This dynamics can have several phases. There is deliberation prior to playing the game, there is information dynamics happening during play, and there are activities of post-play analysis or justification. Indeed, there can even be more drastic dynamics, involving game changers.

To deal with these phenomena, we need a corresponding shift in logic itself. We want dynamic logics of actions and events that make information flow, and more broadly, that change players' beliefs about the game, or their evaluation of its outcomes. Our tool will be *dynamic-epistemic logics* that deal with a variety of update actions, treated as first-class citizens in the system denoting events of model transformation. Dynamic-epistemic logics have been developed for general reasons, not only tied to games. Chapter 7 is a mini-course, explaining the core methodology of epistemic models, actions and postconditions, recursion laws, and program repertoire, covering knowledge update, belief revision, preference change, and other relevant phenomena. In addition to providing tools, it will also give the reader a sensibility to what may be called the dynamic stance. Then in Chapter 8, we take this stance to the deliberation phase, analyzing Backward Induction as a procedure of successive belief revisions that create a stable expectation pattern for a game. This will tie in with the fixed point logics of Chapter 2, suggesting a much broader theory of scenarios with iterated events that provide hard or soft information. Next, in Chapter 9, we discuss the dynamics of actual play, trying to see what actions are relevant there, and pointing out the laws of dynamic logic governing them. As we will see, the same dynamic-epistemic techniques also deal with the postmortem phase of game analysis, and even with game change. Finally, in Chapter 10, we assess what this means for an incipient Theory of Play that can be thought of as the joint offspring of logic and game theory.

7
Logical Dynamics in a Nutshell

The move from Part I to Part II of this book is inspired by a view of games as consisting of many informational events, including deliberation, actual play, and subsequent rationalization that need to be brought to the fore in a Theory of Play. To study this dynamics, we do not have to leave the realm of logic. The tools for making this view of games happen are dynamic-epistemic logics, a family of systems that has been developing over the last decade. These systems are not specific to games. They arise out of a general "Dynamic Turn" toward putting informational acts of observation, inference, and communication at center stage in logical theory. This chapter gives a brief introduction to the main motivations, issues, and systems in this area that will be relevant for the rest of this book. Although compact, this format should be instructive to the reader in making the whole program visible at a glance, and it can be consulted when reading later chapters.[70] In this chapter, general actions are the focus, and games are mainly important as they form a natural stage in the broader logical dynamics program.

7.1 Logic as information dynamics

Logic is often considered the study of valid consequence, but much more is involved in even the simplest informational settings. Consider a common scenario that can be observed all around us.

---

70 For standard monographs on dynamic-epistemic logic, see van Ditmarsch et al. (2007) and van Benthem (2011e). Further references are found at the end of this chapter.

Example 7.1 The Restaurant
In a restaurant, your father has ordered fish, your mother vegetarian, and you have meat. A different waiter arrives carrying the three meals. What happens?

You will see a sequence of informational acts. First, the waiter asks, say, who has the vegetarian dish, and sets that one down. Then he asks who has the fish, and sets that one down as well. And in a final act, the waiter infers that the meat dish is for you, and puts it in front of you without asking. This is logic in action, as we see it around us every day. But note the unity of different informational events: questions, answers, and inference in solving the problem. It seems artificial to separate this into a non-logical part, the questions and answers, and a logical part, the inference. All are crucial, and all belong together. ■

This is a very simple scenario, but the main idea that follows has many sources. We want to bring the natural repertoire of dynamic acts and events of information flow that drive agency inside logic. And this basic repertoire includes, at least, acts of inference, observation, and communication. This view is quite old. The Mohist logicians in ancient China (Zhang & Liu 2007) said that knowledge comes from three sources: communication, demonstration, and perceptual experience.[71]

A basic aspect of much social information flow is information, not just about ground facts, but also about other people. This is a shift from single minds to many-mind problems, in the same way that physics arrives at a true understanding of single objects by studying their interactions with other objects in many-body problems. Simple card games are a useful scenario for studying information flow in action. The following scenario is like the restaurant in a way, but with a further layer of higher-order knowledge.

Example 7.2 The Cards
Three cards *red*, *white*, and *blue* are given to three players ***1***, ***2***, and ***3***, one card each. Each player sees his or her own card, but no others. The actual deal order for ***1***, ***2***, ***3*** is *red*, *white*, *blue*. Now a conversation takes place. ***2*** asks ***1***: "Do you

---

71 The old phrase was *Zhi: Wen, Shuo, Qin* (知 闻 说 亲, know, hear, demonstration, experience). These categories were understood broadly: for instance, hearing stands for learning from communication, and experience covers perception or observation. The Mohist's own example was a dark room where you see an object inside, but not its color. You see a white object outside (observation), and someone tells you the object inside has the same color (communication). You now infer that the object in the room is white.

have the blue card?" *1* answers truthfully: "No." Who knows what then, assuming that the question is sincere?

If the question is sincere, *2* indicates that she does not know the answer, and so she cannot have the blue card. This tells *1* at once what the deal was. But *3* did not learn, since, having the blue card himself, he already knew that *2* does not have blue. When *1* says she does not have blue, this now tells *2* the deal. *3* still does not know the deal; but since he can perform the reasoning just given, he knows that the others know it. Thus, social information flow can be subtle. The systems that follow will analyze this information flow in detail. ■

A famous puzzle of this sort involving longer conversations is the Muddy Children problem, which will appear in Section 7.5.

The variety of information sources highlighted here does not just cover social settings. In science, too, deductive inference and empirical observation are on a par, and it is their interplay that drives progress. And even communicative aspects are crucial to understanding science, which is, after all, one of the oldest human forms of organization dedicated to reliable information flow. To deal with these phenomena, we need dynamic logics of inference, observation, and communication that will be reviewed below.

Beyond mere technicalities, the main theme of this chapter is the dynamic stance as such. It consists of cultivating a sensibility about the acts and events that underlie logic and many other activities. Traditionally, logicians have concentrated on studying only *products* of those activities, such as proofs or propositions. But natural language suggests a richer view. A word such as 'dance' is ambiguous between the actual activity and its products or forms.[72] The same is true for the words 'proof', or 'solution' for games. The latter is both an activity one can engage in, and the end result of such an activity (cf. Chapter 8). Our broader aim in this chapter is to do justice to both activities and products.

7.2 From epistemic logic to public announcement logic

Shrinking information ranges Bringing dynamics into logic requires an account of information states and how these change. Consider the successive updates in the restaurant scenario of Example 7.1. Initially, there are $3! = 6$ possibilities for

---

72 Baltag et al. (2014) takes a dance as its metaphor for the logical essence of epistemology.

distributing three plates over three people, the first question then reduces this uncertainty to two, and the second to just one. The general view here is that of semantic information as a range of options for what the actual world is like, a range that gets updated as new information comes in.

EXAMPLE 7.3 Updates for the Cards
The Cards scenario involves differences in what players know. In diagram form, indexed lines indicate the agents' uncertainties. Informational events then shrink this range stepwise, from six to four to two, as in the following "update video":

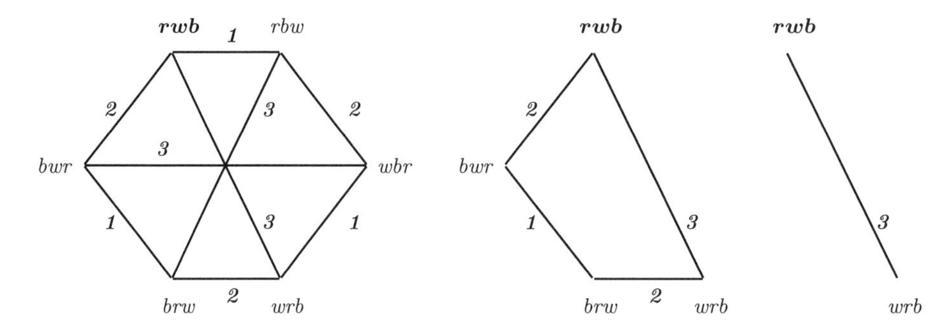

The first update step matches the presupposition of the question, and the second step matches the answer. Note that in the final model to the right, both players *1* and *2* know the cards (they have no uncertainty lines left), but *3* does not, even though *3* can see (in common with us as theorists) that, in both remaining eventualities, *1* and *2* have no uncertainties left. ■

The geometry of the diagram encodes both knowledge about the facts and knowledge about others, such as 3's knowing that the others know the cards. The latter kind is crucial to social scenarios, holding behavior in place. Indeed, at the end of the scenario, everything described above has become *common knowledge* in the group $\{1, 2, 3\}$.[73]

To bring any dynamic activity into logic, we first pick a suitable static base logic, often an existing system, which we then "dynamify" by adding explicit descriptions for the relevant events.

---

73 See Lewis (1969), Aumann (1976), and Fagin et al. (1995) for pioneering uses of this notion in philosophy, game theory, and computer science.

Epistemic base logic Perhaps the simplest example of this methodology involves a well-known base system. Static snapshots of the above update processes are standard models for epistemic logic, a system that we have already used in Chapter 3 for studying imperfect information games.[74] Its language has the syntax rule

$$p \mid \neg\varphi \mid \varphi \wedge \psi \mid K_i\varphi$$

while the corresponding epistemic models were tuples

$$\boldsymbol{M} = (W, \{\sim_i \mid i \in G\}, V)$$

with a set of relevant worlds W, accessibility relations $\sim_i$ and a propositional valuation V for atomic facts. We will mostly take the $\sim_i$ to be equivalence relations, although this is by no means necessary. Knowledge is then defined as having semantic information:

$$\boldsymbol{M}, s \models K_i\varphi \quad \text{iff} \quad \text{for all worlds } t \sim_i s : \boldsymbol{M}, t \models \varphi.$$

Common knowledge $\boldsymbol{M}, s \models C_G\varphi$ is defined as φ's being true for all t reachable from s by finite sequences of $\sim_i$ steps. If necessary, we distinguish an *actual world* in the model, often indicated in bold type when we draw diagrams.

REMARK Third-person versus first-person perspectives
Working with pointed epistemic models $(\boldsymbol{M}, s)$ with an actual world s reflects the external third-person perspective of an outside observer. Epistemic logics can also deal with first-person internal perspectives of agents, but we forego this here.

These epistemic models encode "semantic information," a widespread notion in science, although other logical views exist (van Benthem & Martínez 2008).

Update as model change The key idea now is that informational action is model change. The simplest case is a *public announcement* $!\varphi$, which conveys hard information. Learning with total reliability that φ is the case eliminates all current worlds with φ false from the current model:

---

74 We will not explain epistemic logic and other standard logics in this chapter beyond the barest outlines. The textbook van Benthem (2010b) explains many relevant systems.

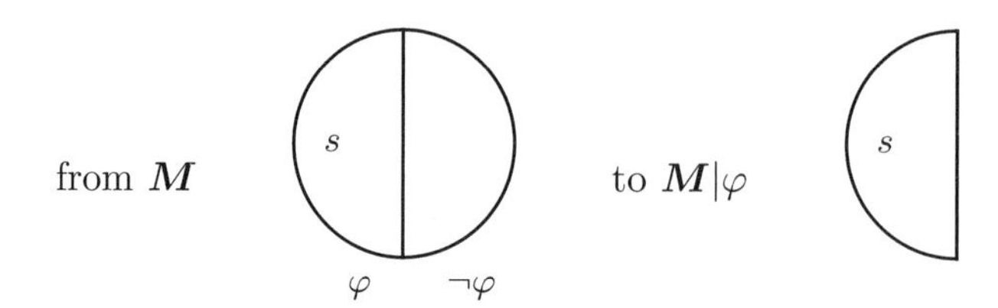

Getting more information by shrinking a current semantic range of options is a common idea that also works for information flow through observation. We call this hard information because of its irrevocable character: counterexamples are eliminated from the model.

It may be thought that this is simple, with actions $!\varphi$ leading to knowledge $K\varphi$. But things are more subtle. Dynamics typically involves truth value change for complex formulas. While an atom p stays true after update (the physical facts do not change), complex epistemic assertions may change their truth values: before the update $!p$, perhaps I did not know that p is the case, but afterward, I do. This results in order dependence of updates.

EXAMPLE 7.4 Order dependence
A sequence $!\neg Kp$; $!p$ makes sense, but the permuted $!p$; $!\neg Kp$ is contradictory. ∎

By contrast, so-called *factual* propositional statements containing no epistemic operators have their truth value invariant under public announcements, and in fact, they are invariant under all update actions that we will consider in this chapter.

Public announcement logic To get clear on such non-trivial phenomena, a logical system is helpful that can serve as a pilot. The system PAL of *public announcement logic* arises by extending the language with a dynamic modality for public announcements, interpreted as follows:

$$\boldsymbol{M}, s \models [!\varphi]\psi \quad \text{iff} \quad \text{if } \boldsymbol{M}, s \models \varphi, \text{ then } \boldsymbol{M}|\varphi, s \models \psi$$

Here, the antecedent "if $\boldsymbol{M}, s \models \varphi$" reflects the assumption that the announcement is truthful. The syntax of this language allows for complex recursions, as even formulas with dynamic modalities can be announced. The system has a complete logic that records some delicate phenomena.

FACT 7.1 The formula $[!p]Kp$ is valid, but $[!\varphi]K\varphi$ is not valid in general.

Proof A popular counterexample is the self-refuting "Moore sentence" $p \wedge \neg Kp$ (which can be read as "p, but you do not know it"). This also shows that the validities of PAL are not closed under substitutions. ∎

The system of public announcement logic PAL combines a standard logic for the static epistemic base plus a "recursion law" for knowledge after update, the basic recursion equation of the dynamic system defined here. In addition, there are recursion laws for the other syntactic forms of statement that can occur after an announcement modality.

THEOREM 7.1 PAL is axiomatized completely by the laws of epistemic logic plus the following recursion axioms

$$\begin{array}{lll} [!\varphi]q & \leftrightarrow & \varphi \to q \\ [!\varphi]\neg\psi & \leftrightarrow & \varphi \to \neg[!\varphi]\psi \\ [!\varphi](\psi \wedge \chi) & \leftrightarrow & [!\varphi]\psi \wedge [!\varphi]\chi \\ [!\varphi]K_i\psi & \leftrightarrow & \varphi \to K_i(\varphi \to [!\varphi]\psi) \end{array}$$

We will not state the needed formal derivation rules precisely, but these include the standard modal rules of Necessitation and Replacement of Provable Equivalents.

Proof Soundness follows by direct inspection. For instance, the key recursion law for knowledge may be understood as follows. In $\boldsymbol{M}, s$, the formula $[!\varphi]K_i\psi$ states the truth of $K_i\psi$ in the model $\boldsymbol{M}|\varphi, s$. But the worlds in that model were the φ-worlds in $\boldsymbol{M}$. Hence, we almost get an equivalence $[!\varphi]K_i\psi \leftrightarrow (\varphi \to K_i(\varphi \to \psi))$, where the prefix $\varphi \to$ just marks that φ needs to be true for a truthful announcement to be possible. However, this is not quite right yet, since complex epistemic formulas φ may change their truth value in passing from $\boldsymbol{M}$ to $\boldsymbol{M}|\varphi$. The correct statement, rather, is that worlds satisfying ψ in $\boldsymbol{M}|\varphi$ satisfy $[!\varphi]\psi$ in $\boldsymbol{M}$. But then we have precisely the stated equivalence.

Conversely, the completeness of PAL can be proved by reducing the innermost announcement modalities to combinations of base cases $[!\varphi]q$ that are then replaced by implications $\varphi \to q$. Working inside out in this manner, and using appropriate matching derivation rules, every statement in the language of PAL is provably equivalent to a pure epistemic formula, and then the completeness theorem for the static base logic applies. ■

In particular, there is no need for a recursion axiom for stacked dynamic modalities, although the following law does in fact hold.[75]

FACT 7.2 $[!\varphi][!\psi]\chi \leftrightarrow [!(\varphi \wedge [!\varphi]\psi)]\chi$ is valid.

---

75 Recent versions of PAL working over generalized update models do require such axioms, as they drop the atomic base reduction (cf. Holliday et al. 2012, van Benthem 2012f).

Dynamic methodology There is a general methodology behind this specific example. Where possible, one dynamifies an existing static logic, making its underlying actions explicit and defining these as suitable model transformations. The heart of the dynamic logic then consists of a compositional analysis of postconditions for the key actions via recursion axioms. In order to make this method work, one needs enough expressive power of "pre-encoding" in the static logic. This puts expressive requirements on the base language, sometimes forcing it to be redesigned.

EXAMPLE 7.5 Static redesign for common knowledge
A recursion law for common knowledge cannot be given using epistemic logic as it stands, and we need to add a new notion of conditional common knowledge looking only inside some definable submodel (see van Benthem et al. 2006a for its precise definition and logic) to obtain the following principle:

$$[!\varphi]C_G^{\psi}\chi \leftrightarrow (\varphi \rightarrow C_G^{\varphi\wedge[!\varphi]\psi}[!\varphi]\chi)$$

Beyond this law, axiomatizing common knowledge in PAL also requires an extension of the static epistemic base logic to modal *S5* plus natural fixed point axioms for conditional common knowledge (cf. Fagin et al. 1995).

General theory The dynamic-epistemic logic of updating with hard information throws fresh light on issues in epistemology, philosophy of language, and philosophy of science. It also raises many technical problems that we cannot address here (see van Benthem 2011d). As to recent samples, the website of the Stanford Logical Dynamics Lab (`http://stanford.edu/~thoshi/ldl/Home.html`) contains new results on two major issues.

(a) The "learning problem" of which syntactic forms of proposition φ always give validity of the formula $[!\varphi]K\varphi$. (All factual formulas without epistemic or dynamic operators have this property, but what about others?).

(b) Decidability and axiomatization for the "schematic validities" of PAL that are valid no matter what formulas we substitute for their proposition letters. (For a start, even though the PAL axiom for atoms p was not schematically valid, all other mentioned recursion axioms are.)

Both issues will return in Chapters 8, 9, and 13. Many further forms of information update can be treated in the same style, as well as other events relevant to agency, such as changes in preference. In what follows, we look at cases that will be needed in our study of games. The most important of these concern beliefs.

7.3 From correctness to correction: Belief change and learning

In addition to knowledge, beliefs are crucial to information-driven agency. These lead to useful inferences going beyond what we know; but beliefs can also be wrong, and when this comes to light under new information, another important rational skill comes into action: acts of belief change. Thus, rational agents are not those who are always correct, but those who have an ability for the dynamics of "correction."

Belief and conditional belief The models for belief that we will use extend the earlier models for epistemic logic by ordering epistemic ranges by a comparison order of *relative plausibility* $\leq_i xy$ between worlds x, y.

REMARK Simplified models
The illustrations to follow are very simple linear plausibility orders. Realistic models can be more complex, but this format suffices for our main points to be made.

EXAMPLE 7.6 A plausibility model
In the following model with three worlds, the black point s to the left is the actual world. Intuitively, as explained in Section 6.9 of Chapter 6, belief is what holds, not in all epistemically accessible worlds, but only in the most plausible ones among these. We drop indices for the ordering $\leq_i$ when we consider only one agent:

$\boldsymbol{s}, p, q$	$\leq$	$t, \neg p, q$	$\leq$	$u, p, \neg q$

In this model, p is true in the actual world s and the agent believes it, and q is also true, while the agent believes erroneously that $\neg q$. ■

In principle, plausibility orders can be ternary, depending on the vantage point of the current world. In this chapter, we will only use binary relations, uniform across epistemic ranges, reflecting an assumption that agents know their beliefs.

DEFINITION 7.1 Epistemic-doxastic models and truth definition
Epistemic-doxastic models are tuples $\boldsymbol{M} = (W, \{\sim_i \mid i \in G\}, \{\leq_i \mid i \in G\}, V)$ with relations as just described. For simplicity, we often ignore the epistemic structure.

The truth condition for belief is as follows, where we immediately add a notion of conditional belief that is crucial for pre-encoding belief changes:[76]

$\boldsymbol{M}, s \models B_i\varphi$ iff $\boldsymbol{M}, t \models \varphi$ for all $t \sim_i s$ maximal in the order $\leq_i$ on $\{u \mid u \sim_i s\}$
$\boldsymbol{M}, s \models B_i^{\psi}\varphi$ iff $\boldsymbol{M}, t \models \varphi$ for all $\leq_i$-maximal t in $\{u \mid s \sim_i u$ and $\boldsymbol{M}, u \models \psi\}$

Clearly, absolute belief $B_i\varphi$ is a special case, being equivalent to $B_i^{\top}\varphi$. ■

The static base logic for the notion of conditional belief is just the standard minimal conditional logic over pre-orders (Burgess 1981, Board 1998). However, a simpler base system will be given below in terms of "safe belief."

Belief changes and model transformers Now we turn to dynamic changes in beliefs under informational events. The simplest triggers are public announcements as before, that shrink current models to definable submodels. But we will soon find other relevant events, corresponding to flows of softer information that modifies plausibility order rather than the domain of worlds. These, too, will be important for analyzing games, and we will therefore explain the total methodology.

Belief change under hard information Clearly, beliefs can change under public announcement of hard information. The key axiom for this change is as follows:

$$[!\varphi]B_i\psi \;\leftrightarrow\; \varphi \rightarrow B_i^{\varphi}[!\varphi]\psi$$

This says essentially that a belief statement $B_i\psi$ holds in the updated submodel consisting of the φ-worlds if and only if the original model verifies a belief in $[!\varphi]\psi$ conditional on φ. Here the prefix $[!\varphi]$ serves as before to allow for truth value changes of ψ under the update action $!\varphi$. To be in harmony, the system also needs a recursion axiom for conditional belief that can be checked in the same manner.[77]

THEOREM 7.2 The complete logic of belief change under public announcements is axiomatized by

(a) the static logic of $B_i^{\psi}\varphi$.
(b) PAL.
(c) $[!\varphi]B_i^{\psi}\chi \leftrightarrow (\varphi \rightarrow B_i^{\varphi\wedge[!\varphi]\psi}[!\varphi]\chi)$.

---

76 We have chosen maximality for the good position here, although other authors prefer minimality. The difference is immaterial, of course, as long as terms are used consistently.

77 Thus, belief revision theory should really be a theory of revising conditional beliefs.

Theorem 7.2 is only the beginning of the dynamics of belief change.

Further notions, safe belief Even the preceding simple setting has its surprises, as we shall see now.

EXAMPLE 7.7 Inducing false beliefs with true information
In the following epistemic-doxastic model:

s, p, q	$\leq$	$t, \neg p, q$	$\leq$	$u, p, \neg q$

giving the true information that q is the case updates to a new model

s, p, q	$\leq$	$t, \neg p, q$

where the agent now has a false belief that $\neg p$. One can mislead with the truth. ■

This scenario from game theory and logics of agency (Shoham & Leyton-Brown 2008), has motivated a new notion of safe, or robust, belief that does not change when true information is received (Stalnaker 1999, Baltag & Smets 2008).

DEFINITION 7.2 Safe belief
The notion of safe belief is the ordinary modality of the plausibility order

$$\boldsymbol{M}, s \models [\leq_i]\varphi \quad \text{iff} \quad \boldsymbol{M}, t \models \varphi \text{ for all } t \text{ with } s \leq_i t$$

This is intermediate in strength between knowledge and belief as defined above. ■

Evidently, safe belief (at least in factual propositions φ) remains true under any update with a true proposition holding at the actual world s. Safe belief can define both absolute belief $B_i\varphi$ and conditional belief $B_i^\psi\varphi$ (Boutilier 1994).

FACT 7.3 Safe belief defines conditional belief by means of

$$K\big(\psi \to \langle\leq_i\rangle(\psi \land [\leq_i](\psi \to \varphi))\big).$$

Thus, analyzing the dynamics of safe belief is simpler and does not lose generality. Its static logic over plausibility pre-orders is just the standard modal logic *S4*. On top of this, the following recursion principle captures the dynamics.

FACT 7.4 The recursion law $[!\varphi][\leq_i]\psi \leftrightarrow (\varphi \to [\leq_i][!\varphi]\psi)$ is valid.

Belief change under soft information In addition to hard information, there is *soft information*, which occurs when we take a signal seriously, but not as gospel truth. There are many sources for this weaker form of information, for instance, in default reasoning (Veltman 1996), in learning theory (Baltag et al. 2011), or just in listening to people telling their vacation stories. In Chapter 8, we will also find a use for it in the deliberation phase of games.

The corresponding updates merely make a proposition more plausible, without eliminating all counterexamples. This happens by changing the plausibility order in the current model, rather than its domain of worlds. One ubiquitous form of soft update works as follows.

DEFINITION 7.3 Radical upgrade
A *radical upgrade* $\Uparrow \varphi$ changes a current belief model $\boldsymbol{M}$ to $\boldsymbol{M} \Uparrow \varphi$, where all φ-worlds become better than all $\neg\varphi$-worlds; within these zones, the old order remains

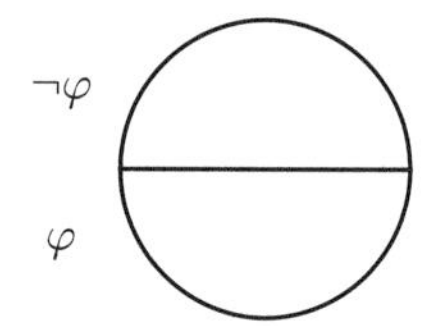

There is a corresponding modality satisfying

$$\boldsymbol{M}, s \models [\Uparrow \varphi]\psi \quad \text{iff} \quad \boldsymbol{M} \Uparrow \varphi, s \models \psi.$$

Reasoning with this notion can be studied in the dynamic logic style of PAL. ∎

EXAMPLE 7.8 Some very simple updates
Consider a model with two worlds encoding an agent's uncertainty about p

$\boldsymbol{s}, p$ ———— $t, \neg p$

An event $!p$ would transform this into the one-world model

$\boldsymbol{s}, p$

By contrast, a radical upgrade $\Uparrow p$ transforms the initial model into

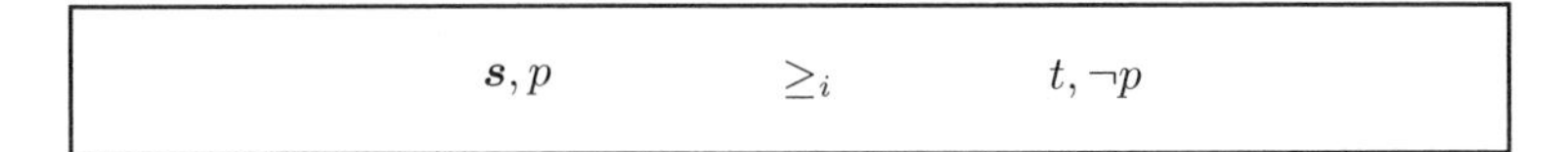

The difference between the two kinds of update, hard and soft, shows when further events retract earlier information. An update sequence $!p$; $!\neg p$ makes no sense. A sequence $!p$; $\Uparrow\neg p$ has the same effect as $!p$, since after the elimination step, there are no more $\neg p$-worlds to be upgraded. However, a sequence $\Uparrow p$; $\Uparrow\neg p$ is perfectly fine, and it transforms the last model depicted into

$s, p \qquad \leq_i \qquad t, \neg p$

Incidentally, the latter scenario shows the overriding effect of radical upgrades: the last one counts. This makes sequences of upgrades in our present sense different from gradual learning rules that accumulate evidence. ■

Of course, further updates are possible; witness the theory of belief revision (Gärdenfors & Rott 1994). For a systematic take on characteristic parameters of variation, see van Benthem (2011a) on analyzing update rules as principles of social choice. These notions support the same sort of logical calculus of reasoning as before. In particular, we state one completeness result in conditional belief style.

THEOREM 7.3 The dynamic logic of radical upgrade is axiomatized by the static base logic of epistemic plausibility models plus a set of recursion axioms including the following key principle (where E is the existential knowledge modality):

$$[\Uparrow\varphi]B^{\psi}\chi \leftrightarrow \big(E(\varphi \wedge [\Uparrow\varphi]\psi) \wedge B^{\varphi\wedge[\Uparrow\varphi]\psi}[\Uparrow\varphi]\chi\big) \vee \big(\neg E(\varphi \wedge [\Uparrow\varphi]\psi) \wedge B^{[\Uparrow\varphi]\psi}[\Uparrow\varphi]\chi\big)$$

A proof is found in van Benthem (2007c). Much simpler, however, is an alternative axiomatization in terms of the recursion law for safe belief under radical upgrade. The latter can be found using the techniques of van Benthem & Liu (2007).

FACT 7.5 The following equivalence is valid on epistemic-doxastic models:

$$[\Uparrow\varphi][\leq_i]\psi \leftrightarrow \big((\varphi \wedge [\leq_i](\varphi \to [\Uparrow\varphi]\psi)) \vee (\neg\varphi \wedge [\leq_i](\neg\varphi \to [\Uparrow\varphi]\psi) \wedge K(\varphi \to [\Uparrow\varphi]\psi))\big)$$

Proof This is a simple verification of cases for the updated plausibility order, using the definition of a radical upgrade $\Uparrow\varphi$ on a current plausibility order $\leq$. ■

Other update policies There are many other revision policies, such as a softer *conservative upgrade* $\uparrow\varphi$ that only makes the best φ-worlds maximal in the ordering. In other words, $\uparrow\varphi$ is radical upgrade with just the most plausible φ-worlds.

EXAMPLE 7.9 Conservative versus radical upgrade
The following example shows the difference between the two kinds of upgrade. Consider three worlds ordered by increasing plausibility as indicated

$$s, p \quad \leq \quad t, p \quad \leq \quad u, \neg p$$

A radical upgrade $\Uparrow p$ turns this into the model

$$u, \neg p \quad \leq \quad s, p \quad \leq \quad t, p$$

making p the winner overall, no matter what past merits put $\neg p$ on top earlier on. In contrast, a conservative upgrade $\uparrow p$ would produce the different model

$$s, p \quad \leq \quad u, \neg p \quad \leq \quad t, p$$

where the agent has acquired a belief in p, but not a deeply entrenched one, since ruling out world 2 would bring the agent back to believing $\neg p$. ■

Comparing soft update policies may be delicate. While conservative upgrade may seem safer, results in learning theory (Gierasimczuk 2010) suggest that the more risky radical upgrade is sometimes the better policy in the long run, since it may shake the agent out of suboptimal local equilibria in beliefs.

Achieving generality There are many further ways of changing plausibility order (see Rott 2006 for a large compendium). To achieve generality, there are several proposals for a generic format for plausibility upgrades as model transformers supporting recursion axioms. One is the program format of van Benthem & Liu (2007), where the new plausibility relations $\leq_{new}$ after upgrade are given by PDL programs $\pi(\leq_{old})$. For instance, radical upgrade $\Uparrow\varphi$ is definable as the PDL program

$$(?\varphi \,;\, \leq \,;\, ?\varphi) \cup (?\neg\varphi \,;\, \leq \,;\, ?\neg\varphi) \cup (?\neg\varphi \,;\, U \,;\, ?\varphi)$$

where U is the universal relation. All upgrades defined in this format lead to automatically derived complete sets of recursion axioms. For another general format of plausibility change called priority upgrade, see Section 7.4. A very general PDL-based format (cf. Chapter 1) was recently proposed in Girard et al. (2012).

Probability Belief revision is often studied quantitatively using Bayesian or other methods (cf. Gärdenfors & Rott 1994). This fits well with the standard uses of probabilistic notions in game theory noted in our Introduction. A complete dynamic-epistemic logic for quantitative probability update can be found in van Benthem et al. (2009b), combined with our earlier qualitative knowledge, showing how the program of this chapter meshes well with probabilistic thinking.[78]

Conclusion We have shown how dynamics extends from informational events with guaranteed correctness to acts of correction where agents learn from errors. Stated in a more philosophical way, logical ability resides in a talent for revising theories as well as building strong foundations. Technically, going beyond the simple world elimination of PAL, this induced a much broader view of update as model change. This raises new issues for logical theory that we will not pursue here.[79]

7.4 Update by general events with partial observation

Let us now move from logics for knowledge and belief change for single agents to genuine multi-agent settings. Most social scenarios do not have the same information for every agent. Parlor games create interest by manipulating information, and

---

78 The logic has an update mechanism merging information from three sources: (a) prior probabilities between worlds, (b) observation probabilities for current events that trigger the update, and, crucially, (c) occurrence probabilities for the chance that a given event will occur under a given precondition. Occurrence probabilities have no direct counterpart in the plausibility models of this book, although they are natural in the presence of procedural information about agent types. Proposals for qualitative plausibility models that match occurrence probabilities are found in van Benthem (2012d), as well as rules for merge of orderings that match numerical probabilistic update rules.

79 One simple open problem is finding good recursion laws for common belief under radical upgrade, a task that seems to require considerable extensions of the static language. A much deeper problem is finding a technical characterization of natural update operations on models that would delimit the realm of dynamics in some principled manner.

so do war games. In the realm of security, an email with a *cc* is a public announcement, but with a *bcc*, it becomes something more subtle. The update mechanism to be discussed here applies to all of these settings, and also, more to the point for us, to the games of imperfect information in Chapters 3 and 6. These typically involve partial observation of events such as drawing cards, resulting in differential information flow for different players.

Event models Many social scenarios are represented in the following structures, which drive a widely applicable general update mechanism (Baltag et al. 1998).

DEFINITION 7.4 Event models and product update
Event models $\mathcal{E} = (E, \{\sim_i \mid i \in G\}, \{PRE_e \mid e \in E\})$ consist of abstract points standing for the relevant events, where indistinguishability relations $\sim_i$ encode agents' knowledge about which event is taking place. Crucially, events e have preconditions PRE_e for their execution, which makes observing them informative.

For any epistemic model $(\boldsymbol{M}, s)$ as defined earlier, and any event model $(\mathcal{E}, e)$, the *product model* $(\boldsymbol{M} \times \mathcal{E}, (s, e))$ is defined as follows:

Domain: $\{(s, e) \mid s$ a world in $\boldsymbol{M}$, e an event in $\mathcal{E}$, $(M, s) \models PRE_e\}$.

Accessibility: $(s, e) \sim_i (t, f)$ iff both $s \sim_i t$ and $e \sim_i f$.

The valuation for atoms p at (s, e) is that at s in $\boldsymbol{M}$.

The latter clause can be generalized to factual change, allowing changes in truth value via postconditions $POST_e$ for events e (van Benthem et al. 2006a). ■

Product update deals with a wide range of informational scenarios. These include misleading actions as well as truthful ones, and belief as well as knowledge. A noteworthy feature is that epistemic models can now get larger as update proceeds, reflecting the fact that treating different agents differently can be a source of complexity. As before, there is a logical calculus keeping all this complexity straight.

Dynamic-epistemic logic Dynamic-epistemic logic has the following syntax and semantics, where $(\mathcal{E}, e)$ stands for any event model with actual event e:

$$p \mid \neg\varphi \mid \varphi \wedge \psi \mid K_i\varphi \mid C_G\varphi \mid [\mathcal{E}, e]\varphi$$

$$\boldsymbol{M}, s \models [\mathcal{E}, e]\varphi \quad \text{iff, if } \boldsymbol{M}, s \models PRE_e, \text{ then } \boldsymbol{M} \times \mathcal{E}, (s, e) \models \varphi$$

THEOREM 7.4 Dynamic-epistemic logic is effectively axiomatizable and decidable.

The proof is similar to that for PAL, and the key recursion axiom extends that for public announcements $!\varphi$ (now one-world event models with precondition φ):

$$[\mathcal{E}, e]K_i\varphi \leftrightarrow (PRE_e \to \bigwedge\{K_i[\mathcal{E}, f]\varphi \mid f \sim_i e \,\mathrm{in}\, \mathcal{E}\})$$

This law works for finite event models, but also, reading the conjunction as an infinite one, for infinite event models.

EXAMPLE 7.10 Another look at the challenge of common knowledge
Common knowledge or common belief in subgroups with secrets may be delicate. The following event model has a pointed arrow indicating that when event e takes place, agent 2 mistakenly believes that it is event f that is happening:

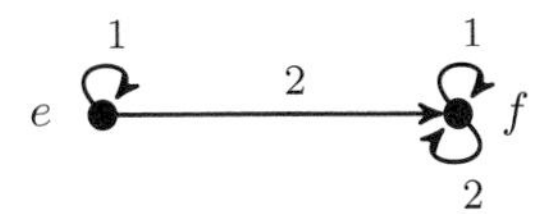

To obtain a recursive analysis of such cases, we have to extend the language to a much richer system of *epistemic PDL* (E-PDL), where arbitrary programs defined over basic relations $\sim_i$ stand for agents viewed as epistemic procedures. ■

THEOREM 7.5 DEL with a recursion axiom for $[\mathcal{E}, e]C_G\varphi$ is axiomatized in E-PDL.

The result is proved in van Benthem et al. (2006a) using finite automata, and another proof proceeds via modal μ-calculus techniques.

This shows something that holds for many formats of dynamic-epistemic update. In the limit, they tend toward propositional dynamic logic (cf. Chapters 1 and 4).

Belief revision and priority The same mechanism can deal with multi-agent belief revision. Now event models consist of signals with doxastic plausibility relations. For instance, an event model for the earlier upgrade $\Uparrow\varphi$ would have two events $!\varphi$ and $!\neg\varphi$, with $!\varphi$ more plausible than $!\neg\varphi$.

DEFINITION 7.5 Priority update
The *priority plausibility order* in epistemic-doxastic product models $\boldsymbol{M} \times \mathcal{E}$ holds if, either old worlds and new events agree, or new events override the prior order:

$$(s, e) \leq (t, f) \text{ iff either } (s \leq t \text{ in } \boldsymbol{M} \;\&\; e \leq f \text{ in } \mathcal{E}) \text{ or } e < f \text{ in } \mathcal{E}$$

with $x < y$ the strict comparison order $x \leq y \wedge \neg y \leq x$. This rule can mimic many others by a suitable choice of signals and plausibility order in event models. ■

There is a complete dynamic logic with modalities for this update mechanism, including the following recursion law for safe belief (Baltag & Smets 2008):

$$[\boldsymbol{\mathcal{E}}, e][\leq_i]\varphi \leftrightarrow (PRE_e \rightarrow (\bigwedge\{[\leq_i][\boldsymbol{\mathcal{E}}, f]\varphi \mid\mid e \leq_i f \text{ in } \boldsymbol{\mathcal{E}}\} \wedge \bigwedge\{K_i[\boldsymbol{\mathcal{E}}, f]\varphi \mid e <_i f \text{ in } \boldsymbol{\mathcal{E}}\}))$$

7.5 Temporal behavior and procedural information

Our next step concerns one more dimension of informational events. All dynamic logics so far were about single steps of information update and attitude change. But such local actions make sense only in global procedures of inquiry over time. A first format are structured programs, as studied in Chapters 1 and 4.

Program operations Action and communication involve complex programs, structured by operations of sequential composition ;, guarded choice IF... THEN ... ELSE, and iteration WHILE... DO. Even parallel composition || occurs when people act or speak simultaneously.

Here is a well-known illustration, which has inspired many authors.

EXAMPLE 7.11 The Muddy Children puzzle
After playing outside, two of a group of three children have mud on their foreheads. They can only see the others, and do not know their own status. Now the Father says: "At least one of you is dirty." He then asks: "Does anyone know if he is dirty?" The children always answer truthfully. What will happen?

As questions and answers repeat, nobody knows in the first round. But in the next round, each muddy child can reason like this: "If I were clean, then the one dirty child I see would have seen only clean children, and so that child would have known that she was dirty at once. But she did not. So I am dirty, too." ■

An initial epistemic model for the puzzle has eight worlds assigning $\boldsymbol{D}$ (dirty) or $\boldsymbol{C}$ (clean) to each child. What children know is reflected in accessibility lines.

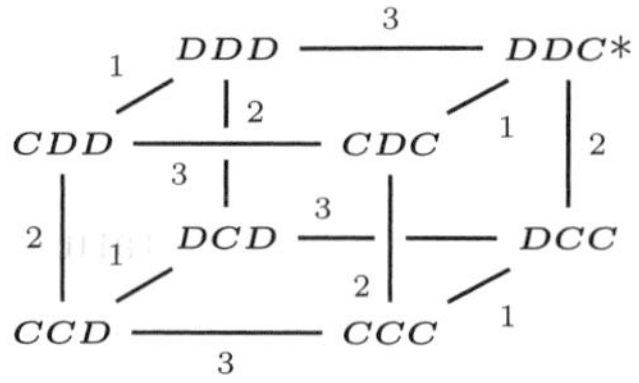

Now, the assertions update this information in the PAL style.[80] The Father's announcement removes ***CCC***, breaking the symmetry of the diagram.

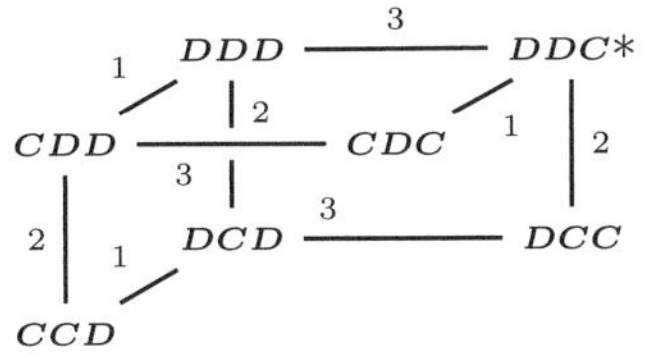

```
            3
      DDD ——————— DDC*
  1  /  | 2       /  |
CDD ——————— CDC  1   | 2
  |     3 |          |
2 |  1   DCD ——————— DCC
  |  /        3
CCD
```

The first public statement of ignorance updates this model to:

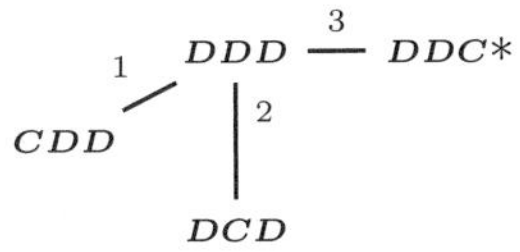

```
         3
  1  DDD — DDC*
   /  | 2
CDD   |
     DCD
```

The final update with an ignorance assertion has common knowledge in the limit:

DDC∗

Thus, beyond single steps, there is a rich structure in temporal information flow, involving complex actions formed out of single update steps. For instance, the puzzle involves a program with sequence, guarded choice, and iteration:

> !"At least one of you is dirty." ; WHILE not know your status
> DO (IF not know THEN "say don't know" ELSE "say know")

One could even add parallel program structure for simultaneous speaking. While the Muddy Children problem used public announcements here, it works equally well for softer upgrades, or mixtures of hard and soft information.

Limit behavior Another interesting temporal feature is the limit behavior in the puzzle, leading to a stable endpoint where updates have no further effect and agents' knowledge is in some sort of equilibrium. This temporal structure will turn out to be highly relevant to games in the following chapters of Part II. In particular, the children have common knowledge of their status in the limit model $\#(\boldsymbol{M}, \varphi)$

---

80 A much sparser agent-internal representation of how the muddy children themselves might view their situation is found in Gierasimczuk & Szymanik (2011).

reached by the iterated updates $!\varphi$ of their ignorance assertion. There is a general phenomenon at work here, having to do with how announcements shrink models.

DEFINITION 7.6 Announcement limits
Starting from an initial model $\boldsymbol{M}$, iterated public announcements of a formula φ always reach a first submodel where they have no further effect, the *announcement limit* $\#(\boldsymbol{M}, \varphi)$, where, in infinite models, iteration takes intersections of all previous stages at limit ordinals. These limits can be of two kinds. If $\#(\boldsymbol{M}, \varphi)$ is non-empty, then φ becomes common knowledge, the "self-fulfilling" scenario, but if $\#(\boldsymbol{M}, \varphi)$ is empty, the negation $\neg\varphi$ has become true, the "self-refuting" scenario. ■

The Muddy Children ignorance scenario is self-refuting in the model obtained after the Father's initial assertion, so self-refutation can yield positive knowledge. But without the Father's announcement, the scenario is self-fulfilling: the initial model itself is the limit, and the ignorance is common knowledge, as it was at the start. In Chapters 8 and 9, we will look at less trivial self-fulfilling scenarios with iterated announcements of rationality in games.

Modal logics Useful systems for describing program structure of epistemic and doxastic events are propositional dynamic logic PDL and the μ-calculus. First consider adding program structure to PAL or other dynamic-epistemic logics. While this is natural, complexity may explode in this setting, going far beyond non-axiomatizability for reasons that we have also seen in Chapters 2 and 3.

THEOREM 7.6 PAL with iteration of updates is Π^1_1-complete.

Proof With this language over universes of epistemic models, geometrical tiling problems of high complexity become encodable (cf. Miller & Moss 2005). ■

We will see in a moment how this might be mitigated in more general models.

Next, turning to limits of updates, again there are links with earlier logics. Limit models for special positive-existential formulas φ are definable in the modal μ-calculus of Chapter 1 (cf. van Benthem 2007d, and Chapter 13), making their theory part of a decidable system. Both the ignorance statement of the Muddy Children puzzle and the rationality assertions to be used for games in Chapter 8 are of this special syntactic kind, and we will revisit some of the issues there. Still, much remains to be understood about the model theory of epistemic limits.

Protocols and extended dynamic logics In the recent literature on dynamic-epistemic logic, there has been an important move toward a generalized temporal picture (van Benthem et al. 2009a; see van Benthem 2012f for a related approach). Our analysis so far took place in the total universe of all possible epistemic, or epistemic-doxastic, models. This universe is what validates the axioms of PAL and subsequent systems. But in real informational scenarios, there may be constraints making not all of these models accessible. Normally there is a "protocol" for a process of communication or inquiry that decides which actions are available at which stage (Fagin et al. 1995).[81] When modeling this procedural information, we can no longer assume that all public announcements of true facts are possible at any stage (if only because telling the truth is sometimes uncivilized). *Protocol models* consist of admissible histories of successive updates starting from some initial model.

This move generalizes our earlier dynamic logics. For PAL, the base axioms change, and so does the key recursion law for knowledge.

FACT 7.6 The following principles are valid on protocol models:

$$[!\varphi]q \leftrightarrow (\langle !\varphi\rangle\top \to q)$$
$$[!\varphi]K_i\psi \leftrightarrow (\langle !\varphi\rangle\top \to K_i(\langle !\varphi\rangle\top \to [!\varphi]\psi))$$

Note the difference with the earlier laws, for instance, that for atomic propositions, $[!\varphi]q \leftrightarrow (\varphi \to q)$. The new axioms crucially contain procedural information encoded in formulas $\langle !\varphi\rangle\top$ telling us whether φ can be announced (or observed, or learned) in the current process. But there is no equivalence between $\langle !\varphi\rangle\top$ and φ, only an implication from left to right. Thus, in proving completeness, the earlier reduction of the full dynamic language to the epistemic base language disappears.

THEOREM 7.7 PAL is axiomatizable and decidable on protocol models.

Proof This goes by a standard modal completeness argument (cf. Hoshi 2009). ■

In this larger class of models, high-complexity results for the standard universe do not transfer automatically. In particular, the earlier result for PAL with iterated assertions becomes open again, although we do not know of a precise result.

Epistemic forests: From dynamic to temporal logic Protocol models as described here are special cases of the branching tree or forest models of Chapter 6 with epistemic links between nodes. In particular, the available histories in such

81 As we have seen in Chapters 5 and 6, this procedural perspective arises even with standard logical systems; witness Beth or Kripke models for intuitionistic logic.

temporal models encode a protocol as above. Thus, one can also bring epistemic temporal logics to bear like those studied in Chapter 5.

This raises the issue of how dynamic-epistemic and temporal logics are related. This issue was clarified in van Benthem et al. (2009a); see Dégremont et al. (2011) for an important amendment. We merely give the upshot here.

The languages of PAL and its multi-agent extension DEL described above lie inside the epistemic-temporal logic of Chapter 5 as a fragment describing agents' knowledge change under one-step actions. For instance, a dynamic modality $\langle !\varphi \rangle$ matches a one-step future operator of the form F_e for some suitable event e. However, there are no counterparts of unbounded future or past operators; that would be more like the iterated program modalities $[(!\varphi)^*]$ discussed earlier.

What about relating the two kinds of model? One way of doing this lets the earlier DEL product update iterate to form epistemic forests in successive stages:

$$\boldsymbol{M}, \boldsymbol{M} \times \boldsymbol{\mathcal{E}}_1, (\boldsymbol{M} \times \boldsymbol{\mathcal{E}}_1) \times \boldsymbol{\mathcal{E}}_2, \ldots$$

A complete characterization exists for the resulting epistemic-temporal models.

THEOREM 7.8 An epistemic forest model can be constructed, up to isomorphism, via iterated DEL updates iff agents satisfy Perfect Recall, No Miracles, and the domains of all events are closed under epistemic bisimulations.

We will not explain these properties further here, and refer to Chapter 9 for the precise representation technique proving this result. Still, the reader will recognize the notion of bisimulation from Chapter 1, while Perfect Recall and No Miracles were discussed in Chapter 3 as possible properties of game players that are endowed with perfect memory, while learning from observations only.

Against this background, the earlier one-step feature explains why dynamic-epistemic logics tend to be simpler than richer epistemic-temporal ones. As noted in Chapter 5, in the latter setting, Perfect Recall and No Miracles are costly, and epistemic temporal logics for models satisfying them can be highly complex. Now, as we just saw, the typical operator interchanges of these properties are built into the very recursion laws for knowledge of PAL and DEL. But in their one-step versions, without an unbounded future, they do no harm.

Belief change over time All that has been discussed here for knowledge over time can be extended to allow for agents' beliefs. This has led to interesting links between our notions of belief update, epistemic protocol models, doxastic-temporal logics, and learning theory (Kelly 1996, Dégremont 2010, Gierasimczuk 2010). As we will see soon, iterated belief update also applies to game theory, for instance in our account of Backward Induction in Chapter 8. Logical foundations become more complex in this area. For instance, instead of limits there can also be cycles in iterated plausibility change (Baltag et al. 2011; see Chapter 8 for more details of this quite realistic phenomenon in social opinions), and richer fixed point logics are then needed to analyze the resulting models.

There is much more to the temporal road than we have discussed here. For instance, our models also suggest taking key ideas of dynamic-epistemic logic from the finite scenarios of classical game theory to the infinite ones in evolutionary game theory, a subject whose logic will return in Chapter 12.

7.6 Inference and variety of information

This concludes our semantic treatment of the most common informational events. But even our original example of the Restaurant already contained further actions, namely, *questions* and *inferences*. While these actions are both crucial to information flow, and also make a lot of sense in games, they are not a major theme in this book, so we will keep our discussion very brief, merely outlining a dynamic-epistemic angle on both.

Starting with the final act, what did the inference do for the waiter? The semantic options did not change, having already reduced to just one. To make sense of inference as changing states, one needs a more fine-grained notion of information. Indeed, it is well-supported that logic involves different kinds of information, called "correlation," "range," and "code" in van Benthem & Martínez (2008). One basic approach to inference dynamics is syntactic, and we now sketch how it works.[82]

Dynamic logic of awareness In the spirit of Fagin & Halpern (1988) and van Benthem & Velázquez-Quesada (2010), we give each world a set of syntactic formulas that the agent is aware of, with a corresponding modality $A\varphi$ saying that φ is in the current awareness set.

This move can deal, for instance, with the usual criticism of the epistemic distribution axiom $K(\varphi \to \psi) \to (K\varphi \to K\psi)$ read as a closure principle for "explicit

82 One might think that logics such as PAL already deal with inference dynamics, since they allow us to reason about updates in a proof calculus. But this is an analysis at the meta-level of reasoning about, not at the object-level of agents engaging in inference.

knowledge" instead of, as we have emphasized throughout, a notion of semantic information. When we think of explicit knowledge, $Ex\varphi$, the distribution axiom really contains a hidden gap for work to be done, an action that one should perform in order to obtain what would be a free gift if we assumed closure:

$$Ex(\varphi \to \psi) \to (Ex\varphi \to [\ldots]Ex\psi)$$

A typical item that fits here is an act $\#\psi$ of "realization" that ψ is the case, dynamically raising awareness by adding the conclusion ψ to the formulas associated with the world. The waiter realized that the last plate was for the remaining person, and could now act on this knowledge.[83] Other updates of fine-grained syntactic information arise from acts of introspection, explicit seeing, and others.

These events can be described with the same methods that we have used before.

THEOREM 7.9 The dynamic logic of awareness raising is completely axiomatizable.

Proof See Velázquez-Quesada (2011) and van Benthem (2011e) for some systems of this kind. Here are two basic recursion laws explaining how awareness may change:

$$[!\varphi]A\psi \leftrightarrow (\varphi \to A\psi) \qquad [\#\varphi]A\psi \leftrightarrow (A\psi \vee \psi = \varphi)$$

∎

Where we stand Different notions of information in logic involve different dynamic acts with logics of their own. Some first dynamic-epistemic logics have been found for inference and related actions, but these were just mentioned here as a proof-of-concept. Richer systems are clearly needed for a more serious analysis of evidence change (van Benthem & Pacuit 2011) or multi-agent argumentation (Grossi 2012). Technically, this may involve finding links between dynamic logics and more combinatorial features of proof theory.

7.7 Questions and direction of inquiry

But there is also one more important informational act that has received less attention in dynamic-epistemic logic, although it does have a distinguished history (see the many strands represented in the special *Synthese* issue, Hamami & Roelofsen 2013). The waiter also asked questions, in a natural combination with inference.

83 For an opposing view entangling actions with unconscious inferences, see Icard (2013).

What is their dynamic informational function? Questions direct inquiry by setting the current *issue*.

This time, not even a standard static logic exists, and one has to be designed. To a first approximation, an issue may be viewed as an equivalence relation $\approx$ partitioning a current epistemic range, where the aim of inquiry is to find out which partition cell we are in.

EXAMPLE 7.12 An epistemic issue model
In the following epistemic model with an additional issue relation, dotted black lines are epistemic links and rectangles are the cells of the issue relation.

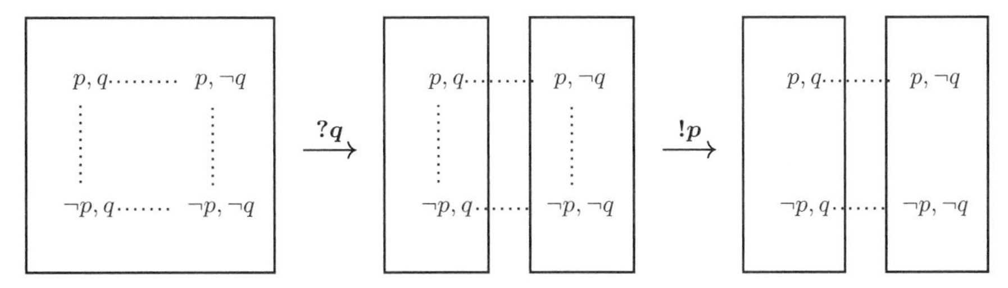

The action $?q$ refines the initial issue relation to the two zones of the q-worlds and the $\neg q$-worlds, while the epistemic relation stays the same. After that, a public announcement $!p$, in a weak "link cutting" version where no worlds get removed (sometimes called testing of the proposition p) refines the epistemic relation while the issue relation stays the same. ■

The static base language of these models has the usual modality $K\varphi$ for knowledge plus two new ones, a somewhat technical universal modality $Q\varphi$ for the issue relation, plus a "resolution modality":

$R\varphi$ says that φ is true in all worlds accessible via both $\sim$ and $\approx$.

This describes what would hold if the issue is resolved given our current knowledge.

On top of this static infrastructure of knowledge and issues, the dynamic methodology works again, and it starts by identifying relevant actions. We will only look at single-agent models, although questions are social acts of communication, and better logics must deal with this essential feature (cf. Minică 2011).

Actions of issue management Let $\boldsymbol{M} = (W, \sim, \approx, V)$ be an epistemic issue model. The *issue update model* resulting from the action of a question action $?\varphi$ on $\boldsymbol{M}$ is the tuple $\boldsymbol{M}^{?\varphi} = (W, \sim, \approx^{?\varphi}, V)$ with $s \approx^{?\varphi} t$ iff $s \approx t$ and $s =_\varphi t$. Here $s =_\varphi t$ says that the worlds s and t agree on the truth value of the formula φ.

Via this refinement, questions update a current issue. Other natural forms of issue management are actions "resolve" ! resetting the epistemic relation to $\sim \cap \approx$, and "refine" ? resetting the issue relation to $\sim \cap \approx$.

Complete dynamic logics Actions of issue management can be axiomatized in our earlier style (van Benthem & Minică 2012). Some key recursion laws for asking and resolving are as follows.

$$[?\varphi]K\psi \leftrightarrow K[?\varphi]\psi \qquad [!]K\psi \leftrightarrow R[!]\psi$$

$$[!\varphi]K\psi \leftrightarrow (\varphi \wedge K(\varphi \rightarrow [!\varphi]\psi)) \vee (\neg\varphi \wedge K(\neg\varphi \rightarrow [!\varphi]\psi))$$

In general, asking questions seems just as important to inquiry as giving answers, and questions will return in a game-theoretic setting in Chapter 22.

Now we have dealt with all actions of the waiter in our initial Restaurant example, but the resulting account of agency may not be at a natural border yet.

7.8 From correct to reasonable: Harmony with evaluation

Rational agency involves more than correctness, in the form of a balance of information and evaluation. Information flow provides the kinematics, but a more explanatory dynamics of behavior also needs a balance with underlying forces linked to agents' *preferences* over outcomes. Preference can be modeled in modal logics, as we have seen in Chapter 2. And the same dynamic-epistemic techniques that represented belief change by transforming plausibility orders can also deal with preference change, as shown in van Benthem & Liu (2007).

For instance, a radical upgrade $\Uparrow\varphi$ of a current preference relation also models accepting a normative command that φ become the case. But many other triggers occur for preference changes, from changes in taste to weak or strong social cues. A typical weaker relation transformer, not discussed so far, works as follows.

DEFINITION 7.7 Suggestion as a trigger for preference change
A *suggestion*, $\sharp\varphi$, in favor of φ changes a current preference relation $\leq$ to

$$(?\varphi\ ;\leq;\ ?\varphi) \cup (?\neg\varphi\ ;\leq;\ ?\neg\varphi) \cup (?\neg\varphi\ ;\leq;\ ?\varphi)$$

removing only those earlier preferences that ran from to a φ-world to a $\neg\varphi$-world. Note the PDL program format employed here: as mentioned earlier, it guarantees the existence of matching recursion laws. ■

Dynamic logics for these systems work via the same techniques as before, relating preferences after update to those the agent had before. Liu (2011) is a systematic study of preference change, connecting it to deontic logic, imperatives in natural language, and ways of rationalizing behavior in games (see also Girard 2008).

Not all of this is standard generalization of our earlier notions. The treatment of reason-based preference in Liu (2011) also includes graphs of prioritized predicates that induce the betterness order of worlds that was still taken as primitive in Chapter 2, with new dynamic operations now emerging on the underlying graphs. This powerful idea of "priority graphs," proposed initially in Andréka et al. (2002), also makes sense in games, whose preference order on end nodes can now be analyzed further as coming from players' prioritized goals.[84]

Preference dynamics has a natural continuation to games (see Chapter 9), and into areas such as social choice theory, where dynamic logics add fine structure of deliberation and decision making by forming new individual and group preferences.

7.9 Conclusion

Logic Dynamic-epistemic logics for information-driven action extend the traditional scope of logic, combining philosophical and computational themes. The dynamic stance has repercussions, too, for interfaces of logic with epistemology and the philosophy of science. But at the same time, the systems defined here satisfy technical standards and raise new technical problems that are in line with mathematical logic. Thus, they fit with two historical trends that have long characterized logic: expanding the agenda, and exploring technical system foundations.

One point remains to be clarified here that may be captured in two prepositions. The conception we have explored is reasoning *with* information dynamics: it is about agents engaged in logical tasks. From that perspective, mathematical activities such as proof and inference are just a small subset of all relevant actions, as we have seen. But these activities were then captured in formal dynamic logics that embody recursion laws and allow for systematic study of general system properties. The

84 This idea is taken further in Grossi & Turrini (2012), exploiting the syntax of the graphs to model players' limited cognitive horizon in actual play of complex games.

latter study might be called reasoning *about* information dynamics. For that meta-study, standard mathematical logic, perhaps even with a special status for proof, is as important as ever.

Games Logical dynamics meets games in two ways. In one sense, we have seen how games are a natural culminating step in the above program for logic, tying together informational, preferential, and long-term strategic aspects of intelligent agency. But in the opposite direction, techniques from our dynamic logics can also illuminate what happens in games, leading to the Theory of Play that will be the further topic of Part II. Many examples will be found in the coming three chapters.

However, this book contains an even more radical sense in which games, logic, and dynamics are linked. The logic games of Parts IV will analyze logic itself as a form of playing games. For the conceptual repercussions of that, we refer to those later parts of this book.

7.10 Literature

A complete development of logical dynamics requires a book of its own, and we refer the reader to van Benthem (1996) and van Benthem (2011e) for more extensive monographs. For a textbook on dynamic-epistemic logics, the reader may consult van Ditmarsch et al. (2007). More specific references for special topics have been given in separate sections, and many more, directed toward games, will appear at specific places in later chapters.

7.11 Recent developments

To conclude this survey, we list a few recent directions in logical dynamics that may eventually have some relevance for the study of games.

Evidence dynamics Relational models provide a somewhat coarse-grained representation of what agents know and believe, suppressing the evidence that took them there. Evidence is analyzed by means of neighborhood models (cf. Chapter 11) in van Benthem & Pacuit (2011), where evidence can cluster in maximally consistent sets, but also contain conflicts when sets are disjoint. They show how evidence models support a rich dynamics of adding and removing pieces of evidence with a delicate underlying language of knowledge and new varieties of conditional belief. This is a much richer setting for many of the themes discussed in this chapter.

Postulational versus constructive approaches Some theories of update and revision do not provide explicit update actions, but rather impose postulates to be satisfied by whatever mechanism is chosen (cf. Gärdenfors 1988, Segerberg 1995, Rott 2001). The two views are connected in van Benthem (2012f) by modal frame correspondences for update postulates on universes of models related by abstract transitions. In particular, the standard recursion axioms turn out to virtually fix the concrete update rules that we have given. As a by-product of this abstract view, one can give a generalized semantics for PAL and DEL where propositions denote sets of pointed epistemic models forming worlds in some larger temporal protocol model. This makes the validities substitution-closed, while the original PAL becomes the special case where atomic propositions are context-independent, depending only on the local states of the pointed model. This is in the sprit of our earlier protocol models, that fit well with games (cf. Dégremont 2010).

Generalized update mechanisms Product update, the PDL program format, and priority update may still not be general enough for modeling complex social scenarios. The discussion of the best technical implementation of logical dynamics continues. Girard et al. (2012) propose a new system combining aspects of all three.

Probabilistic versions Probability is important to games and other social scenarios for several reasons. It provides a refined description of agents in terms of degrees of belief, it extends the range of possible behaviors with mixed strategies (see the Introduction to this book), and a probability value can also be a compact summary of past experience about some process or type of situation. Thus, our systems need to be integrated more systematically with probability theory, and the results in Halpern (2003b) and van Benthem et al. (2009b) suggest that they can.

Mathematical foundations A study of algebraic aspects of dynamic logics with model-changing operators on weaker logical bases has been undertaken in Sadrzadeh et al. (2011) and Kurz & Palmigiano (2012) (see also Holliday et al. 2011 on the laws of the substitution-closed core), and a study of their inference rules and proof-theoretic bases is found in Wang (2011). What such studies can contribute to the analysis of games is a more austere view of the real proof content of basic game-theoretic arguments and structures. A nice example of such a sparse approach is the analysis of Aumann's Theorem in Zvesper (2010) that will be shown in Chapter 13. Sparser approaches fit well with the quasi-empirical look at lean logical structures for common game-theoretic arguments that we have advocated in Part I.

8
Deliberation as Iterated Update

In Chapter 2 of this book, we took the Backward Induction procedure as our pilot example, and described it in a fixed point logic of action and preference, and also, zooming out to a less fine-grained level of detail, in a modal logic of best action. But there is also another way of construing the logical import of solving games. Backward Induction is a *procedure* for creating expectations about how a game will proceed. In the logical dynamics of the present part, it is the procedure itself that deserves attention, as a specimen of what we might call the dynamics of deliberation. Taking this view of game solution as rational procedure also has a broader virtue. We can do justice to an appealing intuition behind the fundamental notion of rationality: it is not a static state of grace, but a style of doing things. Along with this, we get a dynamic focus shift in the epistemic foundations of game theory that may be of interest in itself.

8.1 Backward Induction and announcing rationality

Let us shift attention from the static rationality in Part I to what players do when deliberating about a game. We will see how the relevant procedures reach stable limit models where rationality has become common knowledge. We show this for the method of Backward Induction, as a pilot for our style of analysis. The version of the algorithm we have in mind uses the relational strategies of Chapter 2.[85]

---

85 Indeed, the very word solution has an ambiguity between a reading as a procedure ("Solution of this problem is not easy") and a static product of such a procedure ("Show me your solution").

Our dynamic analysis of Backward Induction takes it to be a process of prior deliberation about a game by players whose minds proceed in harmony. The steps driving the information flow in our first scenario are public announcements $!\varphi$ saying that proposition φ is true, as explained in Chapter 7. These transform an epistemic model $\boldsymbol{M}$ into its submodel $\boldsymbol{M}|\varphi$ whose domain consists of just those worlds in $\boldsymbol{M}$ that satisfy φ.

The driver: Rationality We now explain the driving assertion about games that we will need in what follows. It is closely related to our analysis of best action in Chapter 2, though it differs slightly from the principle RAT used there.

DEFINITION 8.1 Node rationality
At a turn for player i in an extensive game, a move a is *dominated* by a sibling b (a move available at the same node) if every history through a ends worse, in terms of i's preference, than every history through b. Now *rationality* (***rat***, for short) says that "at the current node, no player has chosen a strictly dominated move in the past coming here." ■

This makes an assertion about nodes in a game tree, namely, that they did not arise through playing a dominated move. This is often called playing a "best response." Some nodes will satisfy this, others may not: we only need that ***rat*** is a reasonable local property of nodes. Thus, announcing this formula as a true fact about behavior of players is informative, and it will in general make a current game tree smaller.

But then we get a dynamics as in the earlier scenario of the Muddy Children in Chapter 7, where *repeated true assertions* of ignorance eventually produced enough information to solve the puzzle. In our case, in the new smaller game tree, new nodes may become dominated, and hence announcing ***rat*** again (saying that it still holds after this round of deliberation) makes sense, and so on. This process of iterated announcement always reaches a limit, a smallest subgame where no node is dominated any longer.

EXAMPLE 8.1 Solving games through iterated assertions of rationality
Consider the following extensive game with three turns, four branches, and payoffs for players $\boldsymbol{A}$ and $\boldsymbol{E}$ in that order:

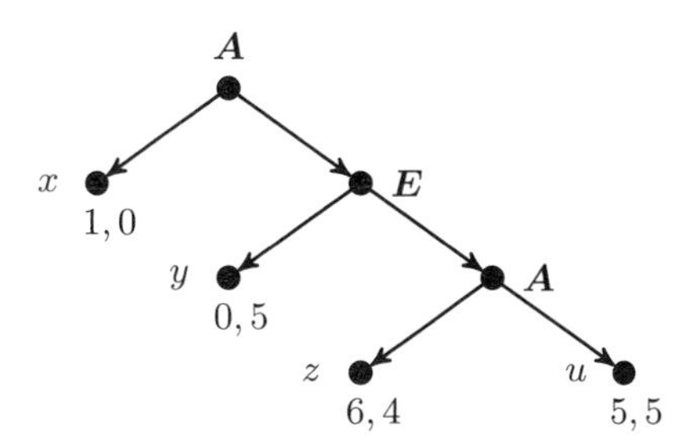

Stage 0 of the procedure rules out point u (the only point where ***rat*** fails), Stage 1 rules out z and the node above it (the new points where ***rat*** fails), and Stage 2 rules out y and the node above it. In the remaining game, ***rat*** holds throughout:

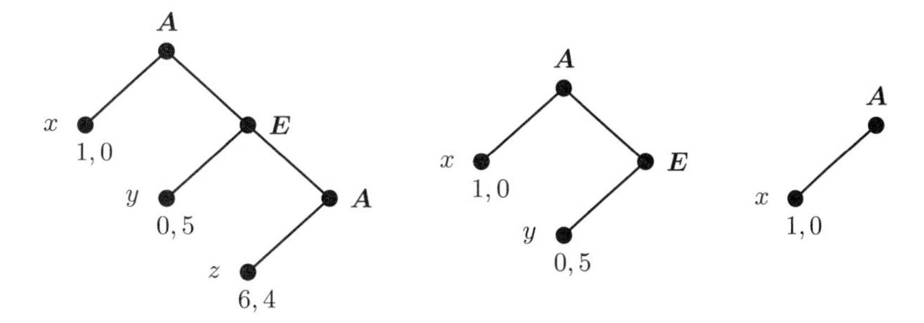

Thus, the Backward Induction solution emerges step by step. ■

It is shown in van Benthem (2007d) that the actual Backward Induction path for extensive games is obtained by repeated announcement of the assertion ***rat*** to its limit. We will now explain this in more detail.

Logical background We reiterate some relevant notions from Chapter 7.

DEFINITION 8.2 Announcement limit
For each epistemic model $\boldsymbol{M}$ and each proposition φ true or false at points in $\boldsymbol{M}$, the *announcement limit* $\#(\boldsymbol{M}, \varphi)$ is the first model reached by successive announcements $!\varphi$ that no longer changes after the last announcement is made. That such a limit exists is clear for finite models, since the sequence of successive submodels is non-increasing, but announcement limits also exist in infinite models, where we stipulate that, at limit ordinals, intersections are taken of all previous stages. ■

There are two cases for the limit model. Either it is non-empty, and ***rat*** holds in all nodes, meaning it has become common knowledge (the "self-fulfilling" case), or the limit model is empty, meaning that the negation $\neg \boldsymbol{rat}$ has become false (the "self-refuting" case). Both possibilities occur in concrete puzzles, although generally speaking, rationality assertions such as ***rat*** tend to be self-fulfilling, while the ignorance statement that drives the Muddy Children was self-refuting: at the end, it held nowhere.

Capturing Backward Induction by iterated announcement With general relational strategies, the iterated announcement scenario produces the relational version of Backward Induction defined in Chapter 2.

THEOREM 8.1 In any game tree $\boldsymbol{M}$, $\#(\boldsymbol{M}, \varphi)$ is the actual subtree computed by Backward Induction.

Proof This can be proved directly, but it also follows from a few simple considerations. For a start, it turns out that it is easier to change the definition of the driving assertion ***rat***. We now only demand that the current node was not arrived at directly via a dominated move for one of the players. This does not eliminate nodes further down, and indeed, announcing this repeatedly will make the game tree fall apart into a forest of disjoint subtrees, as is easily seen in the above examples. These forests record more information.

Now we make some simple, but useful auxiliary observations.

Sets of nodes as relations Two simple facts about game trees are as follows.

FACT 8.1 Each subrelation R of the total *move* relation has a unique matching set of nodes $reach(R)$ being the set-theoretic range of R plus the root of the tree.[86]

FACT 8.2 Vice versa, each set X of nodes has a unique corresponding subrelation of the move relation $rel(X)$ consisting of all moves in the tree that end in X.

These facts link the approximation stages BI^k for Backward Induction (i.e., the successive relations computed by our procedure in Chapter 3) and the stages of our public announcement procedure. They are in harmony all the way.

FACT 8.3 For each k, in each game model $\boldsymbol{M}$, $BI^k = rel((!\boldsymbol{rat})^k, \boldsymbol{M})$.

Proof The argument is by induction on k. The base case is obvious: $\boldsymbol{M}$ is still the whole tree, and the relation BI^0 equals *move*. Next, consider the inductive step. If we announce ***rat*** again, we remove all points reached by a move that is dominated

86 Here, the root of the tree is only added for technical convenience.

for at least one player. Clearly, these are just the moves that were cancelled by the corresponding step of the Backward Induction algorithm. ■

It also follows that, for each stage k,

$$reach(BI^k) = ((!\boldsymbol{rat})^k, \boldsymbol{M}).$$

Either way, we conclude that the algorithmic fixed point definition of the Backward Induction procedure and our iterated announcement procedure amount to the very same thing. ■

One might say that our deliberation scenario is just a way of conversationalizing the mathematical fixed point computation of Chapter 2. Still, it is of interest in the following sense. Viewing a game tree as an epistemic model with nodes as worlds, we see how repeated announcement of rationality eventually makes this property true throughout the remaining model: it has made itself into *common knowledge.*

8.2 Another scenario: Beliefs and iterated plausibility upgrade

Next, in addition to knowledge, consider the equally fundamental notion of belief. Many foundational studies in game theory view rationality as choosing a best action given what one believes about the current and future behavior of the players. Indeed, this may be the most widely adopted view today. We first recall the logical analysis of Backward Induction given in Chapter 2, and relate it to this perspective.

THEOREM 8.2 On finite extensive games, the *BI* strategy is the largest subrelation σ of the total *move* relation that has at least one successor at each node, while satisfying the following property for all players i:

> RAT No alternative move for the current player i yields outcomes via further play with σ that are all strictly better for i than all the outcomes resulting from starting at the current move and then playing σ all the way down the tree.

This rationality assumption was a confluence property for action and preference:

$$CF \quad \forall x \forall y \Big(\big(Turn_i(x) \wedge x\sigma y \big) \rightarrow \forall z \big(x\ move\ z \rightarrow \exists u \exists v (end(u) \wedge end(v) \wedge y\sigma^* u \wedge z\sigma^* v \wedge v \leq_i u) \big) \Big)$$

that could be pictured in the following game tree with additional structure:

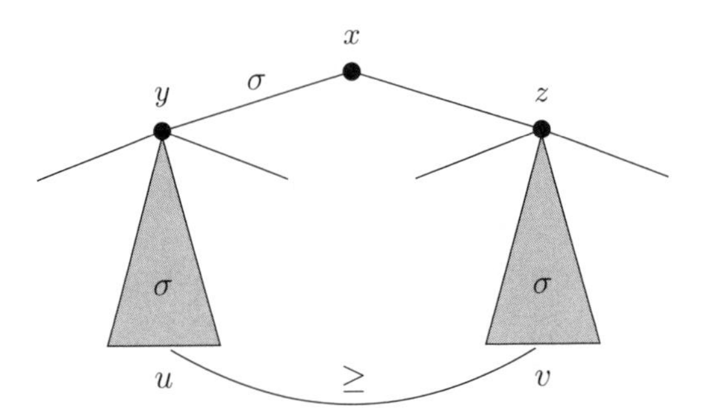

The shaded area is the part that can be reached via further play with our strategy. In the consequent of the syntactic $\forall\forall\exists\exists$ format for CF, all occurrences of σ are positive, and this was the basis for its definability in the standard first-order fixed point logic LFP(FO).

THEOREM 8.3 The *BI* relation is definable in LFP(FO).

This connected game solution with fixed point logics of computation.

The important conceptual point here is that we need not think of the shaded parts as coming from further play by the strategy under consideration. We can also view them as the further histories that players think most plausible, encoding their expectations about the future. This gives us a connection with the dynamic logics of belief in Chapter 7.

Backward Induction in a soft light An appealing take on the relational Backward Induction strategy in terms of beliefs uses soft update that does not eliminate worlds like announcements $!\varphi$, but rearranges the plausibility order between worlds. A typical soft update in Chapter 7 was the radical upgrade $\Uparrow\varphi$ that makes all current φ-worlds best, puts all $\neg\varphi$-worlds underneath, while keeping the old ordering inside these two zones. Now recall our observation that Backward Induction creates expectations for players. The information produced by the algorithm is then in the binary plausibility relations that it creates inductively for players among end nodes in the game, standing for complete histories.

EXAMPLE 8.2 A debatable outcome, hard version
Consider the game that started Part I as a conceptual appetizer. The hard scenario in terms of events $!\boldsymbol{rat}$ removes nodes from the tree that are strictly dominated by siblings as long as this can be done, resulting in the following stages:

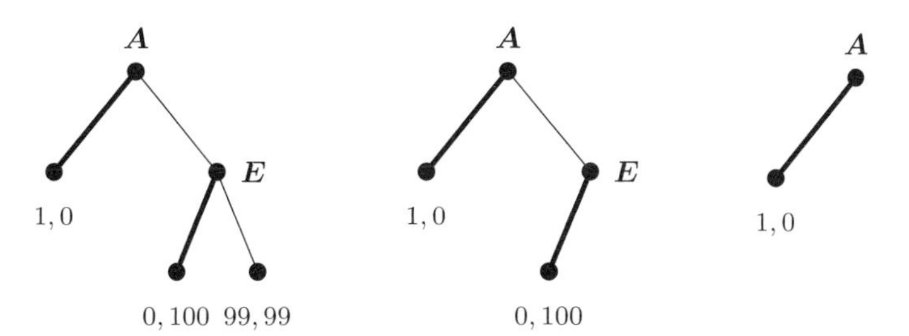

This scenario gives us players' absolute beliefs, but not yet conditional beliefs about what might have happened during off-path play. ∎

By contrast, a soft scenario does not remove nodes but modifies the plausibility relation. We start with all endpoints of the game tree incomparable with respect to plausibility.[87] Next, at each stage, we compare sibling nodes, using an appropriate notion of rationality in beliefs.

DEFINITION 8.3 Rationality in beliefs
A move x for player i *dominates* its sibling y *in beliefs* if the most plausible end nodes reachable after x along any path in the whole game tree are all better for the active player than all the most plausible end nodes reachable in the game after y. *Rationality** (***rat****, for short) is the assertion that no player plays a move that is dominated in beliefs. ∎

Now we perform a relation change that is like an iterated radical upgrade $(\Uparrow\varphi)^*$.

> If x dominates y in beliefs, we make all end nodes from x more plausible than those reachable from y, keeping the old order inside these zones.

This changes the plausibility order, and hence the dominance pattern, so that an iteration can start.[88]

EXAMPLE 8.3 A debatable outcome, soft version
The stages for the soft procedure in the above example are as follows, where we use the letters x, y, and z to stand for the end nodes or histories of the game:

---

87 Other versions of our analysis would have all end nodes initially equiplausible.

88 We omit some details; in general, the plausibility upgrades take place in subtrees only.

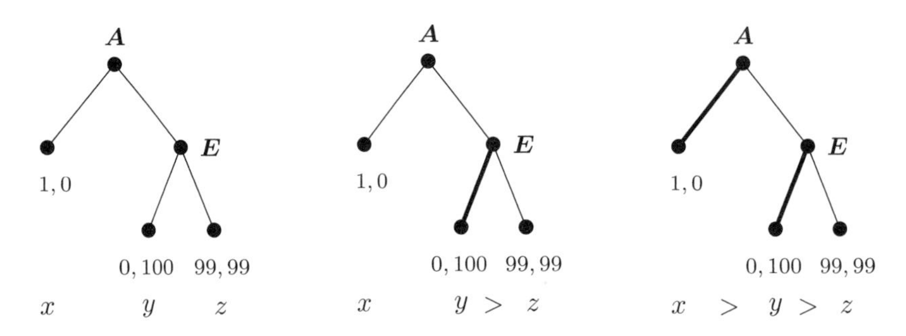

In the first tree, going right is not yet dominated in beliefs for ***A*** by going left. The assertion ***rat****** only has force at ***E***'s turn, and update makes $(0, 100)$ more plausible than $(99, 99)$. After this change, however, going right has become dominated in beliefs, and a new update takes place, making ***A***'s going left most plausible. ■

THEOREM 8.4 On finite trees, the Backward Induction strategy is encoded in the plausibility order for end nodes created in the limit by iterated radical upgrade with rationality-in-belief.

At the end of this procedure, players have acquired *common belief in rationality*. Let us now prove this result, using an idea from Baltag et al. (2009).

Strategies as plausibility relations We first observe that each subrelation R of the total *move* relation induces a total plausibility order $ord(R)$ on leaves x and y of the tree.

DEFINITION 8.4 Leaf order from a sub-move relation
We put $x\, ord(R)\, y$ iff, looking upward at the first node z where the histories of x, y diverged, if x was reached via an R move from z, then so is y. ■

The following property of this order is easy to see by inspection of trees.

FACT 8.4 The relation $ord(R)$ is a total pre-order on leaves.

Moreover, this total order $\leq$ on leaves is "tree-compatible," meaning that, for any two leaves x and y, if z is the first splitting node above x, y as before, all leaves x' reached by taking the move toward x at z stand in the relation $\leq$ to all leaves y' reached by taking the move toward y. This means that there can be no crisscrossing as in the following tree:

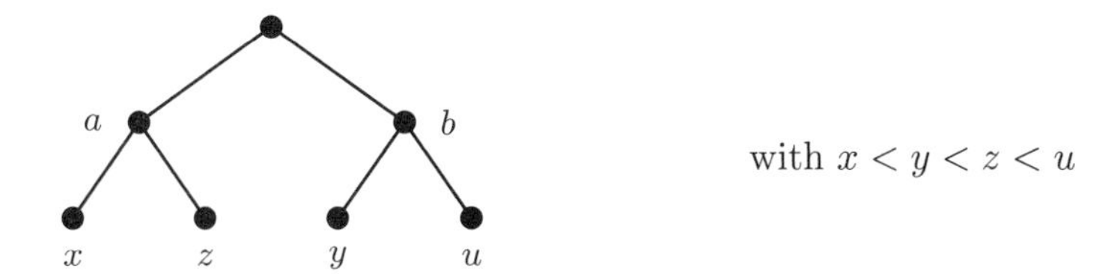

DEFINITION 8.5 Relational strategies from leaf order
Conversely, any tree-compatible total order $\leq$ on leaves induces a subrelation $rel(\leq)$ of the *move* relation, defined by selecting just those available moves at a node z that have the following property: their further available histories lead only to $\leq$-maximal leaves in the total set of leaves that are reachable from z. ■

Together, the maps *rel* and *ord* give a precise meaning to the way in which Baltag et al. (2009) can say that strategies are the same as plausibility relations.[89]

Now we can relate the computation in our upgrade scenario for belief and plausibility to the earlier relational algorithm for Backward Induction. Things are in harmony at each stage.

FACT 8.5 For any game tree $\boldsymbol{M}$ and any k, $rel((\Uparrow \boldsymbol{rat}^*)^k, \boldsymbol{M}) = BI^k$.

Proof The key point was demonstrated in our example of a stepwise solution. When computing a next approximation for the Backward Induction relation according to CF, we drop those moves that are dominated in beliefs by another available one. This has the same effect as making those leaves that are reachable from dominated moves less plausible than those reachable from surviving moves. And that was precisely the earlier upgrade step. ■

Thus, the algorithmic analysis of Backward Induction and its procedural analysis in terms of forming beliefs amount to the same thing. Still, as with iterated announcements, the iterated upgrade scenario has interesting features of its own. One is that, for logicians, it accounts for the genesis of the plausibility orders usually treated as primitives in doxastic logic. Thus, games provide an underpinning for a possible world semantics of belief that is of independent interest.[90]

---

89 Zvesper (2010) relates our dynamic analysis to achieving the sufficient condition for the Backward Induction outcome given in Baltag et al. (2009).

90 We have stated the operations *ord* and *rel* semantically. They can also be viewed as syntactic translations, and then various logical definitions for Backward Induction can be directly transformed into each other (see Gheerbrant 2010 for details).

8.3 Repercussions and extensions

Extensional equivalence, intensional differences Putting together the results in Chapter 2 and those found here, three different approaches to analyzing Backward Induction turn out to amount to the same thing. To us, this means that the notion is stable, and that, in particular, its fixed point definition can serve as a normal form. Still, extensionally equivalent definitions can have interesting intensional differences in terms of what they suggest. For instance, the above analysis of strategy creation and plausibility change illustrates a general conceptual issue: the deep entanglement of agents' beliefs and actions in the foundations of decision and game theory.

Dynamic instead of static foundations As we have said already, a key feature of our dynamic announcement and upgrade scenarios is that they are self-fulfilling: ending in non-empty largest submodels where players have common knowledge or common belief of rationality. Thus, this dynamic style of analysis is a change from the usual static characterizations of Backward Induction in the epistemic foundations of game theory. Common knowledge or belief is not assumed, but produced by the logic.

Announcing rationality, hard or soft, is not the only case of interest. Deliberation can be driven by other statements, and it is not the only activity that falls under this way of thinking.

Other game-theoretic construals We have seen in Chapter 2 how Backward Induction can be viewed as producing various Nash equilibria between functional strategies with unique outputs at nodes. These sharper predictions of behavior corresponded to making assumptions about the player one is against: say, less or more careless about an opponent's interests once the player's own interests have been served. One obvious question is how to extend our current style of analysis to this setting. One way of doing this is by making such assumptions about players explicit, as we will do in Chapters 9 and 10.

Iterated hard announcements Our scenarios have a much broader sweep than may appear from our specific case study. Dégremont & Roy (2009) give a limit analysis for agents that communicate disagreement (cf. Aumann 1976) via iterated hard public announcements of conflicts in belief. They find interesting new scenarios, including the following one.

EXAMPLE 8.4 Stated disagreements in beliefs can switch truth value
Consider two models $\boldsymbol{M}$ and $\boldsymbol{N}$ with actual world s, and accessibility running as indicated. For example, in $\boldsymbol{M}$, $\{s, t, u\}$, $\{v\}$ are epistemic equivalence classes for agent *2* ordered by plausibility. For agent *1*, the epistemic classes are $\{s, v\}$, $\{t\}$, and $\{u\}$. Now, in $\boldsymbol{M}$, $B_1\neg p \wedge B_2 p$ is true at s and t only. Announcing it updates $\boldsymbol{M}$ to $\boldsymbol{N}$, in whose actual world $B_1 p \wedge B_2\neg p$ is true:

$$v, \neg p \quad \leq_1 \quad \boldsymbol{s}, p \quad \leq_2 \quad t, \neg p \quad \leq_2 \quad u, p \qquad\qquad \boldsymbol{s}, p \quad \leq_2 \quad t, \neg p$$

$\boldsymbol{M}$ $\boldsymbol{N}$

Here, truth values are computed entirely following our earlier clauses for belief. ■

This leads to a dynamic-epistemic version of the results in Geanakoplos & Polemarchakis (1982). Any dialogue where agents keep stating whether or not they believe that formula φ is true at the current stage leads to agreement in the limit. If agents share a well-founded plausibility order at the start (their hard information may differ), in the first fixed point, they all believe or all do not believe that φ is true. Dégremont (2010) links these results to syntactic definability of relevant assertions in epistemic fixed point logics.

Iterated soft updates Baltag & Smets (2009) analyze limit behavior of soft announcements, including the radical $\Uparrow \varphi$ and conservative $\uparrow \varphi$ of Chapter 7. Surprises occur, and their flavor is given in the following illustration.

EXAMPLE 8.5 Cycling radical upgrades
Consider a one-agent plausibility model with proposition letters as indicated:

$$s, p \quad \leq \quad t, q \quad \leq \quad \boldsymbol{u}, r$$

Here $\boldsymbol{u}$ is the actual world. Now make the following soft announcement:

$$\Uparrow \big(r \vee (B^{\neg r} q \wedge p) \vee (B^{\neg r} p \wedge q)\big)$$

The formula is true in worlds s and u only, and hence the new pattern becomes:

$$t, q \quad \leq \quad s, p \quad \leq \quad \boldsymbol{u}, r$$

In this new model, the formula $r \vee (B^{\neg r} q \wedge p) \vee (B^{\neg r} p \wedge q)$ is true in t and u only, so radical upgrade returns the original model, starting a cycle:

$$s, p \quad \leq \quad t, q \quad \leq \quad \boldsymbol{u}, r$$

Note that the final world $\boldsymbol{u}$ always stays in place in this scenario. ■

This example provides a formal modeling for two important phenomena. Oscillation in public opinion is a fact of social information flow, which definitely needs further logical study. But the stability of the final world is significant, too. Baltag & Smets (2009) prove that, despite cycles with conditional beliefs, every truthful iterated sequence of radical upgrades stabilizes all absolute factual beliefs. This stabilization result is relevant to the next application of the above ideas.

From belief revision to learning theory Iterated plausibility upgrade also applies to "learning in the limit" as studied in formal learning theory (Kelly 1996), leading to interesting results in Baltag et al. (2011) that throw more light on the earlier mechanism. The set of hypotheses at stake in a learning problem creates an initial epistemic model over which a set of all possible, finite or infinite, histories of signals is then given to the learner, as procedural information about the process of inquiry. The aim of the learning is to find out where the actual history lies in some given partition, corresponding to an issue as discussed in Section 7.7 of Chapter 7. Observing successive signals then triggers either hard or soft information about the actual history.

A key observation is that learning methods can be encoded as plausibility orderings on the initial model that determine agents' beliefs about the issue in response to new input. It then turns out that both iterated public announcement and iterated radical upgrade are universal learning methods, although only radical upgrade still has this feature in the presence of (finitely many) errors in the input stream.[91]

91 Gierasimczuk (2010) has details, including definability results for finite identifiability and identifiability in the limit in the epistemic-temporal language of Chapter 5. One natural question still open is to what extent the initial plausibility order can be mimicked by belief formation on the fly in the process of receiving signals.

8.4 Logical aspects

The preceding topics raise a number of general logical questions. While these may not be relevant to specific games as such, they do provide a broader setting for the kind of analysis that we have proposed in this chapter.

Fixed point logics While relational Backward Induction was definable in the first-order fixed point logic, LFP(FO), this depended on positive occurrence of σ in the syntax of the assertion CF, as noted above. To test the scope of our method, consider a natural maximin variant $\text{BI}^\sharp$ of the Backward Induction algorithm where choices between moves ensure the greatest minimal value. We leave it to the reader to see how the following rule can deviate from our more cautious relational Backward Induction algorithm. This time, the syntactic confluence property $\text{CF}^\sharp$ is

$$\begin{aligned} &\textstyle\bigwedge_i \big(\mathit{Turn}_i(x) \to \\ &\qquad \forall y(x\sigma y \to (x\ \mathit{move}\ y \wedge \forall u((\mathit{end}(u) \wedge y\sigma^* u) \to \\ &\qquad\qquad\qquad\qquad \forall z(x\ \mathit{move}\ z \to \exists z(\mathit{end}(v) \wedge zS\sigma^* v \wedge v \leq_i u)))))\big) \end{aligned}$$

where not all occurrences of the relation symbol S are positive. Hence, $\text{CF}^\sharp$ cannot be used for an immediate fixed point definition in LFP(FO). But we do have a characterization in a slightly extended logical system.

THEOREM 8.5 The relational $\text{BI}^\sharp$ strategy is definable in "first-order deflationary fixed point logic" IFP(FO) using simultaneous fixed points.

Proof A proof can be found in van Benthem & Gheerbrant (2010). ■

Unlike the systems discussed in Chapters 1 and 2, deflationary fixed point logic puts no restrictions on the formulas $\varphi(P)$ used in fixed point operators, but it forces convergence from above by always intersecting the new set with the current approximation. This system is of major interest in understanding computation, but we refer the reader for details to Ebbinghaus & Flum (1999) and Dawar et al. (2004).[92] We will discuss this logic further in Chapter 13 when analyzing solution procedures for strategic games.

---

92 By the results of Gurevich & Shelah (1986) and Kreutzer (2004), $\text{BI}^\sharp$ is still definable in LFP(FO) by using extra predicates. However, their computation no longer matches stages of our algorithm.

Exploiting the well-foundedness of trees Fixed point logics such as LFP(FO) or IFP(FO) work on any model. This generality is attractive when investigating abstract solution procedures for classes of games. However, another approach is possible. Variants of Backward Induction exploit a special feature of finite extensive games, namely, their *well-founded* tree dominance order.[93] Such orders allow recursive definitions without positive occurrence as long as all occurrences of the defined predicate scope under quantifiers looking downward along the ordering.[94] Thus, we get many more recursive definitions on game trees (see Gheerbrant 2010 for matching logics of trees).

Finally, fixed point logics like the above are also conceptually intriguing from the perspective of statics versus dynamics raised in the transition from Part I to Part II. They have a bit of both, since their fixed point operators come with procedures.

Limits in dynamic-epistemic logic We have already seen how limit scenarios for game solution and related tasks such as conversation or learning raise interesting logical issues. We noted in Chapter 7, and again in this chapter, that iterated announcements can end in limit models $\#(\boldsymbol{M}, \varphi)$ where for the first time, a new event $!\varphi$ no longer changes things. These models came in two kinds, non-empty $\#(\boldsymbol{M}, \varphi)$, where φ has become common knowledge, and models where φ has become false in the actual world. Rationality assertions ***rat*** were of the former self-fulfilling kind, while the ignorance statement driving the Muddy Children was of the latter self-refuting kind. Likewise, announcements of disagreement were self-refuting in Dégremont & Roy (2009). Can we say more? Going beyond the few known examples in games and elsewhere, can we say something systematic about the outcome from the syntactic form of the statements and the shape of the initial model?

A simpler related issue is the Learning Problem for public announcement (van Benthem 2011d). Using the notions in Chapter 7, it is easy to show that factual formulas become known upon announcement. But epistemic formulas need not behave in this manner. Moore sentences $p \wedge \neg Kp$ became false when announced truly. Thus, the problem arises of which syntactic shapes of dynamic-epistemic formulas φ guarantee that public announcement makes them known – i.e., $[!\varphi]K\varphi$ is valid. Holliday & Icard (2010) solve this problem.

---

93 Also, all trees allow for recursion over their predecessor ordering toward the root.

94 More precisely, $CF^\sharp$ defines its unique subrelation of the *move* relation by recursion on the well-founded tree order given by the relational composition of the sibling and dominance orderings.

It is an open problem to characterize the self-fulfilling and self-refuting dynamic-epistemic formulas φ syntactically. In fact, this behavior may be so dependent on the initial model, that uniform behavior is rare. Still, van Benthem (2007d) shows how limits of iterated public announcement on epistemic models are definable in a deflationary fixed point extension of the modal μ-calculus. Moreover, behavior gets better for "positive-existential formulas" constructed using this syntax:

$$\text{literals } (\neg)p \mid \wedge \mid \vee \mid \text{existential modalities } \Diamond$$

FACT 8.6 Limit models for positive-existential modal formulas φ have their domain definable by a formula in the modal μ-calculus.

Proof The reason for the fixed-point definability is that positive-existential formulas have monotonic approximation maps in their announcement sequence. This will be covered in more detail in Chapter 13. ■

Both the rationality statement in our Backward Induction analysis and the ignorance statement in the Muddy Children problem are positive-existential. Fact 8.6 then shows that their logic remains simple and decidable. However, the disagreement statement of Dégremont & Roy (2009) is not positive-existential, and yet its limit logic seems simple. We still do not understand in general why rationality is self-fulfilling, and disagreement self-refuting, on the above models.

All of these issues of limit definability and predicting behavior from syntax return with iterated upgrade of plausibility orderings. No general results seem to be known at this interface of dynamic logics and dynamical systems.

Fragments and complexity Moving from definability to proof, which logics are suited for reasoning with our dynamic scenarios? One relevant system seems public announcement logic with Kleene iteration added, PAL*, but this system is highly complex. As noted in Chapter 7, Miller & Moss (2005) prove that validity in PAL* is Π^1_1-complete.

In addition to this source of high complexity, we saw in Chapter 2 that combinations of action and preference satisfying gridlike confluence properties can generate complexity as well. One way out here is that game solution procedures need not use the full power of logical languages for recursive procedures. Which fragments are needed? Moreover, PAL* might be too ambitious, since we may just want to reason about limit models, not all intermediate stages. A second way out, mentioned in Chapter 7, is switching to more general temporal protocol models for dynamic-epistemic logics where complexity may drop.

At the moment, we are not sure what best dynamic logic to use for the theory of game solution, or more generally, a theory of protocols in temporal universes of informational events. The epistemic-temporal and dynamic logics in Fagin et al. (1995), van Benthem et al. (2009a), and Wang (2010) seem relevant, and provide lower-complexity tools for a wide array of tasks.[95]

Infinite models Do our deliberation scenarios extend to infinite games? Infinite ordinal sequences are easy to add to iterations, and fixed point definitions make sense in infinite models. As we saw in Chapter 5, there may be game-theoretic substance to this generalization, since in infinite trees, intuitive reasoning changes direction from backward to forward. An illustration was our recursive analysis of weak determinacy. In this step, the mathematical spirit changed from inductive to co-inductive (Venema 2006), something that also proved attractive for strategies in Chapter 4.

Dynamics in games with imperfect information Many games have imperfect information, with uncertainties for players where they are in the tree. Can our dynamic analysis be extended to this area, where Backward Induction no longer works? We repeat an example from Chapter 3 that the reader may want to try. In the following games, outcome values are written in the order (***A***-value, ***E***-value):

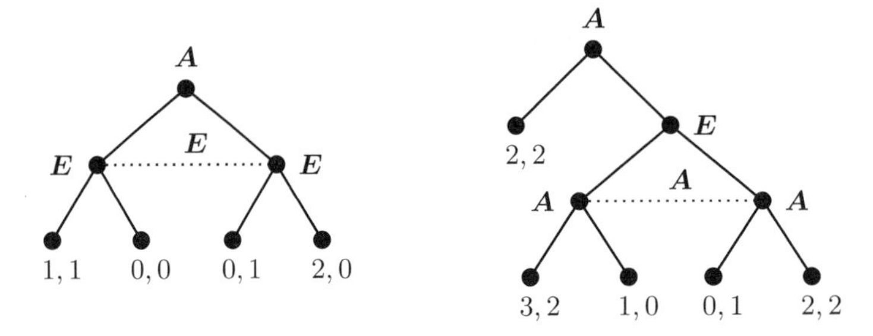

The game to the left yields to our technique of removing dominated moves, but the one to the right raises tricky issues of what ***A*** is telling ***E*** by moving right. Some of these issues will return in Chapter 9 on dynamics in models for games.

95 Incidentally, while we have concentrated on modal formalisms here, all questions raised in this chapter make sense for other logical languages as well.

8.5 Conclusion

We have shown how the deliberation phase of games can be analyzed in terms of dynamic-epistemic iteration scenarios of knowledge update and belief revision. This style of thinking applies more broadly to epistemic puzzles, conversational scenarios, and learning methods. Beyond this conceptual contribution, our analysis raised new technical issues. We extended standard epistemic and doxastic logic with a notion of limit models, an intriguing topic that has hardly been explored yet. Our analysis also creates new bridges between game theory and fixed point logics, a natural mathematics of recursion that fits very well with equilibrium notions.[96]

8.6 Literature

This chapter follows van Benthem (2007d) and van Benthem & Gheerbrant (2010).

Of related work going into more depth, we mention Baltag et al. (2009), Dégremont & Roy (2009), de Bruin (2010), Gheerbrant (2010), Baltag et al. (2011), and Pacuit & Roy (2011).

---

96 We think that our scenarios will also provide a good format for developing alternatives to received views in game theory, although, admittedly, we have not done so here. Chapter 13 will give a few more examples, but even there we do not stray far from orthodoxy.

9
Dynamic-Epistemic Mechanisms at Play

Games involve different sorts of dynamic events. The prior phase of deliberation was studied in Chapter 8 with the help of update mechanisms from Chapter 7 that change beliefs and create expectations. In this chapter, we turn to what happens during actual play. We will use the dynamic techniques of Chapter 7 once more, this time, to look at various kinds of events and information flow as a game proceeds. We do this first by making sense of a given record of a game, in particular, the uncertainty annotations found in imperfect information games. We will make their origins explicit in terms of dynamic-epistemic scenarios that produce these traces, first for knowledge, then also for belief. Next, we discuss updates during play, as the current stage keeps shifting forward. Our vehicle here is the epistemic-temporal forest perspective of Chapter 6 that encodes knowledge and belief for players of a game. While we viewed these models before as complete records of play, they can also serve as information states that can be modified by further events. Next, we show how our techniques can also analyze activities after play, as players ponder a game that has already happened, perhaps rationalizing what they did post facto. We also add some observations on more drastic events such as game change, since the same methods apply. In the course of this analysis, several issues will come to light about how all of these activities and events can work together harmoniously, although we will not present one unified theory. Some further thoughts on the overall program will be found in our concluding Chapter 10 in Part II.

9.1 Retrieving play from a game record

We start with the issue of making the dynamics explicit that lies behind a given game. As we saw in Part I, games annotated with imperfect information links are a record with the traces of some process of actual play. However, we are not told explicitly what relevant events produced the record. How can we tease out what has taken place?

To do this, we use dynamic-epistemic techniques, in particular, the notion of "product update" from Chapter 7. Any finite extensive game arises because events take place to which players may have only partial, and different observational access. This access is described by an "event model" for the moves of the game, perhaps even a sequence of event models, with preconditions on moves encoded explicitly. Through its epistemic accessibilities, this event model encodes the observational powers of players in terms of what they can see of these events. Once we have such an event model, we can decorate the game tree with epistemic links through iterated product update, as happens in dynamic-epistemic update evolution (van Benthem 2001b, Sadzik 2006).[97]

EXAMPLE 9.1 Decorating a game tree by updates

When moving in the following game tree, given as a bare action structure, players can distinguish their own moves, but not all moves of their opponents. Their precise observational powers are described in an accompanying event model for the moves:

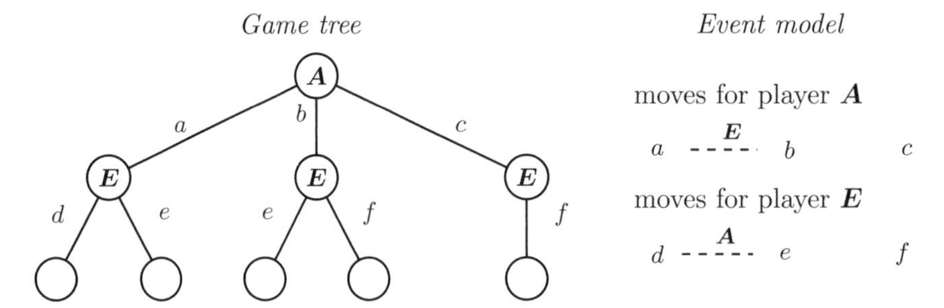

The successive dynamic-epistemic updates that create the uncertainty links in the tree are as follows:

97 Our treatment in this section and the next relies heavily on van Benthem (2011e), Chapter 11, which elaborates on the many interesting technical phenomena that can occur in update evolution.

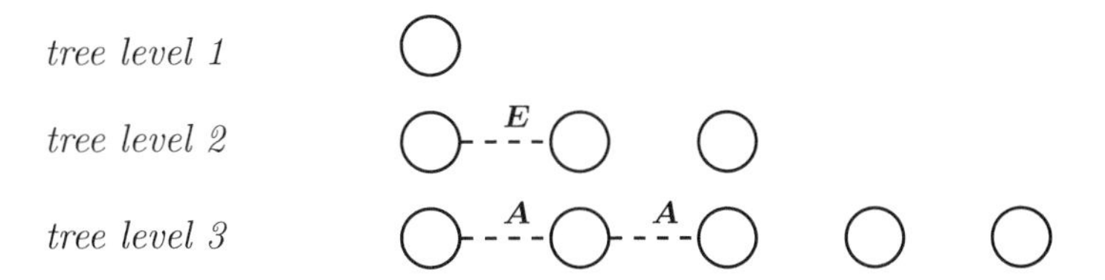

The resulting annotated tree is the following imperfect information game:

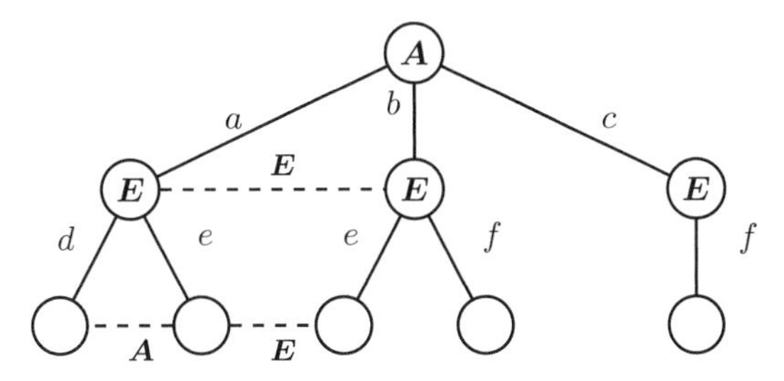

This is a special case of "update evolution," a process that creates successive epistemic models from an initial model by iterated update with some event model, or a sequence of event models. We will soon see in more detail how this works. ■

Game trees that are decorated with uncertainty links in this way are not arbitrary. There are special patterns, if the process worked in this systematic manner. We will now analyze what these are. In order to achieve a higher level of generality, we move from games to the more general epistemic-temporal models of Chapter 5. These models recorded information that agents have about the current protocol.

9.2 A representation for update on epistemic-temporal models

In update evolution, an initial epistemic model $\boldsymbol{M}$ is given, and it then gets transformed by the gradual application of event models $\mathcal{E}_1, \mathcal{E}_2, \ldots$, to form a growing sequence of stages for an epistemic forest model

$$\boldsymbol{M}_0 = \boldsymbol{M}, \;\; \boldsymbol{M}_1 = \boldsymbol{M}_0 \times \mathcal{E}_1, \;\; \boldsymbol{M}_2 = \boldsymbol{M}_1 \times \mathcal{E}_2, \ldots$$

It helps to visualize this in trees, or rather forest models such as the following:

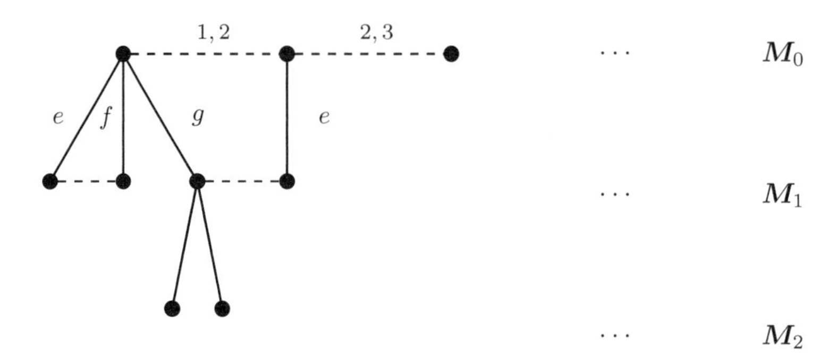

Stages are horizontal, and worlds may extend downward via 0 or more successors. Through product update, worlds in the successive models arise from pair formation, resulting in sequences starting with one world in the initial model $\boldsymbol{M}$ followed by a finite sequence of events that were executable when their turn came. Such worlds are essentially histories in the sense of the epistemic forest models of Chapter 5.

DEFINITION 9.1 Induced epistemic forests
Given a model $\boldsymbol{M}$ and a finite or countable sequence of event models $\mathbb{E}$, the *induced epistemic forest model* $Forest(\boldsymbol{M}, \mathbb{E})$ has as its histories all finite sequences $(w, e_1, \ldots, e_k)$ produced by product update with successive members of the sequence $\mathbb{E}$, with accessibility relations and a valuation defined as in Chapter 7. ■

NOTE We will refer to epistemic forest models as ETL models henceforth, for the sake of brevity. We will also use the abbreviation DEL to remind the reader of dynamic-epistemic product update in the presence of partial observation for different agents.

Induced ETL models have a simple protocol $\mathbb{H}$ (in the sense of Chapter 5) of available histories that determine how the total informational process can evolve, namely, only along the finite sequences that pass the requirements of the DEL update rule. The following three striking properties make these models stand out.

FACT 9.1 ETL models $\boldsymbol{H}$ of the form $Forest(\boldsymbol{M}, \mathbb{E})$ satisfy the following three principles, where quantified variables $h, h', k, \ldots$, range only over histories present in the initial model $\boldsymbol{M}$:

Perfect Recall If $he \sim k$, then there is some f with $k = h'f$ and $h \sim h'$.

Uniform No Miracles If $h \sim k$, and $h'e \sim k'f$, then $he \sim kf$.

Definable Execution The domain of any event e, viewed as a set of nodes in the forest model $\boldsymbol{H}$ is definable in the epistemic base language.

The crucial observation is that these three properties induce the following representation theorem for dynamic-epistemic update evolution in epistemic-temporal forest models.[98]

THEOREM 9.1 For ETL models $\boldsymbol{H}$, the following two conditions are equivalent:

(a) $\boldsymbol{H}$ is isomorphic to some model $Forest(\boldsymbol{M}, \mathsf{IE})$.

(b) $\boldsymbol{H}$ satisfies Perfect Recall, Uniform No Miracles, and Definable Executability.

Proof The direction from (a) to (b) is given by Fact 9.1. Conversely, consider any ETL model $\boldsymbol{H}$ satisfying the three conditions. Define an update sequence as follows:

(a) The initial model $\boldsymbol{M}$ consists of the set of histories in $\boldsymbol{H}$ of length 1, copying their given epistemic accessibilities and valuation.
(b) The event model $\boldsymbol{\mathcal{E}}_k$ is the set of events occurring at tree level $k+1$ in $\boldsymbol{H}$, setting $e \sim f$ if there exist histories s and t of length k with $se \sim tf$ in $\boldsymbol{H}$. The required definability of event preconditions comes from Definable Executability.

We prove by induction that the tree levels $\boldsymbol{H}_k$ at depth k of the ETL model $\boldsymbol{H}$ are isomorphic to the successive epistemic models $\boldsymbol{M}_k = \boldsymbol{M} \times \boldsymbol{\mathcal{E}}_1 \times \ldots \times \boldsymbol{\mathcal{E}}_{k-1}$.

The crucial fact is this, using our definition and the first two properties (writing (s, e) for the history se)

$$(s, e) \sim_{\boldsymbol{H}_k} (t, f) \quad \text{iff} \quad (s, e) \sim_{\boldsymbol{M}_k} (t, f)$$

We first proceed from left to right. By Perfect Recall, $s \sim t$ in $\boldsymbol{H}_{k-1}$, and therefore, by the inductive hypothesis, $s \sim t$ in $\boldsymbol{M}_{k-1}$. Next, by our definition of accessibility, $e \sim f$ in $\boldsymbol{\mathcal{E}}_k$. Then, by the forward half of the DEL product update rule, it follows that $(s, e) \sim_{\boldsymbol{M}_k} (t, f)$.

Next, we proceed from right to left. By the other half of the definition of product update, $s \sim t$ in $\boldsymbol{M}_{k-1}$, and by the inductive hypothesis, $s \sim t$ in $\boldsymbol{H}_{k-1}$. Next, since $e \sim f$, by our definition, there are histories i and j with $ie \sim jf$ in $\boldsymbol{H}_k$. By Uniform No Miracles then, $se \sim tf$ holds in $\boldsymbol{H}$. ■

---

98 Successive versions of this result have appeared in van Benthem (2001b), van Benthem & Liu (1994), and van Benthem et al. (2009a).

This result assumes linguistic definability for preconditions of events e, i.e., the domains of the matching partial functions in the tree $\boldsymbol{H}$.

There is also a purely structural version in terms of a notion from Chapter 1.

THEOREM 9.2 Theorem 9.1 still holds when we replace Definable Executability by Bisimulation Invariance: that is, closure of event domains under all purely epistemic bisimulations of the ETL model $\boldsymbol{H}$.

Proof Two facts from Chapter 1 suffice: (a) epistemically definable sets of worlds are invariant for epistemic bisimulations, and (b) each bisimulation-invariant set has an explicit definition in the infinitary version of the epistemic language.[99] ■

Our results state the essence of DEL update as a mechanism creating epistemic-temporal models. It is about agents with perfect memory, driven by observation only, whose information protocols involve only local epistemic conditions on executability of actions or events.

Caveat Our treatment implies "synchronicity": uncertainty only occurs between worlds at the same tree level. Dégremont et al. (2011) present an important amendment showing how synchronicity is an artifact of the above representation that can be circumvented while keeping the spirit of the other principles, thereby allowing for DEL-induced epistemic forest models in which processes occur asynchronously.

Extended preconditions A mild relaxation of the above definability or invariance requirements for events allows preconditions that refer to the epistemic past beyond local truth. Think of a conversation that forbids repeated assertions: this protocol needs a memory of what was said, which need not be encoded in a local state.

Variety of players In Chapter 3, we looked at different kinds of agents, on a spectrum from perfect recall to memory-free. Our representation can be modified to characterize effects of other update rules, say, for agents with bounded memory.

EXAMPLE 9.2 Update for memory-free agents
A modified product update rule for completely memory-free agents works as follows:

$$(s, e) \sim (t, f) \quad \text{iff} \quad e \sim f$$

Note how the prior worlds play no role at all, only the last event counts. ■

---

99 This only guarantees finite epistemic definitions for preconditions in special models. However, further tightening of conditions has no added value in grasping the essentials.

Alternately, one can think not of agents that are memory-impaired, but of agents following a strategy that uses no memory. This is a general reinterpretation for our results, leaving the nature of the agents open, but capturing their styles of behavior.

9.3 Tracking beliefs over time

The preceding epistemic temporal analysis generalizes to other attitudes that are fundamental to rational agency, and especially, to beliefs that players have, based on their observations. The relevant structures are *epistemic-doxastic-temporal models* (DETL models), that is, branching forests as before, but with nodes in the same epistemic equivalence classes now also ordered by plausibility relations for agents. These expanded forest models interpret belief modalities at finite histories in the manner of Chapter 7. But as before, their belief relations can be very general, and as with knowledge, it makes sense to ask which of them arise as traces of some systematic update scenario.

REMARK Beliefs versus expectations
A clarification may be needed here. As we have seen in Chapter 6, intuitively, beliefs in a game come in two kinds: procedural beliefs about the game and its players, and expectations about the future. The scenario in this section is mainly about the first kind, that is, about beliefs where we stand. We will discuss connections with expectations, as created by Backward Induction or other deliberation methods, later on.

Following van Benthem & Dégremont (2008), we take epistemic-doxastic models $\boldsymbol{M}$ and plausibility event models $\boldsymbol{\mathcal{E}}$ to create products $\boldsymbol{M} \times \boldsymbol{\mathcal{E}}$ whose plausibility relation obeys a notion from Baltag & Smets (2008) introduced in Chapter 7:

$$\text{Priority Rule} \qquad (s, e) \leq (t, f) \;\text{ iff }\; (s \leq t \wedge e \leq f) \vee e < f$$

Let update evolution take place from some initial model along a sequence of plausibility event models $\boldsymbol{\mathcal{E}}_1, \boldsymbol{\mathcal{E}}_2, \ldots$ according to some uniform protocol.[100] The crucial pattern that arises in the forest model created by the successive updates can be described as follows.

---

100 Dégremont (2010) also analyzes pre-orders, and "state-dependent" protocols where the sequence of event models differs across worlds of the initial epistemic model.

FACT 9.2 The histories h, h' and j, j' arising from iterated priority update satisfy the following two principles for any events e and f:

Plausibility Revelation Whenever $je \leq j'f$, then $he \geq h'f$ implies $h \geq h'$.

Plausibility Propagation Whenever $je \leq j'f$, then $h \leq h'$ implies $he \leq h'f$.

Together, these properties express the revision policy in the Priority Rule: its bias toward the last-observed event, but also its conservativity with respect to previous worlds whenever possible given the former priority.

THEOREM 9.3 A DETL model is isomorphic to the update evolution of an epistemic-doxastic model under successive epistemic-plausibility updates iff it satisfies the structural conditions of Section 9.2, with Bisimulation Invariance now for epistemic-doxastic bisimulations, plus Plausibility Revelation and Propagation.

Proof The idea of the proof is as before. Given a DETL forest $\boldsymbol{H}$, we say that

$e \leq f$ in the epistemic plausibility model $\boldsymbol{\mathcal{E}}_k$ if e, f occur at the same tree level k, and there are histories h and h' with $he \leq_{\boldsymbol{H}} h'f$.

One checks inductively, using priority update plus Plausibility Revelation and Plausibility Propagation in the forest $\boldsymbol{H}$, that the given plausibility order in $\boldsymbol{H}$ matches the one computed by sequences of events in the update evolution stages

$$\boldsymbol{M}_{\boldsymbol{H}} \times \boldsymbol{\mathcal{E}}_1 \times \ldots \times \boldsymbol{\mathcal{E}}_k$$

starting from the epistemic plausibility model $\boldsymbol{M}_{\boldsymbol{H}}$ at the bottom of the tree. ■

One can think of the structures described here as generalized imperfect information games, where information sets now also carry plausibility orderings.

Logical languages Languages over these models extend dynamic doxastic logic to a temporal setting as in Chapter 6. In particular, the safe belief modality of Chapter 7 is used in van Benthem & Dégremont (2008) to state correspondences of Plausibility Revelation and Plausibility Propagation with special properties of agents. Dégremont (2010) proves completeness for the logics, relating them to those used in Bonanno (2007), while also connecting doxastic protocols with formal learning theory. We will look at these logics in a different perspective in the next section.

9.4 Witnessing events and adding temporal logic

Having shown how dynamic actions may be retrieved from their traces in a game as usually defined, let us now place the focus directly on actual events that can take place during play.

The simplest example of such an event is the mere playing of a move, and its bare observation. In Chapter 10, we will take up the more sophisticated events discussed at the end of Chapter 6, interpreting moves as intentional, accidental, or otherwise. At this point, we ask how bare observation is dealt with in our dynamic logics. We expect an analysis close to our natural reading of the temporal models of Chapter 5, and this is borne out.

Technically, the topic to come is simpler than the scenarios in the preceding sections, since we focus on perfect information games with events that are publicly observed. Once this is understood, an extension to imperfect information games should be easy to make. Our setting is branching temporal models or epistemic forests, but for many points, the precise choice of models is immaterial. Moreover, we add a theme that was absent from Sections 9.2 and 9.3, namely, axiomatization in logical languages.

Playing a move involves change, as the current point of the forest model shifts, and this can be defined as follows.

DEFINITION 9.2 Updates for moves
An *occurrence* $!e$ of an event e changes the current pointed model $(\boldsymbol{M}, s)$ to the pointed model $(\boldsymbol{M}, se)$, where the distinguished history s moves to se.[101] ■

An equivalence with a standard modality is an immediate consequence.

FACT 9.3 The dynamic modality $\langle !e\rangle\varphi$ is equivalent to $\langle e\rangle\varphi$.

REMARK Existential modalities
Existential modalities often make logical principles in this area a bit easier to state. We will use them for this reason here and later in this chapter. Of course, in contexts with unique events, the difference between existential and universal modalities will be slight. Also, we use $\Diamond$ for the existential knowledge modality, suppressing agent

---

101 Here we assume, as is often done, that moves are fine-grained enough to be unique. Another view of this factual change is that the current history h gets extended to he.

indices for convenience. In the temporal language of Chapter 5 for forest models, the preceding event modality would be $F_e\varphi$, and this is in fact the notation that we will use in this section.

Moves under public observation are a very special case of the potentially much more intricate scenarios provided by DEL product update in preceding sections. Even so, it is illuminating to connect such bare events to earlier topics in this book. First, we show how principles of temporal logic reflect semantic properties of play, as counterparts to earlier laws of dynamic-epistemic logic.

FACT 9.4 The following principle is valid for knowledge and action on forest models with public observation of events:

$$F_e\Diamond\varphi \leftrightarrow (F_e\top \wedge \Diamond F_e\varphi)$$

We have shown the validity of this equivalence in Section 6.5 of Chapter 6. One can view this as a temporal equivalent of the PAL recursion axiom for knowledge, thinking of $!e$ as a public announcement that e has occurred. The precondition for this event is $F_e\top$, which fits with the protocol version of PAL in Chapter 7.

Representation once more The stated principle is not generally valid on all epistemic forest models. In the spirit of Section 9.2, it corresponds to the conjunction of two earlier properties:

Perfect Recall $\forall xyz : ((xR_ey \wedge y \sim_i z) \rightarrow \exists u(x \sim_i u \wedge uR_ez))$
(uncertainty after a move e can only come from earlier uncertainty).

No Miracles $\forall xyuz : ((x \sim_i y \wedge xR_ez \wedge yR_eu) \rightarrow z \sim_i u)$
(uncertainty before a move e must persist after that same move,
i.e., epistemic links can only be broken by different observations).

In slightly modified forms, these properties were also prominent in Chapters 3 and 5. For instance, agents with perfect recall will always know their past history. In their current form, they support a special case of the earlier representation theorem for epistemic forests, where the event models consist of isolated points. We leave the simple details of this specialization to the reader.

Other logical laws Interestingly, for general modal reasons, when events are unique as in this case, further laws will break down postconditions after events:

$$F_e(\varphi \wedge \psi) \leftrightarrow (F_e\varphi \wedge F_e\psi) \qquad F_e\neg\varphi \leftrightarrow (F_e\top \wedge \neg F_e\varphi)$$

General product update What if events happen in a game according to the more general event models of Section 9.2? In that case, the logical axiom is the following, where we assume for convenience that only one event model $\mathcal{E}$ was applied repeatedly, the way things worked in the imperfect information game of Section 9.1. What we get is essentially the characteristic DEL recursion axiom from Chapter 7.

FACT 9.5 On forest models produced by product update, the following is valid:

$$F_e\Diamond\varphi \leftrightarrow (F_e\top \wedge \bigvee\{\Diamond F_f\varphi \mid e \sim f \text{ in } \mathcal{E}\})$$

Beliefs The same analysis applies to beliefs modeled by plausibility relations in DETL forests. We get temporal counterparts to earlier principles of dynamic plausibility change. First, we state some laws governing public events $!e$. These involve absolute and conditional belief, now in existential forms $\langle B\rangle\varphi$, $\langle B\rangle^{\psi}\varphi$, plus a modality $\langle\leq\rangle\varphi$ for safe belief, that can define the other two (cf. Chapter 7).

FACT 9.6 The following principles are valid on doxastic forest models:

(a) $F_e\langle B\rangle^{\psi}\varphi \leftrightarrow (F_e\top \wedge \langle B\rangle^{F_e\psi}F_e\varphi)$

(b) $F_e\langle\leq\rangle\varphi \leftrightarrow (F_e\top \wedge \langle\leq\rangle F_e\varphi)$

Proof We proved (a) in Section 6.9 of Chapter 6 in a model with plausibility running between histories, and (b) is even simpler in that setting. In forest models with plausibility relations between nodes, the argument is similar, using the definition of priority update in the special case when event models just have isolated points. ■

Once more, this result shows an earlier phenomenon: the technical similarity between history- and stage-based models. This shows in the following version for doxastic forests created by priority update with general event models $\mathcal{E}$. (The existential modality $\Diamond$ in our formula is epistemic over all $\sim$-accessible worlds.)

FACT 9.7 The following is valid on forest models created by priority update:

$$F_e\langle\leq\rangle\varphi \leftrightarrow \big(F_e\top \wedge \bigvee(\{\langle\leq\rangle F_f\varphi \mid e \leq f \text{ in } \mathcal{E}\} \vee \{\Diamond F_f\varphi \mid e < f \text{ in } \mathcal{E}\})\big)$$

Note the analogy with the key recursion axiom for belief revision in Chapter 7.

This logic captures the preference propagation and preference revelation that characterized forest models of this sort. For instance, propagation said that, if $je \leq j'f$, then $h \leq h'$ implies $he \leq h'f$. This is expressed by the following temporal formula with an existential modality E over the whole forest, and past modalities:

$$EF_e\langle\leq\rangle P_f^{\cup}\top \rightarrow (\langle\leq\rangle F_f\varphi \rightarrow [e]\langle\leq\rangle\varphi)$$

This can be derived from the preceding recursion principle as a law of the system.

9.5 Help is on the way: Hard information during play

Having reviewed simple events corresponding to official moves of a game, let us now consider a more ambitious scenario. Public moves are not the only events that occur during play. One may also experience events where further information comes in about the structure of the game, or the behavior of other players. There are many such scenarios, and we will discuss a few later on. This means that we now leave the official conception of a game, and we will reflect on this as we proceed.

The simplest new events are public announcements $!\varphi$ of information relevant to play. Here we will apply the standard view of world elimination from $(\boldsymbol{M}, s)$ to $(\boldsymbol{M}|\varphi, s)$ to pointed forest models $\boldsymbol{M}$ with finite histories as worlds. Using the methods of Chapter 7, we can then analyze a wide range of effects on earlier notions of action, knowledge, and belief. We note beforehand that this raises some delicate issues of interpretation, since forest models now become modifiable stages in a dynamic process, rather than universal receptacles of everything that might happen.

Recursion axioms for announcements in forests Effects of basic informational actions can be described explicitly on top of our static game languages.

THEOREM 9.4 The logic of public announcement in forest models is axiomatizable.

Proof As in Chapter 7, the heart of the analysis is finding the right recursion laws for the announcement modality $\langle!\varphi\rangle\psi$. We consider the various postconditions ψ that can occur. The recursion axioms for atoms and Boolean operators are as usual.

Action Consider the pure event structure of forest models (cf. Chapter 1). Here is the law for the atomic modality. For convenience, we use the existential version.

$$\langle!\varphi\rangle\langle a\rangle\psi \leftrightarrow (\varphi \wedge \langle a\rangle\langle!\varphi\rangle\psi)$$

Interestingly, the case of iteration (and hence of future knowledge) $\langle!\varphi\rangle\langle a^*\rangle\psi$ is a bit less obvious, since we now need to make sure that we run along φ-points only.

For the recursion law, we need a system from Chapter 1, PDL with test:[102]

$$\langle !\varphi\rangle\langle a^*\rangle\psi \leftrightarrow (\varphi \wedge \langle(?\varphi\,;\,a)^*\rangle\langle !\varphi\rangle\psi)$$

But then, we really need to show that PDL as a whole has recursion laws for public announcement. This crucially involves the following technical property.

FACT 9.8 The logic PDL with test is closed under relativization.

The simple inductive proof is found at many places in the literature (cf. Harel et al. 2000). In particular, we can now state the following explicit recursion law.

FACT 9.9 $\langle !\varphi\rangle\langle\pi\rangle\psi \leftrightarrow (\varphi \wedge \langle\pi|\varphi\rangle\langle !\varphi\rangle\psi)$ is valid on process graphs.

Here $|\varphi$ is a recursive operation on PDL programs π, surrounding every occurrence of an atomic move a with tests to obtain the program $?\varphi\,;\,a\,;\,?\varphi$. The effect of this transformation can be described as follows.

FACT 9.10 For any PDL program π and formula φ, and any two states s and t in $\boldsymbol{M}|\varphi$, we have that $sR_\pi t$ in $\boldsymbol{M}|\varphi$ iff $sR_{\pi|\varphi}t$ in $\boldsymbol{M}$.

This follows by a straightforward induction on PDL programs, in a simultaneous proof of the standard relativization lemma for formulas.

Knowledge We next consider the epistemic structure of forest models. The only new feature is a recursion law for the epistemic modality. This is just the standard equivalence from PAL, where we write $\langle K\rangle$ for the existential dual modality of K:

$$\langle !\varphi\rangle\langle K\rangle\psi \leftrightarrow (\varphi \wedge \langle K\rangle\langle !\varphi\rangle\psi)$$

Belief Finally, we consider the doxastic structure of forest models with plausibility orderings. The relevant law in this case is one from Chapter 7 for belief change under hard information, stated here for the modality of safe belief:

$$\langle !\varphi\rangle\langle\leq\rangle\psi \leftrightarrow (\varphi \wedge \langle\leq\rangle\langle !\varphi\rangle\psi)$$

This concludes our discussion of all relevant recursion laws for public announcement of facts about a game. A completeness proof clinching Theorem 9.4 now follows on the pattern described in Chapter 7. ∎

102 This is similar to the move to the system E-PDL in van Benthem et al. (2006a).

Thus, updating forest models is an application of standard techniques.[103]

Strategies As we noted in Chapter 4, PDL has the further virtue of explicitly defining strategies for players as programs. Hence the above analysis of PDL programs under public announcement also yields recursion laws for the game modalities $\{\sigma\}\psi$ of Chapters 1, 4, and 11 defined as saying that following strategy σ forces only outcomes in the game that satisfy ψ.[104] This leads to the following result for an extended logic PDL + PAL adding public announcements to PDL.

THEOREM 9.5 PDL + PAL is axiomatized by combining their separate laws while adding the following recursion law: $[!\varphi]\{\sigma\}\psi \leftrightarrow (\varphi \rightarrow \{\sigma|\varphi\}[!\varphi]\psi)$.

Proof One can use Fact 9.10 on relativizing PDL formulas and programs.[105] ■

Strategies involving knowledge In the current information-oriented setting, an interesting kind of strategies are the knowledge programs of Chapter 3, where test conditions have to be known to be true or false by the agent. These programs defined uniform strategies in imperfect information games. How do such programs interfere with getting more information? The above logic PDL + PAL will tell us, but the result is not always what one might expect.

EXAMPLE 9.3 Pitfalls of knowledge-based strategies
One might think that learning more, say by a reliable public announcement, should not affect the effects of a knowledge program. But this is not correct. Consider the knowledge program IF Kp THEN a ELSE b in a model where you do not know if p is the case. It tells you to do b. Now suppose you learn that p is the case, through an announcement $!p$. The knowledge program now switches its recommendation to doing a, which may in fact be disastrous compared to b. ■

---

103 A similar analysis applies to other temporal languages. A recursion law for the earlier branch modality $\exists G$ of Chapter 6 is as follows: $\langle !\varphi\rangle\exists G\psi \leftrightarrow (\varphi \wedge \exists G\langle !\varphi\rangle\psi)$.

104 As with Backward Induction in Chapter 8, we need converse action modalities to define rationality, but our analysis easily extends to PDL with a converse operator.

105 It may be a bit disappointing to see what the preceding result does. The recursion law derives what old plan we should have had in an original game model $\boldsymbol{G}$ to run a given plan σ in the new model $\boldsymbol{G}|\varphi$. But more interesting issues are just the other way around. Let a player have a plan σ in $\boldsymbol{G}$ that guarantees some intended effect φ. Now $\boldsymbol{G}$ changes to a new $\boldsymbol{G}$'. How should σ then be *revised* to get some related effect φ' in $\boldsymbol{G}$'? This seems much harder, as we noted in our discussion of understanding a strategy in Chapter 4. We will encounter similar issues below in our discussion of game change.

Other interesting questions arise when we consider strategies such as the one for Backward Induction that also involves preferences. We will return to this particular issue in Section 9.9.

Conclusion We have shown how forest models support the information dynamics of Chapter 7, allowing us a much richer account of events that can happen during play, from playing moves under uncertainty to receiving extra information beyond observed moves. However, the latter dynamics involved a radical step. Instead of viewing forest models as complete records of everything that has taken place, we now also use them as local states that can be modified when events happen that go beyond the official definition of the game. We will return to this contrast at the end of this chapter.

9.6 Forest murmurs: Soft information during play

The forest dynamics of the preceding section is easily extended to other types of event, since much more can happen in games than just getting hard information.

For a start, continuing with the techniques of Chapter 7, it is easy to add updates with soft information and plausibility change. As we have seen in Chapter 8 on Backward Induction, radical upgrades $\Uparrow\varphi$ may play an important role, and these can come from many sources.

Soft triggers In fact, we usually take information coming to us as soft, unless we trust the source absolutely. This holds for imperfect information games where we can have beliefs about moves that were played on the basis of extra triggers that are suggestive but not conclusive, such as seeing a player draw a card from the stack with a happy smile. But even in public settings like an ordinary conversation, one must take careful note of what is said; but it would be foolish to burn all the bridges of alternative truths.

Complete logics We will not spell out the logic of radical upgrade over forest models, since all the ingredients are in place. Suffice it to say that we need recursion principles for the same structures as above: pure action, knowledge, and belief. For pure action, simple commutations suffice, since plausibility change does not eliminate worlds, and does not affect available actions or epistemic links. There will be axioms showing how pure action affects belief, but these are exactly the same as in Chapter 7, since the models there were fully general.

Forward Induction The issue remains of what soft updates make best sense in play of a game. A case in point is Forward Induction, a way of understanding games that was raised in Chapter 6, and that will be discussed more thoroughly in Chapter 10. Unlike Backward Induction, Forward Induction combines information from two sources: (a) observing the past of the game as played, and (b) analyzing its remaining future part. Note that the way of observing the past need not be a neutral recording of facts, as in our events $!e$ of Section 9.4. There may be more sophisticated intensionally loaded ways, such as

(a) e was played intentionally, on the basis of rationality in beliefs.
(b) e was played by mistake, by deviating from Backward Induction.

When we observe a move e, then, taking e to be rational gives information about the active player i's beliefs: they are such as to make e rational in beliefs. Now we can still view this as public announcement, be it of a more informative statement:

! "move e is rational for i in beliefs."

But since the coloring of the observation, say by rationality, may only be a hypothesis on our part, we may not want to use hard announcements, but the soft information of the radical upgrade $\Uparrow$. This issue will return in Chapter 10.[106]

9.7 Preference change

Many further dynamic events makes sense in games. As we have noted in Chapter 7, rational agency does not just consist of processing information and adjusting knowledge and belief. It also involves maintaining a certain harmony between informational states and agents' preferences, goals, and intentions. Therefore, it makes sense to also study preference change. Triggers for such changes can be diverse. We may obey a command or take a suggestion from some authority, establishing a

106 Using $\Uparrow$ has the additional virtue that we can now make sense of any move, even those that are not rationalizable. A radical upgrade for rationality in beliefs will put those worlds on top where the latter property holds, but when there are no such worlds, it will leave the plausibility order the same. What happens in that case is just a bare observation of the move. Of course, this minimal procedure does not address the issue of how to solve conflicts in our interpretation of behavior, which may involve further updates in terms of changing preferences (see below).

preference where we were indifferent before; we may undergo a spontaneous preference change such as falling in or out of love; or we may adjust our preferences post facto, as in La Fontaine's well-known story of the fox and the sour grapes. We have already seen how to design dynamic logics with recursion axioms for events of preference change that modify betterness ordering between worlds, working on a close analogy with the earlier plausibility changes in models for beliefs. Such logics have been studied in Girard (2008), and especially Liu (2011), to which we refer for details and applications.

Games are a typical instance of the balance between information and preference. Therefore, dynamifying their information content has a natural counterpart in dynamifying their preference structure. There are two kinds of preference change that make sense for games, or models for games. One is the realistic phenomenon of *changing goals*. Intuitively, we often do not play a game for any numerical payoff. Rather, we try to achieve certain qualitative goals, such as winning, or much more refined aims.[107] Things may happen that change players' goals as a game proceeds, and when this structure changes, preferences over outcomes have to be adjusted.[108]

Deontic views Related to this scenario is the natural connection between preferences in games and deontic notions such as obligation and permission. Preferences may come from some moral authority, encoding what one ought to do at the current stage. Normative constraints typically change as moral authorities utter new commands, or pass new laws, and this deontic preference change, too, can be relevant to games. As we noted in Chapter 2 when discussing best action, deontic logics have been applied to games in Tamminga & Kooi (2008), Roy (2011), and in many other publications.

Coda: Preference change or information change? Information and evaluation are not sealed compartments. Sometimes, it is hard to separate preference change from information change. What follows is an example adapted from Liu (2011) studying the entanglement of preference and informational attitudes such as belief. For this entanglement and possible tradeoffs, see also van Benthem (2011e), and Lang & van der Torre (2008).

---

107 The knowledge games of Chapter 22 involve goal structure, and some intriguing results are presented there tying the syntax of goals to the solution behavior of the game.

108 Goal structure suggests the priority graphs of Andréka et al. (2002), used in Girard (2008) and Liu (2011) to model reasons or criteria for preferences (cf. Chapter 7).

EXAMPLE 9.4 Buying a house
A potential buyer likes two houses equally, one located in the Amsterdam neighborhood De Jordaan and the other one in De Pijp. News comes out that a subway line will be built under De Pijp, endangering the house's foundations, and the buyer comes to prefer the De Jordaan house. This starts with an initial model

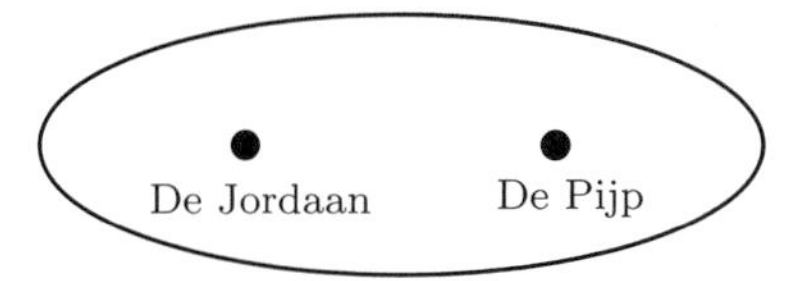

with an indifference relation between the two worlds. The subway line warning triggers a preference change that keeps both worlds, but removes one $\leq$-link, leaving a strictly better De Jordaan house.

Alternately, however, one could describe this scenario purely informationally, in terms of a three-world model with extended options

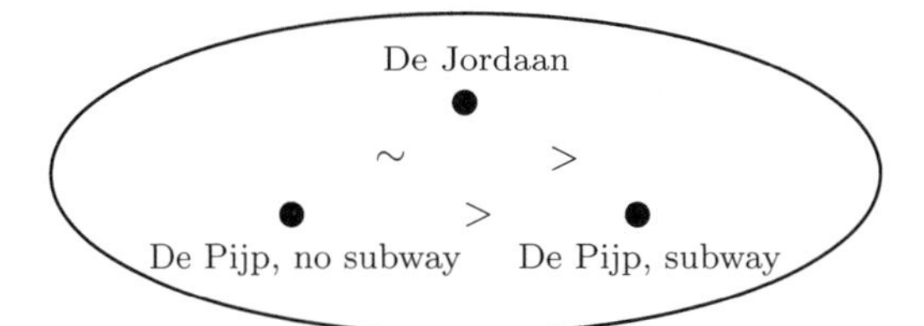

and obvious betterness relations between them. An announcement of "subway" removes the world to the left, yielding the model we got before by upgrading. ■

This example raises issues of choosing worlds in models, the appropriate language to be used in models, and the extent to which one can pre-encode future events. No systematic comparison of the two kinds of dynamics seems to exist so far, but then, we may also view these switches in modeling just as a pleasant convenience.

9.8 Dynamics after play

Preference change also makes sense when we move from playing a game to looking back later at what has happened. Our dynamic logics apply to any activity, from prior deliberation about a game to postmortem analysis of what went wrong.

Rationalization post facto Perhaps the most effective talent of humans is not being rational in their activities, but rationalizing afterward whatever it is that they have done. In this way, a person can even make sense of what looks like irrational behavior. If we just observe the actions of one or more players, constructing preferences on the fly, virtually any behavior can be rationalized. What follows is one of many folklore results that make more precise sense of this.

THEOREM 9.6 Any strategy against the strategy of another player with known preferences can be rationalized by assigning suitable preferences among outcomes.

Proof One algorithm works in a bottom-up fashion. Let $\boldsymbol{E}$ be the player whose moves are to be rationalized. Assume inductively that all subgames for currently available moves have already been rationalized. Now consider the actual move a made by $\boldsymbol{E}$ and its unique subgame G_a. We can make $\boldsymbol{E}$ prefer its outcome more than that of the subgame G_b for any other move b. To do so, we add a number N to all values already assigned to outcomes in G_a. With a large enough N, we can get any outcome in G_a to majorize all outcomes in other subgames G_b.

Here, crucially, adding the same number to all outcome values in G_a does not change any relative preferences in that subgame. Moving upward to turns for the other player $\boldsymbol{A}$, nothing needs to be adjusted for $\boldsymbol{E}$. ■

If we also assume that the player $\boldsymbol{A}$ whose preferences are given never chooses a strictly dominated move, we can even assign preferences to $\boldsymbol{A}$ to match up with Backward Induction play.

EXAMPLE 9.5 Rationalizing a game tree by stipulating preference
Consider the sequence below, where bold arrows are your given moves, and dotted arrows are mine. Numbers at leaves indicate values postulated for you:

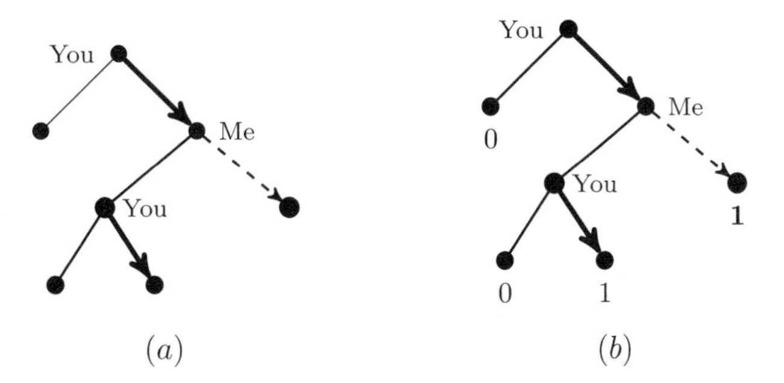

■

Other rationalization algorithms are explored in van Benthem (2007e), adjusting preferences or beliefs, working from the bottom up or from the top down.

Liu (2011) analyzes rationalization scenarios in terms of successive preference changes following observations of moves of a game.[109]

Of course, many other kinds of action make sense in a post-game phase, including updates when new information changes the players' view of what has happened.

9.9 Changing the games themselves

On the road to realism about playing games, there are even more drastic scenarios. Players may not know the game they are playing: a common scenario in daily life. And if they do know the game, they may want to change it. This can happen for several reasons.

Making promises One can break the impasse of a bad Backward Induction solution by changing the game through making promises.

EXAMPLE 9.6 Promises and game change
In the following game from Chapter 8, the Nash equilibrium $(1,0)$ can be avoided by $\boldsymbol{E}$'s promise not to go left. This announcement eliminates histories (we can make this binding by a fine on infractions), and a new equilibrium $(99,99)$ results:

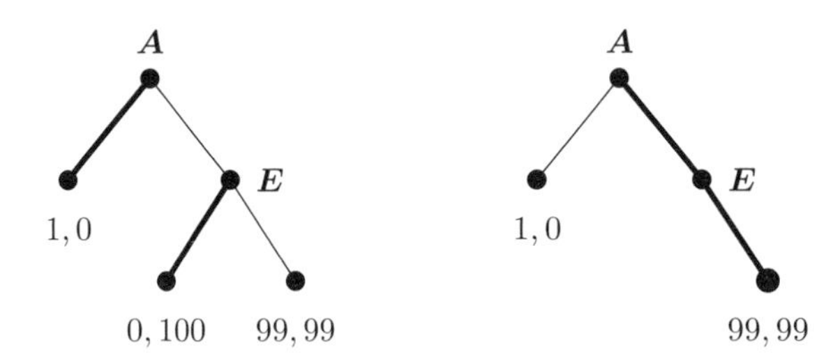

Intriguingly, restricting the freedom of one of the players makes both better off. ■

Game theory has sophisticated analyses of such scenarios, including so-called cheap talk (Osborne & Rubinstein 1994). In principle, these phenomena can be dealt with by our dynamic logics of information, as has been suggested in van Rooij (2004) and van Emde Boas (2011), but little has been done so far. We did see

109 Upgrades are more complex than in the above procedure: see Liu (2011) for details. Adjusting preferences also works if the beliefs of the player are given beforehand, because, as in Chapter 8, a relational strategy or a belief amount to the same thing.

an example of misleading pre-information in our discussion of safe belief in Chapter 7, a phenomenon also known from signaling games. Along a different line, Goranko & Turrini (2012) discusses pre-play negotiation in a logical framework.

Intentions and other scenarios Promises are just one instance of a more general type of event. In the dynamic logic of strategic powers in van Otterloo (2005) games change by announcing intentions or information about players' preferences. This works at the level of the forcing languages of Part III, but one can also write such logics of game change with the above techniques. Roy (2008) uses announcing intentions to obtain simplified procedures for solving strategic games. More special, but more concrete scenarios of extraneous information flow can be found in Parikh et al. (2011) on agents manipulating the knowledge of others during play.

Other kinds of game change The borderline between changing a game or getting more information about a game is tenuous, and in principle, the above dynamic logics on forest models can handle both. Still, later chapters of this book suggest many more kinds of change that have not been studied much, and that could prove tricky. For instance, the sabotage games of Chapter 23 are about deleting moves from a given game. But as we will see there, the corresponding dynamic logic becomes undecidable, and also more complex in other ways. The same might be true with adding moves to games, adding or removing players, or changing individual powers and available coalitions. Changes in admissible sequences of moves also occur in strategy change, a topic studied by PDL techniques in Wang (2010). The general logical study of model-changing transformations relevant to games is still in its infancy (cf. van Benthem 2011e).

The importance of game change Game change may be a drastic scenario. Still, there are many reasons for taking it seriously, as we have seen already in Chapter 4. Very often, we want robust strategies that survive small changes in their current game. If plans cannot be adapted to changing circumstances, they do not seem to be of use. If a student cannot apply a method in new circumstances, the method has not really been grasped.

One more reason for robustness under change arises with the notion of rationality that is so prominent in this book. We have mentioned an analogy with theory structure in mechanics, viewing rationality as an explanatory bridge principle between various observable notions, the way that Newton's axioms postulate forces that tie physical observables together. The main reason why this is such a powerful device in physics is the way this works, not as an ad hoc device on a case-by-case basis, but as a general way of postulating physical forces that make sense across many different mechanical situations. By contrast, many solutions to games are fragile, falling apart as soon as the game changes.[110]

9.10 Conclusion

Chapter 8 showed how the reflection dynamics of pre-play deliberation fits with the dynamic-epistemic logics of Chapter 7. Following up on that, we took dynamic logics of informational events, and preference change, to a wide variety of mid-play processes in and around games. This involved a change from simple annotated game trees to epistemic-doxastic forest models encoding procedural information that players have about the game, and about other players. Our scenarios moved in stages. First we looked at recovering the epistemic and doxastic processes that create forest models, identifying traces satisfying the right conditions of coherence. Our results included several representation theorems. Then we moved to actual dynamics on forest models, providing a number of completeness theorems. This may be viewed as a sort of play over play, and it revealed a space of scenarios where players receive additional information, hard or soft, and even preference changes became legitimate events that can happen during play. These scenarios eventually moved into post-play evaluation, and beyond that, to changing the game itself. In all, we have shown how a wide variety of activities inside or around games can be studied in our dynamic logics.

This picture also raises new issues of its own. How do the various kinds of dynamics studied here fit together? For instance, the deliberation scenarios of Chapter 8 do not move a game forward, whereas the events in this chapter do. This is much like the reality in agency. There is the external time of new information and new events that keep changing the world, but there is also the internal time of rounds of reflection on where we stand at each moment. Somehow, we manage to make both processes happen, without getting stuck in infinite reflection, or in mindless fluttering along with the winds of change. But, just how? We will discuss this issue in Chapter 10, although we cannot promise a final resolution. There are still other general issues raised by this chapter, and we will list a few in the final section on further directions.

---

110 One might object that game change is redundant, since we can put all relevant game changes in one grand "supergame" beforehand, but as said earlier in Chapter 6, this ploy seems entirely unenlightening.

Despite these open problems, we have shown something concrete in this chapter, namely, how dynamic logics help turn play into a rich topic for logical analysis, far beyond the mere description of static game structure found in Part I of this book. Chapter 10 will discuss the resulting contours of a Theory of Play.

9.11 Literature

This chapter is based on van Benthem et al. (2009a), van Benthem & Dégremont (2008), van Benthem (2007e), van Benthem (2011d), and van Benthem (2012b).

In addition, we have mentioned many relevant papers by other authors on logics that are relevant to games, and these can be found throughout the text.

9.12 Further issues

As usual, we list some issues that have not been given their due in this chapter.

Protocols This chapter and the preceding one focused on local events and their preconditions, which ignores the more general protocol structure provided by general epistemic forest models with sets of admissible infinite histories (cf. Fagin et al. 1995). We can use that structure to place global constraints on how local actions can be strung together into longer scenarios. For instance, Backward Induction in Chapter 8 used a strict protocol where only the same assertion can be made at any stage. Other protocols might restrict alternations of hard and soft information, and so on. While this can sometimes still be checked by a local counter, other protocols may need full temporal generality. We know very little about all this, witness the project for a general protocol theory in van Benthem et al. (2009a). Possible frameworks for this include dynamic logic PDL (see Wang 2010 on the related theme of strategy change from Chapter 4). Also, the theory of automata-based strategies for games that has been developed in computational logic (Apt & Grädel 2011, Grädel et al. 2002, and Ramanujam & Simon 2008) seems relevant to a better understanding of these matters (see also Chapter 18).

Histories versus stages Our dynamic logics have mainly taken a local view of temporal structure, with worlds as finite histories, i.e., stages s of some unfolding history that itself might be infinite. But we also saw a use for the temporal logics of Chapters 5 and 6 with histories h themselves as semantic objects, and evaluation of formulas taking place at pairs (h, s). Descriptions of what games achieve often refer to complete histories, as is clear in many chapters of this book. In Chapter 6, we also saw that the two views may bring their own takes on important notions such as belief.[111] The connection between the two perspectives remains to be understood, although we have suggested that they are tantalizingly close under suitable model transformations.

Internalizing external events, thick versus thin models In much of the dynamics of this chapter, epistemic forests change through external events. In particular, external events of public announcement simplified given models, perhaps reducing the forest to just one tree. Another approach is to internalize these external events to events that can happen inside some redesigned "supergame," being an enlarged forest model. For instance, the protocol models of Hoshi (2009) internalize public announcements to actions inside forest models, and this has many benefits, including new protocol versions of PAL and DEL (cf. Chapter 7). On the other hand, internalizing external events blows up model size, going against the spirit of small models advocated in Chapter 6, and against our idea that complexity is best located in the dynamics, rather than in huge static models.

Clearly, there should be general transformations from one kind of model for games to another. Only with these in place would a better understanding arise of the general tradeoffs.[112]

Belief and expectation once more Our update formats for belief in this chapter may still be too poor when we try to really get at beliefs and expectations. An analogy may be helpful with DEL systems for updating probabilities (van Benthem et al. 2009b). The update rule mentioned in Chapter 7 turned out to need three kinds of probabilities. There were prior probabilities among worlds representing our current judgments about their relative weights; and there were observation probabilities expressing uncertainty about which event the agent actually thinks occurred. But

111 Backward Induction was a sort of intermediate scenario here. Even though we focused on its final order for endpoints, the algorithm also creates relative plausibility among stages, namely, among sibling nodes that are successors to a parent node.

112 A related issue is the extent to which update methods on forest models can work on just game trees with plausibility orders, as with Backward Induction. For instance, rough versions of Forward Induction (Chapter 10) can create plausibility order directly on trees. Start from a flat order, and consider successive nodes in the game, where moves partition all reachable outcomes. Then one can upgrade partition cells by radical upgrade for the set of outcomes that majorize, for the active player, at least one outcome for an alternative move (cf. van Benthem 2012b).

in addition to these, there were occurrence probabilities expressing what agents know about the probability that some event occurs, i.e., their knowledge of the process, encoded in probabilistic values of preconditions. The new probabilities for pairs (s, e) of an old world s and an event e then weigh all three factors, and this is important since we need to factor in how probable an event was in a given world to arrive at the right probabilistic information flow in examples such as the Monty Hall problem. Thinking in the same vein, we want a qualitative version of our update logics where we weigh plausibility of current worlds (as in our doxastic models), observation plausibility (as in our event models for priority update), but also, general plausibility reflecting the procedural information encoded in a forest model. At present, no such update systems exist.[113]

Connecting up different dynamics The update methods in the preceding chapters sometimes represent different takes on events. How do they interface? As a concrete example, consider Backward Induction, analyzed in Chapter 8 as a style of deliberation that created plausibility among histories of a game. But in this chapter, we worked with other belief update mechanisms. Can Backward Induction also be obtained via, say, priority upgrade of plausibility models? The answer is negative. The expectation pattern created does not satisfy the characteristic priority upgrade conditions of Preference Revelation and Preference Propagation.[114]

Even so, Backward Induction and our update mechanisms over forest models can live in harmony. We can imagine that Backward Induction has created initial expectations, and we now feed these into the real-time update process as follows. We first create an initial model whose worlds are the histories of the game, ordered

---

113 The challenge of finding plausibility-based update rules for procedural information, even in very simple probabilistic scenarios, is discussed in van Benthem (2012d). The difficulty is in finding qualitative analogues for the two different roles that numbers play in probability: giving degrees of strength for evidence or beliefs, but also weighing and mixing values in update.

114 In a game tree with Backward Induction-style plausibility, there can be a node x with two moves e and f where e leads to more plausible outcomes than f, while at some sister node y of x with the same moves e and f, this order reverses. For instance, assign different payoffs to the two occurrences of e and f. This stipulation also highlights the intuitive difference: Backward Induction looks ahead, while priority upgrade looks at past and present observations.

by the plausibility relation that was created.[115] This is then the starting point for real play. In perfect information games, this consists of publicly observable moves $!e$ as discussed earlier. But with imperfect observation, further information may come in as well, either hard or soft, that can override the initial plausibility order. For instance, if we see a move that could be either an a on the Backward Induction path or an off-path move b, but with highest plausibility it is b, then we will not know where we are in the game tree any more, but the higher plausibility will be for being off-path. We will return to these interface issues in Chapter 10.[116]

Technical issues for model change Our analysis also leads to a more mathematical issue reminiscent of one raised for strategies in Chapter 4. Dynamic-epistemic logics are by no means the last word in studying model change. A standard issue from model theory fits our earlier concerns about robustness across games. When passing from one model to another where objects or facts have changed, one basic question is which statements survive such a transition. A typical example is the Los-Tarski Preservation Theorem: the first-order formulas whose truth is preserved under model extensions are precisely those definable in a purely existential syntactic quantifier form. As another example, in Chapter 1, the first-order formulas that are invariant for bisimulation were precisely those definable by a modal formula. Similar questions make sense for models and languages for games. Which assertions about games in our languages survive the changes that were relevant in this chapter? For instance, as noted in Chapter 4, it is still unknown what a Los-Tarski theorem should look like for PDL, although one has been found for the modal μ-calculus using automata techniques (D'Agostino & Hollenberg 2000).

---

115 The adequacy of this transformation was one of the remodeling issues raised in Chapter 6. Accordingly, we do not claim that this initial model is the optimal encoding of the Backward Induction-annotated game.

116 Similar issues arise when zooming out from Backward Induction to best action (cf. Chapters 2 and 8). Interfacing with dynamics in play involves the relativization used for common knowledge and conditional belief in Chapter 7, with recursion laws $\langle !\varphi\rangle\langle best\rangle^{\alpha}\psi \leftrightarrow (\varphi \wedge \langle best\rangle^{\varphi\wedge\langle !\varphi\rangle\alpha}\langle !\varphi\rangle\psi)$. For instance, in our example of game change by a promise, let p denote all nodes except the end node with $(0, 100)$, and q just the end node with $(99, 99)$. Recomputing Backward Induction in the smaller game, we saw that q resulted. Analyzing the matching assertion $\langle !p\rangle\langle best\rangle\langle best\rangle q$ by the given law, we find that it reduces to $p \wedge \langle best\rangle^{p}(p \wedge \langle best\rangle^{p} q)$. To see that the latter is true, recursion laws no longer help, and we must understand the relativized best-move modality. This shifts the burden to understanding the conditional notion $\langle !\varphi\rangle\langle best\rangle^{\alpha}\psi$ plus its static pre-encoding. This refines our conjecture in Chapter 2 about axiomatizing best action.

Beyond literal preservation, translation of assertions makes sense as well, deriving modified strategies for players, or new descriptions of their powers, in games that have undergone some suitable simple definable change. No general theory seems to exist that can be applied to games without further ado.

Dynamic logic and signaling games This chapter has proposed a richer logical style of analysis for games than the usual static ones. Still, this is a program, and it remains to be applied more systematically in game theory. One obvious area for applying dynamic techniques would be the theory of signaling games (Lewis 1969, Cho & Kreps 1987, Osborne & Rubinstein 1994, Skyrms 1996, van Rooij 2004), where agents send signals about the state of the world in which a game takes place.

10
Toward a Theory of Play

It is time to pull together some threads from the preceding chapters. Chapter 7 showed how we have logical systems at our disposal for modeling actions and events that make information flow, and also for changing preferences in a concomitant stream of evaluation. In Chapters 8 and 9, we explored how this rich array of logical tools applies to various sorts of dynamic events that happen within, or alongside games. But where is all this heading? The purpose of this discursive final chapter of Part II is to combine all of these threads by thinking about a general enterprise that seems to be emerging at the interface of logic and games, which might be called a "Theory of Play." We will not offer an established theory, but a program. Based on the topics in earlier chapters, we will discuss major issues in its design, but also critical points, and further repercussions for the logical study of social activities, gamelike or not.

10.1 Dynamics in games

Many activities happen when people engage in playing a game, with phases running from before to during and after. Chapter 8 was about pregame deliberation, and our emphasis was on procedures that create prior expectations. Taking Backward Induction as our running example, we saw how to construe deliberation as a dynamic process of iterated belief revision, working with extensive game trees expanded with plausibility relations. But this was just a case study. This initial phase contains many natural activities of deliberation and planning. Next, during actual play, many further things can happen. In Chapter 9, we saw a wide variety of events, such as playing moves, public or private observations of moves, but also events of receiving further information about the game and its players, and of changing beliefs or preferences. Again, this list is not complete, as it all depends on the level of detail that is of interest, whether coarser or finer. For instance, many authors assume that players know their own strategies, but this amounts to assuming that a decision has already been taken, whereas, at a finer level, acts of decision themselves might be objects of study.

We have shown how all of these events can be dealt with in dynamic logics, in both hard and soft varieties, where the structures undergoing change were usually epistemic-doxastic forest models encoding procedural and social information that players have about the game and about each other. Finally, turning to the post-game phase of reflection, the same logics dealt with rationalization or other activities that take place afterward. In fact, they even dealt with more drastic events that act as game changers.

This creates a huge space of possibilities. In what follows, we try to get some focus by doing a case study, looking at how one would normally play a game, while identifying a number of issues arising that are of more general interest for the role of logic. Our case is a confrontation of Backward Induction as a style of deliberation to reasoning in actual play, with Forward Induction as an alternative style. Issues that will arise include belief revision, managing hypotheses about other agents, options in modeling update steps, the role of simple cues in normal ways of speaking about social action, and the resulting desiderata on design of logical systems. After that, we will discuss general issues in a Theory of Play, including the role of agent diversity, the tension between sophisticated description and model complexity, and possible repercussions beyond games to other fields interested in a broader theory of social action.

10.2 Problems of fit: From deliberation to actual play

Our extensive discussion in Chapter 8 may have suggested that Backward Induction is the view of games officially endorsed by logic. But this is not true. While its elegant bridge principle of rationality is appealing, and hence should not be given up lightly, exploring alternatives makes sense. Our limit scenarios of iterated hard and soft announcements $(!\varphi)^{\#}$ and $(\Uparrow\varphi)^{\#}$ would work with any formula φ in our language, whether it produces the Backward Induction solution or not, and Muddy Children-style alternatives such as "iterating worries" could well be another option. In Chapter 9, our dynamic logics supported many sorts of events that can override the initial expectations created by a solution algorithm, for instance, in the form of soft updates that change the plausibility ordering. We now turn to what may be the most radical challenge.

From deliberation to actual play In moving from deliberation to reality, some well-known problems arise. Let us accept Backward Induction as a prior deliberation procedure. What about the dynamics of actual play, when these expectations are confronted with what we actually observe? One might think this is a simple matter of following the virtual moves computed in the deliberation phase. But often this makes little sense when run in this opposite direction. We expect a player to follow the path computed by Backward Induction. So, if the player does not, we must perhaps revise our beliefs, and one way of doing that is precisely having second thoughts about the player's style of deliberative reasoning.

This issue has been dramatized in the so-called paradox of Backward Induction. Why would a player who has deviated from the computed path return to Backward Induction later on (Bicchieri 1988)?

EXAMPLE 10.1 Expectations meet facts
Consider the following game, with outcomes evaluated by $\boldsymbol{A}$ and $\boldsymbol{E}$ as indicated:

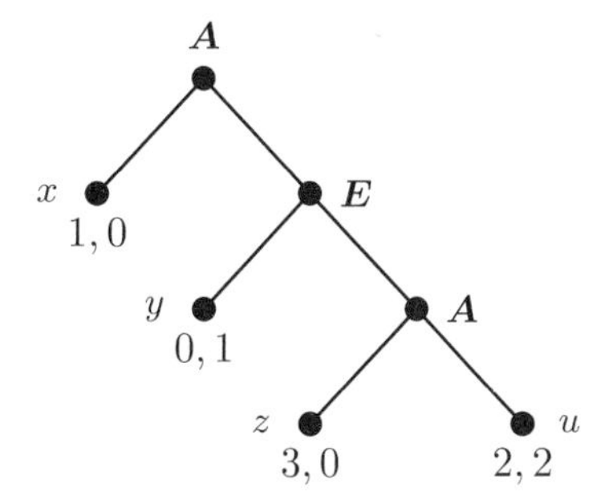

Backward Induction tells us that $\boldsymbol{A}$ will go left at the start. So, if $\boldsymbol{A}$ plays *right*, what should $\boldsymbol{E}$ conclude from this surprising observation? $\boldsymbol{E}$ might still assume that $\boldsymbol{A}$ will play Backward Induction afterward, thinking that a mistake was made. But this is not always a plausible response. In many social scenarios, a participant $\boldsymbol{E}$ might rather think that the observed deviant move refutes the original style of deliberation about $\boldsymbol{A}$, who may now be thought to be on to something else, maybe hoping to arrive at the outcome $(2, 2)$. ■

In other words, although the Backward Induction strategy *bi* is easy to compute, does it make sense when one tries to interpret off-equilibrium parts of the game? The assumption that Backward Induction will prevail later on, no matter how often we observe deviations, is the technical core of the results by Aumann (1995), and within dynamic-epistemic logic, Baltag et al. (2009), that characterize *bi*.[117]

From the game to policies of its players But clearly, this is not the only way of revising beliefs here. Player ***E*** could have reasonable other responses, such as

> ***A*** is saying he wants me to go right, and I will be rewarded if I cooperate.
> ***A*** is an automaton with a rightward tendency, and cannot act otherwise.
> ***A*** believes that ***E*** will go right all the time.

We cannot tell which response is coming, unless we also know at least players' *belief revision policies* (Stalnaker 1999, Halpern 2001). This seems right, but adding this parameter is a momentous move in the foundations of game theory. Halpern (2001) is a masterful logical analysis of how Stalnaker's proposal undermines famous results such as Aumann's Theorem that common knowledge of rationality implies that the Backward Induction path is played (Aumann 1995). This is true on the standard notion of rationality, but it is false on Stalnaker's generalized view, where rationality refers to players' beliefs about the current strategy profile, which can change as surprise moves are played.

What does all this mean for our approach? Our dynamic logics are well-suited to analyzing belief change, so technically, accommodating players' revision policies poses no problems. Our logics are also welcoming: they do not impose any particular policy. The systems of Chapter 7 supported many update methods, of which the one generating Backward Induction is just one. Our subsequent analysis in Chapter 8 may have seemed to favor this scenario, but this was for concreteness and as a technical proof-of-concept, and we identified choice points along the road, for instance, concerning the strong uniformity assumptions underlying the *bi* strategy.

Broader issues in social action But these issues are not just technical. They reflect real challenges for a logic of social action. Our expectations may be based on prior deliberation, including scenarios that we think will not occur. But what if the unexpected happens? A move considered hypothetically may impact us quite differently once it has occurred: cold feet are ubiquitous in social life. How can a priori styles of deliberation and actual play of a game that is constantly being updated with observed events be in harmony? There are often deviations from this harmony in practice, but it is definitely an interesting ideal.

---

117 Baltag et al. (2009) call this attitude an incurable belief in rationality.

Player types Belief revision policies need not be our only focus in studying these phenomena. Social action involves many kinds of hypothesis about other agents. For example, an agent may perform an action based on assumptions about the memory capacity of other agents or their learning behavior. This point is acknowledged in game theory. As we saw in Chapter 6, "type spaces" are meant to encode all hypotheses that players might have about others. But our problem is the distance from reality, where we get by with simpler views of the social scenario we are in. Hence, we have worked with simpler models for games, describing the dynamics of relevant events instead of having things be prepackaged.

To make these new perspectives more concrete, we will now discuss a case study of social reasoning, pursuing a simple view that lends itself to dynamic logical analysis. Our treatment will be light, and we do not offer a final proposal. After the case study, we take stock of some general features of a logical Theory of Play.

10.3 Forming beliefs from observations

Let us discuss the earlier example in a bit more detail, focusing on its update aspects. In the bottom-up Backward Induction computations of Chapter 8, we omitted the history of play: it did not matter for our expectations how we arrived at the present node. But just think about how you yourself operate in social scenarios: the past is normally informative, and we do need to factor it in.

EXAMPLE 10.2 Interpreting a game
Here is a simple example, varying on an earlier game:

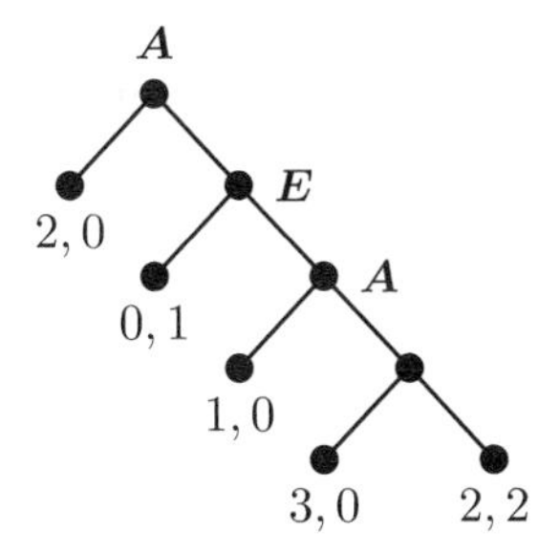

Suppose that ***A*** chooses *right* at the start. Assuming that ***A*** is rational-in-beliefs, this informs ***E*** about ***A***'s beliefs and/or intentions. Clearly, ***A*** does not expect ***E*** to go left at the first turn, because then, playing *left* at the start would have been better. For the same reason, ***A*** does not intend to go *left* after that at the second turn. And we may furthermore assume that ***A*** believes that ***E*** will go *right* at the end, as that is the only way the opening move makes sense. All this might induce ***E*** to go *right* at the first turn, although one hesitates to predict what move will be chosen at the end. ■

The point here is not that we have one simple rule replacing Backward Induction. It is rather that the past is informative, telling us which choices players made or avoided in coming here. Our observations and our expectations work together. Now we do not have one unique way of doing this, but there are clearly intuitive scenarios reflecting our own practice.

Games with a history The change needed is easily pictured. We now look at games ***M*** with a distinguished point s indicating how far actual play has progressed:

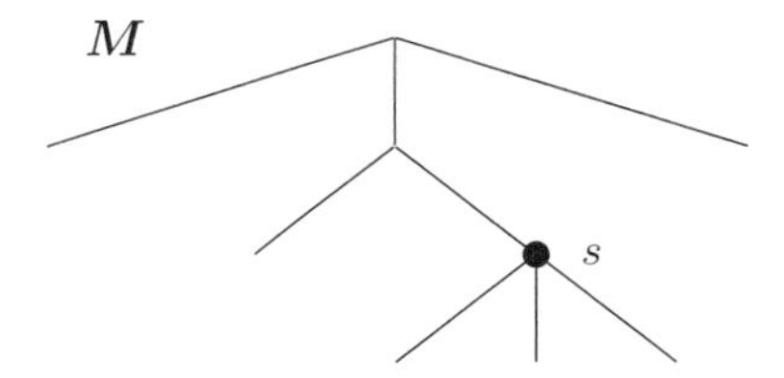

Thus, at least in games of perfect information, at the actual stage, we know what has happened, and players can let their behavior depend on a mixture of two things:

(a) What players have done so far in the context of the larger game.

(b) What the structure of the remaining game looks like.

We will see more complex models later, but for now, we will discuss simple scenarios.

Ways of taking observed behavior How will players change their expectations as the black dot moves along a game? Readers can hone their intuition on simple cases of making decisions.

EXAMPLE 10.3 Making sense of decisions
We picture two simple scenarios:

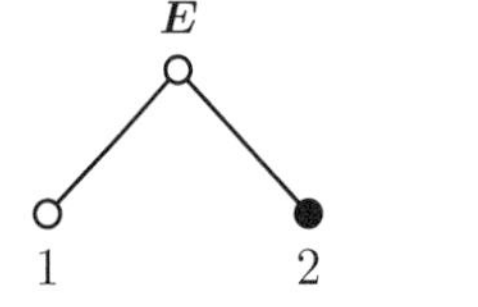

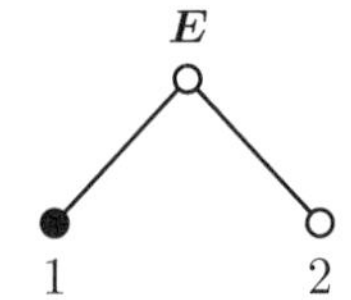

To the left we see the end of a basic rational decision. To the right we see a "stupid move," probably regretted by ***E*** once made.[118] Here are a few more complex cases:

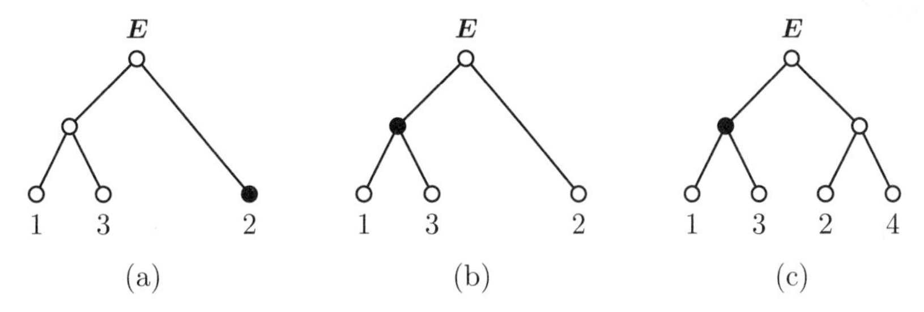

The play observed in game (a) may be considered rational by ascribing a belief to ***E*** that choosing left would have resulted in outcome 1. Game (b) may be rationalized by ascribing a belief to ***E*** that the game will now reach 3. Finally, game (c) suggests that ***E*** thinks that 3 will be reached, while 2 would have been reached if the initial move had been to the right. ■

Rationalizing There are many options for making sense of observed behavior in the preceding example. Actual moves may be considered to be mistakes, or even as signs of stupidity. They may also be taken to be smart, but how smart depends on assumptions. Let us discuss one natural tendency when making sense of what others do. We stay close to rationality, and only drop that assumption about others when forced:

> *Rationalizing* By playing a publicly observable move, a player gives information about beliefs. These beliefs rule out that the player's actual move is strictly dominated in beliefs.

This will only work if the player is minimally rational, not choosing a move that is strictly dominated under all circumstances, i.e., under every possible continuation of the game.[119] Only if rationalization does not work might we consider stupidity or some other obstacle. This suggests a ladder of interpretative hypotheses, where we move to the next step only when forced. But for now, let us stay with rationalization.

---

118 Regret seems central to social life, although it may work best with iterated games.

119 This "weak rationality" avoids stupid things. One could also make the stronger assumption of "strong rationality," where the agent thinks the chosen move is best. Chapter 12 considers counterparts of both notions for strategic form games.

Implementing rationalization still depends on additional assumptions about the belief structure of the agent. In the presence of minimal rationality, and beliefs allowing for ties, one way of rationalizing is simply to assume that the agent considers all continuations to be equally plausible. No observed move can then be strictly dominated in beliefs. But assuming that agents have one unique most plausible history in mind, more information comes out of an observation. Unique belief plus strong rationality were in fact reasons for suggesting that in the above game (c), agent ***E*** believes that 3 will be reached, and that 2 would have been reached after going right. Fixing stipulations, we get various algorithms, all proceeding on the principle that moves reveal beliefs about the future.[120]

Forward Induction Scenarios such as the above are sometimes called Forward Induction (cf. Brandenburger 2007), suggesting a simple computational change to the Backward Induction algorithm, or relatives for strategic games such as Iterated Removal of Weakly Dominated Strategies, that will remedy the earlier-noted deficiencies. Whether this remedy is possible or not, we now explore a relevant question about switching algorithms.

Can dropping Backward Induction be for the best? Sometimes, dropping Backward Induction may be advisable for rational players.

Example 10.4 A Forward Induction scenario
The following game is adapted from Perea (2011):

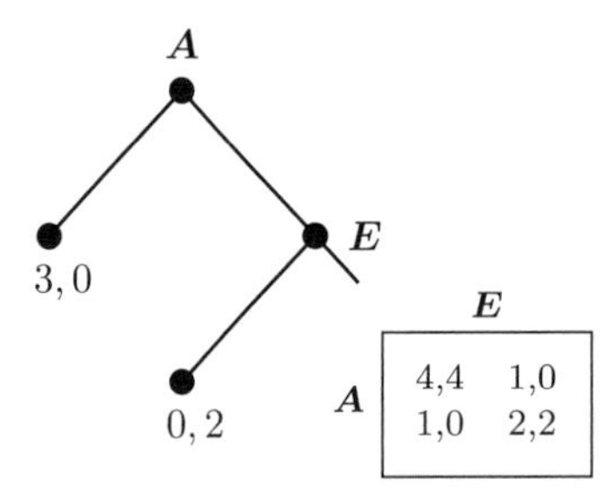

In the matrix game, no move dominates any other. Hence, ***E*** considers all outcomes possible in Backward Induction. Then going left is safer than going right,

120 Incidentally, the beliefs of a player ***E*** do double duty in this setting. At a turn for the other player ***A***, they are real beliefs about what ***A*** will do. But with a turn for ***E***, they are more like intentions.

and hence ***A*** should go left at the start. But if ***E*** rationalizes, and sees ***A*** go right, there is extra information at the next point: ***A*** expects to do better than 3, which can only happen by playing *up* in the matrix game. This tells ***E*** to go there, too, and play *left*, doing better than the Backward Induction move giving ***E*** just 2. ■

Top-level views and fuzzy endings? While this scenario is interesting, it needs a convincing interpretation for the final matrix game. Often we do not know the complete game, or it is beyond our powers to represent it. We only know some top-level structure. The matrix game of imperfect information then gives a rough image of what might happen afterward. This goes against the spirit of analyzing games all the way to rock bottom, the way we did in Chapters 2 and 8. But it is like solution algorithms in AI that work with some top-level game analysis plus general heuristic values for unanalyzed later parts of the game (cf. Schaeffer & van den Herik 2002). This is also closer to the way we often reason in real life.[121]

Coda: Extending rationality Our discussion is not meant to suggest that rationalization in the sense of Forward Induction is the only alternative to Backward Induction. An interesting alternative is minimizing regret, proposed in Halpern & Pass (2012), and studied using dynamic-epistemic logic in Cui (2012). Yet another view is suggested in van Benthem (2007d), based on the social phenomenon of returning favors, compensating agents for risks they have run for improving the payoffs for both of us. This points at perhaps the most general view to emerge from our discussion in this section, that of weighing both future and past. A player acts in a "responsibly rational" way by taking care of that player's own future interests while giving past interactions with the opponent their proper due. Cooperation deserves consideration; lack of cooperation justifies neglect. This is how most of us navigate through life, and it would make sense in many games.

10.4 Logical aspects: Models and update actions

We started by analyzing Backward Induction as a process of deliberation prior to playing a game. We have now discussed how to analyze a game as it is being played, making a case for considering alternatives using information from the past. We now explore a few further logical aspects of this setting, tying in with earlier chapters.

---

121 The general algorithm in Perea (2011) raises further logical issues that we forego here.

Choosing the models Models in Section 10.3 were pointed trees that mark where play currently stands. The pointed forest models of Chapters 5, 6, and 9 were like this, although more general in allowing for variation in what players know or believe, as encoded in epistemic and plausibility relations between nodes. This is the generality that we need. For instance, to describe the earlier rationalization procedure, we need general forest models, since beliefs need not have a simple *bi*-style encoding as best moves. Quite different beliefs for an agent may rationalize an observed move, and these need not have a weakest common case.

EXAMPLE 10.5 Incompatible hypotheses about belief
Consider the following tree. There are two incompatible ways of rationalizing player ***E***'s *left* move arriving at the black dot:

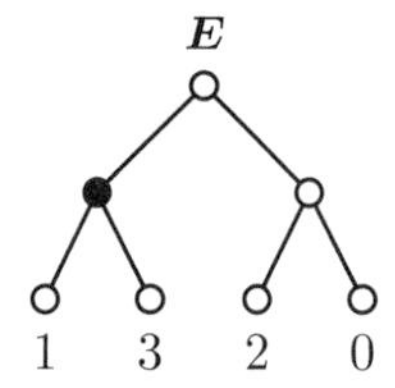

One option assumes that ***E*** expects 3 on the left, and any outcome on the right. Another option lets ***E*** expect any outcome on the left, and 0 on the right. ■

Updating thinner or thicker models We also have different update scenarios for the earlier games, depending on whether we choose thinner or thicker models.

EXAMPLE 10.6 Expectation meets facts, revisited
Consider this earlier game from Example 10.1 in Section 10.2:

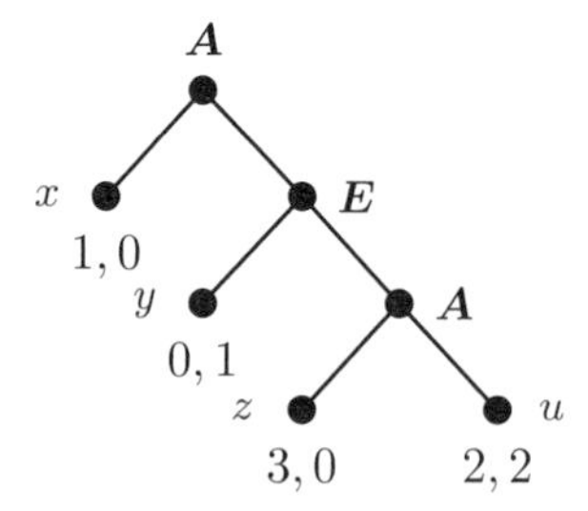

One way of interpreting the *right* moves in the game is as follows. We start with equiplausibility for all branches for all players. ***A***'s first *right* move triggers an upgrade, making histories RRR, RRL more plausible than RL in the ordering $\leq_{\boldsymbol{A}}$. Next, $\boldsymbol{E}$'s *right* move makes RRR more plausible in $\leq_{\boldsymbol{E}}$ than RRL and RL. ■

However, sometimes a simple plausibility shift may not do the job, and we will have to update more complex models. This may be seen in the following example, a variation on a recurrent illustration throughout this book.

EXAMPLE 10.7 Updating thicker models
Consider the following game:

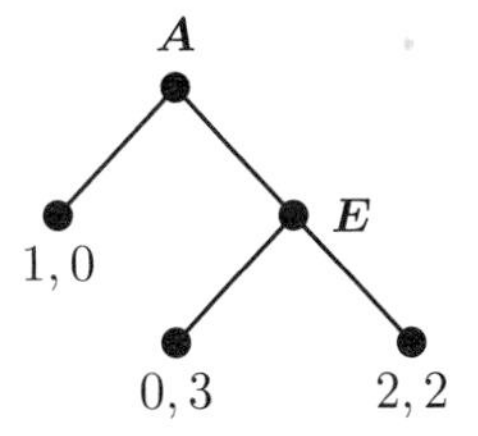

The four possible strategy profiles pre-encode all possible response patterns. These form a forest with epistemic links as indicated, marking a moment when players have decided on their own strategy, but do not yet know what the other will do. We only indicate the top-level links. Lower links will match corresponding lower nodes, reflecting our assumptions of perfect recall and public observation.[122]

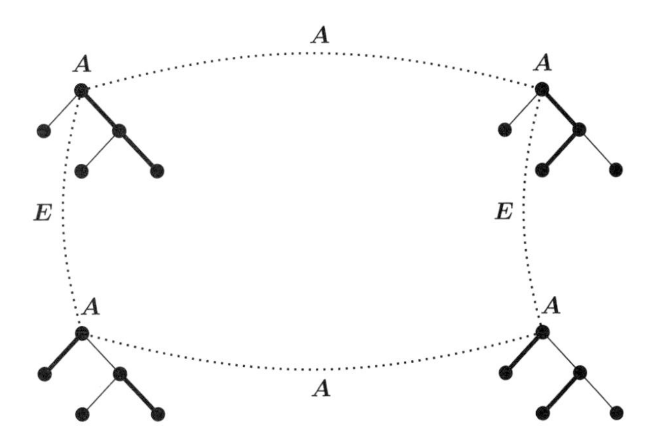

122 Epistemic forests also allow finer distinctions, such as assuming that after the moment of decision, players know their next move, but not their entire strategy yet.

Let the top left tree be the actual one, which means that ***A*** plays *right*. Even then, we still have a choice of update. One view is that moves are played according to players' intentions. That is, an event e takes place with the precondition, "e is a move prescribed by the active player's strategy," and the update would leave only the two topmost models, making ***E***, but not ***A***, know how the game will end. But we can also assume that players make mistakes, with a weaker precondition "e is a move that is available to the active player." Then e may also have been played by mistake, and all four subtrees remain, shifting the current stage of the model, but nothing else. ■

This model can be made more complex to also allow for players' expectations, and their belief changes under public information.

Options galore What we see here is a tradeoff between plausibility upgrade on trees and pre-encoding global information in game models about agent behavior, but then using simpler updates such as public announcement.[123] In addition to this choice of models, there is a variety of choices for updates interpreting events. For instance, we can decide to take moves at face value or with stronger intentional loading, and we can model decision steps explicitly, or leave them implicit. Also, forces can differ, in that rationalization steps need not be public announcements, but could be soft upgrades with rationality, as in the second scenario for Backward Induction in Chapter 8. Likewise, as for the models involved, we can update close to trees, or simplify updates in richer models pre-encoding global information about agent behavior. This variety seems true to the many options that humans have for making sense of behavior.

Now we move to a technical issue that goes slightly beyond our earlier logics.

Ternary plausibility There is also a generalization in the air here of the earlier models of Chapters 7, 8, and 9. While Backward Induction created one uniform binary plausibility relation $x \leq y$ among histories x, y, our discussion of Forward Induction suggested a ternary plausibility relation $\leq_s xy$ (cf. van Benthem 2004c) where the ordering may depend on the current vantage point s. This ternary ordering allows for differences between what players expect hypothetically if another move had been played than the actual one ("that would have been stupid") versus how they would feel if that other move were actually played.

123 A similar tradeoff occurred in Chapter 9 for preference change and information change.

Backward Induction had no distinctions based on current viewpoints in its plausibility ordering, but more general procedures of rationalization need not produce expectations that match up across a game. Our rationalization procedure used off-path expectations in the game as a contrast, allowing us to get more information about relevant beliefs in the future of our current path.[124] Recall Example 10.2, which we now explore more fully.

Example 10.8 Crossing expectations, revisited
This time, read numbers, not as utility values, but as degrees of plausibility:

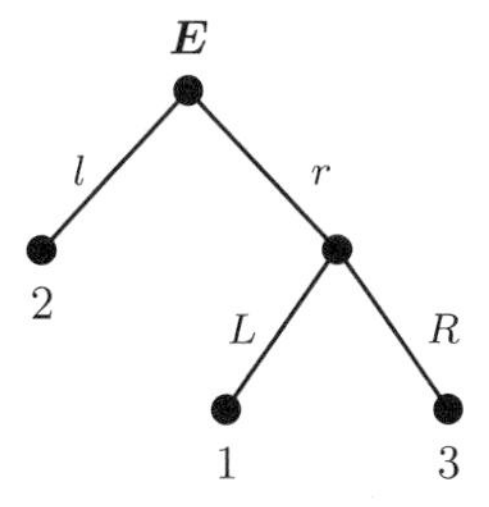

This violates the uniform node-compatible plausibility order for Backward Induction (cf. Chapter 8) that makes one of the moves l, r more plausible than the other, while all their outcomes follow this decision. Yet it is easy to find scenarios where the order depicted is natural, and general models seem more appropriate.[125] ■

Digression Here is a common objection to this move. Is the technical move to ternary orders coherent? Surely, a current plausibility order already determines what is plausible at other nodes by counterfactual reasoning, or does it? A player chooses move a, and says "if I had chosen move b, then the following would have happened." Where is the need for making the latter reasoning dependent on actually being at the node reached by playing b? This objection seems confused. If a player has in fact chosen move a, then that player considers choosing b a mistake, and the counterfactual has a hidden assumption of being in a bad place. But if the

124 There need not even be consistency going down a path: expectations may flip when a player makes a nonrationalizable move. Technically speaking, Backward Induction might be the only uniform consistent algorithm creating expectations.

125 The uniform plausibility relations produced by the Backward Induction algorithm are then a subclass defined by a special axiom in the doxastic-temporal language, whose details we do not spell out here.

player has actually chosen b, judgments will proceed on the assumption that this move was chosen intentionally. Thus, a counterfactual cannot be taken at face value. It can only be evaluated if we make our assumptions explicit about how the move was played: intentionally, by mistake, and so on. These assumptions generate different plausibility orderings, and so the third argument reappears. With Backward Induction, the built-in assumption is that all off-path moves are mistakes.

Dynamic logic over ternary models Ternary ordering relations are a standard tool in conditional logic (Lewis 1973), and ternary versions of our dynamic logics of Chapter 7 were applied in Holliday (2012) to epistemological views that explain knowledge in terms of counterfactual beliefs. Such logics would have to be adjusted to our earlier topics, for instance, tying the limit scenarios of Chapters 7 and 8 to standard fixed point logics such as LFP(FO) or IFP(FO) that easily tolerate ternary predicates.[126]

Update with agent types Our earlier scenarios largely used local updates, where views of the game change as certain events are observed and given a particular interpretation by players. But as we have suggested, there is also room for an additional global view of the type of agent we are dealing with. Backward Induction presupposes a particular kind of future-driven rational agent, while rationalization modified this assumption by allowing agents other beliefs, revealed through their behavior. Still further agent types emerge when such rationalization of behavior fails, perhaps moving to hypotheses of impaired rationality.[127] While these may seem to be pessimistic ladders of agent types descending into inanity, there are also optimistic views, starting with an assumption that the player is a simple machine, and moving up to more complex process views of others as needed. Modeling such ladders fits well with our earlier models, but it requires a more complex structure of update moves than what we have investigated.

EXAMPLE 10.9 Games with different agent types
Assume that players do not know whether the game they are entering is cooperative

126 Our discussion also suggests more radical challenges to the logics used in this book. For instance, one case of Forward Induction looked at extensive games that end in strategic matrix games, or more simply, in open-ended games whose structure we do not have available. What happens to our logical analysis when we allow this sort of hybrid model?

127 Similar ladders occur in social life, where we start by assuming that people are friendly and reasonable, and give up these illusions only when forced to by new facts at hand.

or competitive, a typical situation in daily life. But they do know that there are only two types of player: competitive (playing Backward Induction) and cooperative (striving for a best cooperative outcome). Of course, what the latter means remains to be defined in more detail, but even now, we can see how this may drastically change earlier update scenarios. If we only have these two types, then one observation may tell us the type of the opposing player and all of the future behavior, and the game may reach a stable situation after a few moves. ■

While this scenario looks attractive, and while working with a small set of possible types sounds realistic, it also has an almost static flavor, in that opposing players of fixed types are predictable automata. But there is an interesting asymmetry in much social reasoning. Often, players may see themselves as unique and infinitely flexible, while opponents fall into general types of behavior encountered before, as in Oscar Wilde's famous snub: "I don't know you, but I know the type." The same asymmetry can be seen in many logic puzzles (cf. Liu & Wang 2013): the inhabitants of Smullyan islands are predestined Liars or Truth Tellers, but the visitor is a flexible human trying to design questions so as to detect the inhabitants' type and then profit from it.[128] The most interesting types seem those allowing for sufficient variation in behavior.

I will not develop any logic for these scenarios, but see Wang (2010) and Ramanujam (2011) for some computational logic-inspired analyses of agent types. What might also be of interest is extending standard characterization results in the foundations of game theory. Aumann's Theorem says that if it is common knowledge that all agents are of one particular type, that of standard rationality, then the Backward Induction path is played. What if we relax the assumption, and assume as common knowledge that each agent is either competitive in this sense or cooperative? Could there be an extended logical theory of games in terms of natural families of agent types?

10.5 Theory of Play

We now draw a conclusion from the preceding considerations. There is no unique way of defining the best action in a game. The missing ingredient is information about the types of agent we are interacting with. The structure of a game by itself

128 Logic puzzles would get a lot harder if all participants had flexible types.

does not provide this information, unless we make strong uniformity assumptions. We need more input.

The term coined in van Benthem et al. (2011) is "Theory of Play." To make sense of what happens in a game, we must combine information about game structure plus the agents in play. Game theory allows each player to have personal preferences, but on top of that, say, many solution procedures assume uniformity on how players think and act. But we need much more variety: in cognitive or computational limitations, belief revision policies, and other relevant features. Further, what is true for games is true for social scenarios in general, the players matter. Actually, there are two aspects to this extension: we need to know about the actual play, and also about the players involved. The two are intertwined, but they represent different dimensions. Logic can help with bringing out the variety of scenarios and reasoning styles that go with this.

Of course, this idea is not totally new in the literature. Many developments in game theory tend in this direction, and the same is true for much work on game logics. Still, the current phrasing seems useful as a way of highlighting what is involved. It may also be important here to reiterate a point from our Introduction. Theory of Play is not something that is going to cure the current ills of game theory, or of logic. Rather, it is a joint offspring of ideas from these two fields (and others, such as computer science). Children are often a stronger bond in relationships than are personal transformations.

At present, there is no well-developed Theory of Play but only interesting bits and pieces that might help create one, among which are the rich set of logical tools developed in Part II of this book. This chapter concludes in a discussion of a few major challenges and potential benefits of the enterprise.

10.6 Locating the necessary diversity

One central concern is where to locate the variety that is needed in a Theory of Play. Possibilities are vast: the multi-agent system of the players, or just their strategies, or perhaps their repertoire of interpreting behavior by others, or at least reasoning about it, letting variety be in the eye of the beholder.

Individual events versus general types One can even ask whether we need agents at all, since they are a temporal entity with behavior persisting over time that goes far beyond the few events observed so far. Why not stick to what we observe locally and update based on that? The tension between individual events and postulating general types beyond these seems typical of human language and reasoning. We tend to see each other in generic terms, routinely using adjectives such as friendly or hostile that package a whole style of behavior over time. Cognitively, a genuinely isolated one night stand is about as rare as a free lunch. Presumably this tendency toward instantaneous generalization has a cognitive value, since just sticking to the facts would turn us into mindless signal-recording devices.[129] We therefore turn to agents.

Taxonomy of players There are huge spaces of possible hypotheses about players, but real understanding involves finding smaller manageable sets of relevant options. For instance, Liu (2009) has a nice map of agent diversity from the standpoint of dynamic-epistemic logic. One dimension is the processing properties of agents: what are their powers of memory, observation, and even of inference?[130] Another dimension is the update policies of agents: how will they revise their beliefs, or more generally, what learning methods do they follow? A third dimension might be called "balance types" between information and evaluation: agents can be more optimistic or pessimistic in pursuing their goals, and so on. Finally, also relevant might be social types such as whether players are more competitive or more cooperative, as discussed earlier.

Sophisticated versus simple strategies Agent types are one way of doing things. We might also just consider strategies as partners in interaction. Taxonomy of agents then gives way to taxonomy of strategies. Here lies a challenge to the logical approach in this book with its emphasis on ever greater sophistication in epistemic reasoning. Many studies of social behavior show that simple rules often work best (cf. Axelrod 1984, Gigerenzer et al. 1999). A player may be a sophisticated intellectual full of theory of mind, but perhaps the only thing that matters right now is whether the player is following a simple strategy of Tit for Tat. We are far from a general understanding of when simple strategies suffice, and when sophisticated reasoning about others is really needed.

---

129 This dismissal of individual events may be too hasty, and there is an intermediate option. Ramanujam (2011) is an intriguing exploration of a space of agent types that grow over time in a social process.

130 Along this line of thought, off-path behavior against Backward Induction may indicate the presence of another reasoning style by the relevant player.

Diversity in logics Diversity also abides in our logics. The logical dynamics of Chapter 7 highlighted the diversity of observational access or plausibility shifts by various update mechanisms, greatly extending standard views of logical agents. And even the presuppositions of our logics can be varied. The examples explored in Chapter 9 involved games with dynamic-epistemic update. In line with agent diversity, there were also update rules for memory-bounded agents whose epistemic-temporal forest patterns could be determined.

In other words, our approach is diversity-tolerant. But could it be too much so?

10.7 Some objections

There are certainly objections to what we are proposing. We address them directly.

Messiness Theory of Play comes at the cost of a large space of hypotheses about agents, making models quite complex. How can this explosion be kept in hand?

This objection has a good deal of merit, and it can be a salutary force for keeping things simple. For instance, our study of rationalization used complex updates over complex models from Chapter 9. Perhaps we should instead look for simpler alternatives. In particular, interpreting moves in social scenarios may involve just a small number of ways that are common in practice. In addition to public observations $!e$, these might be "uptake" acts considered earlier such as:

e was played intentionally, on the basis of rationality in beliefs.
e was played by mistake, by deviating from Backward Induction.[131]

Losing the appeal of uniformity Uniformity assumptions such as those embodied in Backward Induction are not just a simplistic modeling view to be replaced by sophisticated diversity. They also represent attractive intuitions of treating people equally, while reflecting a crucial intelligent ability of being able to put oneself in someone else's place (cf. van Benthem 2011d). Moreover, much cognitive behavior is held in place by forms of resonance between similar minds, so we should not give up uniformity lightly.

---

131 Moreover, given the fact that these additional features may be hypotheses on our part, we may want to use these either as hard information or soft information.

Certainly this objection has some force, but perhaps resonance occurs at some higher general levels (for instance, by agreeing to play a game at all), while diversity reigns at more specific levels.

Understanding too much The apparatus developed here can model virtually any behavior. Where is the normative force that is crucial to criticizing behavior, another aspect of taking people seriously, rather than letting them stew in their own juices?

There is no clear response to this quite reasonable objection. Social life is a delicate matter of balance, and perhaps, so is its logic.

10.8 Living with diversity, linguistic cues, and logical design

We end with some more constructive thoughts on the issues raised so far for a viable Theory of Play.

Using information that we have The preceding objections may make things look more complex than they really are. There are also forces that strike a blow for simplicity. Normally, we do not have to produce hypotheses out of the blue. Our expectations about people are based on earlier experience, so we do not enter social scenarios as a tabula rasa. And even though puzzles in the literature seem lifted out of context, often the concrete description of the scenario can be mined for agent types.[132]

Social life and language Coping with diversity is a fact of successful social life that takes several sources of tension in stride, such as the earlier division between thinking in terms of types or just responding to individual events. While this may not be totally convincing (is social life really so successful?), and while appeals to the facts are often the last resort of desperate theorists, looking at empirical evidence may be important at this stage of theorizing. For instance, one underused resource is our natural language. We have a rich linguistic repertoire for talking about individual behavior and social interaction, of which only a tiny part has been studied by logicians. Just think of all of the terms like regret, doubt, hope, reward, revenge, and so on, that structure our lives, while having a clear connection

132 For instance, a famous probabilistic puzzle like the Monty Hall problem can only be solved if we know which protocol the host is following (Halpern 2003b, van Benthem et al. 2009b). Such information can usually be found in the statement of the scenario.

with the balance of information, evaluation, and action that is so crucial to games. This repertoire seems to serve us well, so it might provide a sort of anchor to the logic of social interaction.[133]

Finding unity in all the right places Moving to more technical perspectives, one can also wonder how much unity of methods is needed, and where it resides. For instance, while dynamic logics allowing agent variety may get complex, a counteracting force is "redesign." Consider the powers of observation. One might write different logics for all sorts of agents with varying access to what is happening. But dynamic-epistemic logic packs all of this variety into one relevant event model, and then describes one mother logic for updating with these.[134] The same was true for belief revision. Prima facie, it dissolves into many update policies, with a resulting jungle of logical systems (see van Benthem 2007c on complete dynamic logics for many policies). But Baltag & Smets (2008) let event models encode the variety again, leaving only one rule of priority update obeying a simple complete set of axioms that we saw in Chapters 7 and 9, be it at an intuitive price of having more abstract signal events. And the discussion on best logical formats for social scenarios continues (cf. Girard et al. 2012).

Thus, Theory of Play should acknowledge diversity, while taking full advantage of all available cues, and letting logic do its usual job of abstraction and idealization.

10.9 Connections and possible repercussions

Agent diversity and theory of play make sense beyond games. For instance, computer science has a large body of results on what can be achieved by different kinds of strategies (Chapter 18 surveys some results, generalizing the work on logic games in Part IV). Likewise, behavior in cognitive experiments illustrates the earlier mismatch between deliberative rationality and actual play, because preferences may change in the heat of battle.[135] Our theory should be informed by all of this.

---

133 We have based the logics of this book on calm beliefs and preferences, but what if we base them on the warmer hopes and fears that inform our real decisions?

134 Admittedly, finding that mother logic can be highly non-trivial, witness the discussion in Chapter 7 on recursion laws for DEL with common knowledge.

135 Compare this to McClure (2011) on behavior in auctions deviating from game theory.

Theory of Play may also affect fields such as philosophy that are packed with uniformity assumptions, often based on philosophical intuitions that serve as a uniformizing standard. What happens when we question these assumptions? What is fair play in ethics given the undeniable diversity of agents? Are the usual Kantian ideas about all of us reasoning in the same way the greatest justice, or the greatest form of injustice? Or consider epistemology: what is rationality? Is it doing the best by your own lights, no matter how dim? Similar points apply to the philosophy of language, where the usual models of meaning involve uniform language users that belie the variety of actual communication. There is a tension between the lofty impartiality of uniformity assumptions and the humanity of diversity views, but in any case, both deserve a hearing in our logics.

Theory of Play might even reach logic itself. What about a Theory of Inference describing human or computational agents engaging in deduction and other activities, and their different styles of doing so? Say, different kinds of automata engaging in proof search, or competing in logic games? Can logic get closer to actual reasoning if we relax its standard uniformity assumptions about agents that remain implicit? Might this lead to a new understanding of existing formal systems, when we study them in use?

10.10 Conclusion

This chapter has drawn together the threads of Chapter 7, 8, and 9 toward a conception of game logic as analyzing a Theory of Play rather than just games. We have shown how tools are available for such a program in our dynamic logics of information and preference that help paint a much richer picture of reasoning about social interaction. Still, we also considered objections to the resulting diversity, and problems with drawing natural boundaries. The resulting enterprise stands in need of philosophical reflection as much as technical development, but we hope to have shown the interest of both.

10.11 Literature

This chapter is based on van Benthem (2011b), van Benthem (2011d), van Benthem (2011f), and van Benthem et al. (2011).

Papers with a strong influence on the above views are Bicchieri (1988), Stalnaker (1999), Halpern (2001), and Halpern (2003a), as well as the game-theoretical literature mentioned at several places in this text: see Perea (2012) and Brandenburger et al. (2014) for congenial recent sources.

Conclusion to Part II

In this part, we have set out the general program of logical dynamics as a theory of explicit actions in information-driven agency, and we have introduced the reader to the basic systems and methodology that come with this view. Games are a prime example of intelligent interactive agency, and hence a natural part of a logical dynamics interest. What we achieved with this style of thinking is a dynamic perspective on games as activities, with phases of initial deliberation, actual play, and post-play reflection.

In Chapter 7, we gave a cameo version of dynamic-epistemic logic, broadly conceived. In Chapter 8, we analyzed game solution as exemplified in Backward Induction as a process rather than as a finished product, casting solution algorithms as deliberation procedures tending to limits of iterated hard and soft announcements of rationality. The resulting general framework suggested procedural foundational views of games and social activities. It also established links with fixed point logics of computation that seem of independent interest. Finally, it provided a new take on notions like plausibility that are usually taken as primitive in philosophical logic. Chapter 9 studied a variety of informational processes in actual play, defined mainly over forest models for games. Specific offerings were complete dynamic logics for playing moves, receiving hard or soft information transforming epistemic accessibility or plausibility order, and using the same methods, preference change, and even strategy and game change. Over forest models, representation theorems captured the behavior of agents following the update rules of dynamic-epistemic-doxastic logic, as well as others. Finally, in Chapter 10, we looked at these phenomena as tending toward a joint offspring of logic and game theory that may be called a Theory of Play. We discussed basic issues such as the challenge of Forward Induction, and looked at the contours of what is involved: maintaining the right game models, typology of agents, and repercussions for areas beyond game theory, including philosophy, and even logic itself.

Parts I and II have recorded where things stand in our thinking on game logics. To test the viability of this program, Part III will show how this style of analysis also works for other levels of representing games, namely, players' powers, and strategic matrix forms.

III *Players' Powers and Strategic Games*

Introduction to Part III

In Parts I and II of this book, we have mainly studied games in extensive form at the level of individual actions. But a natural global perspective has come up repeatedly, too: that of players' powers for influencing final outcomes, in terms of their having strategies for making a game produce only states, or histories, that satisfy certain stated properties. This global perspective still fits well with logic, and in Chapter 11, we will use the power view as a test case for earlier techniques, showing how modal languages, bisimulation, and other notions from Part I generalize to this setting. Our tool for this is so-called neighborhood models that have been used in studies of information, concurrency, topology, co-algebra, and games. All this is a stepping stone toward Chapter 12, where we move closer to standard game theory, investigating games in strategic form. We show how these support logical analysis in the same style as we have offered before, but with new twists including topics such as knowledge, equilibrium, rationality, and temporal evolution, while adding themes such as concurrent action. Finally, in Chapter 13, we show how the logical dynamics of Part II applies just as well to studying strategic games.

This part of the book has two faces. It is a natural conclusion to Parts I and II, as it shows how the logical statics and dynamics developed there apply to other views of games. Thus, the first main theme of this book comes to an end here. But at the same time, Part III looks ahead to our other main theme of logic games. As we will see in Part IV, the natural level of grain associated with the logic games of our Introduction is often power bisimulation. And once that is clear, Part V will develop some logical systems of power-based game analysis that owe as much to logic games as to game logics. Thus, while this part of the book can be skipped without loss of continuity, it is actually a nice switching site between the two main strands of our investigation.

11
Forcing Powers

In this chapter, we propose a simple view of global input-output behavior of games through the lens of players' strategic powers. These can be studied with earlier modal techniques, leading to systematic connections with the logics of actions in Parts I and II.

11.1 Forcing and strategic powers

Social powers of agents are an important topic in their own right. In this chapter, we will ignore details of players' moves, and concentrate on their powers for achieving outcomes of a game. Intuitive computations of this sort were already made in the Introduction and in Chapter 1 when discussing different views of games.

DEFINITION 11.1 Forcing relations
The *forcing relation* $\rho^i_G s, X$ in a game tree holds if player i has a strategy for playing game G from state s onward whose resulting end states are always in the set X. When s is known in context (often it is the root of the game), the sets X are called the *powers* of player i. ■

In any finite game tree, we can find forcing relations by enumerating strategies, and sets of final states that can be reached by applying the strategy to all plays by the other player.[136] Determining powers can be done inductively using game algebra, but we defer this until Chapter 19. For now, we compute powers by hand.

136 The concept of powers also applies to infinite games, where outcomes can be identified with the total histories. Much of what we will have to say extends to this setting.

Example 11.1 Computing forcing relations

The following game displays the basic ingredients of strategies and outcome sets:

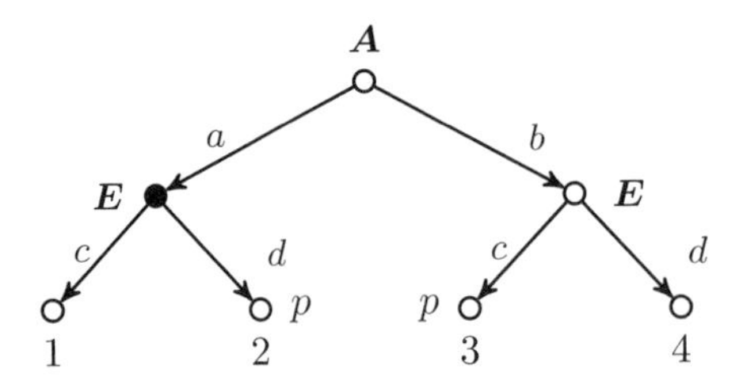

Here player $\boldsymbol{A}$ has two strategies, forcing the sets $\{1,2\}$, $\{3,4\}$, while $\boldsymbol{E}$ has four strategies, forcing the sets $\{1,3\}$, $\{1,4\}$, $\{2,3\}$, $\{2,4\}$. These sets encode players' powers over specific outcomes: a larger set says that the power is not strong enough for controlling a unique outcome, just keeping things within some upper range. ■

For a further perspective, we recall the earlier issue of game equivalence.

Example 11.2 Propositional distribution once again

Consider the following games, discussed in the Introduction and Chapter 1:

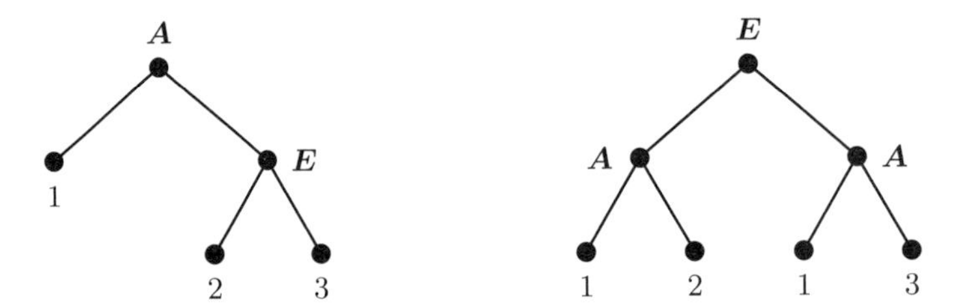

These games were not the same in their local action structure as studied in Chapter 1, but they are the same in terms of global powers. In the left-hand game, player $\boldsymbol{A}$ has two strategies, *left* and *right*, that guarantee outcomes in the sets $\{1\}$, $\{2,3\}$. These are then player $\boldsymbol{A}$'s powers. Player $\boldsymbol{E}$ has two strategies as well, that guarantee outcomes in $\{1,2\}$ and $\{1,3\}$, making these sets $\boldsymbol{E}$'s powers. In the right-hand game, we found the same outcomes. First, player $\boldsymbol{E}$ has two strategies that force the sets $\{1,2\}$ and $\{1,3\}$. Next, $\boldsymbol{A}$ has four strategies LL, LR, RL, RR that yield, respectively, $\{1\}$, $\{1,3\}$, $\{2,1\}$, $\{2,3\}$. Of these four, $\{1,3\}$ and $\{2,1\}$ can be dropped, however. For, on the above understanding of the notion, they represent weaker powers than $\{1\}$, and hence they are redundant. Thus player $\boldsymbol{A}$, too, has the powers $\{1\}$, $\{2,3\}$, just as in the left-hand game. ■

This analysis presupposes some formal properties of players' powers, such as closure under supersets. We will now look at these constraints in some more generality, following a logical line linking up with other themes in this book.[137]

11.2 Formal conditions and game representation

Forcing relations as defined above are evidently closed under supersets:

$C1$ if $\rho^i_G s, Y$ and $Y \subseteq Z$, then $\rho^i_G s, Z$

This property is sometimes called Monotonicity. Another obvious constraint on these relations is Consistency: players cannot force the game into disjoint sets of outcomes, since playing strategies for these powers always produces an outcome:[138]

$C2$ if $\rho^i_G s, Y$ and $\rho^j_G s, Z$, then Y and Z overlap

Recall from Chapter 1 that all finite two-player games are determined: for any winning convention, one of the two players has a winning strategy. In the present abstract terms, this says the following. Let S be the total set of outcome states:

$C3$ if not $\rho^i_G s, Y$, then $\rho^j_G s, S - Y$

One can also think of this determinacy as a sort of Completeness.

It is easy to check these conditions for the earlier examples in this chapter. In fact they hold for players' general powers in our present setting.[139]

FACT 11.1 Any finite perfect information game G between players i and j yields powers in the root s satisfying conditions $C1$, $C2$, $C3$.

Conversely, these conditions are all that hold, by the following representation.

FACT 11.2 Any two families F_1, F_2 of subsets of some set S satisfying conditions $C1$, $C2$, and $C3$ are the root powers in some two-step game.

---

137 An early formal analysis of powers from a game-theoretic perspective was given in Bonanno (1992b).

138 Here and henceforth, we use i and j as generic names for the two players.

139 Interestingly, the stated conditions on powers may be viewed as a two-person version of the standard logical notion of an ultrafilter. This pulling apart of standard notions into many-agent notions will be very typical for the logic games to be studied in Part IV.

Proof Start with player i choosing between successors corresponding to all the inclusion-minimal sets in F_i. At these nodes, player j then gets to move, and can pick any member of the given set. Clearly then, player i has the powers specified in F_i. Now consider the powers of player j. In the game just defined, player j can force any set of outcomes that overlaps with each of the sets in F_i. But by the given constraints, these are precisely the sets in the initial family F_j. For instance, if some set of outcomes $\boldsymbol{A}$ overlaps with all the sets in F_i, then its complement, $S - A$, cannot occur in the latter family, and hence A itself must have been in F_j, by Completeness. ■

This argument gives a bit more than stated, that is to say, a normal form for games related to the usual game-theoretic strategic form that will be studied in Chapter 12. It just has two moves, and it does not matter to the construction which player begins. This is like the disjunctive and conjunctive normal forms of propositional logic. Indeed, the Boolean operations that produce such normal forms are part of a more general logical calculus of game equivalence that will be studied in Parts V and VI of this book.

Implementing social scenarios Our representation theorem may also be viewed in a different light. A list of powers is a specification for how much control we want to give to various agents in a given group, and a game is then the design of a simple social scenario implementing just those powers. For more complex power structures analyzed in logical terms, see Pauly (2001) and Goranko et al. (2013). Social scenarios where ways of giving agents different information may be part of the design are discussed in Papadimitriou (1996).

11.3 Modal forcing logic and neighborhood models

Forcing modalities One can introduce a matching modal language for games at this new level, with proposition letters, Boolean operators, and the following new modal operators:

$\boldsymbol{M}, s \models \{G, i\}\varphi$ iff there exists a set power X for player i in G such that for all $t \in X$: $\boldsymbol{M}, t \models \varphi$

or more in line with the format of modal semantics:

$\boldsymbol{M}, s \models \{G, i\}\varphi$ iff there exists a set X with $\rho^i_G s, X$ and $\forall t \in X$: $\boldsymbol{M}, t \models \varphi$

This is essentially the forcing modality $\{i\}\varphi$ that we have defined in Chapters 1 and 3 as a global perspective on some given extensive game G. In what follows, for convenience, we often write $\{G\}\varphi$ for the modality since the player i is understood from context.

This semantics involves a generalization of the modal models of Chapter 1, where accessibility took worlds to worlds, since we now work with world-to-set relations. The main effect of this change at the level of validities is reminiscent of the base logic of temporal forcing in Chapter 5.

FACT 11.3 Modal logic with the forcing interpretation satisfies all principles of the minimal modal logic of $\Diamond$ except for distribution of $\{\}$ over disjunctions.

That is, $G\varphi$ is upward monotonic:

$$\text{if} \models \varphi \to \psi, \text{ then} \models \{G\}\varphi \to \{G\}\psi$$

But the following is not valid:

$$\{G\}(\varphi \vee \psi) \to \{G\}\varphi \vee \{G\}\psi$$

The latter failure is precisely the point of forcing. Other players may have powers that keep me from determining results precisely. I may have a winning strategy, but it may still be up to you just where my victory will take place. For instance, in the very first game of this chapter, player $\boldsymbol{A}$ can force the set of outcomes $\{1, 2\}$, but neither $\{1\}$ nor $\{2\}$. It is also easy to check that the forcing modality $\{\}$ does not distribute over conjunction.

The basic modal neighborhood logic is decidable, with lower computational complexity than the minimal modal logic K over standard relational models.

Neighborhood modal logic Models with world-to-set accessibility relations for modal logics are called neighborhood models (Chellas 1980, van Benthem 2010b). The term reflects a connection with topology: see Section 11.5 below. These structures have many applications (Hansen et al. 2009), including fine-grained evidence models for the information dynamics of Chapter 7 (van Benthem & Pacuit 2011).

11.4 Bisimulation, invariance, and definability

Here is a key model-theoretic feature of neighborhood models that goes back to the issues of game equivalence discussed in the Introduction and later on in this book.

The invariance notion of bisimulation from Chapter 1 lifts to this setting without major effort. Consider any game model $\boldsymbol{M}$ plus forcing relations as defined above.

DEFINITION 11.2 Power bisimulation

A *power bisimulation* between two game models G, G' is a binary relation E between states in G, G' that satisfies the following two conditions:

(a) If xEy, then x and y satisfy the same proposition letters.
(b) If xEy and $\rho^{i}_{G}x, U$, then there exists a set V with $\rho^{i}_{G}y, V$ and $\forall v \in V \exists u \in U$: uEv; and vice versa. ■

Power bisimulation is a natural notion. It was proposed independently in concurrent dynamic logic (van Benthem et al. 1994), topological modal logics (Aiello & van Benthem 2002), and co-algebra (Baltag 2002). It also has a general model theory that is a natural extension of that in Chapter 1 for ordinary relational bisimulation. Without going into great detail, we give two illustrations.

FACT 11.4 The modal forcing language is invariant for power bisimulation.

Proof Consider two models $\boldsymbol{M}$ and $\boldsymbol{N}$ with $\boldsymbol{M}, s \models \{\}\varphi$ and sEt. By the truth definition, there is a set U with $\rho^{i}_{\boldsymbol{M}}s, U$ and for all $u \in U$: $\boldsymbol{M}, u \models \varphi$. Now by the back-and-forth clause (2), there is a set V in $\boldsymbol{N}$ with $\rho^{i}_{\boldsymbol{N}}t, V$ and $\forall v \in V \exists u \in U$: uEv. So, every $v \in V$ is E-related to some $u \in U$, and by the inductive hypothesis, we then have that $\boldsymbol{N}, v \models \varphi$. But then by the truth definition, $\boldsymbol{N}, t \models \{\}\varphi$. ■

FACT 11.5 If finite models $\boldsymbol{M}, x$ and $\boldsymbol{N}, y$ satisfy the same forcing formulas, then there is a power bisimulation E between them with xEy.

Proof Define a relation E between states in the two models as follows:

$$uEv \text{ iff } \boldsymbol{M}, u \text{ and } \boldsymbol{N}, v \text{ satisfy the same modal forcing formulas.}$$

CLAIM E is a power bisimulation.

The atomic clause is clear from the definition. Now, suppose that sRt, while also, for some subset U of $\boldsymbol{M}$, $\rho^{i}_{\boldsymbol{M}}s, U$. We need to find a set V with

$$\rho^{i}_{\boldsymbol{N}}t, V \text{ and } \forall v \in V \exists u \in U : uEv$$

Suppose that no such set exists. That is, for every set V in $\boldsymbol{N}$ with $\rho^{i}_{\boldsymbol{N}}t, V$, there is a state $v^{V} \in V$ that is not E-related to any $u \in U$. Let us analyze the latter statement further. By the definition of the relation E, this means that for each

$u \in U$, v^V disagrees with u on some forcing formula ψ^u: say, it is true in u, and false in v^V. But then, the disjunction Ψ^V of all of these formulas is true in every member of U, and still false in v^V. Now let Ψ be the conjunction of all the latter formulas, where V runs over all sets satisfying $\rho^{\boldsymbol{i}}_{\boldsymbol{N}} t, V$. Evidently, we have

$$\boldsymbol{M}, u \models \Psi \text{ for each } u \in U$$

and hence also

$$\boldsymbol{M}, s \models \{\}\Psi$$

But then, by the above definition of E, also $\boldsymbol{N}, t \models \{\}\Psi$. This says that there is a set V with $\rho^{\boldsymbol{i}}_{\boldsymbol{N}} t, V$ each of whose members satisfies the formula Ψ. This contradicts the construction of Ψ, since v^V certainly does not satisfy its conjunct Ψ^V. ■

These results also hold over more general process models, not just game trees.[140]

11.5 Digression: Topological models and similarity games

Let us now take a more spatial look at the preceding notions, injecting some fresh air into what might otherwise be a rather dry exercise in generalized modal logic.

Topological semantics A special case of neighborhood models are *topological models* $M = (\boldsymbol{O}, V)$ consisting of a topological space $\boldsymbol{O}$ and a valuation V for proposition letters. The forcing relation $\rho\ x, U$ is now read as follows: x belongs to the open set U. The semantics for modal languages on these mathematical structures dates back to the 1930s.

DEFINITION 11.3 Topological interpretation of modal logic
We say that $\Box\varphi$ is true at a point s in a topological model $\boldsymbol{M}$ in case s has an open neighborhood U all of whose points satisfy φ. ■

We refer to the literature for details on the topological reading of the modalities (see the survey van Benthem & Bezhanishvili 2007 or the textbook van Benthem 2010b). For instance, the base logic is $S4$, and there are many deep technical results tying further modal logics to significant mathematical structures.

140 More results lie down the road, generalizing those stated in Chapter 1 for standard modal logic. For instance, we can find power invariants for any game G by infinitary modal forcing formulas defining the class of all models power bisimilar with G.

Here, however, our interest is in the relevant notion of structural invariance, connecting neighborhood bisimulation with the logic games of Part IV in this book. As we will see in Chapter 15, each notion of bisimulation invites a corresponding logic game for comparing models, where two players probe differences and similarities up to a specified number of rounds. This yields a fine structured version of bisimulation measuring degrees of similarity. Instead of doing this for power bisimulation in its abstract version on neighborhood models, we demonstrate the idea in a more concrete topological setting, taken from Aiello & van Benthem (2002).

Comparison games One can analyze the expressive power of our modal language with games between two players called *spoiler* (the difference player) and *duplicator* (the analogy player), comparing points in two given topological models.

DEFINITION 11.4 Topo-games
Rounds in *topo-games* proceed as follows, starting from some current match $s - t$. Spoiler takes one of these points, and chooses an open neighborhood U in its model. Duplicator responds with an open neighborhood V of the other current point. Still in the same round, Spoiler chooses a point $v \in V$, and then Duplicator chooses a point $u \in U$, making $u - v$ the new match. Duplicator loses if the two points differ in atomic properties. ■

This looks abstract, but it can be made very concrete.

EXAMPLE 11.3 Comparing points on spoons
In the spoons shown below, compare the following intuitively different points:

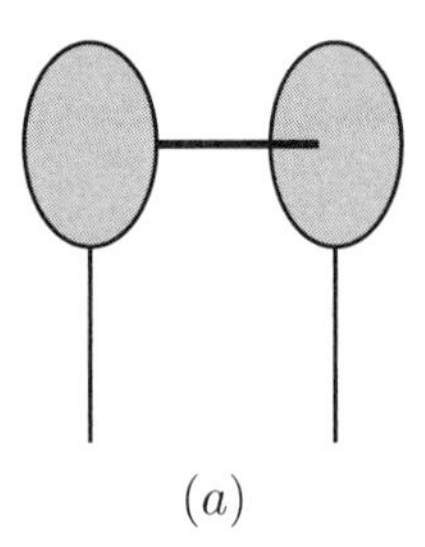

(*a*)

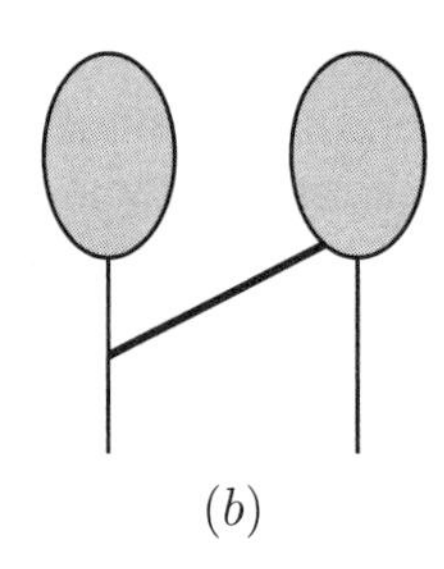

(*b*)

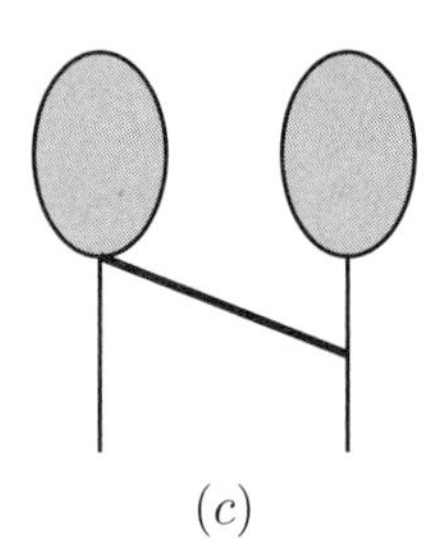

(*c*)

It is helpful to observe that, in these games, it does not matter whether players choose small or large open neighborhoods.[141]

141 To see this, recall the standard observation that, on relational models, evaluation of modal formulas at a current point s only needs to look at R-closed generated submodels

Case (a) If spoiler chooses a neighborhood to the left, then duplicator chooses a small interior disk to the right, and whatever spoiler chooses, there will be an inside point that duplicator can match in the open to the left. Therefore, this is a bad idea. If spoiler starts with a small disk on the right, however, then duplicator must respond with a disk on the edge to the left, which then allows spoiler to choose an object outside of the spoon, and every response by duplicator is losing, since it must be inside the spoon. Therefore, spoiler has a winning strategy in one round. This is reflected in the simple observation that there is a modal difference formula of operator depth 1 distinguishing the two positions. As a concrete example, $\Box p$ is true on the right, but not on the left.

Case (b) Spoiler's winning strategy starts by taking an open set on the handle to the left, after which duplicator must choose an open on the rim of the oval. Now spoiler chooses an object inside the spoon, and duplicator can only respond by either choosing outside of the spoon (an immediate loss), or on the handle. But the latter choice reduces the game to Case (a), which spoiler could already win in one round. A suitable difference formula this time has modal depth 2, say, $\Diamond\Box p$.

Case (c) is the most complex, as the point connecting rim and handle is much like an ordinary rim point. Spoiler has a winning strategy in three rounds, matching a modal difference formula of modal depth 3, say, $\Diamond(p \wedge \neg\Diamond\Box p) \wedge \Diamond\Box p$. ■

The main result governing this game is similar to the "success lemma" for model comparison games found in Chapter 15.

FACT 11.6 Duplicator has a winning strategy in the comparison game over k rounds starting from two models $\boldsymbol{M}, s$ and $\boldsymbol{N}, t$ iff these two pointed models satisfy the same modal forcing formulas up to modal operator depth k.

The matching global relation for topological spaces plus propositional valuations is *topo-bisimulation*, that is, power bisimulation restricted to topological models.[142] Conversely, the topo-game is also easily adapted to general power bisimulation.

---

around s. Likewise, for any pointed topological model $\boldsymbol{M}, s$ and modal formula φ, and for any open neighborhood U of s, the following two assertions are equivalent: (a) $\boldsymbol{M}, s \models \varphi$, and (b) $\boldsymbol{M}|U, s \models \varphi$, where $\boldsymbol{M}|U$ is the model $\boldsymbol{M}$ restricted to the subset U.

142 The back-and-forth clauses resemble the topological definition of a continuous open function, for which both images and inverse images of open sets are open.

11.6 Compositional computation and game algebra

Powers in games can often be computed compositionally, as the reader may have done already when checking earlier examples. For instance, when a game starts with a move for player $\boldsymbol{E}$, the powers of $\boldsymbol{E}$ are just the union of the powers in the separate subgames that can be chosen. More abstractly, writing $G \cup H$ for an operation of choice between games G and H, we have the following valid equivalence:

$$\rho^{\boldsymbol{E}}_{G \cup H} s, Y \quad \text{iff} \quad \rho^{\boldsymbol{E}}_{G} s, Y \text{ or } \rho^{\boldsymbol{E}}_{H} s, Y$$

What we see here is an incipient algebra of game-forming operations. Natural operations forming new games are choice, but also role switch, sequential and parallel composition, that support identities of their own.[143] This algebra will emerge in Part IV of this book, and it will be explored further in Chapter 19, using logics with forcing modalities $\{G, i\}\varphi$ that can talk about composite game structure in their G argument.

11.7 Forcing intermediate positions

Powers tell us what players can achieve in the end. However, sometimes we also want to describe intermediate stages, getting closer to the earlier action level.

EXAMPLE 11.4 Intermediate forcing

The local dynamics of the two distribution games in Example 11.2 are different. In the root on the left, but not on the right, player $\boldsymbol{A}$ can hand player $\boldsymbol{E}$ a choice between achieving q and achieving r. This might be expressed in a simple notation:

$$\{A\}^+(\{E\}^+q \wedge \{E\}^+r)$$

true in the left root, and false in the root on the right. The modified forcing modality $\{E\}^+\varphi$ in this formula says that player $\boldsymbol{E}$ has the power to take the game to some state, final or intermediate, where the proposition φ holds. ■

---

143 A simple example of a valid identity is that for role switch: $\rho^{\boldsymbol{E}}_{G^d} s, Y \leftrightarrow \rho^{\boldsymbol{A}}_{G} s, Y$.

This is a natural new intermediate level of game description: unlike bisimulation, intermediate forcing does not track specific actions, but it does care about choices of players on the move. It is easy to find a matching notion of bisimulation, and thus, we have one more level in the ladder of game equivalences discussed in Chapter 1.

11.8 Interfacing modal logics of forcing and action

Defining powers by actions The modal action languages of Chapter 1 were at least as expressive as the present forcing languages.

FACT 11.7 In any finite game model, $\{E\}\varphi$ is definable as a greatest fixed point of the recursion

$$\{E\}\varphi \leftrightarrow ((\boldsymbol{end} \wedge \varphi) \vee (\boldsymbol{turn_E} \wedge \bigvee_a \langle a \rangle \{E\}\varphi) \vee (\boldsymbol{turn_A} \wedge \bigwedge_a [a]\{E\}\varphi))$$

In specific finite models, we can even unwind $\{E\}\varphi$ as one modal action formula, whose depth depends on the size of the model. This is connected to the following general observation.

FACT 11.8 If there exists an action bisimulation between two models connecting points s and t, then there also exists a power bisimulation connecting s and t.

Games and game boards The two levels of description for games, in terms of moves and of powers, coexist when we make a natural distinction between games G themselves and associated structures that may be called "game boards" $\boldsymbol{M}(G)$ marking relevant external aspects of game states only. Game boards include the physical board states of a game like Chess, but they can also be abstract structures, such as the models where one plays the logical evaluation games of Chapter 14. Game boards will be studied in more depth in Chapters 18, 19, and 24, and a few simple observations will suffice for now.

Let G be a game of finite depth over a game board $\boldsymbol{M}(G)$, using moves available on the board, assigning turns to players, and determining a winning convention. This gives us an obvious projection map F from internal game states to external states on the board. Also, let the winning convention only use external facts on the game board, definable by a formula α:

$$\boldsymbol{win_E}(s) \text{ iff } \boldsymbol{M}, F(s) \models \alpha_{\boldsymbol{E}}$$

This sort of external winning convention holds for many games. Under these assumptions, the two levels are in harmony.

FACT 11.9 There is an effective equivalence between forcing statements about outcomes in a game G and modal properties of the associated game board $\boldsymbol{M}(G)$.

Proof If the game structure over atomic games is given by choices, switches, and compositions, we translate forcing statements at nodes into matching disjunctions, negations, and substitutions. At atomic games, we plug in the given definition. ■

We will see instances of this harmony in Chapters 14, 19, and 25.[144]

11.9 Powers in games with imperfect information

Having dealt with perfect information games, we now look at imperfect information. Chapter 3 studied such games at their action level, using a modal-epistemic language. Let us now consider the more global power level. Our treatment will be more sketchy than in preceding sections, showing merely how forcing views apply.

Powers First, the definition of players' powers is adapted as follows. Recall the basic notion of a uniform strategy assigning the same moves to nodes in the game tree that the relevant player cannot distinguish.

DEFINITION 11.5 Powers in imperfect information games
At each node of an imperfect information game, a player can *force* those sets of outcomes that are produced by following one of the player's uniform strategies. ■

EXAMPLE 11.5 Diminished powers in imperfect information games
Consider the two games drawn in Section 11.1. In the game of perfect information, player $\boldsymbol{A}$ had 2 strategies, and player $\boldsymbol{E}$ had 4, producing the following set powers:

$$\boldsymbol{A} \quad \{1,2\}, \{3,4\} \qquad\qquad \boldsymbol{E} \quad \{1,3\}, \{1,4\}, \{2,3\}, \{2,4\}$$

144 A useful example are graph games in computational logic (Libkin 2004), where players make alternating steps in their given move relations on a board. The first player to get stuck loses. There is an equivalence here between the assertions (a) player $\boldsymbol{E}$ has a winning strategy in the game at (G, s), and (b) $G, s \models (\langle R \rangle [S])^n \top$, with n the size of G.

With imperfect information added for $\boldsymbol{E}$ in the second game, $\boldsymbol{A}$'s powers are not affected, but $\boldsymbol{E}$'s are. $\boldsymbol{E}$'s two uniform strategies only yield two set powers:

$$\boldsymbol{A} \quad \{1,2\}, \{3,4\} \qquad\qquad \boldsymbol{E} \quad \{1,3\}, \{2,4\}$$

This loss of control may be seen as a weakness, but it really shows something else: the much greater power for representing varieties of social influence in imperfect information settings. ■

We can measure this power by extending the representation of Section 11.2.

Representation theorem The earlier conditions of monotonicity $C1$ and consistency $C2$ still hold for powers in imperfect information games. What fails, however, is the completeness $C3$. As we just saw: player $\boldsymbol{A}$ lacks the power $\{1,4\}$, while $\boldsymbol{E}$ lacks its complement $\{2,3\}$.

THEOREM 11.1 Any two finite families of sets satisfying conditions $C1$ and $C2$ can be realized as the powers in a two-step imperfect information game.

Proof Instead of stating the procedure in complete formal detail, we go through an example displaying all the necessary tricks. We are given two lists:

minimal powers for player $\boldsymbol{A}$	$\{1,2,3\}$, $\{3,4\}$
minimal powers for player $\boldsymbol{E}$	$\{2,3\}$, $\{1,4\}$

Start with player $\boldsymbol{A}$ and put successor nodes for all of $\boldsymbol{A}$'s power sets, with possible duplications of these sets (as explained below), where player $\boldsymbol{E}$ gets to move. Here $\boldsymbol{E}$'s uniform strategy must prescribe the same move at each node, depicted below as lines with the same slope. $\boldsymbol{E}$'s actions, too, may involve duplications. First, take action types for each set of player $\boldsymbol{E}$, making sure all of these get represented via uniform strategies. There may still be excess outcomes in the powers for $\boldsymbol{E}$, and then dilute these by copying and permuting so that they end up in supersets of $\{2,3\}$, $\{1,4\}$. A picture conveys the idea better than words:

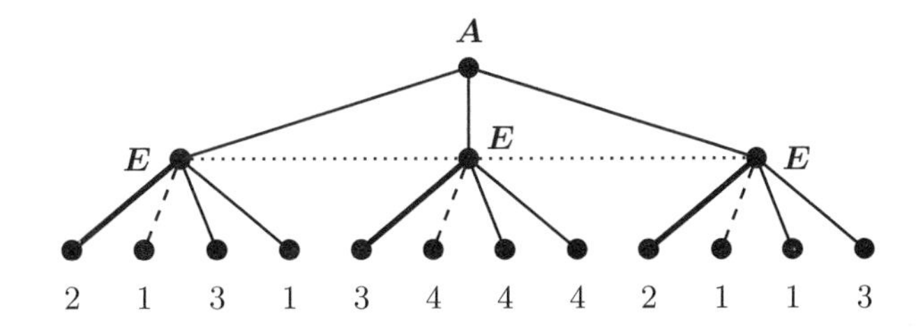

The reader may want to construct this representation by starting with just $\boldsymbol{A}$'s two left-most choices and three actions for $\boldsymbol{E}$, and then seeing why the additional actions have to be there. The sets for the third and fourth uniform strategy are all supersets of $\{2, 3\}$, $\{1, 4\}$.

A complication arises when $\boldsymbol{A}$'s sets involve outcomes not occurring in $\boldsymbol{E}$'s list. For concreteness, consider this case

minimal powers for player $\boldsymbol{A}$	$\{1, 2, 3\}$, $\{3, 4\}$
minimal powers for player $\boldsymbol{E}$	$\{2, 3\}$, $\{1, 4, 5\}$

We must dilute still more, adding $\boldsymbol{A}$-moves to cases where $\boldsymbol{E}$ can choose between 1, 4, and 5 so that the resulting powers of $\boldsymbol{E}$ are all extensions of $\{1, 2, 3\}$, $\{3, 4\}$.

The recipe A more precise description of the representation method is as follows. First, make sure that all atomic outcomes in players' sets occur for each of them, by adding redundant supersets if needed. Then create a preliminary branching for $\boldsymbol{A}$, making sure that all of $\boldsymbol{A}$'s sets are represented. Now represent each outcome set for $\boldsymbol{E}$ via a uniform strategy, by choosing enough branchings at midlevel, starting from the left. This step may involve some duplication of nodes, as illustrated by the case of initial lists $\boldsymbol{A}$ $\{1, 2\}$, $\boldsymbol{E}$ $\{1, 2\}$. Next, if there are still left over outcomes, repeat the following routine:

Suppose outcome i at midlevel point x was not used in our third step so far. Fix any outcome set for $\boldsymbol{E}$ produced by some uniform strategy σ. Say, σ chose outcome j at x. Duplicate this node x, and add two new branchings throughout (for all midlevel nodes considered so far) to get two further uniform strategies: one choosing j at x, and i in its duplicate, the other doing the opposite, and both following σ at all other nodes. This addition makes outcome j appear as it should for $\boldsymbol{A}$, while generating only a harmless superset of outcomes for $\boldsymbol{E}$.[145] ■

A variation One source of complexity in our construction is the fact that intersecting powers of the two players need not result in a singleton set with a unique outcome. If we add the latter condition, then we arrive at the matrix games of Chapter 12, where the forcing powers arise from simultaneous action, where players know their own choice, but can only observe the other player's choice afterward. We will discuss the latter scenario in connection with STIT logics of agency.

---

145 This method is very messy, and it would be nice to have a uniform construction, say, with pairs of actions and appropriate equivalence relations over these for the players.

Game algebra As with powers in games of perfect information, the present setting suggests an algebra of game operations that respect power equivalence. For imperfect information games, however, compositionality with respect to subgames is much less obvious, and we postpone the topic until Chapters 20, 21, and 24.

11.10 Game transformations and extended logic

There are also direct links between our analysis and a classical topic in game theory. Let us take up a thread first started in Chapter 1.

Game transformations and power invariance The four "Thompson transformations" of game theory (Thompson 1952, Osborne & Rubinstein 1994) turn normal form equivalent extensive games with imperfect information into each other. We show how this relates to our power analysis, returning to further logical background in Chapter 21.

EXAMPLE 11.6 Addition of a superfluous move
The first transformation, called addition of superfluous moves, tells us how to tolerate a new uncertainty by duplicating a move, switching game trees as follows:

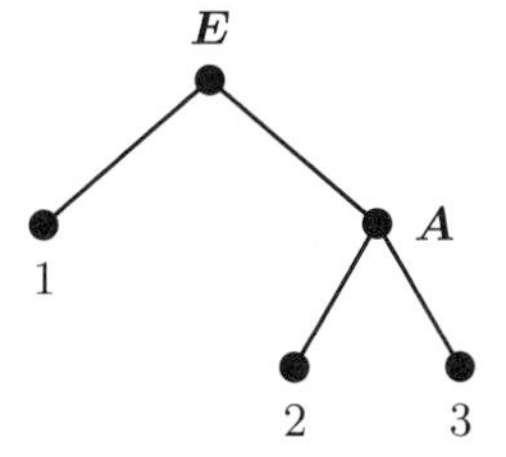

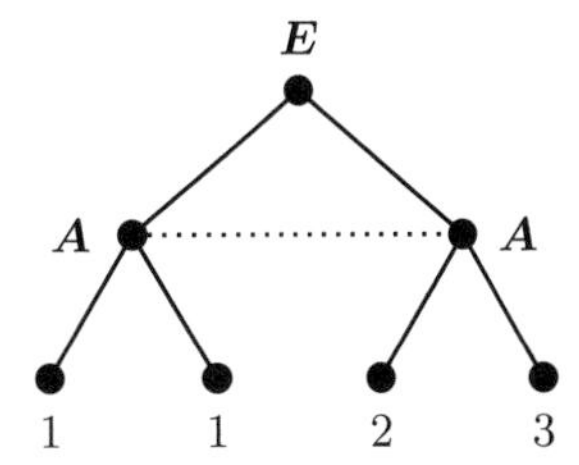

Computing powers, nothing changes from left to right at the root. In the game to the right, the duplication adds no new outcomes for player $\boldsymbol{E}$, while the two uniform strategies for player $\boldsymbol{A}$ on the right capture the same outcome sets $\{1, 2\}$, $\{1, 3\}$ as on the left. ■

EXAMPLE 11.7 Interchanging moves
The key transformation interchanging moves of players occurred in the Introduction and elsewhere. We give a slight variant here, with some imperfect information added. In the following two games, the set powers for the players are the same:

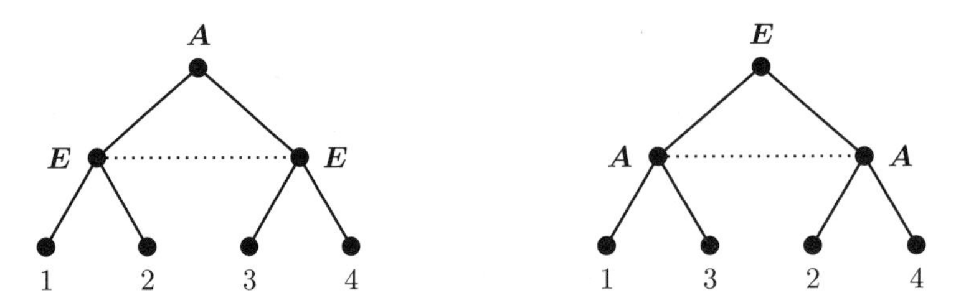

Player $\boldsymbol{A}$ has powers $\{1,2\}$, $\{3,4\}$, while $\boldsymbol{E}$ has powers $\{1,3\}$, $\{2,4\}$. ■

This second Thompson transformation makes sequential play under imperfect information similar to simultaneous play, a topic that will return in Chapter 12.

The remaining two Thompson transformations are

(a) Inflation Deflation Adding a move that will not materialize in play.

(b) Coalescing Moves Re-encoding a zone where a player is to move successively as just one set of choices.

These reflect an emphasis on powers as the criterion of identity for a game, rather than imperfect information as such.

All of our observations so far are summarized in the following result.

THEOREM 11.2 Powers of players in imperfect information games remain the same under the Thompson transformations.'

This analysis suggests an analogue to Thompson's result that his transformations are precisely those that leave the reduced form of a game unchanged. See Dechesne (2005) for further discussion, and connections with the independence-friendly logic of Chapter 21.

Game transformation as generalized logic A fascinating alternative interpretation is possible. The Thompson transformations induce a calculus for a generalized propositional logic. The above pictures of equivalent games relate to logical laws, the way we saw earlier with games and propositional distribution laws in the Introduction.

FACT 11.10 Addition of a Superfluous Move is propositional idempotence

$$A \wedge (B \vee C) \leftrightarrow (A \bigvee A) \wedge (B \bigvee C)$$

where the larger disjunctions indicate an uncertainty link.

We are into a nonstandard use of propositional logic now, where duplications matter, and operators can be linked. Another valid principle in this realm may be even more surprising.

FACT 11.11 Interchanging Moves is a propositional distribution law:

$$(A \bigvee B) \wedge (C \bigvee D) \leftrightarrow (A \bigwedge C) \vee (B \bigwedge D)$$

with large operators indicating dotted links, again a nonstandard feature.[146]

Thus, finite game trees with imperfect information are a natural extension of propositional formulas, leading to an intriguing calculus of equivalence in extended notation.[147] This theme will return in Parts IV and V of this book, showing how logical formulas themselves can act as game forms. But even without that more general background, it seems an interesting open problem to find a complete propositional logic of finite imperfect information games viewed as extended logical forms.

11.11 Forcing languages, uniform strategies, and knowledge

Again, there is a linguistic counterpart to power equivalence and outcome thinking. We need an epistemic version of the earlier forcing language (cf. Chapters 3 and 4), including knowledge operators, but now also with modified modalities

$\{G, i\}\varphi$ player i has a uniform strategy forcing only outcomes satisfying φ

This seems straightforward, but the epistemic setting hides a subtlety with outcomes. Indeed, the meaning of our modality in imperfect information games is not crystal clear.

EXAMPLE 11.8 Subtleties of interpretation
Suppose a player is in an end node x satisfying p, but cannot distinguish this from another end node y satisfying $\neg p$. Does the uniform strategy of doing nothing force p in the end node x, as we have always assumed so far? Yes, if we look at the

---

146 In the independence-friendly IF games of Chapter 21, this equivalence is expressed by $\forall x \exists y/\boldsymbol{x}\, Rxy \leftrightarrow \exists y \forall x/\boldsymbol{y}\, Rxy$, where slashes indicate informational independence.

147 Also striking is the occurrence character of the above rules. One occurrence of A need not have the same effect as several occurrences when dealing with uniform strategies. This is reminiscent of linear logic, whose game content will be explored in Chapter 20.

actual outcomes produced (just x), but not, if we think of the outcomes that are epistemically relevant to the player's deliberations, being both x and y. ■

Thus, before inquiring into this extended logic, we may have to worry about its intended interpretation. We will not propose an epistemic modal forcing logic here, but merely point to a connection with the logics of Chapter 3 for action and knowledge. If the truth of the assertion $\{G, i\}\varphi$ at game state s is to imply knowledge for player i at s that the forced proposition φ will hold, we need a broader notion of relevant outcomes. This also ties up with the earlier issue of knowing one's strategy and the interplay of know-how and know-that, as discussed in Chapter 4.

11.12 Conclusion

The power perspective on players of games has surfaced at several places in this book, and it will continue to do so. We have made this level of analysis precise, and we have shown how it can be studied by generalized modal logics over neighborhood models, lifting earlier notions such as bisimulation to power equivalence for games, and adding concrete spatial visualizations via topological games. Of course, there is an issue of adding further crucial game structure from Part I, such as knowledge and imperfect information. This can be done, and we found a strong connection between power invariance at this level and the classical Thompson transformations of game theory, although some non-trivial issues of interpretation arose with designing an epistemized logic. Another task that remains is adding preference logic as in Chapter 2, and studying a forcing language for what might be called players' optimal powers. Finally, moving beyond Part I, the logical dynamics of Part II applies just as well to the power view of games, but for evidence of this, we refer the reader to some recent literature on dynamic logics of acts that change neighborhood models.[148]

Even so, the style of analysis and the results in this chapter give ample reasons to conclude that our logical study in Parts I and II of this book generalizes smoothly to a natural global perspective on games that focuses on players' strategic powers.

---

148 In particular, we noted in Chapter 10 that neighborhood models can generalize dynamic-epistemic logic to deal with more fine-grained notions of evidence (van Benthem & Pacuit 2011). Another study of dynamics in neighborhood models is found in Zvesper (2010) for the purpose of analyzing epistemic game theory.

11.13 Literature

This chapter is based on van Benthem (2001b) and van Benthem (2002a). Use was also made of Aiello & van Benthem (2002).

We have pointed at the earlier work of Bonanno (1992b) on set-theoretic forms for extensive games. Also, we listed several good sources on neighborhood semantics. Finally, the results in Pauly (2001) on dynamic game logic and coalition logic are highly relevant, while Goranko et al. (2013) presents amendments and extensions.

12
Matrix Games and Their Logics

The best known format in game theory is different from both the extensive game trees that we have studied so far in this book and their associated power structures. The so-called *strategic form* gives players' complete strategies as single actions, where it is assumed that players choose these strategies simultaneously, resulting in a profile that can be evaluated at once. This is pictured in matrices for standard games such as Hawk versus Dove, or Prisoner's Dilemma, which were discussed briefly in our Introduction. While strategic forms are not a major focus in the book, this chapter and the next will show how they can be studied by means of the logical techniques at our disposal. Our main focus will be the additional perspectives on social action emerging in this way.

12.1 From trees and powers to strategic matrix forms

In Chapter 1, we defined extensive games as an object of study, and in Chapter 6, there was an extensive discussion of models for such games, ranging from game trees themselves to forest models and beyond to more abstract modal models for games. The following standard notion suffices for our purposes in what follows.

DEFINITION 12.1 Strategic games
A *strategic game* G for a set of players N is a tuple $(N, \{A_i\}, \{\leq_i\})$ with a non-empty set A_i of actions for each $i \in N$, and a preference order $\leq_i$ for each player on the set of *strategy profiles*, tuples of actions for each player. As usual, given a strategy profile $\sigma = (a_1, \ldots, a_n)$, σ_i denotes the ith projection and σ_{-i} denotes the remaining profile after deleting player i's action σ_i. ■

Strategic games are given by the familiar matrix pictures of game theory books. Extensive game trees resemble existing models in computational or philosophical logic, but these matrix structures, too, provide a direct invitation for logical analysis of action, preference, knowledge, and freedom of choice. We will show how this can be done, using notions and techniques from Parts I and II, and then develop portions of this logic having to do with independent and correlated action, best response, equilibrium, and various rationality assertions that occur naturally. We will mainly discuss finite two-player games, but most results extend to more players.

Strategic games come with an extensive theory of their own (Osborne & Rubinstein 1994, Gintis 2000, Perea 2011, Brandenburger et al. 2014), and toward the end of the chapter, we briefly discuss some possible further interfaces, such as evolutionary games and the role of probability. We will also show how the logics of this chapter fit with existing topics in philosophical logic such as the STIT analysis of deliberate action, where STIT stands for "seeing to it that".

12.2 Models for strategic games

To a logician, a game matrix is a semantic model that invites the introduction of modal languages, with the strategy profiles themselves as possible worlds. As in Chapter 6, we could also use more abstract worlds, carrying strategy profiles without being identical to them. This is common in epistemic game theory, but it is not needed in what follows. Over strategy profiles, players have three natural kinds of structure, through preference, epistemic view, and action freedom.

Preference The definition of strategic games contained preference relations:

preference $\quad \sigma \leq_i \sigma'$ iff player i prefers profile σ' at least as much as σ

This is the standard notion of preference that we began studying using the logics of Chapter 2. It is usually taken to run between outcomes of a game, and this is what we will do. However, our strategy models would also support another, more ambitious generic use of preference, namely, that players prefer using one strategy over using another.

Epistemic view Matrices also support other natural relations. Here is another basic one from a logical perspective:

epistemic view $\quad \sigma \sim_i \sigma'$ iff $\sigma_i = \sigma'_i$

This epistemic relation represents player i's view of the game at the interim stage where i's choice is fixed, but the choices of the other players are unknown.[149]

Action freedom There is yet a third natural relation for players i in game models:

action freedom $\quad \sigma \approx_i \sigma'$ iff $\sigma_{-i} = \sigma'_{-i}$

This relation of freedom gives the alternative choices for player i when the other players' choices are fixed.[150] The three relations are entangled in various ways, a topic that we will return to.

This structure can all be packaged in the following notion.

DEFINITION 12.2 Game models

Game models $\boldsymbol{M} = (S, N, \{\leq_i\}_{i \in N}, \{\sim i\}_{i \in N}, \{\approx_i\}_{i \in N}, V)$ are relational structures as described above, where a valuation V evaluates atomic propositions p on strategy profiles, viewed as special properties V(p) of these. The *full model* over a strategic game G is the game model $\boldsymbol{M}(G)$ whose worlds are all strategy profiles, with the three relations as defined above. Finally, a *general game model* is a submodel of a full game model. ■

We will find a logical and a game-theoretic use for general game models below.

What the models mean In this minimalist view, matrix games model a moment where players know their own action, but not that of the others, which happens at the time of decision when all the relevant available evidence is in. Richer models of games in the style of Chapter 6 would allow for knowledge or ignorance concerning other features, and also for more fine-grained beliefs that players can have about what will happen. Despite their intuitive importance, we will ignore these extra features at this point, as well as any issues having to do with probabilistic combinations of strategies.

Standard game matrices correspond directly to models in our style.

---

149 This is just one take on matrix games, albeit a common one. Bonanno (2012) provides an illuminating discussion of different construals of games in terms of deliberation versus actual choice. Likewise, Hu & Kaneko (2012) contains a careful disentanglement of various views of matrix games, in terms of predictions versus decisions.

150 The term freedom is taken from Seligman (2010), who points out that these three-relation structures make sense far beyond games. See Guo & Seligman (2012) for some further development.

EXAMPLE 12.1 Models from game matrices
Consider the following matrix, with this order of writing utility values for outcomes: ($\boldsymbol{A}$-value, $\boldsymbol{E}$-value). It will be discussed in more detail in a later section.

		$\boldsymbol{E}$		
		a	b	c
$\boldsymbol{A}$	d	$2,3$	$2,2$	$1,1$
	e	$0,2$	$4,0$	$1,0$
	f	$0,1$	$1,4$	$2,0$

The associated model is as follows with regard to epistemic view and freedom:

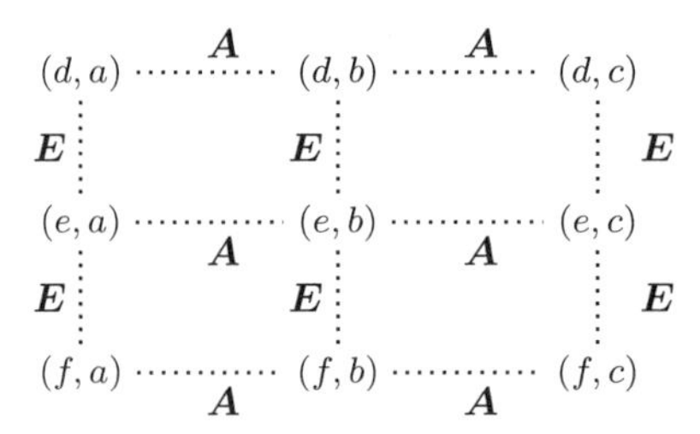

Here the uncertainty relation $\sim_{\boldsymbol{E}}$ runs inside columns, because $\boldsymbol{E}$ knows the action that $\boldsymbol{E}$ will take, but not that of $\boldsymbol{A}$. The uncertainty relation of $\boldsymbol{A}$ is symmetric, running inside the rows. ■

These concrete pictures will help understand later arguments. In a sense, the gestalt switch from matrix games to logical models is the main point of this chapter.

Alternative perspectives There are other ways of interpreting our pictures. We have already mentioned the diversity of interpretations in the game-theoretic literature. But also, one person's freedom can be another person's ignorance. Freedom is read in van Benthem (2007d) as "distributed knowledge" for player i in the sense of Fagin et al. (1995), stating what all other players, that is, the group $N - \{i\}$, know implicitly about the game.

12.3 Matching modal languages

Syntax The above three kinds of structure invite matching modalities. Game models obviously support the preference language of Chapter 2 with modalities $\langle pref_i \rangle$ or $\langle \leq_i \rangle$, since the latter can be interpreted on any model consisting of worlds with a betterness order. For matrix games, it also turns out to be useful to add modalities $\langle <_i \rangle$ for strict preference order as in van Benthem et al. (2009c). Likewise, game models interpret the epistemic language of Chapter 3 with knowledge operators K_i (or their existential variants $\langle know_i \rangle$) where a few special-purpose proposition letters may carry information about which player plays what action in a given world, or other relevant properties of strategy profiles. Finally, freedom can be described in the same style by modal operators $\langle free_i \rangle$.

DEFINITION 12.3 Logical language for matrix games
The syntax of a combined *logic of matrix games* can be given as follows:

$$p \mid \neg\varphi \mid (\varphi \vee \psi) \mid \langle pref_i \rangle\varphi \mid \langle <_i \rangle\varphi \mid \langle know_i \rangle\varphi \mid \langle free_i \rangle\varphi$$

We will often write $\langle \leq_i \rangle\varphi$ for $\langle pref_i \rangle\varphi$, $\langle \sim_i \rangle\varphi$ for $\langle know_i \rangle\varphi$, and $\langle \approx_i \rangle\varphi$ for $\langle free_i \rangle\varphi$ to emphasize the underlying accessibility relations for these modalities. ■

Expressive power With a little expressive boost, languages such as these can define basic features of social scenarios and of games in particular. A benchmark is Nash equilibrium of profiles that has been much studied by logicians (cf. van der Hoek & Pauly 2006). As it happens, no formula of our basic modal language can define this notion. The reason is that it is not invariant under bisimulation (cf. Chapter 1, and van Benthem 2010b). But we can help ourselves to one more technical device, namely, modalities for intersections of relations (see Dégremont 2010, Zvesper 2010 for its uses in analyzing extensive games). Recall from our Introduction that action σ_i is a *best response* for player i against given behavior σ_{-i} of the others if no alternative action for i will lead to an outcome more preferred by i. Then, enough expressive power is available.

FACT 12.1 (a) The modal definition BR_i for best response is $\neg\langle <_i \cap \approx_i \rangle\top$. (b) The strategy profiles in Nash equilibrium are those satisfying the formula $\bigwedge_{i \in N} BR_i$.

Thus we see how game models suggest the use of mild extensions of modal logic, often from the area of hybrid logics (Areces & ten Cate 2006).[151]

---

151 Van Benthem (2004b) has a hybrid logic with preference modalities, nominals for specific worlds, a universal modality, and distributed group knowledge, that can formalize the rationality principles found in Section 12.7. Hybrid preference logics are explored in van Benthem et al. (2006b), Guo & Seligman (2012) have further game examples.

12.4 Modal logics for strategic games

Basic modal logics Next, we explore a calculus of reasoning for strategic games. For convenience, we restrict our attention to two-player games. First, given the nature of our three relations, the separate logics are standard systems: modal $S4$ for preference, and modal $S5$ for both epistemic outlook and action freedom. What is of greater interest is the interaction of the three modalities. In general, logics of knowledge and action need not support strong bridge principles, as we have seen in Chapter 3; or, at the very least, such bridge principles express non-trivial assumptions about agents. But our models for strategic games are full of mutual connections, so we expect much more to be valid.

One strong connection involves the following natural combination of modalities.

FACT 12.2 $[\sim_i][\approx_i]\varphi$ makes a proposition φ true in each world of a game model.

Proof On matrix game models, the sequential composition of the earlier relations for epistemic view and freedom is precisely the universal relation. ■

This gives the base language a universal modality $U\varphi$ (with an existential dual $E\varphi$), a device used widely in Part I. This modality can even be defined in two ways.

FACT 12.3 The equivalence $[\sim_i][\approx_i]\varphi \leftrightarrow [\approx_i][\sim_i]\varphi$ is valid in matrix game models.

Proof This validity depends on the geometrical confluence property of matrices that, if one can go $x \sim_i y \approx_i z$, then there exists a point u with $x \approx_i u \sim_i z$. To visualize this, compare the confluence property for perfect recall in Chapter 3. ■

This system can prove interesting principles, including one that will serve later as a law of independence for simultaneous actions in STIT logics.

FACT 12.4 The formula $(\langle\approx_i\rangle[\sim_i]\varphi \wedge \langle\approx_j\rangle[\sim_j]\psi) \rightarrow E(\varphi \wedge \psi)$ is derivable.

Proof First we prove an auxiliary formula in our logic:

$$\langle\approx_i\rangle[\sim_i]\varphi \rightarrow [\sim_i]\langle\approx_i\rangle\varphi$$

The steps are as follows:

(a)	$\varphi \to [\approx_i]\langle\approx_i\rangle\varphi$	(in modal $S5$)
(b)	$[\sim_i]\varphi \to [\sim_i][\approx_i]\langle\approx_i\rangle\varphi$	(from (a) in basic modal logic)
(c)	$[\sim_i]\varphi \to [\approx_i][\sim_i]\langle\approx_i\rangle\varphi$	(from (c) by the commutation axiom)
(d)	$\langle\approx_i\rangle[\sim_i]\varphi \to \langle\approx_i\rangle[\approx_i][\sim_i]\langle\approx_i\rangle\varphi$	(from (c) in basic modal logic)
(e)	$\langle\approx_i\rangle[\sim_i]\varphi \to [\sim_i]\langle\approx_i\rangle\varphi$	(from (d) in modal S5)

Now consider the formula $\langle\approx_i\rangle[\sim_i]\varphi \wedge \langle\approx_j\rangle[\sim_j]\psi$. The left-hand side implies $[\sim_i]\langle\approx_i\rangle\varphi$. Also, given that $\approx_j$ equals $\sim_i$ for opposite players, the right-hand side $\langle\approx_j\rangle[\sim_j]\psi$ is equivalent to $\langle\sim_i\rangle[\approx_i]\psi$. But then a standard formal proof in the basic modal logic derives $\langle\sim_i\rangle(\langle\approx_i\rangle\varphi \wedge [\approx_i]\psi)$, and this again implies $\langle\sim_i\rangle\langle\approx_i\rangle(\varphi \wedge \psi)$, that is, $E(\varphi \wedge \psi)$. ■

Complexity of grid structure Confluence may look like a pleasant feature of matrices, but its effects on logical systems can be dangerous. As we have seen in Chapters 2 and 3, bimodal logics on grid models are often not decidable, and not even axiomatizable, once they have a universal modality available. The reason was that such logics can encode tiling problems of high complexity, and hence our simple-looking strategic matrix games can have a Π^1_1-complete logical theory encoding a large amount of geometry.

Finite games These high-complexity results might be circumvented for finite games. The latter realm validates special laws that fail on general game models. For instance, finiteness implies upward well-foundedness of the two-step relation $\sim_A ; \sim_E$, i.e., it has only ascending sequences of finite length. Well-foundedness validates a Grzegorczyk axiom for the epistemic modalities (cf. Blackburn et al. 2001). The logic of finite matrix games may be less complex than for all games, as infinite tiling problems cannot be encoded.

12.5 General game models, dependence, and correlation

There is another, more rewarding avenue toward defusing the complexity danger. It is well known that the complexity of bimodal logics may go down drastically when we allow more models. In particular, the earlier general game models left out some strategy profiles to create less grid-like patterns. Then the logic changes drastically to something much simpler.

THEOREM 12.1 The complete epistemic logic of general game models is multi-S5.

Proof This result is from van Benthem (1996), which proves the key fact that every multi-$S5$ model has a bisimulation with a general game model.[152] ■

While this may look ad hoc, general game models have their uses. For instance, in Chapter 13, we will see how they arise in solving games when external information may come in ruling out strategy profiles in the manner of Chapter 7.

But there is a deeper interpretation of the change to general game models, having to do with *dependence* and independence of actions. Leaving out profiles from a full matrix model creates intuitive dependencies between actions. If we change the choice for one player, we may have to change that for the other to stay inside the given universe of available profiles.

Such models with gaps have been studied by logicians (van Benthem 1997, and in a different, much more systematic vein, Väänänen 2007). The motivation is that dependence is a key notion in many areas of reasoning (see the independence-friendly logics of Chapter 21), as well as in studies of information as correlation among situations (van Benthem & Martínez 2008).

Game theorists, too, have studied general game models through an interest in "correlated behavior" (Aumann 1987, Brandenburger & Friedenberg 2008). Once we omit strategy profiles, what one player does may have repercussions for another, and this leads to new views of games.[153] Thus, complexity of logics for games matches interesting decisions on how we view players: as acting independently, or as being correlated.

12.6 Special logics of best response

Let us now look more concretely at the logic of matrix game models, in both full and general versions. In what follows, to simplify matters, we will not use the full generality of our earlier three relations, choosing a different approach instead. We treat some basic game-theoretic properties involving preference and action as atomic propositions, to keep the logic simple. While somewhat nonstandard, this is an acceptable modus operandi. In our views on logical zoom, there is no moral duty to formalize everything beyond what is perspicuous and convenient.

---

152 It is an open problem to extend this to game models with all three base relations.

153 Sadzik (2009) analyzes games with correlated equilibria in terms of distributed knowledge and a notion of bisimulation. Isaac & Hoshi (2011) discuss game equivalence with correlations in play.

Absolute best For a start, observe that best response as defined earlier is an absolute property: the conjunction over alternatives runs over all actions available in the original strategic game G, whether these occur in the general game model $\boldsymbol{M}(G)$ or not. Because of absoluteness, B_j may be viewed as an atomic proposition that keeps its value when models change from larger to smaller domains of profiles, or vice versa. This context independence simplifies notation considerably.

EXAMPLE 12.2 Expanded game models

Well-known games generate models of interest here. Consider the well known game of Battle of the Sexes with its two Nash equilibria (Osborne & Rubinstein 1994). The abbreviated diagram to the right has best-response atomic propositions at worlds where they are true:

		E	
		a	b
A	c	$2,1$	$0,0$
	d	$0,0$	$1,2$

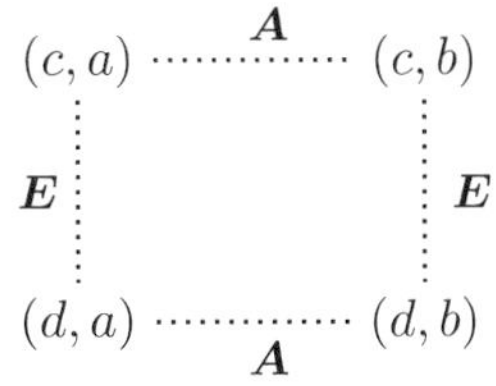

(c,a) ···· ***A*** ···· (c,b)

E ⋮ ⋮ ***E***

(d,a) ···· ***A*** ···· (d,b)

$B_{\boldsymbol{A}}, B_{\boldsymbol{E}}$	–
–	$B_{\boldsymbol{A}}, B_{\boldsymbol{E}}$

REMARK Mixed equilibria

This game also has an equilibrium in mixed strategies that we will not discuss here. This would require a probabilistic extension of the logics in this chapter that we do not provide, although we think it should be a straightforward generalization.

Likewise, Example 12.1 yields a full epistemic game model with nine worlds:

	a	b	c
d	$B_{\boldsymbol{A}}, B_{\boldsymbol{E}}$	–	–
e	$B_{\boldsymbol{E}}$	$B_{\boldsymbol{A}}$	–
f	–	$B_{\boldsymbol{E}}$	$B_{\boldsymbol{A}}$

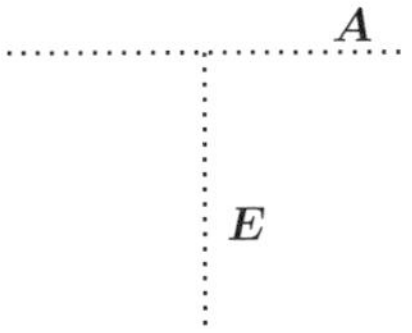

···· ***A*** ····

⋮ ***E***

As for the distribution of B_j atoms, by the above definition, every column in a full game model must have at least one occurrence of $B_{\boldsymbol{A}}$, and every row one of $B_{\boldsymbol{E}}$. ■

EXAMPLE 12.3 Evaluating epistemic statements

(a) The following formula of our modal language expresses that all players think their current actions might be best for them:

$$\langle \boldsymbol{E} \rangle B_{\boldsymbol{E}} \wedge \langle \boldsymbol{A} \rangle B_{\boldsymbol{A}}$$

As we shall see in Section 12.7, this expresses a form of rationality related to those that were discussed in Chapters 3, 7, and 8. In the preceding model, this proposition is true in exactly the six worlds occurring in the a and b columns.

(b) The same model also highlights an important epistemic distinction. B_j states that j's current action is in fact a best response at ω. But j need not know that, as j need not know what the other player is doing. Indeed, the formula $K_{\boldsymbol{E}}B_{\boldsymbol{E}}$ is false throughout the above model, even though $B_{\boldsymbol{E}}$ is true at three worlds. A fortiori, then, common knowledge of rationality in its most obvious sense is often false in the full model of a game, even one with a unique Nash equilibrium. ■

With this enriched language, the logic of game models gains interest. An example are valid laws for best response such as the following principle, subtly different in syntax from the one discussed above.

FACT 12.5 The formula $\langle \boldsymbol{E} \rangle B_{\boldsymbol{A}} \wedge \langle \boldsymbol{A} \rangle B_{\boldsymbol{E}}$ is valid in all full game models.

Proof Use the fact that all rows and columns have entries with maximal values. ■

Relative best There are also alternative views on game models. In particular, the word "best" in the phrase best response is context-dependent. A natural relative version of best response in a general game model $\boldsymbol{M}$ looks only at the strategy profiles available inside $\boldsymbol{M}$. Inside a model, players know these are the only action patterns that will occur.

DEFINITION 12.4 Relative best response
The *relative best response* proposition B_j^* in a general game model $\boldsymbol{M}$ is true at only those strategy profiles where j's action is a best response to that of the opponent when the comparison set is all alternative strategy profiles inside $\boldsymbol{M}$. ■

With relative B_j^*, best profiles for j may change as a model changes. For instance, in a one-world model for a game, the single profile is relatively best for all players, although it may be absolutely best for none.[154]

REMARK What others know
Relative best response has independent interest. With two players, it says that the other player knows that j's current action is at most as good for j as j's action at ω. More generally, continuing with an earlier observation, B_j^* says that

154 For a similar idea in an abstract computational view on game solution, see Apt (2007).

the proposition "j's current action is at most as good for j as j's action at ω" is distributed knowledge at ω for the rest of the players $G - \{j\}$.[155]

Absolute best obviously implies relative best, but the converse is not true.

EXAMPLE 12.4 All models have relative best positions
To see the difference between the two notions, compare the two models

$1, 1(B_{\boldsymbol{A}})$	$0, 2(B_{\boldsymbol{E}})$
$0, 2(B_{\boldsymbol{E}})$	$1, 1(B_{\boldsymbol{A}})$

$1, 1(B_{\boldsymbol{A}}, B^*_{\boldsymbol{E}})$	
$0, 2(B_{\boldsymbol{E}})$	$1, 1(B_{\boldsymbol{A}})$

Removing one entry left no absolute $B_{\boldsymbol{E}}$ in the first row, but $B^*_{\boldsymbol{E}}$ adjusted. ■

It would be of interest to axiomatize the modal logic of best response completely. In this chapter, we will only pursue one particular form of reasoning with the notions introduced here, involving two basic notions of rationality.

12.7 A case study: Rationality assertions, weak and strong

The notion of rationality was studied extensively in Parts I and II of this book. In strategic games, rationality means playing a best response given what one knows or believes. But this is not the whole story. Our game models support further distinctions, such as absolute versus relative best response. Also, even when players in fact play their best action, they need not know that they are doing so. Thus, if rationality is a self-reflexive property, as commonly assumed, what can players know? This issue will return in our conversation scenarios for solving games in Chapter 13, where players can only communicate things they know to be true.

Weak rationality Players may not know that their action is best, but they can know *there is no alternative action that they know to be better*. They are no fools.

155 In the dynamic terms of Part II of this book, the other players might learn this fact about j by "pooling" their information. This observation is used in van Benthem et al. (2006b) for defining Nash equilibrium in an extended epistemic preference logic.

Definition 12.5 Weak rationality
Weak rationality of player j at world ω in a model $\boldsymbol{M}$ is the assertion that, for each available alternative action, j thinks the current one may be at least as good:

$$WR_j \qquad \bigwedge_{a \neq \omega(j)} \langle j \rangle \text{ "}j\text{'s current action is at least as good for } j \text{ as } a\text{"}$$

The index set runs over worlds in the current model, as for relative best B^*_j.[156] ■

Weak rationality WR_j for j fails at those rows or columns in a two-player general game model that are strictly dominated for j. Unpacking quantifiers, $WR_{\boldsymbol{E}}$ says for a matrix column x that for each other column y, there is at least one row where $\boldsymbol{E}$'s value in x is at least as good as that in y. Such columns always exist, by a simple combinatorial argument.

Fact 12.6 Each finite general game model contains worlds where $\bigwedge_j WR_j$ is true.

Proof For convenience, look at games with just two players. We show something stronger, namely, that the model has "*WR* loops" of the form

$$s_1 \sim_{\boldsymbol{A}} s_2 \sim_{\boldsymbol{E}} \ldots \sim_{\boldsymbol{A}} s_n \sim_{\boldsymbol{E}} s_1 \qquad \text{with } s_1 \models B^*_{\boldsymbol{E}},\ s_2 \models B^*_{\boldsymbol{A}},\ s_3 \models B^*_{\boldsymbol{E}},\ \ldots$$

By way of illustration, a Nash equilibrium by itself is a one-world *WR* loop.

First, taking maxima on the available positions (column, row) in the full game matrix, we see that the following two statements must hold everywhere:

$$\langle \boldsymbol{E} \rangle B^*_{\boldsymbol{A}} \qquad \langle \boldsymbol{A} \rangle B^*_{\boldsymbol{E}}$$

For example, the first says that, given a world with some action for $\boldsymbol{E}$, there must be some world in the model with that same action for $\boldsymbol{E}$ where $\boldsymbol{A}$'s utility is highest. (This need not hold with the above absolute $B_{\boldsymbol{A}}$, as its witness world may have been left out.) Repeating this, there is a never-ending sequence of worlds

$$B^*_{\boldsymbol{E}} \sim_{\boldsymbol{E}} B^*_{\boldsymbol{A}} \sim_{\boldsymbol{A}} B^*_{\boldsymbol{E}} \sim_{\boldsymbol{E}} B^*_{\boldsymbol{A}} \cdots$$

that must loop since the model is finite. Thus, some world in the sequence with, say, $B^*_{\boldsymbol{A}}$ is $\sim_{\boldsymbol{A}}$-connected to some earlier world w. Now, either w has $B^*_{\boldsymbol{E}}$, or w has a successor with $B^*_{\boldsymbol{E}}$ via $\sim_{\boldsymbol{A}}$ in the sequence. The former case reduces to the latter by the transitivity of $\sim_{\boldsymbol{A}}$. But then, looking backward along such a loop, using

156 An alternative version lets the index set run over all strategy profiles in the whole initial game, as with absolute best assertions B_j. It can be dealt with similarly.

the symmetry of the relations, we have a WR loop as defined above. Its worlds evidently validate weak rationality for both players: $\langle \boldsymbol{E} \rangle B^*_{\boldsymbol{E}} \wedge \langle \boldsymbol{A} \rangle B^*_{\boldsymbol{A}}$. ■

FACT 12.7 Weak rationality is epistemically introspective.

Proof By the accessibility in epistemic game models, if WR_j holds at some world ω in a model, it also holds at all worlds that j cannot distinguish from ω. Hence, the epistemic principle $WR_j \rightarrow K_j WR_j$ is valid on general game models. ■

Thus, $WR_j \rightarrow K_j WR_j$ is a logical law of game models with best response and rationality. This makes weak rationality suitable for public announcement, ruling out worlds on strictly dominated rows or columns every time when uttered.

Strong rationality Weak rationality is a logical conjunction of epistemic possibility operators: $\bigwedge \langle j \rangle$. A stronger form of rationality assertion inverts this order, expressing that players think that *their actual action may be best.* Instead of merely being no fools, they now reasonably hope they are being clever.

DEFINITION 12.6 Strong rationality
Strong rationality for player j at a world ω in a model $\boldsymbol{M}$ is the assertion that j thinks that the current action may be at least as good as all others:

$$SR_j \qquad \langle j \rangle \bigwedge_{a \neq \omega(j)} \text{“}j\text{'s current action is at least as good for } j \text{ as } a\text{”}$$

This time we use the absolute index set, running over all action profiles in the game. Hence, the assertion can be written equivalently as the modal formula $\langle j \rangle B_j$. Strong rationality for the whole group of players is the conjunction $\wedge_j SR_j$. ■

By the $S5$-law $\langle j \rangle \varphi \rightarrow K_j \langle j \rangle \varphi$, SR_j is something that players j will know if it is true. Thus, it behaves like WR_j. Strong and weak rationality are related as follows.

FACT 12.8 SR_j implies WR_j, but not vice versa.

Proof Consider the following game model, with B atoms indicated:

		$\boldsymbol{E}$	
	a	b	c
$\boldsymbol{A}$ d	1, 2	1, 0	1, 1
e	0, 0	0, 2	2, 1

$B_{\boldsymbol{A}}, B_{\boldsymbol{E}}$	$B_{\boldsymbol{A}}$	–
–	$B_{\boldsymbol{E}}$	$B_{\boldsymbol{A}}$

No column or row dominates any other, and WR holds throughout for both players. But $SR_{\boldsymbol{E}}$ holds only in the two left-most columns. This is because it rejects actions that are never best, even though there need not be one alternative in the model that is better overall. ■

One advantage of SR_j over WR_j is the absoluteness of its proposition letters B_j. In the dynamics of Chapter 13, this feature underlies the monotonicity of the set transformation defined by announcing *SR*. Also, strong rationality has a direct game-theoretic meaning. It says that the current action of the player is a best response against at least one possible action of the opponent. This is the key assertion in rationalizability views of game solution, due to Bernheim and Pearce (cf. de Bruin 2004, Apt 2007), where one discards actions for which a better response exists under all circumstances.

Strong rationality need not be satisfiable in all general game models. But it is satisfiable in all full game models, thanks to the maximal utility values in rows and columns. The following result explains the logical import.

THEOREM 12.2 Each finite full game model has worlds verifying strong rationality.

Proof Much as with Theorem 12.2, there are "*SR* loops" of the form

$$s_1 \sim_{\boldsymbol{A}} s_2 \sim_{\boldsymbol{E}} \cdots \sim_{\boldsymbol{A}} s_n \sim_{\boldsymbol{E}} s_1 \qquad \text{with } s_1 \models B_{\boldsymbol{E}},\ s_2 \models B_{\boldsymbol{A}},\ s_3 \models B_{\boldsymbol{E}},\ \ldots$$

Instead of a proof, we give an illustration showing that each finite full game model has three-player *SR* loops. In such models, by the earlier observations about maxima on rows and columns, the following is true everywhere:

$$\langle \boldsymbol{B}, \boldsymbol{C} \rangle B_{\boldsymbol{A}}, \quad \langle \boldsymbol{A}, \boldsymbol{C} \rangle B_{\boldsymbol{B}}, \quad \langle \boldsymbol{A}, \boldsymbol{B} \rangle B_{\boldsymbol{C}}$$

Here the special modalities $\langle i, j \rangle$ have an accessibility relation $\sim_{\{i,j\}}$ keeping the coordinates for both i and j the same, that is, the intersection of $\sim_i$ and $\sim_j$.

But then, repeating this, by finiteness, we must have loops of the form $B_{\boldsymbol{A}} \sim_{\{\boldsymbol{A},\boldsymbol{C}\}} B_{\boldsymbol{B}} \sim_{\{\boldsymbol{A},\boldsymbol{B}\}} B_{\boldsymbol{C}} \sim_{\{\boldsymbol{B},\boldsymbol{C}\}} B_{\boldsymbol{A}} \cdots$ returning to the initial world with $B_{\boldsymbol{A}}$. Any world in such a loop satisfies $\langle \boldsymbol{A} \rangle B_{\boldsymbol{A}} \wedge \langle \boldsymbol{B} \rangle B_{\boldsymbol{B}} \wedge \langle \boldsymbol{C} \rangle B_{\boldsymbol{C}}$. For example, if the world itself has $B_{\boldsymbol{A}}$, by reflexivity, it satisfies $\langle \boldsymbol{A} \rangle B_{\boldsymbol{A}}$. Looking back at its parent $B_{\boldsymbol{C}}$ via $\sim_{\{\boldsymbol{B},\boldsymbol{C}\}}$, by symmetry, it has $\langle \boldsymbol{C} \rangle B_{\boldsymbol{C}}$. And looking at its grandparent $B_{\boldsymbol{B}}$ via $\sim_{\{\boldsymbol{B},\boldsymbol{C}\}}$ and $\sim_{\{\boldsymbol{A},\boldsymbol{B}\}}$, by transitivity, it also satisfies $\langle \boldsymbol{B} \rangle B_{\boldsymbol{B}}$. ■

In infinite game models, *SR* loops need not occur, and irrationality may prevail.

EXAMPLE 12.5 Irrationality in regress
Consider a grid of the form $N \times N$. Suppose that the best-response pattern has the predicates $B_{\boldsymbol{A}}$ and $B_{\boldsymbol{E}}$ occurring only on the two marked diagonals:

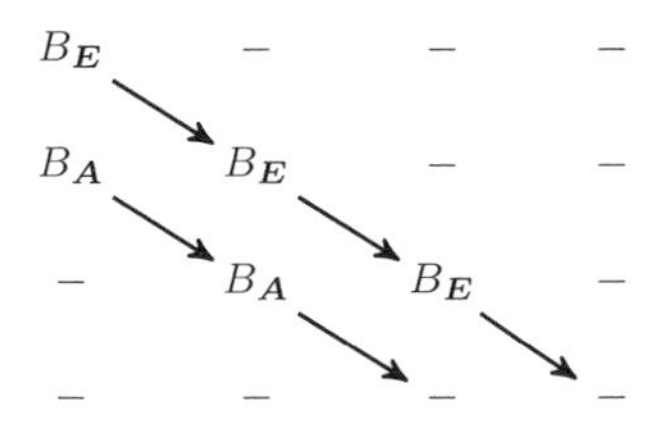

Every putative sequence $B_E \sim_A B_A \sim_E B_E \sim_A B_A \cdots$ must break off. ■

Our case study will have shown how much surprising logical structure there is to reasoning with rationality once we look at models and syntax. It would be of interest to axiomatize the complete logic of general and full game models expanded with B_j, B_j^*, WR_j, and SR_j, either in our epistemic base language, or with the expressive additions from hybrid logic that we have mentioned occasionally. But we leave matters at this stage here, returning to rationality assertions only in the dynamic-epistemic setting of Chapter 13.

12.8 STIT logic and simultaneous action

The modal systems in this chapter may also be viewed as basic logics of simultaneous action, outside of the setting of games. In this coda to our main theme, we present an illustration, in the form of a brief excursion toward a well-known philosophical paradigm for agency called STIT (seeing to it that). For details of what follows here, the reader may consult van Benthem & Pacuit (2012). We also refer to classical sources for more extensive development of STIT, such as Horty & Belnap (1995) and Belnap et al. (2001), while Horty (2001) connects the framework with deontic logics of agency.

Models, language, and logic Models for STIT are branching time structures as defined in Chapter 5, with the following new feature. At each stage (h, t) of a point t on history h, each agent i is assigned a partition $C_i(h, t)$ of all future histories passing through t. Partition cells are choices that the agent has at time t. These choices satisfy one important constraint, expressing the independence of actions chosen by agents:

Independent action   Any two partition cells for different agents overlap.

These models support a temporal language as in Chapter 5, but we now add modalities for agents who see to it that some stated proposition is achieved:

STIT modality $\boldsymbol{M}, h, t \models [stit, i]\varphi$ iff there exists a partition cell X of $C_i(h, t)$ such that, for all histories h' in X, $\boldsymbol{M}, h', t \models \varphi$

The logic of this system is upward monotonic in these modalities, but not distributive over conjunction or disjunction: this can be seen by analogy to the forcing modalities of Chapter 11. The independence constraint validates the following:

Product Axiom $([stit, i]\varphi \wedge [stit, i]\psi) \rightarrow \Diamond(\varphi \wedge \psi)$

where $\Diamond$ is the existential modality over current branches introduced in Chapter 5.

Intuitively, one might think that actions achieve effects, not right now, but later in time. This can be incorporated using modal combinations $[stit, i]O\varphi$, where O stands for at the next moment (Broersen 2011) or $[stit, i]G\varphi$.

Two views of action STIT embodies a view of action as *choice* plus *control* of agents over outcomes, where it is crucial that actions may not give unique control over outcomes. This slack may be because other agents have control as well over what happens, or even with a single agent, multiple histories in a partition cell may arise from actions of the environment. Prima facie, this view of choosing an action looks different from the one in this book, which comes from the dynamic and temporal logics in Part I. There, as emphasized in Chapter 6, branchings between events or moves model possible choices for the agent, and labels provide explicit names for transitions.

The two styles of thinking are not mutually exclusive, however. Here is a way of combining insights from both. Despite the loose talk interchanging events, moves, and actions in this book (and most of the literature), modal logics of action, computation, and games really describe a structure of possible *events*. What STIT then adds is the idea of control that agents can have over events. Thus, a richer view of action doing justice to both would be this:

$$Action = Events + Control$$

Embedding STIT into matrix game logic One way of benefiting from this junction is by embedding the basic STIT logic in the logics of this chapter. The following result comes from van Benthem & Pacuit (2012), but Herzig & Lorini (2010) contains essentially the same insight.

Recall the matrix game models of Section 12.3, where we think of one player i choosing an action a while others choose theirs independently to create a complete strategy profile, that is, a tuple of individual actions. We can think of these composite simultaneous actions as labeling transitions in a temporal STIT model. The following connection then works at the level of languages.[157]

DEFINITION 12.7 Modal game translation
In terms of the modalities of Section 12.4, one can translate as follows

$$[stit, i]\varphi = [\sim_i]\varphi \qquad \Diamond\varphi = \langle\sim_i\rangle\langle\approx_i\rangle\varphi$$

Arbitrary complex formulas are then translated compositionally. ■

THEOREM 12.3 The preceding translation embeds the basic STIT logic faithfully into the modal logic of full matrix games.

Proof We look at only two agents. Our translation validates all STIT axioms, where the freedom modality refers to choices, and the action modality to simultaneous actions compatible with them. The Product Axiom holds on all game models, and indeed we derived it in modal game logic in Section 12.4.

To prove that the embedding is faithful, we must refute each non-valid STIT formula φ in our matrix models. To do so, consider any STIT temporal counter-model, and note that it suffices to look at the current moment where $\neg\varphi$ holds and its successor moments, since our basic STIT language does not contain temporal modalities. Thus, one can simply derive a two-agent basic STIT $S5$ model out of the temporal structure by letting histories be worlds, and defining agent's equivalence relations from their choice partitions. Now, the temporal box $\Box$ is the universal modality, while the two STIT modalities match the two equivalence relations. In principle, this might just lead to a general game model as defined earlier. However, the structure of the STIT model ensures that any two equivalence classes intersect.

Crucially, we can "unravel" this $S5$-model into a bisimilar model of a special form where the intersection of any two equivalence classes is a singleton. Then, a representation in matrix form is obvious: the actions are the equivalence classes for the agents, the unique outcomes are their intersections.[158]

---

157 One might want to prefix an existential modality over choices for i in the translation that follows, but the STIT modality is about the choice made on the actual history.

158 We suppressed important details here. A complete proof, including a bisimulation unraveling by means of a product construction, is in van Benthem & Pacuit (2012).

It follows in a straightforward manner that, under our translation, any STIT-satisfiable formula is also satisfiable in matrix games. Thus, our translation is both correct and faithful.[159] ■

The role of knowledge Our connection between STIT and matrix games introduced a notion of knowledge for agents that have decided, but do not know yet what the others have chosen. This is just one of several notions of knowledge that are important to games, as discussed in Chapters 3 and 6, when looking at information that players have about the past, present, and future of play. Knowledge is not made explicit in the STIT framework, but it is behind the scenes.[160] One can make this feature explicit by moving in the opposite direction of the above, involving the key system used in Part II.

Extending DEL with control Consider a dynamic-epistemic language as in Chapter 7 with knowledge modalities $K_i\varphi$ and dynamic event modalities $[\boldsymbol{\mathcal{E}}, e]\varphi$. Control fits very easily in this setting. First, we endow event models with equivalence relations for control by the relevant agents. Then we can define a dynamic STIT-style "control modality" as follows:

$\boldsymbol{M}, s \models [\boldsymbol{\mathcal{E}}, e, control_i]\varphi$ iff in all product models $(\boldsymbol{M}, s) \times (\boldsymbol{\mathcal{E}}, f)$
for all events f that are control equivalent in $\boldsymbol{\mathcal{E}}$ to e for agent i,
$(\boldsymbol{M}, s) \times (\boldsymbol{\mathcal{E}}, f), (s, f) \models \varphi$.

This is like a DEL operator stating the knowledge that an agent has acquired after product update with the current event model $\boldsymbol{\mathcal{E}}$.[161] The complete dynamic logic of this expanded system lies embedded in DEL in an obvious manner. Its laws for the new control operator are essentially those of STIT. Thus, we get a combined logic of events and agent's choices that might enrich our analysis of games in Part II.[162]

---

159 If we make our modal language stronger, however, the singleton intersection property would give rise to a special axiom beyond basic STIT.

160 A recent proposal for adding knowledge to STIT while making a junction with the modal logic of games in this chapter is in Ciuni & Horty (2013).

161 Control in private-information scenarios is harder to interpret intuitively.

162 However, this richer logic of agent's choices has special features. It lacks the characteristic recursion laws of Chapter 7 directly for the STIT control modality, since this operator does not distribute over conjunction or disjunction. Also, unlike most DEL logics, this new system does not have a modality reflecting its dynamic control relations in the static epistemic base models. The latter feature makes event models really sui generis.

Simultaneous action in extensive games Let us now generalize to interactive behavior over time, as in the extensive games of Parts I and II. Our presentation of STIT emphasized single moments of simultaneous choice. Next, consider players' consecutive choices in longer games, represented by the strategic powers of Chapter 11. We cannot quite adopt the simplest set of constraints here, since determinacy is lost in a setting of simultaneous action. Yet the STIT constraints on choice seem close to those in the representation results for powers in imperfect information games in Section 11.9 that required only monotonicity and consistency.

Instead of pursuing this, we make an observation in the main line of this chapter. What happens to the key constraints on choice when we consider iterated simultaneous action? Most importantly, the crucial "partition property" for histories under single choices disappears. When we make consecutive choices, the space of available strategies grows. In a one-step game, agents can only choose one of their actions ab initio. But now, they can let the next action depend on observed behavior of other agents. A famous case is the Tit for Tat strategy in evolutionary game theory (see also our Introduction): it copies the opponent's preceding move. Hence, strategies available in extensive games need not just choose one action uniformly; they can depend on the behavior of others. It is easy to see that disjointness for sets of outcomes, that is, the powers matching players' strategies, may fail now, and what is left is just monotonicity and consistency.

REMARK Public observation
This richer set of strategies does depend crucially on public observation of moves. Without this, players cannot make their choices dependent on what others have done, and we get a DEL product model of two consecutive actions that does satisfy the partition condition. In terms of Section 12.5, one-step simultaneous action does not allow for sequential dependence of actions, but it may allow for correlation.

Dynamifying STIT Finally, the DEL perspective of Part II also suggests a more radical move, reanalyzing the scenarios that motivated STIT in the first place. What are the main events that occur in a choice scenario? The main stages would seem to be: deliberation, decision, action, and observation. In a first stage, we analyze our options, and find optimal choices. Next, at the decision stage, we make up our mind and choose an action. Then everyone acts publicly, and this gets observed, something we can also model as a separate stage, although things happen simultaneously. All of these stages can be analyzed using the DEL-style models of Part II: Chapter 8 discussed deliberation, while Chapter 9 had many systems for mid-game scenarios. Taken together, this might give a much richer view of simultaneous action than those suggested above.

Further directions Much more can be said about merges of control-based action with preference structure and temporal evolution. We refer to the cited literature, as well as Broersen et al. (2006) and Xu (2010) for some recent developments.

12.9 Conclusion

In this chapter, we have connected the logical approach of this book to the standard format of game theory, that is, games in strategic form. The link arose by sensitizing the reader to a gestalt switch. The simple matrix pictures in any textbook on game theory are models for sophisticated modal logics of action, knowledge, and preference. We introduced these logics originally for extensive games in Part I, but they turn out to make equal sense for strategic games, although with a number of new twists. We have developed a bit of the resulting reasoning about knowledge, freedom, best response, and rationality, showing a number of interesting valid patterns, and finding sources of complexity in assumptions of dependence and independence.

Many new research problems come to light in this view, which has not been as well-investigated as the extensive games of Parts I and II. These have occurred at various places in this chapter, concerning complexity of validity, and complete axiomatizations of various logics, modal or extended hybrid, of full and general matrix game models. In addition, strategic form games raise issues beyond those investigated in earlier chapters, including correlations between behavior of players, linking with a current interest in dependence logics, and with reasoning about simultaneous action.

12.10 Literature

This chapter is based on van Benthem (2007d), van Benthem et al. (2011), and van Benthem & Pacuit (2012).

For further links between logic and classical game theory, offering a wide spectrum of motivations and tools, see Battigalli & Bonanno (1999a), Stalnaker (1999), Bonanno (2001), Halpern (2003a), Brandenburger & Keisler (2006), Kaneko (2002), Lorini et al. (2009), Kaneko & Suzuki (2003), and Fitting (2011). Bonanno (2012) surveys ways of construing logics for games under different scenarios of exploratory and in-game thinking, with an insightful analysis of the role of strategies. Surveys of recent contacts between modal logic and games are found in van der Hoek & Pauly (2006), de Bruin (2010), Dégremont (2010), and Zvesper (2010).

12.11 Further directions

This chapter proposed a modal logic perspective on games in strategic form, but many further directions remain to be explored. Here are a few.

From knowledge to belief The preceding discussion has been couched entirely in terms of knowledge. But as in Part II, agency is usually driven by beliefs, and so one would like to redo things in that style. One simple format would add the plausibility orderings of Chapter 7 to our game models, modeling players' beliefs as what is true in all most plausible epistemically accessible strategy profiles. Again, one could think of these expectations as produced by a dynamic process, either one of deliberation, as explored in Chapter 8, or of external information signals about behavior of other players, as discussed in Chapter 9. There is no obstacle in principle to such an extension of our logics for strategic games, and in fact, doing so may get closer to belief-based models of games used by game theorists (cf. Battigalli & Bonanno 1999a, Perea 2011, Brandenburger et al. 2014).

From games to general social action While our matrix logics stay close to strategic games, we can also view them as a way of analyzing more general social scenarios. Hu & Kaneko (2012) give a case study, tying logic to the general postulates for competitive games in Johansen (1982), that extract the basics of styles of social interaction at a higher abstraction level. Moving beyond games, the authors also make interesting connections with philosophical views on determinism and free will as guiding our intuitions about social action, using logics with simultaneous fixed points, such as those in Parts I and II of this book. Finding general postulates characterizing major styles of social interaction is an intriguing challenge for our logical approach as well.

Evolutionary game theory Matrix games are usually seen as one-shot scenarios, but, infinitely repeated, they are also the standard building blocks for evolutionary games (Maynard-Smith 1982, Hofbauer & Sigmund 1998, Gintis 2000). Evolutionary game theory is a major pillar of modern game theory, and in the form of signaling games whose equilibria set up matches between real world situations and code signals, it has found uses in fields such as linguistics and philosophy, witness the examples in our Introduction (cf. Skyrms 1996, van Rooij 2004, Clark 2011).

Some key notions of evolutionary game theory fall directly under the style of analysis in this chapter. For instance, Kooistra (2012) shows, following Gintis (2000), how to collapse players and strategies in evolutionary games, leading to matrix models with a relation $\tau \leq_\sigma \tau'$ saying that strategy τ' leads to better results than strategy τ from the perspective of the σ player. This reflects our earlier remark about preferences between strategy profiles rather than outcomes. Over these models, evolutionarily stable equilibria can be defined using the above modal logics plus some hybrid gadgets.

Another approach to evolutionary games would use the temporal logic of forcing in infinite games developed in Chapter 5, now in a version including simultaneous moves by all players. It seems quite feasible to generalize most of the results that we obtained for sequential action, perhaps exploiting analogies with alternating-time temporal logic (Alur et al. 2002).

However, the main feature of evolutionary games is the temporal progression of a dynamical system driven by recursion equations of fitness as affected by encounters described in a matrix game. The logical dynamics of such a system calls for a representation of the resulting *population changes* rather than belief changes for individual players. Therefore, we need systems in the spirit of Chapter 7 that update numbers of players following the given strategies. In addition to this local dynamics, evolutionary games typically show long-term temporal behavior that is sui generis. The logics of Chapters 4 and 5 may have something to contribute here – witness our brief discussion of typical evolutionary strategies such as Tit for Tat (Axelrod 1984) – and the same may be true for the learning scenarios via limits of updates introduced in Chapter 7 (cf. Leyton-Brown & Shoham 2008, Hutegger & Skyrms 2012 for evolutionary views on learning).

There have hardly been any logical studies of general dynamical systems per se (but see Kremer & Mints 2007 for a dynamic topological logic capturing some key recurrence patterns in limit behavior). It should be easy to add more expressive temporal languages capturing further features of infinite long-term behavior.

Evolutionary games are a natural extension of the concerns in this book, describing phenomena of mass behavior or public opinion that reign when details of deliberation have long been forgotten. They also pose a challenge of interfacing the dynamic logics in Part II, driven by discrete recursion laws for short episodes, with the differential equations driving the long-term behavior of dynamical systems. The current lack of a natural connection between these two worlds seems to be one of the major open ends in this book.

Probability, belief, and behavior Another major open end is the missing analysis of the use of probability of game theory. Probabilities enter in various guises in the study of agency. They may represent strengths of belief for individual agents, but they may also stand for frequencies of behavior in entire populations, or aggregated experiences in memory (Bod et al. 2003). There is no obstacle in principle to combining the logics in this book with either kind of probability. As noted in Chapter 7, van Benthem et al. (2009b) show how to extend update in dynamic-epistemic logics with both individual strengths of belief and more frequency-like process probabilities. But as pointed out in the Introduction, probabilities play one more important role in game theory, as they extend the set of possible behaviors making every matrix game have a Nash equilibrium in mixed strategies. The closest logical counterpart to such a construction in this book may be the imperfect information games of semantic evaluation in Chapter 21, where the logical universe of objects and functions can get probabilized. A lot more can and should be said on these interfaces of logic and probability in the study of social behavior, but, perhaps unwisely, this topic is outside the scope of this book.

13
Rational Dynamics for Strategic Games

Chapter 12 has shown how the logics in Part I of this book fit well with games in strategic form, while raising new issues about simultaneous action. In this chapter, we turn from static game structure to the dynamics of game solution, which brings us to the logical dynamics of information-driven agency studied in Part II of this book. We show how the techniques of Chapter 7 and later ones provide a new take on solution procedures for games in strategic form, by using processes of deliberation in terms of iterated public announcement. In the course of the exposition, new contacts emerge with dynamic-epistemic logics, fixed point logics, and other earlier topics. While our main focus is on iterated removal of strictly dominated strategies,, the reader will see a style of thinking that works more generally, as we will show in a number of examples.[163]

13.1 Reaching equilibrium as an epistemic process

Iterative solution Solving games often uses an algorithm for finding optimal strategies, as Backward Induction did for extensive games. We now demonstrate a classic method that works for solving, or at least pruning, games in strategic form.

EXAMPLE 13.1 Iterated removal of strictly dominated strategies (SD^{ω})
Consider the following matrix, with this legend for pairs ($\boldsymbol{A}$-value, $\boldsymbol{E}$-value):

163 This chapter depends heavily on Chapter 15 of van Benthem (2011e).

			E	
		a	b	c
	d	$2,3$	$2,2$	$1,1$
A	e	$0,2$	$4,0$	$1,0$
	f	$0,1$	$1,4$	$2,0$

First remove the dominated right-hand column: ***E***'s action c. After that, the bottom row for ***A***'s action f has become strictly dominated, and after its removal, ***E***'s action b becomes strictly dominated, and then ***A***'s action e. In the end, these successive removals leave just the Nash equilibrium (d, a). ■

In this example, SD^ω reaches a unique equilibrium profile. In general, it may stop at some larger solution set of matrix entries where it performs no more eliminations. There is an extensive literature analyzing such game-theoretic solution concepts in terms of epistemic logic, defining the optimal profiles in terms of common knowledge or common belief of rationality. Instead, we will give a dynamic take.

Solution methods as epistemic procedures For convenience, we repeat the puzzle of the Muddy Children, already discussed in Chapter 7, whose game-theoretic importance has been highlighted in Fagin et al. (1995) and Geanakoplos (1992).

EXAMPLE 13.2 The puzzle of the Muddy Children
After playing outside, two of three children have mud on their foreheads. They can only see the others, so they do not know their own status. Now their father says: "At least one of you is dirty." He then asks, "Does anyone know if he is dirty?" The children answer truthfully. As questions and answers repeat, what happens?

Nobody knows in the first round. But in the next round, each muddy child can reason like this: "If I were clean, the one dirty child I see would have seen only clean children, and so she would have known that she was dirty at once. But she did not. So I must be dirty, too!" ■

An initial epistemic model for this puzzle has eight possible worlds assigning ***D*** (dirty) or ***C*** (clean) to each child. What children know is reflected in the accessibility lines in the diagrams below. The given assertions update this information.

EXAMPLE 13.2, CONTINUED Updates for the muddy children
Updates start with the father's public announcement that at least one child is dirty. As with the public announcements of Chapter 7, this eliminates all worlds from the initial model where the stated proposition is false. That is, CCC disappears:

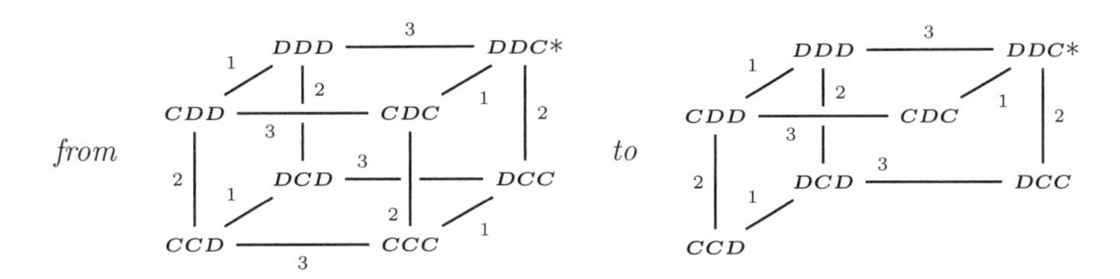

When it is announced that no one knows his status, the bottom worlds disappear:

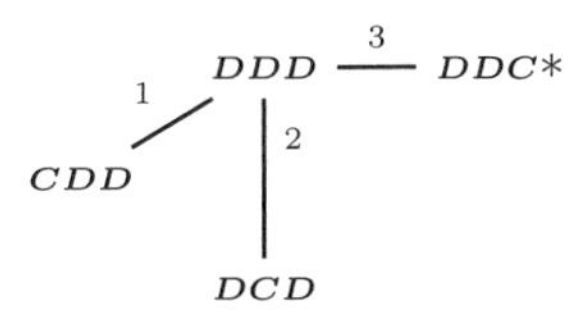

The final update is to DDC* ■

With k muddy children, k rounds of stating the ignorance assertion "I do not know my status" by everyone yield common knowledge about which children are dirty. Thus, the puzzle revolves around iterated announcement of one key proposition.

The same setting also analyzes effects of changes in the procedure, bringing to light an order dependence. If only the first child says he or she does not know, DCC is eliminated. Then the second child knows his or her status. Announcing this takes out all worlds but DDC and CDC. In the final model, it is common knowledge that *2* and *3* know, but here things stop: ***1*** never finds out who is clean or dirty by further epistemic assertions.

This same idea of *repeated announcement* applies to our basic solution algorithm.

EXAMPLE 13.2, CONTINUED Updates for SD^{ω} rounds

Here is the sequence of updates for the rounds of the preceding algorithm:

1	2	3
4	5	6
7	8	9

1	2
4	5
7	8

1	2
4	5

1
4

1

Each box may be viewed as an epistemic game model. Each step increases players' knowledge, until an equilibrium sets in where they know as much as they can. ■

The static logic of these matrix models was studied at length in Chapter 12, giving us a flying start in what follows.

13.2 Iterated announcement of rationality and game solution

Deliberation as conversation We now cast iterative solution as a virtual conversation. Take some game model $\boldsymbol{M}$ with actual world s. Players "deliberate in synch" by thinking of what they know about the scenario. As with the Muddy Children problem, we use generic assertions in the language of knowledge and best response. Our first result recasts SD^{ω} in terms of a notion from Chapter 12.

THEOREM 13.1 The following are equivalent in full game models $\boldsymbol{M}(G)$:

(a) World s is in the SD^{ω} solution zone of $\boldsymbol{M}(G)$.

(b) Repeated successive announcement of weak rationality for players stabilizes at a submodel $\boldsymbol{N}, s$ whose domain is that solution zone.

The harmony in this result is immediate, once we recall what weak rationality said. Announcing it prunes successive general models of G in just the right way.

The general program The theorem restates what we already knew. But its main point is the style of analysis. We can now match games and epistemic logic in two directions. Going from games to logic, given some algorithm defining a solution concept, we look for epistemic actions driving its dynamics. Going from logic to games, any type of epistemic assertion defines an iterated solution process that may have independent interest. Of course, this generality will not do much good unless we can find other cases.

Other scenarios: Announcing strong rationality Instead of weak rationality (WR), we can announce another driving formula in deliberation, namely strong rationality (SR). An announcement that player j is strongly rational leaves only states s where SR_j holds. But this may eliminate worlds, invalidating SR_k at s for other players, as their existential modalities now lack a witness. Thus, repeated announcement of strong rationality makes sense.

EXAMPLE 13.3 Iterated announcement of strong rationality
Our running example gives the same model sequence for SR as with SD^{ω}. But the strong rationality sequence differs from that for weak rationality in the following modified setting:

		E		
		a	b	c
A	d	2, 3	1, 0	1, 1
	e	0, 0	4, 2	1, 1
	f	3, 1	1, 2	2, 1

$B_{\boldsymbol{E}}$	–	–
–	$B_{\boldsymbol{E}}, B_{\boldsymbol{A}}$	–
$B_{\boldsymbol{A}}$	$B_{\boldsymbol{E}}$	$B_{\boldsymbol{A}}$

WR does not remove anything; *SR* removes the top row and rightmost column. ■

EXAMPLE 13.4 Ending in *SR* loops

In the following model, iterated announcement of strong rationality gets stuck in a four-loop of worlds:

$B_{\boldsymbol{E}}$	–	$B_{\boldsymbol{A}}$
–	$B_{\boldsymbol{A}}$	$B_{\boldsymbol{E}}$
$B_{\boldsymbol{A}}$	–	$B_{\boldsymbol{E}}$

$B_{\boldsymbol{E}}$	$B_{\boldsymbol{A}}$
–	$B_{\boldsymbol{E}}$
$B_{\boldsymbol{A}}$	$B_{\boldsymbol{E}}$

$B_{\boldsymbol{E}}$	$B_{\boldsymbol{A}}$
$B_{\boldsymbol{A}}$	$B_{\boldsymbol{E}}$

Nash equilibria in an initial game survive into the limit of the *SR* announcement procedure, but sometimes, more remains. ■

EXAMPLE 13.5 Battle of the Sexes, revisited

In the matrix model of Example 12.2 for Battle of the Sexes, announcing strong rationality would get stuck: the defining formula $\langle \boldsymbol{E} \rangle B_{\boldsymbol{E}} \wedge \langle \boldsymbol{A} \rangle B_{\boldsymbol{A}}$ for *SR* is already true everywhere. ■

Finally, repeated announcement of strong rationality is a "self-fulfilling" process.

FACT 13.1 Strong rationality is common knowledge in its limit model.

This is a difference with the Muddy Children scenario, where the children's ignorance statement has become false in its limit model. We will return to this difference in the eventual behavior below.

Other rational things to say Our method is general: many other assertions can drive the scenario. If the initial game has a Nash equilibrium, and players have decided on one, they can also keep announcing something stronger than strong rationality, namely, the "Nash statement"

$$\langle \boldsymbol{E} \rangle NE \wedge \langle \boldsymbol{A} \rangle NE$$

The limit model of this scenario contains all Nash equilibria plus all the worlds that are both $\sim_{\boldsymbol{E}}$ and $\sim_{\boldsymbol{A}}$ related to one.

This concludes our description of the public announcement view of game solution. It offers a dynamic alternative to standard epistemic foundations of game theory. In particular, common knowledge of rationality is not presupposed but actually produced by the logic.

13.3 From epistemic dynamics to fixed point logic

Many key features of the dynamic-epistemic logics of Part II return in these scenarios. For instance, the order of assertions may matter in game solution, as can be seen with the Muddy Children problem, where the scenario changes considerably when the children speak in turn. Such changes reflect the importance of procedure in social settings. Also, earlier technical notions return. For instance, different game models may be bisimilar in the sense of Chapter 1, inviting simplification during the update process (van Benthem 2011d). In what follows, we concentrate mainly on the update steps and the procedure.[164]

Program syntax and iterated announcement limits Take any formula φ in our epistemic language. For each initial model $\boldsymbol{M}$ we can keep announcing φ as long as it is true, retaining just those worlds where it holds. This yields a sequence of nested decreasing sets. In infinite models, we take the sequence across limit ordinals by taking intersections of all stages so far. This process always reaches a first fixed point, a submodel where taking just the worlds satisfying φ no longer changes things.

DEFINITION 13.1 Announcement limits in a model
For any model $\boldsymbol{M}$ and formula φ, the *announcement limit* $\#(\boldsymbol{M}, \varphi)$ is the first submodel in the iteration sequence where announcing φ has no further effect. If $\#(\boldsymbol{M}, \varphi)$ is non-empty, then φ is common knowledge there (φ is self-fulfilling in $\boldsymbol{M}$). Otherwise, φ is self-refuting in $\boldsymbol{M}$, and its negation has become true. ■

Rationality assertions for games were self-fulfilling, resulting in common knowledge of rationality. But as we have noted, the joint ignorance statement of the muddy children was self-refuting.

What we find here are interesting new problems in dynamic-epistemic logic. Determining syntactic shapes of self-fulfilling or self-refuting formulas generalizes

164 To make this chapter self-contained, we will repeat a few issues from Chapter 8.

the Learning Problem of finding out which formulas become known upon announcement (Holliday & Icard 2010). As for questions of axiomatizing the new epistemic notions emerging here, one can add announcement limits to our language:

$$\boldsymbol{M}, s \models \#(\psi) \quad \text{iff} \quad s \text{ belongs to } \#(\boldsymbol{M}, \psi)$$

Is PAL with this operator # still decidable? Finally, in terms of Chapter 7, iterations take us from public announcement logic to a version with PDL program constructions over basic announcement actions $!\varphi$.[165]

Equilibria and fixed point logic To motivate our next step, here is a different take on the SD^ω algorithm. The original game model need not shrink, but we compute a new property of worlds in stages, zooming in on a new subdomain that is a fixed point. Fixed point operators added to first-order logic give the system LFP(FO) of Chapters 2 and 14. However, for present purposes, we will use an epistemic version of the modal μ-calculus from Chapters 1 and 4.

THEOREM 13.2 The limit set of worlds for repeated announcement of *SR* is defined inside the full game model by $\nu p \bullet (\langle \boldsymbol{E} \rangle (B_{\boldsymbol{E}} \wedge p) \wedge \langle \boldsymbol{A} \rangle (B_{\boldsymbol{A}} \wedge p))$.

Proof By the definition of greatest fixed points, any world in the set P defined by $\nu p \bullet (\langle \boldsymbol{E} \rangle (B_{\boldsymbol{E}} \wedge p) \wedge \langle \boldsymbol{A} \rangle (B_{\boldsymbol{A}} \wedge p))$ satisfies $\langle \boldsymbol{E} \rangle (B_{\boldsymbol{E}} \wedge p) \wedge \langle \boldsymbol{A} \rangle (B_{\boldsymbol{A}} \wedge p)$. Thus, the formula $\langle \boldsymbol{E} \rangle B_{\boldsymbol{E}} \wedge \langle \boldsymbol{A} \rangle B_{\boldsymbol{A}}$, being a logical consequence of this, also holds throughout P, and a further public announcement of strong rationality has no effect. On the other hand, the announcement limit for *SR* is by definition a subset P of the current model that is contained in the set $\langle \boldsymbol{E} \rangle (B_{\boldsymbol{E}} \wedge p) \wedge \langle \boldsymbol{A} \rangle (B_{\boldsymbol{A}} \wedge p)$. Thus, it is contained in the greatest fixed point for the monotonic operator matching this formula. ■

Announcement limits as deflationary fixed points More generally, an announcement limit $\#(\boldsymbol{M}, \varphi)$ arises by iterated application of the following map:

$$F_{\boldsymbol{M},\varphi}(X) = \{s \in X \mid \boldsymbol{M}|X, s \models \varphi\}, \text{with } \boldsymbol{M}|X \text{ the restriction of } \boldsymbol{M} \text{ to } X$$

In general, the function $F_{\boldsymbol{M},\varphi}$ is not monotonic with respect to set inclusion. The reason is this: when $X \subseteq Y$, an epistemic statement φ may change its truth value from $\boldsymbol{M}|X$ to the larger model $\boldsymbol{M}|Y$. We do not compute in a fixed model, as with formulas $\nu p \bullet \varphi(p)$, but in ever smaller ones, changing the range of the modal

165 This is a momentous move: Miller & Moss (2005) showed that public announcement logic with iteration PAL* is non-axiomatizable, and indeed of very high complexity.

operators in φ. Still, announcement limits can be defined in "deflationary fixed point logic" (Ebbinghaus & Flum 1999).

THEOREM 13.3 The iterated announcement limit is a deflationary fixed point.

Proof Take any φ, and relativize it to a fresh proposition letter p, yielding $(\varphi)^p$. Here, p need not occur positively (it becomes negative when relativizing box modalities), and no μ-calculus fixed point operator is allowed. But a standard relativization technique applies. Let P be the denotation of p in $\boldsymbol{M}$. Then for all s in P:

$$\boldsymbol{M}, s \models (\varphi)^p \quad \text{iff} \quad \boldsymbol{M}|P, s \models \varphi$$

Therefore, the above definition of $F_{\boldsymbol{M},\varphi}(X)$ as $\{s \in X \mid \boldsymbol{M}|X, s \models \varphi\}$ equals

$$\{s \in M \mid \boldsymbol{M}[p := X], s \models (\varphi)^p\} \cap X$$

But this computes a greatest fixed point of a generalized sort. Consider any formula $\varphi(p)$, without restrictions on the occurrences of p. Now define the map

$$F^{\#}_{\boldsymbol{M},\varphi}(X) = \{s \in \boldsymbol{M} \mid \boldsymbol{M}[p := X], s \models \varphi\} \cap X$$

This map need not be monotonic, but it always takes subsets. Thus, it finds a greatest deflationary fixed point, starting with $\boldsymbol{M}$, and iterating as needed, taking intersections at limit ordinals. If $F^{\#}$ is monotonic, this coincides with the usual fixed point procedure. ∎

Dawar et al. (2004) show that definitions for announcement limits can go beyond the epistemic μ-calculus, while the extension of this μ-calculus with inflationary fixed points is undecidable.[166]

Monotonic fixed points after all Fortunately, some announcements are better behaved. The limit domain for the assertion SR of strong rationality was definable in the epistemic μ-calculus. The reason is that its update function $F_{\boldsymbol{M},SR}(X)$ is monotonic for set inclusion. This has to do with syntactic form. Consider "existential modal formulas" built with only existential modalities, proposition letters or

---

166 By contrast, as we have mentioned in Chapter 8, deflationary first-order logic is equivalent in expressive power to LFP(FO) (Kreutzer 2004), making the differences in the styles of definition intensional, rather than extensional, in this more expressive language.

their negations, plus conjunction and disjunction. Semantically, such formulas are preserved under extensions of models.

THEOREM 13.4 $F^{\#}_{\boldsymbol{M},\varphi}(X)$ is monotonic for existential modal formulas φ.

Proof The $F^{\#}_{\boldsymbol{M},\varphi}$ are monotonic by extension preservation: if $\boldsymbol{M}|X, s \models \varphi$ and $X \subseteq Y$, then also $\boldsymbol{M}|Y, s \models \varphi$. Hence, a μ-calculus greatest fixed point exists. ■

Thus, due to the syntax of strong rationality, reasoning about it escapes the high complexity results for PAL* by living inside the epistemic μ-calculus. The ignorance announcements of the muddy children were existential, too.[167]

A final application of our main theorem uses the decidability of the μ-calculus.

FACT 13.2 Dynamic-epistemic logic plus $\#(\psi)$ for existential ψ is decidable.

13.4 Variations and extensions

Muddy Children revisited The ignorance assertion driving the puzzle suggests alternative "self-defeating" solution methods. Players might say "My action may turn out badly" until their worries dissolve. Also, the Muddy Children problem has a crucial "enabling action," viz. the father's announcement that broke the symmetry of the initial model. This, too, makes sense. It is easy to find matrix games where the SD^{ω} algorithm has no effect, but it will start pruning the model if we first break some symmetries by a prior announcement. The art here is to find well-motivated enabling statements that set conversation going: Roy (2008) shows how to use players' intentions for this purpose. The logical setting for all these scenarios are the general game models defined in Chapter 12.

Changing beliefs and plausibility Many logical analyses of games use players' beliefs instead of knowledge. The main results in this chapter generalize then, using

---

167 For other uses of monotonicity, recall the earlier order dependence for announcements (van Benthem 2007d). With monotonic assertions, this can often be avoided. For instance, it can be shown that the announcement limit of $SR_{\boldsymbol{E}}\,;\, SR_{\boldsymbol{A}}$ is the same as that of the simultaneous SR. An application relevant to games is that, for any model $\boldsymbol{M}$, $\#(\boldsymbol{M}, SR) \subseteq \#(\boldsymbol{M}, WR)$. With non-existential formulas, even when φ implies ψ, $\#(\boldsymbol{M}, \varphi)$ need not be included in $\#(\boldsymbol{M}, \psi)$. Apt (2007) gives a general analysis of confluence and non-confluence in game solution procedures using abstract rewriting systems.

the plausibility models of Chapters 7 and 8. For instance, with the SD^ω algorithm, in the limit, players will now have a common belief that they are in the solution set. In addition to hard information about beliefs, one can also have scenarios with soft information, changing plausibility patterns over strategy profiles (cf. Zvesper 2010). As was explained briefly in Chapter 7, soft upgrades have been used to make a junction with truth tracking and temporal limit learning in formal learning theory (Kelly 1996, Gierasimczuk 2010, Baltag et al. 2011).

13.5 Iteration, limits, and abstract fixed point logic for games

The techniques presented here are not just about algorithms for solving games. They apply to any interactive scenario where iterated statements play a role, driven by positive or negative assertions. As we saw in Chapter 8, Dégremont & Roy (2009) use iterated public announcement of disagreement in belief to analyze the seminal result of Aumann (1976) that disagreement cannot be maintained forever. Similar analyses apply to skeptical scenarios in philosophy when doubts keep getting raised about knowledge claims. We conclude with a more abstract look at limit procedures for solving games.

Fixed point logics and abstract game solution At a higher abstraction level, Zvesper (2010) links dynamic-epistemic logic with the computational framework of Apt (2007). The setting is abstract models for games (cf. Chapter 6), with belief operators $B(X)$ sending sets of worlds X to the worlds where the agent finds X most plausible, and optimality operators $O(X)$ selecting those worlds from a set X that are best for the agent in an absolute global sense.[168] A counterpart of the solution concept SD^ω is the greatest fixed point O^∞ of repeatedly applying the optimality operator O to the model. In this setting, a suitably chosen rationality formula *rat* denotes the set $O(\boldsymbol{B})$ where agents do their best given their most plausible belief worlds $\boldsymbol{B}$. These models support a modal fixed point language for basic epistemic notions. For instance, $CB\varphi$ (common belief in φ) is defined by the greatest fixed point formula $\nu p \bullet \bigwedge_{i \in I} B_i(\varphi \wedge p)$. This syntax supports perspicuous proofs, using the basic rules of the modal μ-calculus for greatest fixed points:

---

168 The use of absolute best (as in the atoms $B_{\boldsymbol{A}}$ of Chapter 12), makes the optimality operator monotonic: if $X \subseteq Y$, then $O(X) \subseteq O(Y)$.

$\nu p \bullet \psi(p) \to \psi(\nu p \bullet \psi(p))$ unfolding axiom

if $\vdash \alpha \to \psi(\alpha)$, then $\vdash \alpha \to \nu p \bullet \psi(p)$ inclusion rule

EXAMPLE 13.6 Proving an epistemic characterization syntactically
First, for all φ, (a) $rat \to (B\varphi \to O\varphi)$, since $B\varphi$ implies $\boldsymbol{B} \subseteq [\![\varphi]\!]$, using the monotonicity of O and the definition of *rat*. Simplifying $CB\varphi$ to the notation $\nu p \bullet B(\varphi \wedge p)$, one fixed point unfolding yields: (b) $CBrat \to B(rat \wedge CBrat)$. Using (a) with $\varphi = rat \wedge CBrat$ yields (c) $rat \wedge CBrat \to O(rat \wedge CBrat)$. Finally, by the introduction rule for greatest fixed points, (d) $rat \wedge CBrat \to \nu q \bullet Oq (= O^{\infty})$. ■

Thus, a few proof lines in an abstract modal fixed point logic capture the essence of famous results in the epistemic foundations of game theory (cf. Tan & Werlang 1988, Aumann 1999). This abstract analysis also works on the neighborhood models of Chapter 11. Further applications of formal modal proofs in game theory are found in de Bruin (2004).

13.6 Literature

This chapter is based on van Benthem (2007d), the first source for this dynamic-epistemic style of analyzing games.

For recent work in this line that links up more closely with the foundations of game theory, see also Dégremont & Roy (2009), Baltag & Smets (2009), Zvesper (2010), and Pacuit & Roy (2011).

Conclusion to Part III

In the three chapters of Part III, we have shown how the game logics of Parts I and II apply to more global views of games. More specifically, we have introduced

(a) Power representations, modal logics over neighborhood models, generalized bisimulation, and topological comparison games.
(b) Combined modal logics of preference, knowledge, and freedom in full and general matrix models, with a special case study of the syntax and logic of rationality, and applications to a more general logic of simultaneous action.
(c) Scenarios for the dynamics of game solution as a case study of iterated public announcement, leading to new perspectives on the epistemic foundations of games and other interactive scenarios.

The power perspective of Chapter 11 will remain important in the following two parts of this book: Part IV on logic games, and Part V on logics for game construction. As for strategic games, perhaps the most important insight in Chapters 12 and 13 is how this different face of game theory fits naturally with logic, just as well as our earlier extensive games. This fit was evident both for the statics of Part I and the dynamics of Part II, and thus we have found a second source of evidence for our earlier program.

IV Logic Games

Introduction to Part IV

This is a turning point where this book changes both content and style. Parts I, II, and III have completed our general analysis of games and play in a logical style. But, as we have seen in the Introduction, while logic can be applied to anything, in the case of games, there is a more intimate connection. Many basic logical notions can themselves be cast in the form of games. We now proceed to explore this perspective of logic *as* games versus the earlier logic *of* games. In this part, we survey some basic logic games for evaluating formulas, comparing models, constructing models, and carrying on dialogue. Many logicians use these games solely as didactic tools; and indeed, they are powerful for this purpose, since game intuitions pack a lot of complex structure into something easily imagined. But there is more to it. Logic games contain lots of general ideas and themes with a broader thrust, as will become clear in the coming chapters, while we will also discuss some broader themes explicitly in the concluding Chapter 18. In particular, in an appendix to that chapter, we provide a window onto some related work on games in computational logic, where a flourishing theory has emerged in recent years. Our general themes will then be taken further in Parts V and VI of this book, where the latter also closes the circle with the game logics of Parts I, II and III.

A note on sources The material in this part comes largely from my lecture notes, *Logic in Games* (van Benthem 1999), that have circulated until now in several printed and electronic versions. A compact compendium is given in van Benthem (2011c), while an excellent regularly updated survey can be found in Hodges (2001). The educational origin of much of the coming material shows in the style of the chapters. Even without the grand design of this book, the reader might appreciate what follows as an unusual tour of basic logic.

14
Formula Evaluation

Logical languages usually have a model-theoretic semantics defining when a formula φ is true in a model $\boldsymbol{M}$, perhaps with an auxiliary setting. The paradigm is first-order logic, with its notion $\boldsymbol{M}, s \models \varphi$ where s is an assignment of objects in $\boldsymbol{M}$ to variables. Now, stepwise evaluation of first-order assertions can be cast dynamically as a game of evaluation for two players. "Verifier" claims that φ is true in the setting $\boldsymbol{M}, s$, "falsifier" that it is false. This is our most basic logic game. In this chapter we explain first-order evaluation games, establish their adequacy with respect to truth and falsity, explore their more general game-theoretic character, demonstrate how other logics can be gamified in the same style, and identify some general issues of game logic behind first-order games, including the role of players' strategies and game operations.[169]

14.1 Evaluation games for predicate logic

Two parties disagree about a proposition φ in some situation $\boldsymbol{M}, s$: *verifier* $\boldsymbol{V}$ claims that it is true, *falsifier* $\boldsymbol{F}$ that it is false. Here are the natural moves of defense and attack in the first-order evaluation game, that we will indicate henceforth as $\boldsymbol{game}(\varphi, \boldsymbol{M}, s)$.

DEFINITION 14.1 Moves in evaluation games
The moves of *evaluation games* follow the inductive construction of formulas. They

169 Games like this occur in Hintikka (1973). Since then, evaluation games have been given for many logics. Hintikka & Sandu (1997) has a game-theoretical semantics for natural language, and Chapter 21 will pursue the resulting independence-friendly logic.

involve typical notions in the dynamics of games, such as choice, switch, and continuation, in dual pairs with both players allowed the initiative once:

atoms $Pd, Rde, \ldots$	$\boldsymbol{V}$ wins if the atom is true, $\boldsymbol{F}$ if it is false
disjunction $\varphi \vee \psi$	$\boldsymbol{V}$ chooses which disjunct to play
conjunction $\varphi \wedge \psi$	$\boldsymbol{F}$ chooses which conjunct to play
negation $\neg\varphi$	role switch between the two players, play continues with respect to φ

Next, the quantifiers make players look inside $\boldsymbol{M}$'s domain of objects:

existential $\exists x\varphi(x)$	$\boldsymbol{V}$ picks an object d, play continues with $\varphi(d)$
universal $\forall x\varphi(x)$	the same move, but now for $\boldsymbol{F}$

Here the clause for atoms may look circular, but one might think of it as the players consulting the model to see whether it supports such a bottom-level statement. As for complex structure, the schedule of the game is determined by the form of the statement φ. ■

Example 14.1 Formulas and schedule of play
To see how this works, consider a model $\boldsymbol{M}$ with two objects s, t. Here is a game for $\forall x\exists y\, x{\neq}y$, pictured as a tree of moves, with the scheduling from top to bottom:

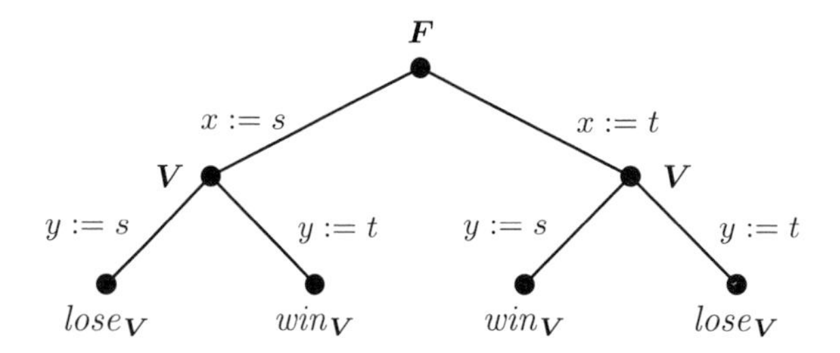

We interpret this as a game of perfect information: players know throughout what has happened. Falsifier starts, and verifier must respond. There are four possible plays, with two wins for each player. But verifier has a winning strategy, in the standard sense of our earlier chapters. ■

Trees such as this are not a complete definition of the game yet, but for many purposes, we are better off without further detail. Evaluation games for slightly more complex formulas in richer models have proved attractive in teaching logic.

EXAMPLE 14.2 Find noncommunicators

Consider the following communication network with arrows for directed links, and with all self-loops present but suppressed in the drawing:

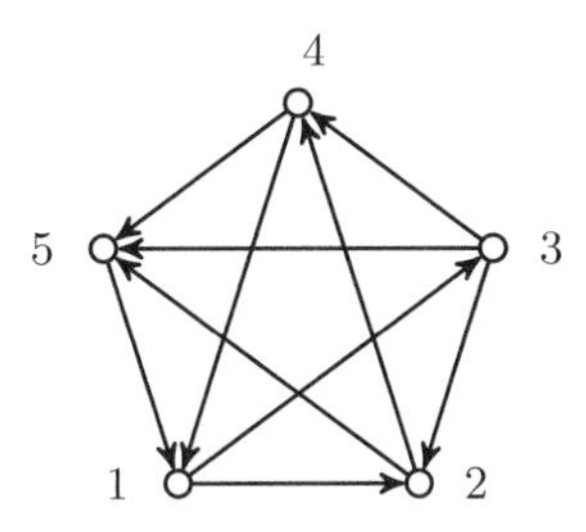

The formula $\forall x \forall y(Rxy \vee \exists z(Rxz \wedge Rzy))$ says that every two nodes in this network can communicate in at most two steps. Here is a run of the evaluation game:

player	*move*	*next formula*
$\boldsymbol{F}$	picks 2	$\forall y(R\boldsymbol{2}y \vee \exists z(R\boldsymbol{2}z \wedge Rzy))$
$\boldsymbol{F}$	picks 1	$R\boldsymbol{2}\,\boldsymbol{1} \vee \exists z(R\boldsymbol{2}z \wedge Rz\boldsymbol{1})$
$\boldsymbol{V}$	chooses	$\exists z(R\boldsymbol{2}z \wedge Rz\boldsymbol{1})$
$\boldsymbol{V}$	picks 4	$R\boldsymbol{2}\boldsymbol{4} \wedge R\boldsymbol{4}\,\boldsymbol{1}$
$\boldsymbol{F}$	chooses	$R\boldsymbol{4}\,\boldsymbol{1}$
test	$\boldsymbol{F}$ loses	

Falsifier started with a threat by picking object 2, but then picked 1. Verifier chose the true right conjunct, and picked the witness 4. Now, falsifier loses with either choice. Still, falsifier could have won, by choosing object 3 that 2 cannot reach in $\leq$2 steps. Falsifier even has another winning strategy, namely, x=5, y=4. ■

In this way, each formula φ is a game form of fixed depth but indefinite branching width, with a schedule of turns and moves. It becomes a real game when a model $\boldsymbol{M}$ is given that supplies possible quantifier moves and outcomes for atomic tests, while an assignment s to the free variables in φ sets the initial position of the game.

14.2 Truth and winning strategies of verifier

In our first example, participants were not evenly matched. Player $\boldsymbol{V}$ can always win: after all, a verifier is in line with the truth of the matter. More precisely, $\boldsymbol{V}$ has a *winning strategy*, a map from $\boldsymbol{V}$'s turns to moves following which guarantees, against any play by $\boldsymbol{F}$, that the game ends in outcomes where $\boldsymbol{V}$ wins. $\boldsymbol{F}$ has no winning strategy, as this would contradict $\boldsymbol{V}$'s having one.[170] Even more can be said. $\boldsymbol{F}$ does not have a losing strategy either: $\boldsymbol{F}$ cannot force $\boldsymbol{V}$ to win, but in our example, player $\boldsymbol{V}$ does have a losing strategy. Thus, players' powers of controlling outcomes in a game may be quite different.

Here is the key to the behavior of evaluation games, the "success lemma."

FACT 14.1 The following are equivalent for all models $\boldsymbol{M}, s$ and formulas φ:

(a) $\boldsymbol{M}, s \models \varphi$, (b) $\boldsymbol{V}$ has a winning strategy in $\boldsymbol{game}(\varphi, \boldsymbol{M}, s)$.

Proof The proof is a direct induction on formulas. One shows simultaneously:

If a formula φ is true in $(\boldsymbol{M}, s)$, then verifier has a winning strategy.
If a formula φ is false in $(\boldsymbol{M}, s)$, then falsifier has a winning strategy.

The steps show the close analogy between logical operators and ways of combining strategies.[171] The following typical cases will give the idea. (a) If $\varphi \vee \psi$ is true, then at least one of φ or ψ is true, say, φ. By the inductive hypothesis, $\boldsymbol{V}$ has a winning strategy σ for φ. But then $\boldsymbol{V}$ has a winning strategy for the game $\varphi \vee \psi$: the first move is *left*, after which the rest is the strategy σ. (b) If $\varphi \vee \psi$ is false, both φ and ψ are false, and so by the inductive hypothesis, $\boldsymbol{F}$ has winning strategies σ and τ for φ and ψ, respectively. But then the combination of an initial wait-and-see step plus these two is a winning strategy for $\boldsymbol{F}$ in the game $\varphi \vee \psi$. If $\boldsymbol{V}$ goes left in the first move, then $\boldsymbol{F}$ should play σ, while, if $\boldsymbol{V}$ goes right, $\boldsymbol{F}$ should play strategy τ. (c) If the formula φ is a negation $\neg\psi$ we use a role switch.

EXAMPLE 14.3 Role Switch
Consider the game for a formula $p \vee q$ in a model where p is true and q is false, as well as its dual game $\neg(p \vee q)$, that switches all turns and win markings:

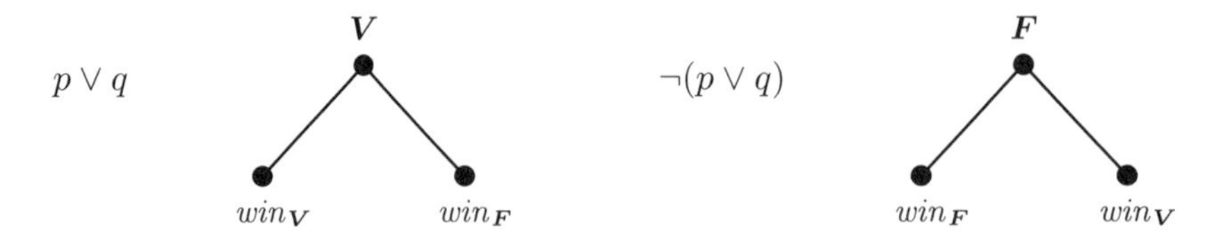

170 Playing two winning strategies against each other yields a contradiction at the end.

171 This inductive proof is virtually the argument for Zermelo's Theorem in Chapter 1.

The second game works out to that for the De Morgan equivalent $\neg p \wedge \neg q$. ■

Thus, strategies for $\boldsymbol{V}$ in a game for $\neg\psi$ are strategies for $\boldsymbol{F}$ in the game for ψ, and vice versa. Now we prove case (c). Suppose that $\neg\psi$ is true. Then ψ is false, and by the inductive hypothesis, $\boldsymbol{F}$ has a winning strategy in the ψ-game forcing an outcome in the set of $\boldsymbol{F}$'s winning positions. But this is a strategy for $\boldsymbol{V}$ in the $\neg\psi$-game, and indeed one forcing a set of winning positions for $\boldsymbol{V}$. The other direction is similar. ■

This is our first link between a key notion in logic (truth) and one in game theory (strategy). We will broaden the interface as we continue. Some critics see the success lemma as showing how games yield nothing new. To them, a game-theoretic analysis is good only if it captures some pre-existing logical notion. Our focus is the opposite: what new themes are intrinsic to games, and might enrich the old agenda of logic?

14.3 Exploring the game view of predicate logic

Simple as it is, there is more to the success lemma than meets the eye. In particular, this result suggests new perspectives on what makes standard predicate logic tick. Many technical distinctions to be formulated in the following discussion will recur in subsequent chapters.

Different winning strategies Truth occurs if and only if there is a winning strategy for player $\boldsymbol{V}$, and likewise for falsity and $\boldsymbol{F}$. But there can be more than one such strategy. For instance, $\boldsymbol{F}$ had two winning strategies in our Example 14.2, using two different counterexamples to the claim. Thus, winning strategies are more refined semantic objects than standard truth values, that we might call reasons for truth or falsity.

Games and game boards The success lemma compares two semantic settings. One is the model $\boldsymbol{M}$, or its associated space of assignments s of individual objects $s(x)$ to all relevant first-order variables x. Here a notion from Chapter 11 returns. This space serves as a "game board," a setting where evaluation games can be played, or even other games. Compare a Chess board with possible positions. Chess expands this with conventions, defining turns for players, as well as their winning positions. The latter are game-internal: there is nothing intrinsic to the distribution of pieces on the board that makes it a win for White or Black.

Comparing two different languages The success lemma compares the game and its board using expressions from different languages appropriate to them:

$$\boldsymbol{V} \text{ has a winning strategy in } \boldsymbol{game}(\varphi, \boldsymbol{M}, s) \quad \text{iff} \quad \boldsymbol{M}, s \models \varphi$$

The expression on the left can be rewritten in a game language referring to forcing powers of players (cf. Chapter 11), while that on the right-hand side is best viewed as a modal formula referring to actions on the board, as in Chapter 1:

$$\boldsymbol{game}(\varphi, \boldsymbol{M}, s) \models win_{\boldsymbol{V}} \quad \text{iff} \quad \boldsymbol{M}, s \models \varphi$$

This dual perspective can be generalized. On the left, one can talk about both players' powers for forcing any set of positions in the game. This corresponds to nested substitutions in modal assertions about the game board on the right.

The general topic of matching games and game boards will be pursued in greater depth in Chapters 19 and 24.

Defining the games formally Defining complete trees for logic games is largely routine. Still, formalization brings out interesting twists to understanding first-order logic. Let us define the tree for $\boldsymbol{game}(\varphi, \boldsymbol{M}, s)$ as follows. Nodes are all pairs

$$(s, \psi) \qquad \text{where } s \text{ is an } \boldsymbol{M}\text{-assignment, and } \psi \text{ is a subformula of } \varphi$$

Game moves reflect the earlier ones, changing one or both components of a state. In particular, atomic tests do not change the state, while choices only change its formula, moving from a current node $(s, \varphi \vee \psi)$ to one of its daughters (s, φ) and (s, ψ). But formalizing the other rules leads to departures from received views in predicate-logical semantics. Consider assignment change with quantifiers. Starting at $(s, \exists x\psi)$, verifier chooses an object d from the domain of $\boldsymbol{M}$, and s is set to $s[x := d]$. Play then continues with $\psi(d)$: that is, it starts afresh from the formal game state $(s[x := d], \psi)$. But this analysis suggests that, unlike in standard logical syntax, we can view the quantifier symbol $\exists x$ by itself as a separate interpretable entity, and more specifically, that *quantifiers are atomic games* of object picking. Standard thinking assimilates quantifiers to Boolean disjunctions or conjunctions. By contrast, here, the real game operation involved in $\exists x\psi$ is *sequential composition*, gluing the game for ψ after the independent atomic game for $\exists x$. On this view,

> *Predicate-logical semantics is really a system of games of object picking and fact testing, related by suitable game operations.*

Next, we need game-internal predicates of turn taking and winning. A formula φ tells us who is to move at which stage, although we need to take care with role switches for negations.

DEFINITION 14.2 Formal game trees
We define $\boldsymbol{game}(\varphi, \boldsymbol{M}, s)$ inductively, for any assignment s, starting from an initial state (s, φ). The first two clauses are for the two kinds of atomic game:

(a) $\boldsymbol{game}(\varphi, \boldsymbol{M}, s)$ for atomic φ is a one-node end game, which is a win for verifier if $\boldsymbol{M}, s \models \varphi$, and for falsifier otherwise.

(b) $\boldsymbol{game}(\exists x, \boldsymbol{M}, s)$ is a one-move game starting at s where it is $\boldsymbol{V}$'s turn, with possible moves to any state $s[x := d]$, always ending in a win for $\boldsymbol{V}$.

Next we turn to game constructions:

(c) $\boldsymbol{game}(\varphi \vee \psi, \boldsymbol{M}, s)$ is the disjoint union of the two games $\boldsymbol{game}(\varphi, \boldsymbol{M}, s)$ and $\boldsymbol{game}(\psi, \boldsymbol{M}, s)$ put under a common root $(s, \varphi \vee \psi)$ that is $\boldsymbol{V}$'s turn.

(d) $\boldsymbol{game}(\varphi \wedge \psi, \boldsymbol{M}, s)$ is defined likewise, but with an initial turn for $\boldsymbol{F}$.

(e) $\boldsymbol{game}(\neg\varphi, \boldsymbol{M}, s)$ is $\boldsymbol{game}(\varphi, \boldsymbol{M}, s)$ with turn and win markings reversed.

The negation switch was illustrated with the success lemma. Finally, to deal with quantifiers, we add a clause for an operation of composition for evaluation games:

(f) $\boldsymbol{game}(\varphi\,;\,\psi, \boldsymbol{M}, s)$ is the tree arising by first taking $\boldsymbol{game}(\varphi, \boldsymbol{M}, s)$ and continuing at end states with assignment t with a copy of $\boldsymbol{game}(\psi, \boldsymbol{M}, t)$.

These tree constructions for games will return in Chapters 20 and 25. ■

Emancipation of syntax The above semantics interprets more than just the usual well-formed formulas. For instance, it makes perfect sense to play a game for the string

$$Px\ ;\ \exists x$$

which translates to "First test that P holds for object $s(x)$ under the current assignment s, then change the value of x, and stop." In contrast with this, the game for $\exists x\,;\,Px$ would first change the value of $s(x)$ to obtain an object with the property P. With composition as a new syntax construction, predicate logic extends to a language of discourse chunks $\alpha\,;\,\beta\,;\,\ldots$ that might be interesting to axiomatize.[172] Even not-so-well-formed formulas now get their chance.

Propositions versus activities Finally, the reader should beware of a common confusion. Reading formulas φ in their ordinary notation as evaluation games makes them serve a double role: as a static *proposition* in truth-conditional semantics, or as a *game* in our new semantics. A game is not a statement, but a dynamic activity. To be sure, one can state propositions about games, say, that verifier has a winning strategy in the φ-game. But this statement is not that game itself, and it does not exhaust the latter's content. Much of the original literature on game semantics conflates these two readings of first-order syntax, resulting in the prejudice that existence statements about verifier's winning strategies are all there is to know to the game-theoretic meaning of φ.[173]

14.4 Game-theoretic aspects of predicate logic

Now we explore connections with game theory, a theme raised in the Introduction. First, via the success lemma, logical laws acquire game-theoretic import.

Determinacy Our first encounter between logic and game theory was the following simple but telling observation.

FACT 14.2 Excluded middle $A \vee \neg A$ expresses determinacy of evaluation games; that is, one of the two players must always have a winning strategy.

Evaluation games turned out to be determined because of Zermelo's Theorem that all zero-sum two-player games of finite depth are determined (cf. Chapter 1).

---

172 We also get new distinctions. Consider a "bounded quantifier" $\exists y(Rxy \wedge \varphi)$, inducing a game where verifier chooses an object d for y, then falsifier chooses a conjunct, and one either tests the atom $R^{\boldsymbol{M}} s(x)d$, or play continues at $(s[y := d], \varphi)$. It seems more natural to package this as one atomic game: verifier picks an R-successor of $s(x)$. This is won by verifier if she can produce such an object, and lost otherwise. In extended first-order syntax, this would be the game for $\exists y\,;\, Rxy$. This is the view that underlies our later evaluation games for modal logic, and it will return in the theoretical analysis of logic games as a format for game algebra in Chapter 24.

173 Significantly, two similar levels, of actions and propositions, were carefully kept separate in the syntax of propositional dynamic logic PDL (Chapters 1 and 4), with *programs* and *formulas*.

Infinite evaluation games will also be relevant later in this chapter, and there, we need to appeal to further results such as the Gale-Stewart Theorem of Chapter 5.

Game equivalence In the Introduction, we also saw an issue of game equivalence.

FACT 14.3 The propositional distribution law states that the evaluation games for its formulas are equivalent in terms of players' forcing powers.

EXAMPLE 14.4 Switching games with invariant powers
Switching between games $A \wedge (B \vee C)$ and $(A \wedge B) \vee (A \wedge C)$ transforms turns for players without affecting their strategic powers concerning outcomes:

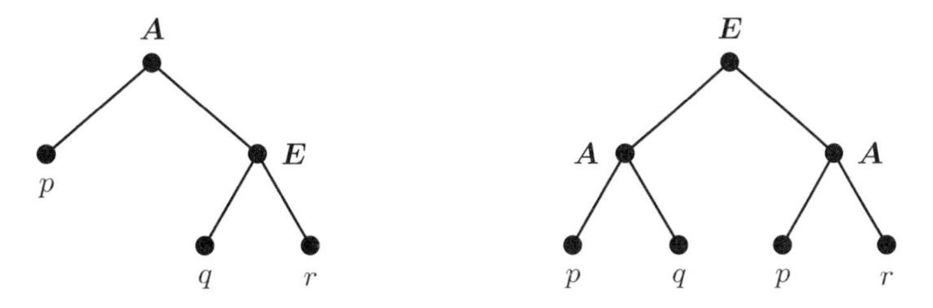

The relevant computations were given in the Introduction and Chapter 1. ■

Chapter 11 contained an abstract perspective on powers that fits this equivalence precisely. In particular, Boolean operations turn into general operations on games. A matching algebra of power equivalence will be studied in Chapters 19 and 21.

Syntactic normal forms Propositional normal forms now serve as normal forms for games. The same is true for first-order quantificational prenex forms such as

$$\forall x \exists y \forall z \, Q(x, y, z)$$

with all quantifiers moved in front, dividing object picking moves into alternating blocks for each player, fixing scheduling without affecting players' powers. Also relevant are Skolem forms, taking first-order formulas to second-order equivalents

$$\exists f_1 \cdots \exists f_k \, \forall x_1 \cdots \forall x_m$$

followed by a quantifier-free propositional part, which will be used in Chapter 22.

Here, an earlier caveat applies. Skolem forms are better read as statements about games, viz. the existence of strategies, rather than as games in themselves. Still, one can view second-order formulas such as $\exists x \exists g \forall x \forall y \, Q(x, f(x), y, g(x, y))$ as defining a new evaluation game, where verifier picks a strategy at the start, after which falsifier has a go, and finally a regular propositional game is played.[174]

14.5 Gamification: Variations and extensions

Evaluation games exist for many logical languages. The above explanations provide almost automatic gamifications, provided that the truth conditions employ quantifiers and connectives. For instance, the Skolem game mentioned at the end of the preceding section was an evaluation game for a second-order language, that goes as before, letting players now also choose functions for second-order quantifiers $\exists f$ (or, in other second-order games, sets), in addition to objects for first-order quantifiers $\exists x$. In this section, we discuss a few further illustrations with additional points.

Basic modal logic Consider the basic modal language over models $\boldsymbol{M} = (W, R, V)$ used in Parts I and II of this book. We start with a simple variation on first-order evaluation games, showing how they transfer to modal languages. Accessibility encodes moves that can be made to get from one world or state to another.

EXAMPLE 14.5 A modal model
Consider the following graph with four worlds and accessibilities as indicated:

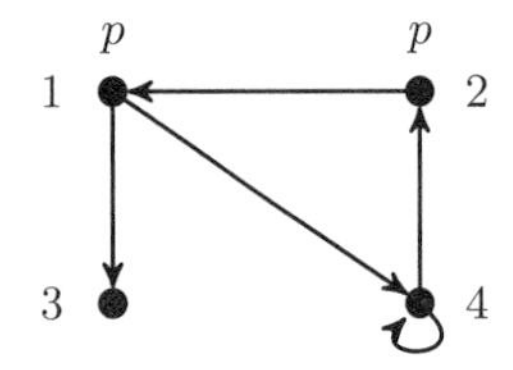

The formula $\Diamond\Box\Diamond p$ is true in states 1 and 4, but false in states 2 and 3. ■

Modal evaluation games search through such a model, with two key moves:

$\Diamond$	verifier chooses a successor of the current world
$\Box$	falsifier chooses a successor of the current world

174 Redescriptions are frequent with logic games. Switching between games raises general issues of game equivalence (cf. Chapter 1). We will discuss this theme in Chapter 18.

Game states are pairs (*state, formula*). Players lose when defending an atom that fails at the current state, or when they must choose a successor but cannot. This is like the bounded first-order quantifier $\exists y(Rxy \wedge Py)$ discussed earlier.

Again we have a modal success lemma:

$$\boldsymbol{M}, s \models \varphi \quad \text{iff} \quad \boldsymbol{V} \text{ has a winning strategy for the } \varphi\text{-game in } \boldsymbol{M} \text{ from } s.$$

EXAMPLE 14.6 A modal evaluation game
The graph of Example 14.5 induces the following game tree for $\Diamond\Box\Diamond p$ at state 1:

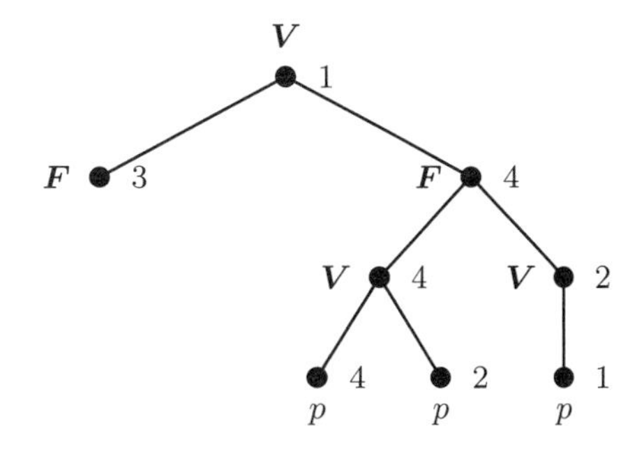

Here, $\boldsymbol{V}$ has two winning strategies: *left*, and *right* ; ⟨*right*, *down*⟩. These are the two ways of verifying $\Diamond\Box\Diamond p$ in the given model at world 1. The game also illustrates the well-known feature of losing when a player must move but cannot. ■

The same games work for polymodal languages with indexed operators $\langle a \rangle$.

Modal μ-calculus A more drastic change in evaluation games is needed for the fixed point logics in Parts I and II of this book. To define their games, we need to explain their central idea of recursion in a bit more detail than before, starting with an important special case. Recall the modal fixed point logic providing the recursive definition used to analyze Zermelo's algorithm in Chapter 1, which returned in many places, including Chapter 13 on solving games in strategic form. We present the basics here; interested readers may look to van Benthem (2010b) and, especially, Venema (2007) for more didactic detail.

DEFINITION 14.3 Modal μ-calculus
The modal *μ-calculus* extends basic modal logic with a syntactic operator $\mu p \bullet \varphi(p)$

in which all occurrences of p in φ are positive.[175] In any model $\boldsymbol{M}$, the new operator $\mu p \bullet \varphi(p)$ defines the *smallest fixed point* with respect to set inclusion for the following set operator $\varphi^{\boldsymbol{M}}$ on the model associated with the formula $\varphi(p)$:

$$\varphi^{\boldsymbol{M}}(X) = \{s \in \boldsymbol{M} \mid \boldsymbol{M}[p := X], s \models \varphi\}$$

Here a fixed point of a function F is an argument X such that $F(X) = X$. ■

By the positive syntactic occurrence of p in $\varphi(p)$, we can easily show the following important property.

FACT 14.4 The map $\varphi^{\boldsymbol{M}}$ is monotonic for inclusion: $X \subseteq Y \Rightarrow \varphi^{\boldsymbol{M}}(X) \subseteq \varphi^{\boldsymbol{M}}(Y)$.

The definition relies on the following fact, called the Tarski-Knaster Theorem.

FACT 14.5 Every monotonic operator F on a power set has a smallest and a greatest fixed point in the inclusion order.[176]

A greatest fixed point operator $\nu p \bullet \varphi(p)$ is defined analogously. This powerful new formalism allows us to define properties of states in a model by recursion.

EXAMPLE 14.7 Fixed point evaluation
Consider the formula $\mu p \bullet (q \vee \langle a \rangle p)$ in the following model:

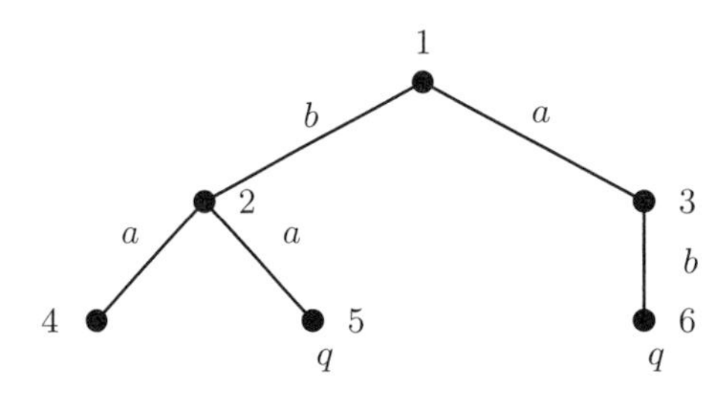

175 Positive occurrence means that, counting from the outside of φ, p lies in the scope of an even number of negations. Some formulas positive in p are $\neg q \vee \langle a \rangle p$, $\neg \langle a \rangle \neg p$, and $\neg \mu q \bullet (\neg \langle a \rangle \neg q \wedge \neg p)$.

176 The proof of the Tarski-Knaster Theorem shows how the smallest fixed point $F_* = \bigcap\{X \subseteq A \mid F(X) \subseteq X\}$, while the greatest fixed point $F^* = \bigcup\{X \subseteq A \mid X \subseteq F(X)\}$. One can also think of these as reached by approximation through the ordinals. F_* is the first set where F reaches a fixed point in the sequence $\varnothing, F(\varnothing), FF(\varnothing), \ldots, F^{\omega}(\varnothing), FF^{\omega}(\varnothing), \ldots$, taking the union of all previous stages at limit ordinals.

Here is the approximation sequence for the associated set function of $q \vee \langle a \rangle p$:

stage	*set*	*defining formula*
0	$\varnothing$	$\bot$
1	$\{5, 6\}$	q
2	$\{5, 6, 2\}$	$q \vee \langle a \rangle p$
3	$\{5, 6, 2\}$	the fixed point

What $\mu p \bullet (q \vee \langle a \rangle p)$ describes in general is the set of all points in the given model that can reach a q-world by a finite sequence of a-steps. In this way, the μ-calculus defines the typical modality $\langle a^* \rangle \varphi$ of propositional dynamic logic (PDL) used at many places in this book. ■

By a similar analysis, the formula $\mu p \bullet [a]p$ holds in points with only finite a-sequences coming out, the so-called "well-founded part" of the relation R_a. But there is an interesting difference. The approximation process for $\mu p \bullet (q \vee \langle a \rangle p)$ always stops in ω steps, while that for $\mu p \bullet [a]p$ can go on to any ordinal, depending on the size of the model. This has to do with the syntax of these fixed point formulas (cf. Fontaine 2010).

Related to the latter smallest fixed point formula is the greatest fixed point formula $\nu p \bullet \langle a \rangle p$ defining the s at which an infinite sequence $sR_a s_1 R_a s_2 R_a \cdots$ starts. Actually, $\nu p \bullet \varphi(p)$ is equivalent to $\neg \mu p \bullet \neg \varphi(\neg p)$.

EXAMPLE 14.8 Evaluating greatest fixed point formulas
Consider the formula $\nu p \bullet \langle a \rangle p$ in the following model:

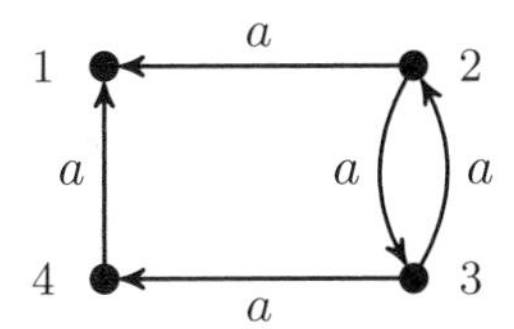

The computation stabilizes at the set of worlds $\{2, 3\}$. ■

Our syntax even allows inhomogeneous nested fixed point formulas of shapes such as $\nu p \bullet \mu q \bullet \varphi(p, q)$, whose intuitive meaning can be harder to decode.

Infinite evaluation games Niwinski & Walukiewicz (1996) and Stirling (1999) define evaluation games for the μ-calculus (Venema 2007 has a lucid presentation). These involve a significant change reflecting the ordinal approximation process, taking us into the realm of Chapter 5: runs may become infinite. Accordingly, the winning convention changes in a delicate manner that goes back to fundamental results connecting logic and automata theory (Rabin 1968, Thomas 1992).

DEFINITION 14.4 Evaluation games for the μ-calculus
In *μ-calculus evaluation games*, verifier and falsifier play by the earlier rules when the main operator is modal. When a fixed point formula $\mu p \bullet \varphi(p)$ or $\nu p \bullet \varphi(p)$ is reached in the game, the next formula is $\varphi(p)$, with the following understanding. Whenever later play hits an atom p, no test takes place (p is a bound variable), but $\mu p \bullet \varphi(p)$ (or $\nu p \bullet \varphi(p)$, as the case may be) is substituted back in.[177]

Next, we note that in infinite runs, some fixed point subformula of the finite initial formula φ must have been called infinitely often. Indeed, it is easy to see that there is a unique recurrent fixed point subformula occurring in the highest syntactic position in φ. We say that an infinite run is a *win for* $\boldsymbol{V}$ in the φ-game if the syntactically highest recurrent fixed point subformula is a greatest fixed point. If it is a smallest fixed point, then $\boldsymbol{F}$ wins. ■

The winning convention displays a general pattern from graph games with a "parity condition" (see Section 18.6 in Chapter 18, and Venema 2007).

As in our earlier analyses, a success lemma connects our language and these infinite games.

FACT 14.6 A formula in the modal μ-calculus is true if and only if verifier has a winning strategy in the game just described.

We refer to the cited literature for a proof, and its further background in automata theory. The μ-calculus is a good framework for studying general interactive processes, and its game aspects will return in Chapter 18.

We now move to a more general formalism showing the same ideas at work, which has also been used earlier on in this book, for instance, in Chapters 2, 8, and 13.

First-order fixed point logic First-order logic with fixed point operators LFP(FO)(Moschovakis 1974, Ebbinghaus & Flum 1999) was used for analyzing game-theoretic solution methods in Part II. Again the language has smallest fixed point operators

$$\mu P \bullet \varphi(P) \qquad \text{with } P \text{ occurring only positively in } \varphi$$

---

177 Here we assume, without loss of generality, that occurrences of fixed point variables have been made unique.

More precisely, with $\boldsymbol{x}$ (or $\boldsymbol{d}$) standing for finite tuples of variables (or objects),

$[\mu P, \boldsymbol{x} \bullet \varphi(P)](\boldsymbol{d})$ says that $\boldsymbol{d}$ is in the smallest set X with $\varphi^{\boldsymbol{M}}(X) = X$

$[\nu P, \boldsymbol{x} \bullet \varphi(P)](\boldsymbol{d})$ says that $\boldsymbol{d}$ is in the largest set X with $\varphi^{\boldsymbol{M}}(X) = X$

Example 14.9 Transitive closure

For instance, $\mu P, xy \bullet (Rxy \vee \exists z(Rxz \wedge Rzy))$ is a definition for the transitive closure of a relation R, showing the typical recursive behavior of this notion:

$$trans(R)(x, y) \leftrightarrow R(x, y) \vee \exists z(R(x, z) \wedge trans(R)(z, y))$$

As with the μ-calculus, this recursive unwinding shows how games for LFP(FO) differ from first-order ones: they can *cycle*. For, if $\boldsymbol{V}$ chooses the right-hand disjunct, taking an R-successor, the original formula returns, and play loops. For certain types of formula, infinite games then become indispensable. ■

The games that we need are as before, with some notational adaptations.

Definition 14.5 First-order fixed point games

In games for LFP(FO), verifier and falsifier play by earlier rules when the main operator is first-order. When a fixed point formula $[\mu P, \boldsymbol{x} \bullet \varphi(P)](\boldsymbol{d})$ is reached, the next formula is $\varphi(P)](\boldsymbol{d})$, with the following understanding. Whenever subsequent play reaches an atom $P\boldsymbol{e}$, no test takes place in the model (P is a bound variable), but $[\mu P, \boldsymbol{x} \bullet \varphi(P)](\boldsymbol{e})$ is substituted back in. Greatest fixed point formulas $[\nu P, \boldsymbol{x} \bullet \varphi(P)](\boldsymbol{d})$ are treated analogously.

Again, in infinite runs, some fixed point subformula of the finite initial formula φ must have been called infinitely often, and there is a unique subformula of this kind occurring in the highest syntactic position in φ. As was stipulated for the μ-calculus, an infinite run is a *win for* $\boldsymbol{V}$ in the φ-game if the syntactically highest recurrent fixed point subformula is a greatest fixed point. If it is a smallest fixed point, then $\boldsymbol{F}$ wins.[178] ■

Again, the following success lemma shows how games and models connect.

178 Consider the smallest fixed point for transitive closure. To show that Pde, verifier may first choose the second disjunct, take some object f for z, and claim that Pfe. But verifier may only do that finitely often; otherwise, a loss results. Greatest fixed points are dual: falsifier must put up in some finite number of cycles, while it is fine for verifier to keep the cycle going.

FACT 14.7 A formula in first-order fixed point logic is true if and only if verifier has a winning strategy in the game just described.

See Ebbinghaus & Flum (1999) and Doets (1999) for proofs. There is a general result in the background here. While these games do not have an open winning condition for players in the sense of the Gale-Stewart Theorem of Chapter 5, their winning condition is still Borel, making the games determined by Martin's Theorem.

EXCURSION Second-order reformulation
An alternative finite version of this game translates formulas $[\mu P, \boldsymbol{x}] \bullet \varphi(\boldsymbol{d})$ inside out by equivalent second-order statements. The Tarski-Knaster definition of a smallest fixed point says that $\boldsymbol{d}$ satisfies every predicate that is a smallest pre-fixed point of the monotonic set operation for the positive formula φ, or in second-order terms, $\forall Q\,(\forall x(\phi(Q, \boldsymbol{x}) \rightarrow Q\boldsymbol{x}) \rightarrow Q\boldsymbol{d})$. A standard evaluation game for these second-order formulas is of finite depth, although at the cost of letting players choose sets of (tuples of) objects. As before, there is an issue in which sense this is the same as our earlier fixed point game, and this time, a non-trivial proof is needed to show that players' powers do not change.

There is a deep literature on evaluation games for fragments of second-order logic, of which Rabin (1968) is a classic. Walukiewicz (2002) is an influential format for automata-based games for monadic second-order logic (first-order logic with added quantifiers over sets), which plays an important role in games and computation, witness our discussion of forcing logics in Chapter 5. These and other automata-based games are explained (with many relevant results) in an accessible manner in Zanasi (2012). A short overview will be found in Chapter 18.

Games with changing models Even for first-order logic, other variations are possible on standard evaluation games, making players perform different tasks. In contrast to object-picking moves, one may consider removing objects from a domain without replacement, or moves that change a domain or interpretation function, changing the playground by adding or subtracting objects and facts. Such variations mix the process of logical evaluation with model construction that will be considered in Chapter 16.

14.6 Conclusion

This concludes our tour of two-player evaluation games as a dynamic view on logical semantics. Most of these games had finite depth, but some are naturally infinite when the language contains fixed points, a natural correlate to the notion of game-theoretic equilibrium in this book. In addition, we found a number of general game-theoretic themes illustrating general topics in this book: namely, determinacy, game equivalence, game algebra of choice, switch, or composition, and the systematic importance of calculus of strategies. Many of these themes will return in the integrative Parts V and VI of this book.

14.7 Literature

The presentation in this chapter is largely from van Benthem (1999).

Hintikka (1973) is a source of the ideas, and an introduction supported by software is found in Barwise & Etchemendy (1999). A recent non-trivial extension to many-valued logics is Fermüller & Majer (2013). Mann et al. (2011) has a more formal development, continuing on to the more complex games with imperfect information found in Chapter 21. We have also pointed at the flourishing literature on related games for fixed point logics in computer science, that link up with automata theory, with references on the μ-calculus such as Venema (2006), and Janin & Walukiewicz (1995). Chapter 18 will provide more details.

15 Model Comparison

Logical formulas express properties of semantic structures. Different languages have different expressive strength over models, showing in powers of distinction. A poor language with only "Yes" and "No" distinguishes few situations, while a rich language can distinguish a whole spectrum. Model comparison can be cast as a game between a "duplicator" $\boldsymbol{D}$ who claims that two given models $\boldsymbol{M}$ and $\boldsymbol{N}$ are similar, and a "spoiler" $\boldsymbol{S}$ who claims they are different. In this chapter, we define comparison games for first-order logic, prove their adequacy for model equivalence, give correspondences between players' winning strategies and logical difference formulas or potential isomorphisms, discuss general game-theoretic aspects of the games, and show how to create variations and extensions. These games go back to Fraïssé (1954) and Ehrenfeucht (1961). Thomas (1997) uses them in computer science, and for a mathematics-oriented slant, see Doets (1996) or Väänänen (2011).

15.1 Isomorphism and first-order equivalence

Expressive power and invariances The expressive power of a language shows in its power of distinguishing situations, as we saw in Chapter 1. The notions of transformations and invariants from 19th century geometry make precise sense of this. In logic, this requires two things: a relation of structural *invariance* between models, and a *language* expressing the properties of those models. The analysis aims to show that the invariance matches those differences between models that the language cannot detect. About the most important structural invariance relation is the following widespread notion in mathematics.

Definition 15.1 Isomorphism
Two models $\boldsymbol{M}$ and $\boldsymbol{N}$ are *isomorphic* (written $\boldsymbol{M} \cong \boldsymbol{N}$) if there exists a bijection f between the objects in their domains that preserves all the relevant structure: atomic properties, relations, distinguished objects, and operations. Thus, we have

$R^{\boldsymbol{M}}de$ iff $R^{\boldsymbol{N}}f(d)f(e)$	for all binary predicates R, and objects d, e in $\boldsymbol{M}$
$f(G^{\boldsymbol{M}}(d)) = G^{\boldsymbol{N}}(f(d))$	for all unary functions G, and objects d in $\boldsymbol{M}$

These two clauses show the general pattern of structure preservation. ■

Coarser invariants may be just the right comparison level for some other purpose: witness the notions of bisimulation between process models in Chapters 1 and 11.[179]

First-order expressiveness For convenience, in this chapter, we only use a first-order logic whose vocabulary has finitely many predicate letters and individual constants. The linguistic notion of model comparison is elementary equivalence $\boldsymbol{M} \equiv \boldsymbol{N}$: that is, $\boldsymbol{M}$ and $\boldsymbol{N}$ satisfy the same sentences. How close is this to a structural similarity? Let us look at the basic Isomorphism Lemma to find out.

Fact 15.1 For all models $\boldsymbol{M}$ and $\boldsymbol{N}$, if $\boldsymbol{M} \cong \boldsymbol{N}$, then $\boldsymbol{M} \equiv \boldsymbol{N}$.

Proof An easy induction on first-order formulas φ shows that, for all tuples of objects $\boldsymbol{a}$ in $\boldsymbol{M}$, and any isomorphism f sending the latter to the model $\boldsymbol{N}$, we have that $\boldsymbol{M} \models \varphi[\boldsymbol{a}]$ iff $\boldsymbol{N} \models \varphi[f(\boldsymbol{a})]$. ■

This implication holds for any well-behaved logical language. The converse is by no means true for first-order logic. What does hold is full harmony in the special case of *finite* models.

Fact 15.2 For all finite models, the following two assertions are equivalent:

(a) $\boldsymbol{M}$ is isomorphic with $\boldsymbol{N}$.

(b) $\boldsymbol{M}$ and $\boldsymbol{N}$ satisfy the same first-order sentences.

Proof From (b) to (a). Write a first-order sentence $\delta^{\boldsymbol{M}}$ describing $\boldsymbol{M}$. Let there be k objects. Then quantify existentially over $x_1, \ldots, x_k$, enumerate all true atomic statements about these in $\boldsymbol{M}$, plus the true negations of atoms, and state that no other objects exist. Since $\boldsymbol{N}$ satisfies $\delta^{\boldsymbol{M}}$, it can be enumerated just like $\boldsymbol{M}$. The isomorphism is immediate. ■

179 For much more on invariance and logical definability, see van Benthem (1996, 2002b).

This proof does not extend to infinite models, as first-order logic cannot define finiteness. Nor, for instance, can it tell the rationals $\mathbb{Q}$ apart from the reals $\mathbb{R}$ in their order $<$.

EXAMPLE 15.1 Natural versus supernatural numbers
Elementary equivalence cannot even distinguish the natural numbers $\mathbb{N}$ from the model $\mathbb{N} + \mathbb{Z}$ that continues with the integers as supernatural numbers:

$\mathbb{N}$	versus	$\mathbb{N} + \mathbb{Z}$
$0, 1, 2, \dots$		$0, 1, 2, \ \dots \ \dots \infty + 1, \infty, \infty - 1, \dots$

We will see the reason for this indistinguishability in Section 15.5. ■

With a richer vocabulary, however, a language may see differences that used to be invisible. But there is no need to always extend our systems. In fact, weak expressive power can also be a good thing, as it yields transfer of properties across different situations, say, between standard models and nonstandard models.

15.2 Ehrenfeucht-Fraïssé games

The fine structure of the above invariance analysis is brought out by playing a certain type of logic games. These will work for any models, finite or not.

Playing the game Consider two models $\boldsymbol{M}$ and $\boldsymbol{N}$. A player called "duplicator" claims that $\boldsymbol{M}$ and $\boldsymbol{N}$ are similar, while a player called "spoiler" maintains that they are different. Players agree on some finite number k of rounds for the game, the severity of the probe.

DEFINITION 15.2 Comparison games
A *model comparison* game works as follows, packing two moves into one round. *Spoiler* (also written $\boldsymbol{S}$ for brevity) chooses one of the models, and picks an object d in its domain. *Duplicator* (also written $\boldsymbol{D}$ for brevity) then chooses an object e in the other model, and the pair (d, e) is added to the current list of matched objects. After k rounds, the object matching is inspected. If it is a partial isomorphism, duplicator wins; otherwise, spoiler does. Here, a "partial isomorphism" is an injective partial map f between models $\boldsymbol{M}$ and $\boldsymbol{N}$ that is an isomorphism between its own domain and range. ■

This alternating schedule $(DS)^*$ occurs in many games. We now present some sample plays of our comparison games. As in Chapter 14, players may lose, even when they have a winning strategy. We use a language with a binary relation symbol R only, mostly disregarding identity atoms $=$.

EXAMPLE 15.2 Comparing integers and rationals
The linear orders of integers $\mathbb{Z}$ and rationals $\mathbb{Q}$ have different first-order properties: the latter is dense, the former discrete. Here is how this will surface in the game:

$$\begin{array}{ccccccc} & -1 & 0 & +1 & & & \\ \cdots & \circ & \circ & \circ & \cdots & & \mathbb{Z} \\ & & & & & & \\ \cdots\cdots\cdots & \circ\cdots\cdots & \circ\cdots\cdots & \circ\cdots\cdots & \cdots & & \mathbb{Q} \\ & -1 & 0 & +1 & & & \end{array}$$

By choosing objects well, duplicator has a winning strategy for the game over two rounds. But spoiler can always win the game in three rounds. Here is a typical play:

Round 1	***S*** chooses 0 in $\mathbb{Z}$	***D*** chooses 0 in $\mathbb{Q}$
Round 2	***S*** chooses 1 in $\mathbb{Z}$	***D*** chooses 1/3 in $\mathbb{Q}$
Round 3	***S*** chooses 1/5 in $\mathbb{Q}$	any response for ***D*** is losing

These moves will convey the typical strategic flavor of the game. ■

Difference formulas and spoiler's strategies In playing the games, winning strategies for spoiler are correlated with first-order formulas φ that bring out a difference between the models. The correlation is tight. The quantifier syntax of φ triggers the moves for spoiler.

EXAMPLE 15.2, CONTINUED Exploiting definable differences
Spoiler can use the first-order definition of density for a binary order, written as $\forall x \forall y(x < y \rightarrow \exists z(x < z \wedge z < y))$, to distinguish $\mathbb{Q}$ from $\mathbb{Z}$. We spell this out, to show how there is an almost algorithmic derivation of a strategy from a first-order difference formula. For convenience, we use existential quantifiers only. The idea is for spoiler to maintain a difference between the two models, of stepwise decreasing syntactic depth. Spoiler starts by observing that

$$\exists x \exists y(x < y \wedge \neg\exists z(x < z \wedge z < y)) \text{ is true in } \mathbb{Z}, \text{ but false in } \mathbb{Q} \qquad \#$$

Spoiler then chooses an integer d for $\exists x$, making $\exists y(d < y \wedge \neg\exists z(d < z \wedge z < y))$ true in $\mathbb{Z}$. Now duplicator can take any rational number d' in $\mathbb{Q}$: the first-order formula $\exists y(d' < y \wedge \neg\exists z(d' < z \wedge z < y))$ will be false for it, by #:

$$\mathbb{Z} \models \exists y(d < y \wedge \neg\exists z(d < z \wedge z < y)), \text{ not } \mathbb{Q} \models \exists y(d' < y \wedge \neg\exists z(d' < z \wedge z < y))$$

In the second round, spoiler continues with a witness e for the new outermost quantifier $\forall y$ in the true existential formula in $\mathbb{Z}$: making $d < e \wedge \neg\exists z(d < z \wedge z < e)$ true there. Again, whatever object e' duplicator now picks as a response in $\mathbb{Q}$, the formula $d' < e' \wedge \neg\exists z(d' < z \wedge z < e')$ will be false there. In the third round, spoiler analyzes the mismatch in truth value. If duplicator kept $d' <' e$ true in $\mathbb{Q}$, then, since $\neg\exists z(d < z \wedge z < e)$ holds in $\mathbb{Z}$, $\exists z(d' < z \wedge z < e')$ holds in $\mathbb{Q}$. Spoiler then switches to $\mathbb{Q}$, chooses a witness for the existential formula, and wins. ■

Thus, even the right model switches for the strategy of spoiler are encoded in the difference formulas. Such switches are mandatory whenever there is a syntactic change from one type of outermost quantifier (existential, universal) to another.[180]

15.3 Adequacy and strategies

As with evaluation games, the interesting information is in players' strategies. In the results to follow, we think of winning strategies for duplicator, although spoiler's strategic point of view will return later. For the sake of brevity, let us write $WIN(\boldsymbol{D}, \boldsymbol{M}, \boldsymbol{N}, k)$ for: "duplicator has a winning strategy against spoiler in the k-round comparison game between the models $\boldsymbol{M}$ and $\boldsymbol{N}$."

Comparison games can start from any initial match of objects in $\boldsymbol{M}$ and $\boldsymbol{N}$. In particular, if models have distinguished objects named by individual constants, these are matched automatically. In the proofs to come, we think of all initial matches in the latter way. We now look at an analogue of the success lemma from Chapter 14.

THEOREM 15.1 For all models $\boldsymbol{M}$ and $\boldsymbol{N}$, and $k \in \mathbb{N}$, the following two assertions are equivalent:

(a) $WIN(\boldsymbol{D}, \boldsymbol{M}, \boldsymbol{N}, k)$: duplicator has a winning strategy in the k-round game.

(b) $\boldsymbol{M}$ and $\boldsymbol{N}$ agree on all first-order sentences up to quantifier depth k.

---

180 Our examples may also suggest a correlation: "winning strategy for spoiler over n rounds $\sim$ difference formula with n quantifiers altogether." But as we shall soon see, the right measure is different, being the maximum length of a quantifier nesting in a formula.

This improves on the Isomorphism Lemma in two ways. Our adequacy result matches up a language-dependent and a language-independent comparison relation. And it provides fine structure not available before, which helps in applications.

Proof From (a) to (b) is an induction on k. We start with the base step. With 0 rounds, the initial match of objects must have been a partial isomorphism for $\boldsymbol{D}$ to win. So $\boldsymbol{M}$ and $\boldsymbol{N}$ agree on all atomic sentences, and hence on their Boolean combinations, the formulas of quantifier depth 0. We proceed with the inductive step. The inductive hypothesis says that, for any two models, if $\boldsymbol{D}$ can win their comparison game over k rounds, the models agree on all first-order sentences up to quantifier depth k. Now let $\boldsymbol{D}$ have a winning strategy for the $k+1$ round game on $\boldsymbol{M}$ and $\boldsymbol{N}$. Consider any first-order sentence φ of quantifier depth $k+1$. Such a φ is equivalent to a Boolean combination of (i) atoms, (ii) sentences of the form $\exists x\psi$, with ψ of quantifier depth at most k. Thus, it suffices to show that $\boldsymbol{M}$ and $\boldsymbol{N}$ agree on the latter forms.

The essential case is this. Let $\boldsymbol{M} \models \exists x\psi$. Then for some object d, we get $\boldsymbol{M}, d \models \psi$. Think of $(\boldsymbol{M}, d)$ as an expanded model with a distinguished object d to which we assign a new name $\underline{d}$. In this way $(\boldsymbol{M}, d)$ verifies the sentence $\varphi(\underline{d})$. Now, $\boldsymbol{D}$'s winning strategy has a response for whatever $\boldsymbol{S}$ can do in the $k+1$-round game. For instance, let $\boldsymbol{S}$ start with $\boldsymbol{M}$ and object d. Then $\boldsymbol{D}$ has a response e in $\boldsymbol{N}$ to this move such that $\boldsymbol{D}$'s remaining strategy still gives a win in the k-round game played from the given link $d - e$. This yields an expanded model $(\boldsymbol{N}, e)$, with e as its interpretation of the name $\underline{d}$. The remainder is an ordinary k-round game starting from the models $(\boldsymbol{M}, d)$ and $(\boldsymbol{N}, e)$. By the inductive hypothesis, these models agree on all sentences up to quantifier depth k: and hence also on $\varphi(\underline{d})$. Therefore, $\boldsymbol{N}, e \models \varphi(\underline{d})$, and so $\boldsymbol{N} \models \exists x\psi$.

The converse direction from (b) to (a) requires another induction on k. This time we need a small auxiliary result about first-order logic in a finite relational vocabulary, the so-called Finiteness Lemma.

LEMMA Fix variables $x_1, \ldots, x_m$. Up to logical equivalence, there are only finitely many first-order formulas $\varphi(x_1, \ldots, x_m)$ of quantifier depth $\leq k$.[181]

181 The proof is by induction on k, analyzing formulas of quantifier depth $k+1$ in the same way as above, and then using the fact that Boolean combinations of any finite set of formulas are finite modulo logical equivalence.

Now we do the inductive proof from (b) to (a). The base step is trivial: doing nothing is a winning strategy for $\boldsymbol{D}$. As for the inductive step, we give the first move in $\boldsymbol{D}$'s strategy. Let $\boldsymbol{S}$ choose one of the models, say $\boldsymbol{M}$, plus some object d in it. Now, $\boldsymbol{D}$ looks at the set of first-order formulas true of d in $\boldsymbol{M}$, which may refer to distinguished objects available through their names in the language. By the Finiteness Lemma, this set is finite modulo logical equivalence, and so one existential formula $\exists x\psi^d$ true in $\boldsymbol{M}$ summarizes all this information. Now, because the models $\boldsymbol{M}$ and $\boldsymbol{N}$ agree on all first-order sentences of depth $k+1$, and $\exists x\psi^d$ is such a sentence, it also holds in $\boldsymbol{N}$. Therefore, $\boldsymbol{D}$ can choose a witness e for it in $\boldsymbol{N}$. Then the expanded models $(\boldsymbol{M}, d)$, $(\boldsymbol{N}, e)$ agree on all sentences up to quantifier depth k, and so, by the inductive hypothesis, $\boldsymbol{D}$ has a winning strategy in the remaining k-round game between them. $\boldsymbol{D}$'s initial response plus the latter further strategy gives $\boldsymbol{D}$ an overall winning strategy over $k+1$ rounds. ∎

15.4 An explicit version: The logic content of strategies

Theorem 15.1 still leaves out the precise match we found earlier between spoiler's winning strategies and first-order formulas. Thus, it displays a phenomenon that we discussed in Chapter 4, under the heading of *∃-sickness.* In logic (but also elsewhere), one often rushes to formulating notions and results with an existential quantifier when more constructive information would be available if we made the witnesses for that existential quantifier explicit. The symptoms of this disease are overuse of indefinite articles such as "a" or modal affixes such as "-ility." Why have a theory of lovability when we can have one of love?

Fortunately, ∃-sickness can often be cured with a little further effort:[182] The following result is the earlier adequacy theorem made explicit.

THEOREM 15.2 There exists an explicit correspondence between

(a) Winning strategies for $\boldsymbol{S}$ in the k-round comparison game for $\boldsymbol{M}$ and $\boldsymbol{N}$.

(b) First-order sentences φ of quantifier depth k with $\boldsymbol{M} \models \varphi$, not $\boldsymbol{N} \models \varphi$.

---

182 Another strain of the disease occurs in standard completeness theorems, that link provability to validity, instead of seeking a more direct match between proofs and semantic verifications. Remedies include the full completeness theorems discussed in Chapter 20.

Proof We first look at the direction from (b) to (a). Every φ of quantifier depth k induces a winning strategy for $\boldsymbol{S}$ in a k-round game between any two models. Each round $k-m$ starts with a match between objects linked so far that differ on some subformula ψ of φ with quantifier depth $k-m$. By Boolean analysis, $\boldsymbol{S}$ then finds some existential subformula $\exists x \bullet \alpha$ of ψ with a matrix formula α of quantifier depth $k-m-1$ on which the models disagree. $\boldsymbol{S}$'s next choice is a witness in that model of the two where $\exists x \bullet \alpha$ holds.

Our next direction is from (a) to (b). Any winning strategy σ for $\boldsymbol{S}$ induces a distinguishing formula of proper quantifier depth. To obtain this, let $\boldsymbol{S}$ make the first choice d in model $\boldsymbol{M}$ according to σ, and now write down an existential quantifier for that object. Our formula will be true in $\boldsymbol{M}$ and false in $\boldsymbol{N}$. We know that each choice of $\boldsymbol{D}$ for an object e in $\boldsymbol{N}$ gives a winning position for $\boldsymbol{S}$ in all remaining $(k-1)$-round games starting from an initial match $d-e$. By the inductive hypothesis, these induce distinguishing formulas of depth $k-1$. By the Finiteness Lemma, only finitely many such formulas exist. Some of these will be true in $\boldsymbol{M}$ (say $A_1, \ldots, A_r$), and others in $\boldsymbol{N}$ (say $B_1, \ldots, B_s$). The total difference formula for strategy σ is then the $\boldsymbol{M}$-true assertion

$$\exists x \bullet (A_1 \wedge \cdots \wedge A_r \wedge \neg B_1 \wedge \cdots \wedge \neg B_s)$$

whose appropriateness is easy to check. ■

Thus, spoiler's winning strategies in a comparison game correspond to formulas, that is, logical objects of independent interest.[183] A similar match exists for the other player. One might call duplicator's strategies analogies of some finite quality measured by the number k. Technically they are cut-off versions of the "potential isomorphisms" that will be defined in Section 15.6.

REMARK Explicitness versus computability

We gave an explicit definable match of strategies with other objects, but it need not be computable. Also, strategies in evaluation or comparison games need not be effective. They range from history-free (with all next moves read off from the current state) to dependent on a complete record of the game so far. The strategic invariants in the next section illustrate the kind of memory to be maintained.[184]

---

183 We have a caveat. The formulation of Theorem 15.2 is still $\exists$-sick. Can you cure it?

184 On the more computational side, Chapter 18 will present an important theorem on the adequacy of history-free strategies in parity games.

15.5 The games in practice: Invariants and special model classes

In practice, comparison games involve not just logic, but also combinatorial analysis of the models involved. Facts 15.1 through 15.3 provide some examples.

FACT 15.3 The rationals $(\mathbb{Q}, <)$ are elementarily equivalent to the reals $(\mathbb{R}, <)$.

It suffices to show that duplicator can win the comparison game for every k. A good method is to identify an *invariant* for duplicator to maintain at intermediate game states. In this particular case, the invariant is simply that all matches so far form a finite partial isomorphism. All further choices of spoiler can then be countered using the unboundedness and density of the orders. More complicated invariants may depend on the number of rounds still to go.

FACT 15.4 $(\mathbb{N}, <)$ is elementarily equivalent with $(\mathbb{N} + \mathbb{Z}, <)$.

This time, if the length of the game is known in advance, duplicator can counter choices of spoiler from the supernatural numbers in $\mathbb{Z}$ by matching them with large natural numbers in $\mathbb{N}$.[185]

Invariants are concrete descriptions of positions where players have a winning strategy. Some descriptions of solutions in the game logics of Part I had this character, witness our brief discussion in Chapter 4.

Finally, comparison games also work on model classes where standard methods of first-order logic fail. Fact 15.3 gives an example of such a negative use of games.

FACT 15.5 Even or odd are not first-order definable on the finite models.

The usual proof for this nondefinability on all models is a compactness argument that fails on finite models. But now, using games, suppose that even size had a first-order definition on finite models, of quantifier depth k. Then any two finite models for which duplicator can win the k-round comparison game are both of even size, or both of odd size. But this is refuted by any two finite models with k versus $k+1$ objects in their domains.

185 The invariant maintains suitable distances between objects. Duplicator makes sure that with k rounds to go, the two sequences $d_1, \ldots, d_m$ in $\mathbb{N}$ and $e_1, \ldots, e_m$ in $\mathbb{N} + \mathbb{Z}$ chosen so far have the following properties: (a) $d_i < d_j$ iff $e_i < e_j$, (b) if d_i, d_j have distance $< 2^k - 1$, then $\mathit{distance}(e_i, e_j) = \mathit{distance}(d_i, d_j)$; else, d_i, d_j and e_i, e_j both have distance $\geq 2^k - 1$ (finite or infinite).

15.6 Game theory: Determinacy, finite and infinite games

Comparison games are two-player zero-sum games of some finite depth k. Therefore, Zermelo's Theorem applies, and either duplicator or spoiler has a winning strategy.

FACT 15.6 Model comparison games are determined.

But these games also have a natural version that goes on forever, say over ω rounds. A natural winning convention in that case is the safety property that duplicator wins the infinite game by not losing at any finite stage, maintaining a partial isomorphism all the time. This is stronger than being able to win all finite-round games. With this understanding, $\mathbb{N}$ and $\mathbb{N}+\mathbb{Z}$ can be distinguished by spoiler in an infinite game: it suffices to start with a supernatural number and keep descending. But when comparing $\mathbb{Q}$ with $\mathbb{R}$, duplicator can hold out indefinitely.

These games still fall under earlier results from Chapter 5. In particular, the winning set for spoiler is open (failure of partial isomorphism always occurs by some finite stage), and hence the Gale-Stewart Theorem applies.

FACT 15.7 The infinite comparison game is determined.

For infinite games, duplicator's winning strategies do correspond to a notion of independent interest.

DEFINITION 15.3 Potential isomorphism
A *potential isomorphism* between two models $\boldsymbol{M}$ and $\boldsymbol{N}$ is a non-empty family I of finite partial isomorphisms between $\boldsymbol{M}$ and $\boldsymbol{N}$ satisfying the following back-and-forth property:

(a) If $f \in I$ and $a \in \boldsymbol{M}$, then there exists a $b \in \boldsymbol{N}$ with $f \cup \{(a,b)\} \in I$.

(b) If $f \in I$ and $b \in \boldsymbol{N}$, then there exists an $a \in \boldsymbol{M}$ with $f \cup \{(a,b)\} \in I$.

This is like a bisimulation, but now for much richer non-modal languages. ■

FACT 15.8 The potential isomorphisms between two models correspond to duplicator's winning strategies in the infinite comparison game.

By contrast, in infinite games, spoiler's winning strategies are methods blocking each attempt at establishing potential isomorphism by some finite stage, guided by a difference formula.

Potential isomorphism implies elementary equivalence. If duplicator can win the infinite game, then duplicator can win every finite cut-off, and the success lemma applies. But the models $\mathbb{N}$ and $\mathbb{N}+\mathbb{Z}$ refuted the converse. It is easy to see directly that the partial isomorphisms in a potential isomorphism I satisfy the same first-order formulas, even with infinite conjunctions and disjunctions added. In fact, Karp's Theorem says that two models are potentially isomorphic iff they satisfy the same sentences in infinitary first-order logic.[186]

15.7 Modifications and extensions

Model comparison games capture a wide variety of logics. In this section we explore some illustrations.

Modal games Restricting players' choices of objects to local successors of currently matched objects leads to basic modal languages, and a link with the notion of bisimulation in Chapter 1.[187] The back-and-forth clauses in a bisimulation between two models strongly suggest a game where one player mentions a challenge, letting one process make an available move, while the other player must then respond with an appropriate simulating move. This might go on forever, but there is also a natural finite variant restricting the number of rounds. More precisely, the fine structure of bisimulation suggests the following games between duplicator and spoiler, comparing successive pairs (m, n) in two models $\boldsymbol{M}$ and $\boldsymbol{N}$.

DEFINITION 15.4 Bisimulation game
Fix a finite number of rounds. In each round of the *bisimulation game*, spoiler chooses a state x in one model that is a successor of the current m or n, and duplicator responds with a matching successor y in the other model. Spoiler wins if, at any stage, x and y differ in atomic properties, or if duplicator cannot choose a matching successor. Infinite bisimulation games have no finite bound, while all other conventions remain the same. ■

186 More on infinitary first-order logic games is found in Barwise & van Benthem (1999), where they are used to prove new kinds of interpolation theorems, as well as a Lindström-type characterization for infinitary modal logic in terms of bisimulation.

187 A more general case of the approach is the Guarded Fragment (Andréka et al. 1998).

Example 15.3 Modal comparison games
Spoiler can win the game between the models depicted below (cf. Example 1.3) starting from their roots:

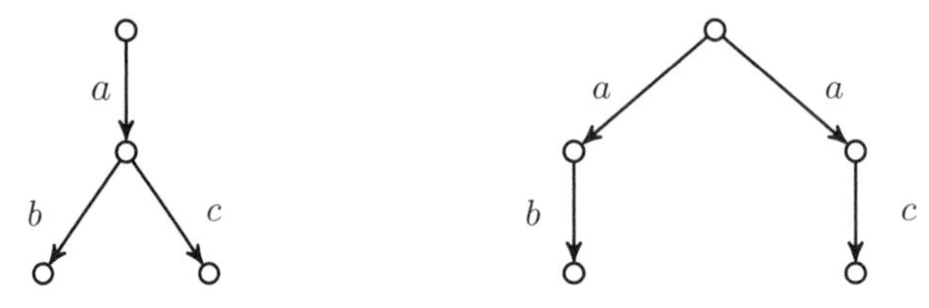

Spoiler needs two rounds, and different strategies do the job. One stays on the left, exploiting the modal difference of depth 2, with three existential modalities:

$$\langle a\rangle(\langle b\rangle\top \wedge \langle c\rangle\top)$$

Another strategy switches models, using the smaller formula

$$[a]\langle b\rangle\top$$

where the type of modality switches between universal and existential. ■

A success lemma can be proved for the finite bisimulation game like for first-order logic (cf. van Benthem 2010b). Analyzing the games further, the following two relevant observations emerge.

Fact 15.9

(a) Spoiler's winning strategies in a k-round game between $(\boldsymbol{M}, s)$ and $(\boldsymbol{N}, t)$ match the modal formulas of operator depth k on which s and t disagree.

(b) Duplicator's winning strategies in an infinite game between $(\boldsymbol{M}, s)$ and $(\boldsymbol{N}, t)$ match the bisimulations between $\boldsymbol{M}$ and $\boldsymbol{N}$ that link s to t.

Clause (b) reveals the close connection between our games and bisimulations.

Pebble games One can also add structure to the games, for instance, in the way that players operate. For instance, to make memory a concern, one can let objects be chosen only by using a finite resource that has been supplied to the players, marking them with one of a finite set of pebbles (Immerman & Kozen 1989). In this setting, duplicator has a winning strategy for the n-round k-pebble game between two models $\boldsymbol{M}$ and $\boldsymbol{N}$ iff $\boldsymbol{M}$ and $\boldsymbol{N}$ agree on all first-order sentences of quantifier depth $\leq k$ that use at most the variables $x_1, \ldots, x_m$ (free or bound), a so-called "finite-variable fragment."[188]

Other languages Other comparison games capture first-order logic with generalized quantifiers (Keenan & Westerståhl 1997), or the first-order fixed point logic LFP(FO) of Chapter 14. This may raise new perspectives. For LFP(FO), for instance, it is not known whether there exists a model comparison game that would be more analogous to the elegant fixed point evaluation game of Chapter 14.

15.8 Connections between logic games

Now that we have seen two major logic games, one for evaluation and one for comparison, questions arise of general architecture, and connections between games. We end with three suggestive observations.

Parallel game operations Comparison games suggest new operations on games in addition to the earlier choices and switch: most obviously, parallel composition. As we will see in Chapter 18, model comparison games are interleaved evaluation games. A general study of parallel game operations will be found in Chapter 20.

Model comparison as evaluation Comparison games sometimes reduce to evaluation games. Through their definition in Chapter 1, bisimulations E may be viewed as non-empty greatest fixed points for a first-order operator between models $\boldsymbol{M}$ and $\boldsymbol{N}$ defined by the equation:

$$\mathsf{E}xy \leftrightarrow \bigwedge_P \left((Px \leftrightarrow Py) \wedge \forall z(R_a xz \rightarrow \exists u(R_a yu \wedge \mathsf{E}zu)) \wedge \forall u(R_a yu \rightarrow \exists z(R_a xz \wedge \mathsf{E}zu))\right)$$

with the right-hand side taken over all atomic predicates P and actions R_a.

Thus, existence of a bisimulation between states s and t amounts to the truth of some LFP(FO) formula in a suitable disjoint sum model $\boldsymbol{M} + \boldsymbol{N}$. Such a formula can be checked by the fixed point evaluation game of Chapter 14. The latter can be infinite; but so can model comparison games. We will highlight the broader significance of facts such as this in Chapter 18.

---

188 This result is proved in Immerman & Kozen (1989). As an important genre of further results, these authors also show that three pebbles suffice as a working memory for winning all comparison games over linear orders. Finite-variable fragments in general play an important role in finite model theory (cf. Ebbinghaus & Flum 1999, Libkin 2004).

Game equivalence The literature often switches between supposedly equivalent formulations without explanation. For instance, Barwise & van Benthem (1999) define the following comparison game starting with a finite partial isomorphism between two models:

> In each round, duplicator selects a set F^+ of partial isomorphisms satisfying the back-and-forth property that, for every object a in one model, there is an object b in the other with $f \cup \{(a, b)\} \in F^+$; and vice versa. In the same round, spoiler then chooses a match in F^+ again, and so on.

In each round, duplicator offers spoiler a complete panorama of all choices that spoiler could make in the former game, plus duplicator's own responses to them. Spoiler then makes a choice of both spoiler's own move as well as duplicator's prepackaged response, setting the new stage. The two games are obviously power-equivalent in the sense of Chapters 11 and 19, but their internal structure also matches closely. Indeed, the above transformation exemplifies a general turn switch such as we saw in the Thompson transformations that we discussed in Chapter 11.

Chapter 18 has further discussion of equivalence levels for logic games.

15.9 Conclusion

Comparison games are a concrete and powerful way of thinking about the interplay of logic and structure. By now they are widely used for many purposes, and the reader will have understood their appeal, while we have also proved their basic properties. Besides being successful special logical activities, comparison games also demonstrate interesting general features, as we have seen. Just to mention one of these, they perform a striking new sort of parallel combination of evaluation games in different structures.

Further, the games of this chapter have a direct impact on the game logics in Parts I and II of this book. As we already noted in Chapter 1 when discussing invariance, playing comparison games offers a systematic way of adding fine structure to existing notions of simulation between processes or games, revealing further information about invariants. Thus, comparison games can be a useful tool in the general study of the right levels at which to analyze general games, and the design of their languages. In line with the integrative spirit of this book, they can be used as games about games.

15.10 Literature

Two highly accessible textbooks are Doets (1996) and Väänänen (2011). Kolaitis (2001) is an excellent presentation geared more toward computer science.

16
Model Construction

So far, we looked at games with given models, evaluating formulas for truth or comparing for similarity. The next basic logical task asks for finding models that make given assertions true. This consistency issue occurs when making sense of a conversation, or creating a structure meeting some specifications. This is more difficult in general than earlier tasks, in that satisfiability is undecidable for first-order logic. Still, there are techniques for testing existence of models, in particular, the semantic tableaus of Beth (1955). In this chapter, we turn the tableau method into a game between a "builder" $\boldsymbol{B}$ and a "critic" $\boldsymbol{C}$ disagreeing about a construction making certain assertions true, and others false. We will explain tableau games for first-order logic, prove their adequacy for satisfiability, analyze correspondences between builder's winning strategies and models, and critic's winning strategies and proofs, explain game-theoretic aspects of tableau games, and discuss variations and extensions for several languages.[189]

16.1 Learning tableaus by example

This section works by example leading up to some general results. Readers interested in more detailed coverage should consult any standard logic textbook explaining the tableau method.

189 Tableau games are not established techniques, such as those in Chapters 1 and 2. They are a test on gamification helping us see new issues. More sophisticated games for model building have been developed in model theory and universal algebra, witness Hodges (2006), and Hirsch & Hodkinson (2002).

At the start, two sets of formulas are given: Σ and Δ. Semantic tableaus test satisfiability in the following sense: Does a model exist that makes all assertions in Σ true, and all those in Δ false?

EXAMPLE 16.1 Disjunctive syllogism
We test the validity of from $A \vee B$, $\neg A$ to B by asking whether this inference could have a *counterexample* making the premises true, and the conclusion false. Thus, we put

$$\Sigma = \{A \vee B, \neg A\}, \qquad \Delta = \{B\}$$

and analyze the situation as follows, leading to progressively less complex tasks.

(a) Making $\neg A$ true is the same as making A false in classical logic, and hence we switch to the equivalent problem

$$\Sigma = \{A \vee B\}, \qquad \Delta = \{B, A\}$$

(b) Making $A \vee B$ true can be done in two ways, and so we split into cases

$$\text{(b1) } \{A\},\ \{B, A\} \qquad \text{(b2) } \{B\},\ \{B, A\}$$

Both these tasks are infeasible, as some formula occurs on both sides, and would have to be both true and false. We call these cases "closed," and say that the tableau is closed. All ways toward a counterexample have failed; thus, the original inference is valid. ■

The general idea of tableaus is to reduce satisfiability problems to simpler ones using a number of decomposition rules for the logical operators.

EXAMPLE 16.2 $A \wedge (B \vee C)$ implies $(A \wedge B) \vee C$
We write the tableau as a tree of unfolding potential counterexamples, noting the rules. Formulas to be true stand on the left, those on the right are to be false:

$$A \wedge (B \vee C) \quad \bullet \quad (A \wedge B) \vee C$$

First, conjunctions on the left are just unpacked, as both conjuncts must be true:

$$A, (B \vee C) \quad \bullet \quad (A \wedge B) \vee C$$

Next, disjunctions on the right are also unpacked, as both disjuncts must be false:

$$A, (B \vee C) \quad \bullet \quad A \wedge B, C$$

Again, we split the disjunction to the left, as at least one disjunct must be true:

$$A, B \quad \bullet \quad A \wedge B, C \qquad\qquad A, C \quad \bullet \quad A \wedge B, C$$

The node to the right closes, as C occurs on both sides. No counterexample lies that way. Finally, as with the left-split for disjunctions, we split the conjunction to the right in our remaining case, since at least one conjunct must be false:

$$A, B \quad \bullet \quad A, C \qquad\qquad A, B \quad \bullet \quad B, C$$

Both of these close, because some formula occurs on both sides. ■

But tableau branches may also remain open, leading to counterexamples.

EXAMPLE 16.3 Creating a counterexample
Consider a tableau for the invalid converse from $(A \wedge B) \vee C$ to $A \wedge (B \vee C)$:

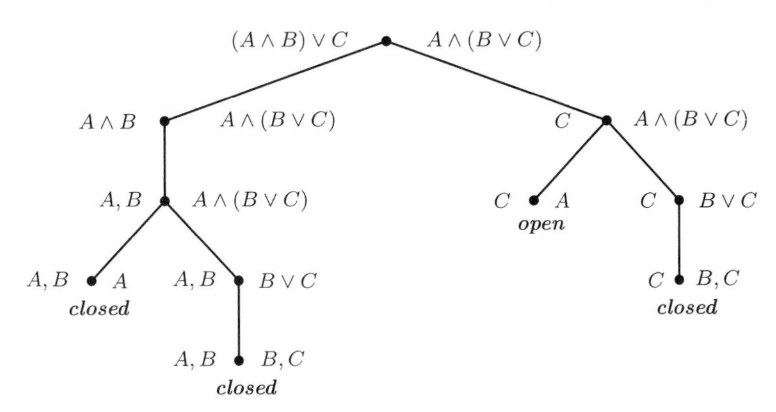

The branch ending in atoms C and A remains "open." Now the stipulation

$$\text{Set } V(C) = 1,\ V(A) = 0,\ V(B) \text{ arbitrary}$$

encodes an obvious counterexample to the initial inference. ■

The method extends to quantifiers. This time, branches will also have a growing domain of objects that witness true existential formulas and false universal formulas. True universals and false existentials act as general requirements on all objects, without asking for new ones.

EXAMPLE 16.4 From $\exists x \forall y\, Rxy$ to $\forall y \exists x Rxy$
Here is a closed tableau showing the validity of this inference:

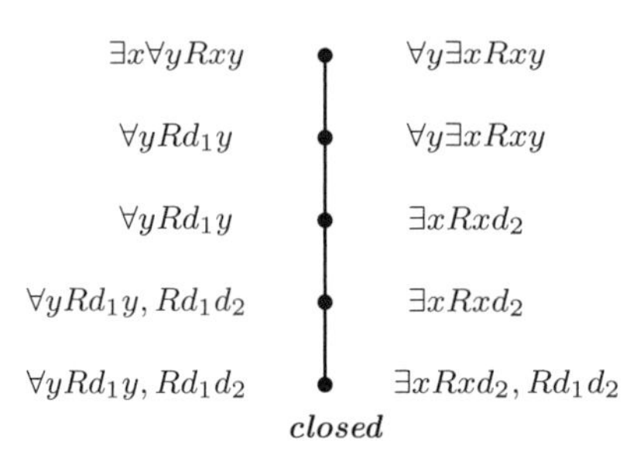

Note how the tableau performs its creation of objects as needed. ■

EXAMPLE 16.5 A counterexample to a quantifier inference
The following table refutes the inference from $\forall x(Ax \vee Bx)$ to $\forall xAx \vee \forall xBx$:

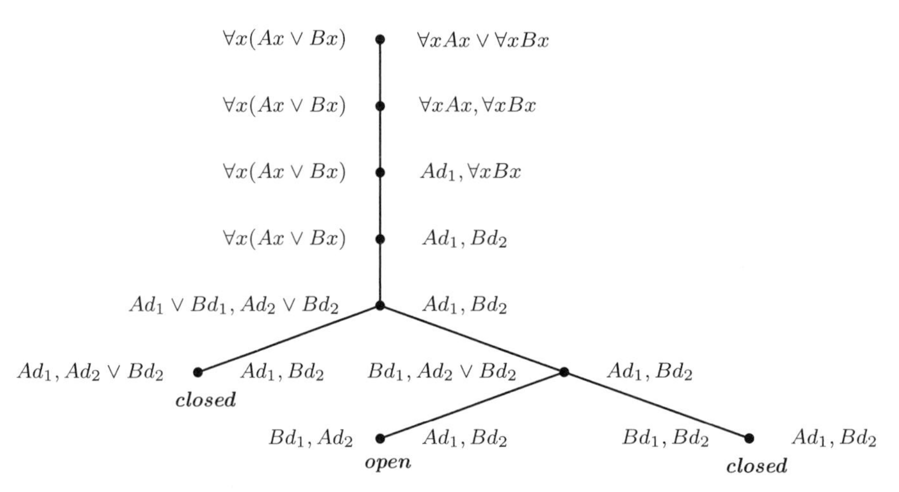

The open branch in the middle describes the simplest counterexample to the above rule. It has two objects, one of which satisfies B only, and the other A only. ■

Not all tableaus are finite. Some satisfiable formulas have only infinite models.

EXAMPLE 16.6 Schütte's formula
The first-order formula $\forall x \neg Rxx \wedge \forall xyz((Rxy \wedge Ryz) \rightarrow Rxz) \wedge \forall x \exists y Rxy$ holds only for irreflexive transitive orders R without endpoints, and these are infinite.

Here, on the left in a tableau, the quantifier combination $\forall\exists$ triggers an endless supply of new objects. Via $\forall x \exists y Rxy$, each object d yields a requirement $\exists x Rdy$, that creates a new object e with Rde, that can again be plugged into the universal quantifier, and so on. In some cases, earlier objects can be reused to create the necessary loop, but with Schütte's formula, the infinity is inevitable. ■

16.2 Tableaus, some general features

A complete tableau tree displays all possible models meeting the initial requirements, making sure that each complex formula on a branch is subjected to decomposition rules. This requires scheduling to make sure that true universals and false existentials are applied to all available objects. Also, on infinite branches, we need a tracking device to make sure that no formula gets omitted. Once things have been set up in this way, the tableau method is correct and complete for testing consistency and validity.

THEOREM 16.1 The following two assertions are equivalent in first-order logic:

(a) The set of formulas $\{\varphi_1, \dots, \varphi_k, \neg\psi_1, \dots, \neg\psi_m\}$ is satisfiable

(b) There is an open tableau with top node $\varphi_1, \dots, \varphi_k \bullet \psi_1, \dots, \psi_m$

One can also state this in terms of "closed tableaus" and valid consequence. A closed tableau has closed branches only. The following assertions are equivalent:

(a′) $\bigwedge\{\varphi_1, \dots, \varphi_k\}$ logically implies $\bigvee\{\psi_1, \dots, \psi_m\}$

(b′) There is a closed tableau with top node $\varphi_1, \dots, \varphi_k \bullet \psi_1, \dots, \psi_m$

Proof From (a) to (b), we reason as follows. Any model $\boldsymbol{M}$ for the given set induces an open branch $\beta_{\boldsymbol{M}}$ in the tableau. At propositional splits, a choice is made by checking which disjunct is true (or conjunct false) in the model, taking objects as required from the model. This cannot lead to closure. Now consider the direction from (b) to (a). Any open branch β induces a model $\boldsymbol{M}$ whose domain consists of all objects introduced on β, where we make all atomic statements to the left on this branch true, and those on the right false. For other atoms, the stipulation is free. Using the tableau decomposition rules, it follows by a straightforward formula induction that all formulas to the left on the branch β are true in $\boldsymbol{M}_\beta$, while all those to the right are false.

The version with closed tableaus involves an additional argument. Clause (a)′ is the negation of (a). The negation of (b), however, says there is no tableau for the initial situation with an open branch. This implies that any tableau must be closed, and hence one exists, i.e., (b)′. But conversely, why does (b)′ imply that there is no tableau at all with an open branch? After all, we have freedom in scheduling the rules, leading to tableaus with possibly quite different shapes for the initial problem. The reason is in the following two observations.

FACT 16.1 Closed tableaus are finite.

The reason is as follows. Any infinite tableau is an infinite finitely branching tree. Now *König's Lemma* says that every finitely branching infinite tree has an infinite branch. But that means that an infinite tableau cannot be closed.

The implication from (b)′ to (a)′ is proved by induction on the size of closed tableaus. The following fact is the crucial observation that we need.

FACT 16.2 Closed tableaus correspond to proofs of the initial implication from $\bigwedge\{\varphi_1, \ldots, \varphi_k\}$ to $\bigvee\{\psi_1, \ldots, \psi_m\}$.

Here we see two faces of tableaus. Read top-down, they are attempts at finding a countermodel, bottom-up (when closed), they are proofs.[190] ■

The same two faces of tableaus play in general argumentation. In court, all the evidence is in, and now the prosecutor's burden of proof is to show that the accused has committed the crime. But the defendant's attorney seeks a scenario in which all the evidence is compatible with the defendant's being innocent. Thus, administrative scheduling features of tableaus become procedural rules for dialogue (cf. Chapters 17 and 23).

16.3 Model construction games

Semantic tableaus invite gamification. Pulling their tasks apart into two roles, tableaus can be cast as a game between two players, *builder* and *critic*.

DEFINITION 16.1 Model construction games

Each stage of a *model construction game* (also called tableau game) has two finite boxes of formulas representing builder's current tasks. The box *YES* contains the formulas to be made true, and *NO* those to be made false. The game moves decompose complex formulas or tasks. Some rounds are automatic, say, in an alternating schedule with the active rounds to follow:

---

190 Beth saw this as vindicating the historical intertwinement of mathematical analysis, decomposing some initial problem, and synthesis, putting together a proof.

- if $\neg\varphi$ is in one box, it changes to φ in the other box
- if $\varphi \wedge \psi$ is in *YES*, it is replaced by φ, ψ separately
- if $\varphi \vee \psi$ is in *NO*, it is replaced by φ, ψ separately
- if $\exists x\varphi$ is in *YES*, it is replaced by $\varphi(d)$ for some new object d not yet used in any formula in *YES* or *NO*
- if $\forall x\varphi$ is in *NO*, it is replaced by $\varphi(d)$ for some new object d not yet used in any formula in *YES* or *NO*

Next come active rounds in which critic schedules some formula for treatment.

(a) Critic can schedule a disjunction in *YES* or a conjunction in *NO*, and builder must choose a subformula replacing that formula.

(b) For existential formulas $\exists x\varphi$ in *NO*, critic mentions some object d in the domain under construction so far, and adds $\varphi(d)$ to the *NO* box. For universal formulas $\forall x\varphi$ in *YES*, critic mentions some object d in the domain under construction so far, and adds $\varphi(d)$ to the *YES* box.[191]

Here is the winning convention. A stage is a loss for builder if some formula occurs in both boxes, while builder wins a run of the game if no such loss occurs at any stage. Note that no model can make a formula both true and false: these are indeed conflicting tasks.

Finally, one can stipulate procedural conventions on admissible runs of the game, say, restricting critic's scheduling to always issuing fresh challenges. ■

Every tableau for a predicate-logical satisfiability problem may be viewed as a record of a game such as this, although see below for some divergences. Note, though, that strife need not be a helpful way of understanding logic games. Critic might be a manager making sure builder forgets no tasks, and creates a house of the highest quality.[192]

One can turn tableaus into trees for the construction game. Still, tableaus are not enough, as the game must list all scheduling orders for critic. So, consider scheduling. First, players must keep moving when they still can. This forces critic

---

191 In standard tableaus, formulas $\exists x\varphi, \forall x\varphi$ treated once remain available for later calls. There is no advantage to carefully choosing a one-time challenge, and thus, case (b) becomes automatic.

192 One can give builder more control, also allowing choices from the available objects. Moreover, as an incentive, we might give builder a higher payoff for smaller models.

to eventually examine every task in any finite game. This explains why critic's scheduling order is not crucial to the propositional game. But predicate-logical tableaus may be infinite, and they need task scheduling making sure that each formula gets considered on every branch. Otherwise, a branch might remain open with a hidden inconsistency. Here, we leave this fine structure of games, as well as natural equivalences between different formats, as an open end, as we are mostly interested in players' winning strategies.

16.4 The success lemma and some game theory

From the above description, one might think that there is not much strategic content to tableau games. The following results should correct that mistaken impression. The above adequacy theorems for tableaus induce a success result for model construction games.

THEOREM 16.2 The following two assertions are equivalent:

(a) The set of formulas $\{\varphi_1, \ldots, \varphi_k, \neg\psi_1, \ldots, \neg\psi_m\}$ is satisfiable.

(b) Builder has a winning strategy in the above construction game starting with the φ's put in the *YES* box and the ψ's in *NO*.

A proof is given in van Benthem (2006b), but the idea is obvious from the usual adequacy proofs for tableaus showing that they test their initial problem correctly and completely for satisfiability.

Theorem 16.2 implies a familiar game-theoretic property. A tableau is either closed or open: critic or builder has a winning strategy in the game.

FACT 16.3 Model construction games are determined.[193]

As we saw in Chapter 15 for model comparison games, this theorem, like so many adequacy results, is still $\exists$-sick. As with evaluation and comparison games, what do winning strategies encode as independent logical objects?

The answer is clear. Builder's winning strategies are tied to open branches that generate models satisfying all initial requirements. By contrast, critic's winning strategies allow critic to end every branch of the game tree at some finite stage in a

193 This also follows from the Gale-Stewart Theorem of Chapter 5: the winning condition for critic (finite failure by builder) defines an open set of runs.

failure for builder. Thus, the subtree of the full game tree played according to that strategy has no infinite branches. But then, critic must have a finite procedure for achieving this, by König's Lemma. This explains why winning strategies for critic are associated with finite closed tableaus, being proofs of the initial implication

$$\bigwedge\{\varphi_1, \ldots, \varphi_k\} \rightarrow \bigvee\{\psi_1, \ldots, \psi_m\}$$

Summing up, we have found the following explicit connection.

THEOREM 16.3 Construction games have an explicit correspondence between

(a) winning strategies for builder, (b) models for the given formulas.

Construction games also have an effective correspondence between

(a)′ winning strategies for critic, (b)′ proofs for the initial sequent.

Note that this is not a one-to-one correspondence. Different open branches in a tableau may encode the same model, but also, one open branch may stand for several models by underspecification of truth values for atoms. Even so, the single notion of strategy for two players in one construction game unifies the notions of model and proof. Normally vastly different things meet here directly.

Our own view of tableaus is as task decomposition games. Their properties assume little about the details of these tasks, but we shall now discuss some specifics.

16.5 Making critic more essential: Fragments and variations

Propositional tableau games seem lifeless, nothing depends on critic's precise choice of moves. This is borne out by the *NP* complexity of propositional satisfiability. In order to win, builder must choose some witness of polynomial size in the length of the input formulas, whose goodness can be checked in polynomial time: an open branch, or a valuation satisfying all initial *YES* and *NO* demands. Things are different with predicate logic, as critic has to schedule in the right way to make every branch end at some finite stage, if closure is possible at all. But if we impose strict scheduling as a constraint on runs, then again, nothing depends on critic's precise selection of moves. These observations bring us to extensions and variations changing these features.

Modal logic Consider modal logic again, as a bounded-quantifier fragment of first-order logic. The complexity of satisfiability is *Pspace* for the basic modal language, which is the complexity of many genuine two-player games (Papadimitriou 1994). Here is the reason. In modal tableaus, modalities are treated in a special fashion. For the universal modality, we need the following instruction.

Definition 16.2 Modal tableau rules
Analyze the tableau at any world stage w as far as possible by Boolean rules. If no closure occurs, the *modal operator rule* looks at the remaining box formulas $\Box\varphi$ in *YES*, and box formulas $\Box\alpha$ in *NO*. For each $\Box\alpha$, it picks a new R-successor t of w, starts there with α in *NO*, and puts all φ of the first kind in *YES*. ■

This fan-out of successors is a universal requirement for builder, and so the matching model construction game for modal logic gives critic a choice of which new world to open. Here, builder's being able to cope with any challenge by critic is essential.[194]

One-shot predicate logic First-order tableaus, too, may give critic real work to do. Consider a natural variant where rules become one-shot. In particular, for existential quantifiers in *NO* or universal ones in *YES*, critic gets to issue one challenge: one object to be plugged in. The existential formula is then discarded.

Tableaus are now finite. In propositional or modal tableaus, this change to the extensive game makes no difference to players' powers in terms of validity and consistency. But in first-order logic, it does. Our earlier quantifier shift depended essentially on critic's choices for objects plugged in for the universal quantifiers to the left. By modifying an earlier tableau, we see how critic loses the one-shot game.

Example 16.7 Plato's Law
Plato's Law is a classically valid principle that fails in the one-shot regime:

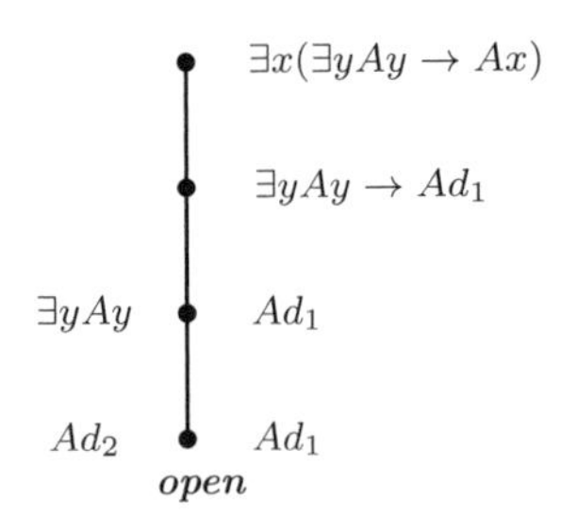

194 Also crucial to computational complexity is that each task can be handled on its own.

In standard tableaus, one repeats the instruction for $\exists x(\exists y Ay \rightarrow Ax)$, plugging in the new object d_2 to generate more requirements that close the tableau. ■

The following property of the new one-shot games is easy to see.

THEOREM 16.4 The validities of one-shot predicate logic are decidable.

REMARK Contraction
Proof-theoretically, the classical repetitive character of the quantifier rules reflects the structural inference rule of

$$\text{Contraction} \quad \text{from } \Sigma, A, A \Rightarrow \varphi, \text{ infer } \Sigma, A \Rightarrow \varphi$$

This apparently innocent rule is crucial to standard inference.

EXAMPLE 16.8 Plato's Law once more
Consider an earlier example, now duplicated. Unlike the earlier variant, the following tableau does have a winning strategy for critic:

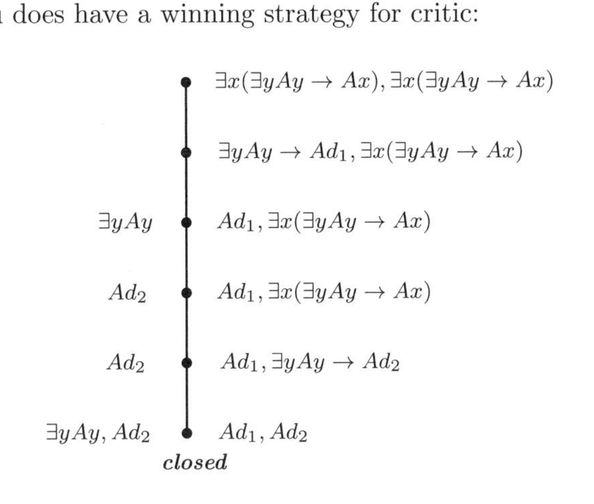

The difference with the preceding tableau speaks for itself. ■

One-shot valid sequents are axiomatized by a Gentzen sequent calculus with the usual logical rules of first-order logic, plus all structural rules minus the Contraction Rule. For more general ideas along these lines, see the substructural linear game logic of Chapter 22.

New logical operators The new game also suggests a more radical redesign of a first-order language. If a task can be scheduled infinitely often, this can be marked by an "iteration operator" $!\varphi$ allowing critic to schedule φ any finite number of times. This makes a decidable system such as one-shot predicate logic undecidable, as the new system can embed full predicate logic by prefixing all quantifiers with **!**, enforcing the old tableau rules. But still, the language extension throws some new light on the earlier first-order games. Further iteration operators for games will be considered in Chapter 19.

16.6 Conclusion

We have turned the well-known method of testing satisfiability via semantic tableaus into a model construction game. Doing so provides a unified framework for two basic logical tasks that seem very different, but are in fact intuitively intertwined: finding proofs, and constructing models. Moreover, it suggests interesting new variations on logics, such as one-shot predicate logic. This theme of finding variant logics via games will return in later chapters. But there was also a more general game-theoretic thrust to what we did. The idea of reasoning as a potentially infinite game seems suggestive in itself, and also, the format of tableau games seems well suited for absorbing more realistic game-theoretic structure that was studied in Part I, such as preferences, imperfect information, and strategic equilibria between participants in discourse.

16.7 Literature

This chapter is based on van Benthem (2006b).

Much more sophisticated model construction games with many mathematical applications are found in Hodges (2006), and Hirsch & Hodkinson (2002). Hamami (2010) uses semantic tableaus together with the dynamic-epistemic logics of Part II to model inquiry in the sense of Hintikka (1973) as a process combining information coming from acts of inference and questions.

17 Argumentation and Dialogue

Argumentation is perhaps the oldest logic game, as we saw in the Introduction to this book. In this chapter, we briefly discuss the dialogue games of Paul Lorenzen, influential mainly in the 1970s, but with a new lease of life in Barth & Krabbe (1982) and Rahman & Rückert (2001). Our main interest this time is not in the logical rules of the system but in its procedure, seeing how varying players' procedural powers in argumentation may generate different logics, something that makes the earlier discussions of game equivalence in Chapters 1 and 11 quite vivid. Eventually, dialogue games may also be a strong suit for logic in joining forces with argumentation theory, for which we give references at the end of this chapter (cf. also Chapter 23). But there is also a historical line from Lorenzen dialogues to the linear logics of Chapter 20.

There is more to this chapter than just one discourse practice. Argumentation is a good testing ground for a Theory of Play as proposed in Chapter 10.

17.1 Dialogue games and actual debate

Dialogue games The dialogues of Lorenzen (1955) are related to the construction games of Chapter 16,[195] but they are closer to real argumentation. A formal debate takes place between a *proponent* $\boldsymbol{P}$ who defends a claim against an *opponent* $\boldsymbol{O}$ who grants initial concessions. Moves are attacks and defenses on assertions according to logical and procedural rules. Logical rules involve choices, switches, and picking

---

195 There is a letter from Lorenzen to E. W. Beth (cf. van Ulsen 2000) that explains his games as records of arguments, but then concedes: *"So entstehen eben Ihre Tableaux ..."*.

instances, but no external world determines who wins or loses, only internal criteria such as consistency. Likewise, in actual debate, people often lose by incoherent positions, rather than the judgment of an external arbiter. Procedural conventions are real, too: they can be observed in a court of law.

Logical rules for defense and attack Conjunctions $A \wedge B$ can be attacked in either conjunct, while asserting the attacked A or B is the matching defense. A disjunction $A \vee B$ is attacked by asking for a choice. The defense consists of choosing a disjunct A or B, and then defending that. A negation $\neg A$ is attacked by defending the statement A itself: no defense exists.[196] We also give a rule for implications $A \to B$. The attack consists of defending A, the matching defense is B. To complete the system, consider the two quantifiers. Attacks on an existential quantifier $\exists x \varphi$ ask for a witness, the defense then chooses some name d of an object, and defends $\varphi(d)$. Attacks on universal quantifiers $\forall x \varphi$ consist of mentioning a challenge: that is, any object d, for which the appropriate defense is $\varphi(d)$.

These attacks and defenses are summarized in the following definition.

DEFINITION 17.1 Dialogue rules of attack and defense
Here are the above rules summarized in a table:

Operator	*Attack*	*Defense*
$A \wedge B$	$?L$	A
	$?R$	B
$A \vee B$	$?$	A, B
$\neg A$	A	$-$
$A \to B$	A	B
$\exists x \varphi$	$?$	$\varphi(d)$
$\forall x \varphi$	d	$\varphi(d)$

Each move is either a defense to a preceding attack, or an attack on something the other player has asserted. ■

Successive moves require a "score," a record of what is said for what purpose. Such context sensitivity is entirely realistic for dialogue.

Winning and losing A player loses if there is nothing legitimate left to say at that player's turn. In dialogue games, this way of ending reflects a self-contradiction.

---

196 Crucially, A itself can now become the target of a counterattack.

Procedural conventions The scheduling of the dialogue is $(\boldsymbol{PO})^*$: players move in turn, as in earlier logic games. Next, we constrain debate by stipulating the rights and duties of players:

(a) Proponent may only assert an atomic formula after opponent has asserted it.

(b) If one responds to an attack, this has to be to the latest still open attack.

(c) An attack may be answered at most once.

(d) An assertion made by proponent may be attacked at most once.

17.2 Learning by playing

Some examples will demonstrate the framework. It is important to record at each stage which previous line is being attacked, or which attack is being responded to.

EXAMPLE 17.1 Defending $p \wedge \neg(p \wedge q) \to \neg q$

Here is a fairly typical dialogue for a classically valid implication:

1	$\boldsymbol{P}$	$p \wedge \neg(p \wedge q) \to \neg q$	
2	$\boldsymbol{O}$	$p \wedge \neg(p \wedge q)$	[A, 1]
3	$\boldsymbol{P}$	$\boldsymbol{P}$ can respond to the first attack, or counterattack. Say:	
	$\boldsymbol{P}$	$?L$	[A, 2]
4	$\boldsymbol{O}$	p	[D, 3]
		this is the only thing $\boldsymbol{O}$ can do at this stage	
5	$\boldsymbol{P}$	$?R$	[A, 2]
6	$\boldsymbol{O}$	$\neg(p \wedge q)$	[D, 5]
7	$\boldsymbol{P}$	$\neg q$	[D, 2]
8	$\boldsymbol{O}$	$\boldsymbol{O}$ has no further attacks to respond to, so $\boldsymbol{O}$ must attack:	
	$\boldsymbol{O}$	q	[A, 7]
9	$\boldsymbol{P}$	$p \wedge q$	[A, 6]
10	$\boldsymbol{O}$	$?L$	[A, 9]
11	$\boldsymbol{P}$	p	[D, 10]
		this is admissible, as p has been asserted before by $\boldsymbol{O}$	
12	$\boldsymbol{O}$	$?R$	[A, 10]
13	$\boldsymbol{P}$	q	[D, 12]
		this is admissible, as q has been asserted before by $\boldsymbol{O}$	

$\boldsymbol{O}$ has nothing legitimate to say at this stage, and loses. ■

EXAMPLE 17.2 Excluded middle

The games show some striking differences with classical logic:

1	***P***	$q \vee \neg q$	
2	***O***	?	[A, 1]
3	***P***	$\neg q$	[D, 2]
4	***O***	q	[A, 3]

This time, ***P*** has nothing legitimate to say at this stage, and loses.

If we want ***P*** to win this game for a classical law after all, we have to change the procedural rules stated above, and allow ***P*** to reply again to an earlier attack:

5	***P***	q	[A, 3]

Now ***O*** is the one who has nothing legitimate left to say, and loses. ■

Some people love these quirks of procedure with logical import, while others hate them. In Chapter 20, we will see a way of moving beyond such gut responses, in terms of resources that players have available.

Winning strategies and logical validity We define game validity for a formula φ as the proponent's having a winning strategy for defending φ against any opponent. This should correspond with the assertion being valid or provable in some natural logical system. But which system, now that it is not necessarily classical logic?

17.3 Constructive versus classical logic

Three accounts of valid inference Historically, there are several accounts of valid inference. The model-theoretic view sees it as transmission of truth from premises to conclusions. The inferential view sees it as derivability via elementary proof steps in some natural system. Right now, we have a third game-theoretic account of validity as guaranteeing wins in argumentation. Should all three accounts of validity capture the same premise/conclusion transitions? Rather, they seem to be independent stances. Games have something of their own to say.

The provability interpretation To appreciate dialogue games, one must understand the difference between classical and intuitionistic logic. The latter arises on a constructive view of reasoning in terms of proofs (Troelstra & van Dalen 1988).

DEFINITION 17.2 The constructive proof-theoretic view of meaning
The explanation of the logical operators now goes inductively in terms of proofs:

Proof of	
$A \wedge B$	a pair of proofs: one for A, one for B
$A \vee B$	a proof for A or for B
$A \to B$	a method for turning arbitrary proofs of A into proofs for B
$\neg A$	a method for turning proofs of A into proofs of a contradiction $\bot$
$\exists x \varphi$	an object d plus a proof of $\varphi(d)$
$\forall x \varphi$	a method for producing proofs of $\varphi(d)$ for every object d

Complex proofs are explained by successive combination of these steps. ■

Classical and intuitionistic logic agree on many principles, but they diverge on non-constructive laws that reflect truth without reason. The prime example of the latter is the law of the excluded middle $A \vee \neg A$. Without further information, we neither have a proof, nor a refutation for for A.[197]

One learns the difference between classical and constructive reasoning by developing a more discriminating palate than usual.

EXAMPLE 17.3 Double negation
The implication $A \to \neg\neg A$ is constructively provable. Given any proof p for A, here is one for $\neg\neg A$, being a method for proving a contradiction from any proof q of $\neg A$. Just combine p and q to prove $A \wedge \neg A$. But the converse $\neg\neg A \to A$, valid classically, fails constructively: there is no way of transforming proofs of $\neg\neg A$ into direct proofs of A. Related to this example is another crucial difference. In classical logic, there are two equivalent forms of the key principle of reductio ad absurdum:

to prove $\neg A$,	assume A and derive a contradiction
to prove A,	assume $\neg A$ and derive a contradiction.

Of these two, only the former is constructively valid. The latter would produce a proof of $\neg\neg A$, which is not enough as a proof for A itself. ■

197 This is not the end of the matter. The linear logic of Chapter 20 drops excluded middle when reading disjunction as a once-only choice between games as in Chapter 14. But it does have a valid law of excluded middle for another disjunction, referring to games played in parallel that can be revisited, giving proponent more chances to win.

Constructive proofs are often given as natural deduction trees.

EXAMPLE 17.4 A constructive natural deduction proof
We display a derivation for $(A \wedge \neg(A \wedge B)) \to \neg B$ that proponent could win:

$$\dfrac{\dfrac{\dfrac{\dfrac{A \wedge \neg(A \wedge B)\ (1)}{A} \qquad B\ (2)}{A \wedge B} \qquad \dfrac{A \wedge \neg(A \wedge B)\ (1)}{\neg(A \wedge B)}}{\neg B}\ \textit{withdraw assumption (2)}}{(A \wedge \neg(A \wedge B)) \to \neg B}\ \textit{withdraw assumption (1)}$$

Another good exercise are the four implications in the De Morgan laws:

$$\neg(A \wedge B) \leftrightarrow (\neg A \vee \neg B), \quad \neg(A \vee B) \leftrightarrow (\neg A \wedge \neg B)$$

of which only the implication $\neg(A \wedge B) \to (\neg A \vee \neg B)$ is invalid constructively. Thus, conjunction and disjunction are not interdefinable in intuitionistic logic. ■

This is an instance of a more general phenomenon. The weaker a base logic, the greater the variety of non-equivalent logical operations it supports.[198]

Information models A semantic viewpoint illuminates these issues. Intuitionistic logic describes trees of stages in a process of acquiring knowledge, classical logic is only about endpoints where all evidence is in. The order of the stages is reflexive and transitive. Intuitionistic conjunctions and disjunctions are then read as usual, locally at any stage. But an intuitionistic negation $\neg\varphi$ constrains further inquiry:

$\neg\varphi$ holds at a stage if φ never holds later on in the tree.

Likewise, an intuitionistic implication $\varphi \to \psi$ says that

ψ holds at every later stage where φ holds.

EXAMPLE 17.5 Refuting classical laws
The following simple model illustrates many phenomena of interest:

198 Likewise, the classical interdefinability of the quantifiers disappears. We have constructive validity of $\neg\exists x\varphi \leftrightarrow \forall x\neg\varphi$ and $\exists x\neg\varphi \to \neg\forall x\varphi$, but not of $\neg\forall x\varphi \to \exists x\neg\varphi$.

In the black dot, $p \vee \neg p$ is false. The model also refutes the classical law $\neg\neg p \to p$ ($\neg\neg p$ is true on the left, p is not), as well as Peirce's Law $((p \to q) \to p) \to p$. ■

Intuitionistic models have something of the flavor of the dynamic-epistemic models studied in Part II of this book (cf. van Benthem 2009). We will not provide further details here, but the reader can glean a lot from the illustrations to follow.

17.4 The logic of the games

To show the harmony of intuitionistic logic with natural deduction and winning strategies in dialogue games, we will just discuss an example.

The use of proofs The De Morgan law $(\neg A \vee \neg B) \to \neg(A \wedge B)$ is valid in information models: if A is excluded from now on after the current stage, then so is $A \wedge B$. Indeed, this implication has the following natural deduction proof:

$$\dfrac{\dfrac{\neg A \vee \neg B \qquad \dfrac{\neg A \quad \dfrac{A \wedge B}{A}}{\neg(A \wedge B)} \qquad \dfrac{\neg B \quad \dfrac{A \wedge B}{B}}{\neg(A \wedge B)}}{\neg(A \wedge B)}}{(\neg A \vee \neg B) \to \neg(A \wedge B)}$$

In dialogues, proponent has a winning strategy reflecting this natural deduction:

1	***P***	$(\neg A \vee \neg B) \to \neg(A \wedge B)$	
2	***O***	$\neg A \vee \neg B$	[A, 1]
3	***P***	$\neg(A \wedge B)$	[D, 2]
4	***O***	$A \wedge B$	[A, 3]
5	***P***	$?L$	[A, 4]
6	***O***	A	[D, 5]
7	***P***	$?R$	[A, 4]
8	***O***	B	[D, 7]
9	***P***	$?$	[A, 2]
10	***O***	has two options for defense: both go the same way. Say, ***O*** picks	
		$\neg A$	[D, 9]
11	***P***	A	[A, 10]:

This is admissible, as ***O*** said A before. Now ***O*** has nothing to say and loses.

The use of models Now consider a dual search for counterexamples.

EXAMPLE 17.6 Refuting nonconstructive principles
The invalid De Morgan law $\neg(A \wedge B) \rightarrow (\neg A \vee \neg B)$ fails on information models:

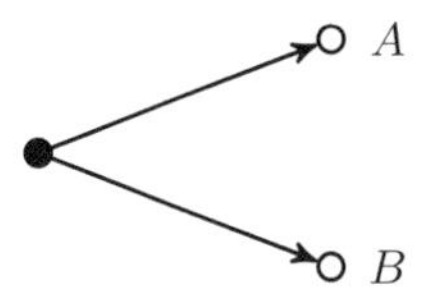

At the black stage to the left, $\neg(A \wedge B)$ is true, but at the same time, $\neg A \vee \neg B$ is false there, since both A and B might still be reached via further investigation. In accordance with this finding, there is no constructive proof for this law. ■

But the perspective of models is interesting in its own right for the other player. In a dialogue game, it is easy to see that they guide behavior in the following way.

FACT 17.1 Opponent can use any intuitionistic model as a winning strategy.

Proof The idea is to maintain the following invariant. Since the formula stated by proponent is false in some state, opponent can choose moves so that all concessions are true in the current state in mind, while opponent's attacks on proponent's last statement can be read off from that state. For instance, in the above situation, it is safe for opponent to concede $\neg(A \wedge B)$, bearing in mind the initial state. If proponent then answers with a counterattack $A \wedge B$, opponent can attack that, and proponent has no response (neither A nor B holds in the black state). And if proponent answers with the (false) disjunction $\neg A \vee \neg B$, opponent can attack that, being sure that both answers are false. But there is also a crucial dynamic process at work of changing stages. Suppose that proponent chooses to defend $\neg A$. Then, opponent can exploit its falsity by moving *upward* in the information model, going to the upper right-hand state where A is true, and then continuing play with that world in mind.[199] ■

Determinacy and adequacy theorems Now we turn to general game-theoretic aspects of dialogue games, in the same spirit as for earlier logic games in Chapters 15

199 The driving force behind such strategies is that opponent can always force proponent to address concerns in the current world opponent has in mind, because proponent must reply to the last attack made.

through 17. For a start, dialogue games allow infinite runs, as proponent might keep repeating the same attack, allowing opponent to repeat the defense. Such infinite runs act as draws, neither winning nor losing. Still, the games are determined, and this can be proved from an analysis of the game tree, or from the main result about dialogue games (Felscher 2001).

THEOREM 17.1 The following are equivalent for first-order formulas $A_1, \ldots, A_k, B$:

(a) $\boldsymbol{P}$ has a winning strategy in the dialogue game for $A_1 \wedge \cdots \wedge A_k \rightarrow B$

(b) $A_1 \wedge \cdots \wedge A_k \rightarrow B$ is valid in intuitionistic logic.

This result also has the following explicit strategy version.

THEOREM 17.2 There exist effective procedures taking intuitionistic proofs to winning strategies for proponent in L-games, and vice versa.

Proofs for both versions can be found in Felscher (2001). More sophisticated results in this line are Abramsky's full completeness theorems (cf. Chapter 20) that relate proofs and strategies in a category-theoretical setting.

17.5 Extensions and variations

Other logics We saw how proponent can win an excluded middle game, if allowed to switch defenses. Manipulating procedure can influence the logic of dialogues.

FACT 17.2 The dialogue game yields classical logic for proponent's winning strategies if the earlier procedural restrictions (b) and (c) apply only to opponent.

Now proponent can repeat attacks, shifting the focus of play. Using such modulations, dialogues model a variety of logics (cf. Rahman & Rückert 2001).

Resource logics In Chapter 20, we will see related ideas in linear logic. There, we will also see the innovative line running from dialogue logic to modern game semantics (Blass 1992, Abramsky & McCusker 1999) where procedural rules have been re-encoded in a greater range of parallel game-forming operations. In particular, as we have noted earlier, linear logic games make excluded middle invalid as a Boolean choice game $A \vee \neg A$, but valid as a parallel choice game $A + \neg A$.

Linear logic is a pioneering instance of modern resource logics that systematically keep track of the resources (linguistic, computational, cognitive) that agents have available in performing reasoning (cf. van Benthem 1991, Girard 1993, Restall

2000). These systems change the standard notion of logical consequence to make constructive feasibility, rather than mere truth, the hallmark of validity.

Other applications Dialogue games have been used as an operational semantics in physics (Mittelstaedt 1978). They have also influenced argumentation theory as a model for commitment and discourse dynamics (Walton & Krabbe 1995, Gabbay & Woods 2002). They also fit legal procedure with its standards of proof (Prakken 1997). Such applications may shift the emphasis from competition to cooperation, a useful alternative stance. We will explore some of these angles in Chapter 23.

Formalizing procedure The flexibility of the procedural component, often considered a nuisance, is in fact a key attraction of dialogue games. There is no general theory of procedure in the framework operating at an illuminating higher abstraction level. One promising line is a junction with "argumentation systems" in the sense of (Dung 1995): directed graphs of propositions with a binary relation of "p attacks q" supporting a theory of stable assertability. A link between dialogue games and abstract argumentation systems was suggested in van Benthem (1999), and Grossi (2010) is a promising implementation using games for modal fixed point logics. More detail will be given in the appendix to this chapter.

Game equivalence Behind our topics so far, some of the big issues in this book reappear. For instance, our discussion of procedure is typically concerned with extensive games, and whether different procedures have the same effect is closely related to notions of equivalence one might have for argumentation, finer or coarser. This is reminiscent of the discussions of game equivalence in Chapters 1 and 11, and others throughout this book. In particular, getting different logics as sets of validities matching our games, whether classical, intuitionistic, or linear, suggests a new view of outcomes of games that we will discuss further in Chapter 18.

Theory of Play Another earlier theme that seems close to the surface in this chapter is the notion of a Theory of Play developed in Chapter 10. The guiding idea there was to make players themselves into explicit first-class citizens of the logical analysis, including their diverse informational powers and habits of play. Looking at players' resources, as done above, is one step in this direction, and so is taking a more finely-grained view of their possibly different procedural rights and duties. We will discuss this sort of merge between logic games and game logics in more detail in Chapter 25.

17.6 Conclusion

Dialogue games offer an interesting dynamic view of logical reasoning, suggesting new handles on procedural options in argumentation. They have not had an undivided success as a paradigm for logic games, but their resilience since the 1950s has been remarkable, especially when dialogues are linked to proof-based constructivist logics and resource logics. Chapter 20 will offer examples from computation and linear logic, and a glimpse of other interfaces will be found in the appendix.

17.7 Literature

This chapter is based on the lecture notes in van Benthem (1999).

Good sources for dialogue games in their original sense are Kamlah & Lorenzen (1967), Barth & Krabbe (1982), and Rahman & Rückert (2001). Sources for broader related areas will be listed at the end of the following appendix.

17.8 Appendix on argumentation

Dialogue games are still far removed from daily argumentation, an area where we all have vivid intuitions and experiences, and perhaps even the area where logic originally started. This appendix adds a few thoughts on this empirical practice.

Real argumentation differs from our games so far in the richer role of procedure in shaping behavior or the dynamics of arguments and counterarguments that help or hinder each other. Even in the history of logic, there have been richer argumentation games, in terms of maintaining focus and relevance: see the study of Obligatio and related medieval practices in Dutilh-Novaes (2007) and Uckelman (2009). Many such aspects are investigated in argumentation theory and related fields.

Modeling realistic argumentation Decisions involve a balance: we weigh arguments, and decide when the balance tips in favor of a proposal. Weighing per se is a noninteractive procedure. However, experience in actual meetings tells us that strategic action plays a role, too: things depend on how we play our cards. There are diverging trends here. Arguments once put forward lose force with time. This is why it is often useful to refer to your opponents' considerations ahead of time.

But while familiarity breeds contempt, success also breeds success: the trend in a debate can work to your advantage. And finally, it is often important not to throw your best arguments on the table at once, keeping some trumps up your sleeve for when you know more about your opponents' strength. Let us model some of this.

DEFINITION 17.3 Strategic argumentation games
A *strategic argumentation game* has an issue P on the table. $\boldsymbol{A}$ has k arguments in favor of P, with values in natural numbers, and $\boldsymbol{E}$ has k arguments against P. In each round, $\boldsymbol{A}$ places a new argument for P on the table from $\boldsymbol{A}$'s remaining stock, and $\boldsymbol{E}$ places a new argument against. The player with the largest total of values on the table wins the round, but draws are possible. If you win the round or draw, your current total goes on to the next, but if you lose, your total diminishes by 1 (until you drop to 0). The goal is to win the final round, after k steps. ■

We can think of this game as simultaneous or sequential. This time, mere weights of arguments do not say everything: their distribution matters.

EXAMPLE 17.7 A balance game
Let $k = 4$, where $\boldsymbol{A}$ has $2, 0, 0, 0$ and $\boldsymbol{E}$ has $1, 1, 1, 1$. $\boldsymbol{A}$ can win the game by first putting 2 on the table, and then winning intermediate rounds. But if $\boldsymbol{A}$ were to play the 2 later, this would not work, and $\boldsymbol{A}$ can lose. ■

Thus, timing is critical. This is demonstrated clearly in the following scenario.

EXAMPLE 17.8 A more complex game tree
$\boldsymbol{A}$ has $4, 1, 0$ and $\boldsymbol{E}$ has $3, 2, 1$. It is easy to see that $\boldsymbol{A}$ can win by playing 4 first:

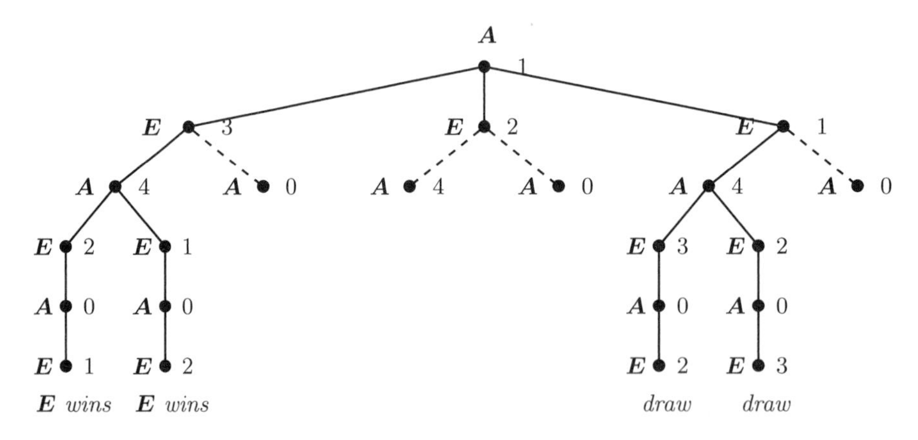

The tree also has corners where $\boldsymbol{E}$ has a winning strategy, or a draw occurs. ■

The sequential version of this game falls under Zermelo's Theorem of Chapter 1, modified for draws. Still, it does not do justice to our experience of waiting with our best arguments, since starting with the highest value is never bad for a player. But it is easy to add such features by modifying argument deterioration.

> Let the value of arguments in each round deteriorate, not by losing points, but by halving. Anyone who wins or draws gets a bonus point.

In this case, playing one's stronger arguments later can be useful after all.

EXAMPLE 17.9 Arguments with half-life time
Let $\boldsymbol{A}$ have $6, 2$, while $\boldsymbol{E}$ has $5, 3$. Consider the new game regime, and compute points. If $\boldsymbol{A}$ plays 6 first, then $\boldsymbol{E}$ can play 3: resulting in an intermediate score of $(4, 1.5)$. After that, $\boldsymbol{A}$ plays 2 and $\boldsymbol{E}$ plays 5: resulting in a final score of $(3, 4.25)$ where $\boldsymbol{E}$ wins. By contrast, if $\boldsymbol{E}$ were to play the argument with value 5 first, the intermediate score would be $(4, 2.5)$, and the final score $(4, 2.75)$, and $\boldsymbol{A}$ wins. We leave other possible runs to the reader. ∎

Many variations are possible. One can have more sophisticated formulas for familiarity effects, allow approval points from other participants, and so on. While these features say little about logic (though initial values might well be grounded in logical content), argumentation of this sort is clearly a rational skill. Moreover, it is easy to refine the game so as to factor in logical implications or other dependencies between propositions on the table.[200]

Our second example is about a specialized professional area of argumentation.

Juridical procedure Many authors have claimed that there are deep differences between logical and juridical reasoning. According to the classic Toulmin (1958): "Logical proof turns on abstract mathematical form, whereas legal reasoning is all about procedure: the formalities." Indeed, a legal setting has procedural features such as timing, parties with different aims, limited resources, and the aim for, not absolute, but reasonable certainty. But these features also make sense in our logic games, providing a more realistic view of the strategies involved, and the way they can be in equilibrium for a fair outcome.

---

200 There is much more to debate that should be of interest to logicians. A more systematic study would fit well with themes in this book, such as the dynamics of deliberation in Chapter 8 and the underlying notions of procedural fairness (van Benthem 2011d).

Single versus repeated play A problematic feature here might be the one-shot character of specific plays. Say that verifier won one run of an evaluation game for φ. This does not prove that φ is true. Having a winning strategy means reacting successfully to every play by falsifier. But here is where falsifier has an interesting role. It is in falsifier's interest to offer the best possible counterplay to have the best shot at winning, but at the same time, this also offers the best guarantee that verifier really has a winning strategy. This seems a realistic point in legal procedure, but also in real argumentation and debate. Setting up the game correctly maximizes the chances that a single run will be informative.

Meshing different tasks In fact, some logic games have features that are highly reminiscent of a legal setting in court. Consider the model construction games of Chapter 16. Builder's role is not to check whether things did happen as described, but whether they could have happened that way. This satisfiability task is the object of a lawyer's strategy, pointing out scenarios that are consistent with the innocence of the client. But in the same game, critic had the proof task of showing that the given requirements are inconsistent, or more positively, that some claim denied by builder follows from the evidence on the table. This seems like the role of a prosecutor. The consistency and the proof task are of different logical complexity, an interesting insight by itself, but they do belong together.

Procedure dependence Next, consider the role of procedure in our dialogue games. Depending on how we set the balance of rights and duties for proponent and opponent, different logics came out, classical or intuitionistic. This might be relevant in legal or argumentative settings, too.[201] Procedure dependence is well-attested in social choice or judgment aggregation (Anand et al. 2009). And logic may have something to say about the different reasoning styles that go with it.

Our final point is that, as in Chapters 19 and 20, procedure may involve playing the same game repeatedly. We end with a logical evergreen with a neat recent juridical twist.

---

201 Consider an, admittedly contrived, example. Classical dialogue games allowed players to repeat a defense, if one was not successful the first time around. Let the evidence be $\neg(A \wedge B), \neg A \rightarrow C, \neg B \rightarrow C$. Does it follow that C is true? Yes, if proponent can argue in classical logic, but not with the stricter intuitionistic procedure. Then the first premise does not give the disjunction of $\neg A$ and $\neg B$, as one may be unable to say which one obtains, and hence unable to proceed, when challenged at some stage of the argumentation game. Should criminals go free on intuitionistic technicalities?

EXAMPLE 17.10 Corax versus Euathlos
The sophist Corax had taught Euathlos debating: the fee would be the first money his student made in winning a lawsuit. But Euathlos never entered court. So, Corax brought a case against him, arguing: "Either I win, and you have to pay me by the verdict, or I lose, and you have to pay by our contract." But the well-taught Euathlos produced a counter dilemma: "Either I lose and need not pay by our contract, or I win, and need not pay by the verdict." A lawyer once pointed out a line that logicians had never considered. The student wins the first case, since he had not promised to practice the law. But then, the teacher can bring a second lawsuit to collect his fee, and he will win that. ■

Argumentation theory There are many further links between logic and argumentation theory today, most of them collected in Gabbay & Woods (2002). We mention only two strands here.

One conspicuous program at the interface is that of Dung (1995) mentioned briefly in Section 17.5. "Argumentation systems" are directed graphs whose nodes stand for propositions, connected by a binary relation "p attacks q." One can classify propositions for their status in a graph, saying, for instance, that p is eventually accepted iff it is not attacked by any accepted proposition.

EXAMPLE 17.11 Defining attacks and defenses in logical systems
Formulated as a fixed point equation for accept (A) and reject (R), the preceding stipulation can be written with a modality over the attack relation:

$$A \leftrightarrow \Box R, \qquad R \leftrightarrow \Diamond A$$

Accordingly, Grossi (2010) studies argumentation systems by means of modal fixed point logics such as those in our Chapters 1, 11, and 14, where equilibrium equations like the above arise from a game over the graph. ■

This modal approach may be the correct abstraction level for something that logic games lack so far, an abstract theory of scheduling and general procedure.

Another congenial line is Gabbay's networks approach (Barringer et al. 2012). This generalizes Dung's argumentation systems considerably, introducing algebraic and numerical techniques for solving argumentation networks, and providing logics for the temporal dynamics of their evolution over time. The approach links logic to dynamical systems, a natural interface that we also noted in Chapters 5 and 12 in connection with infinite games and evolutionary game theory.[202]

Conclusion Logic games can get much closer to practical reasoning than has been done so far. This appendix has only opened a small window to the lively interfaces of logic and argumentation theory (see Walton & Krabbe 1995, Gabbay & Woods 2002, Barringer et al. 2012) and of logic and law (see Prakken 1997, Vreeswijk 2000), that include many further topics such as default reasoning, effective solution methods, or deontic reasoning with norms.[203]

Literature This appendix is a condensed summary of material from van Benthem (2001a), van Benthem (2004a), van Benthem (2010a), and van Benthem (2012c).

---

202 A comparison between this network dynamics and the dynamic logic-based approach in this book can be found in van Benthem (2012c).

203 In all this, there is a clear move toward the empirical realm of reasoning. Logic games might even interface with cognitive science, if we focus on the strategic interactive practices that constitute intelligent behavior, rather than single inferences.

18
General Lines through Logic Games

The traditional core tasks of logic concern expressive and inferential power, with notions such as evaluation, proof, model construction, or model comparison. In Chapters 14 through 17, we have shown how all of these tasks can be cast as games. This gives a view of logic as a family of multi-agent activities enriching the received view of abstract formal patterns. Games in this family differ in many respects, as they serve different purposes. Still, we found shared structure, such as determinacy or the unifying role of strategies. This points toward the existence of general themes behind the diversity. In this chapter, we will identify a few such themes, although without any pretense at finding a uniform theory. Our aim is mainly to show how the general viewpoints on games in the earlier Parts I through III of this book apply to logic games, suggesting some interesting questions that make sense for this special class of games, but by implication also for logic itself. Logic shows unfamiliar features through a game-theoretic lens, but it also acquires a new unity. Some of the themes raised in this way will return in Parts V and VI.

18.1 What is gamification?

Behind the specific games of this part, there lies a more general practice that might be called gamification. While it may not be routine to design a logic game, there are some clear patterns. If a concept has alternating quantifiers at some level of structure, these can be turned into different players. Of course, the art is to pull apart different roles in a way that yields new insights, and finding these may take some reflection. For instance, the evaluation games of Chapter 14 are often considered a routine way of restating the truth definition. By then, thinking a bit harder, these games changed our view of first-order logic, deconstructing it into a decidable core algebra of test and object picking, plus principles reflecting less central specifics of assignment models. Likewise, the model construction games of Chapter 16 recast familiar semantic tableau rules in a way that revealed new choice points, suggesting a decidable one-shot first-order logic. Pulling apart classical notions can also happen in other ways. The comparison games of Chapter 15 took partial maps between models, and gave one component of each ordered pair to one player, and the remaining component to the other.

Similar phenomena occurred elsewhere in this book. Our analysis of players' powers in Chapter 11 took a standard ultrafilter, but then distributed its sets over two owners, such that, if one person does not own some set, the other owns its complement. We will see further instances in the knowledge games of Chapter 22 and the sabotage games of Chapter 23. While all this is suggestive, there seems to be no general algorithm for gamification (cf. van Benthem 2008). Hopefully, by now, with our examples, the reader can perform this task solo.

18.2 Calculus of strategies

Strategies were a constant across logic games, connecting very different notions: models and proofs in Chapter 16, and difference formulas and bisimulations in Chapter 15. Sometimes, they represent new notions that have been studied less in logic: winning strategies in the evaluation games of Chapter 14 offered a finer level of reasons for truth than truth values. Still, there are many shared features. What we would like to have is a calculus of reasoning about strategies that contains the many basic arguments that are shared between all of these different games.

This topic was discussed in Chapter 4, and no conclusive answer was found. We merely repeat an earlier illustration, showing how elementary arguments about logic games that come naturally to us contain interesting cues toward such a calculus.

EXAMPLE 18.1 Exploring basic strategy calculus
Consider the following simple sequent derivation for a propositional validity:

$$
\begin{array}{ccc}
A \Rightarrow A & B \Rightarrow B & \\
A, B \Rightarrow A & A, B \Rightarrow B & C \Rightarrow C \\
\multicolumn{2}{c}{A, B \Rightarrow A \wedge B} & A, C \Rightarrow C \\
\multicolumn{2}{c}{A, B \Rightarrow (A \wedge B) \vee C} & A, C \Rightarrow (A \wedge B) \vee C \\
\multicolumn{3}{c}{A, B \vee C \Rightarrow (A \wedge B) \vee C} \\
\multicolumn{3}{c}{A \wedge (B \vee C) \Rightarrow (A \wedge B) \vee C}
\end{array}
$$

Here is a corresponding richer form indicating strategies:

$$
\begin{array}{c}
x : A \Rightarrow x : A \qquad\qquad y : B \Rightarrow y : B \\
x : A, y : B \Rightarrow x : A \qquad x : A, y : B \Rightarrow y : B \qquad\qquad z : C \Rightarrow z : C \\
x : A, y : B \Rightarrow (x, y) : A \wedge B \qquad\qquad x : A, z : C \Rightarrow z : C \\
x : A, y : B \Rightarrow \langle l, (x, y) \rangle : (A \wedge B) \vee C \qquad x : A, z : C \Rightarrow \langle r, z \rangle : (A \wedge B) \vee C \\
x : A, u : (B \vee C) \Rightarrow IF\, head(u) = l\ THEN\, \langle x, tail(u) \rangle\, ELSE\, tail(u) : (A \wedge B) \vee C \\
v : A \wedge (B \vee C) \Rightarrow IF\, head((v)_2) = l\ THEN\, \langle (v)_1, tail((v)_2) \rangle\, ELSE\, tail((v)_2) : (A \wedge B) \vee C
\end{array}
$$

The format can be read as a construction of strategies with operations of

storing strategies for a player who is not to move	$\langle\, ,\, \rangle$
using a strategy from a list	$(\ )_i$
executing the first action of a strategy	*head*( )
executing the remaining strategy	*tail*( )
choosing depending on a test	IF THEN ELSE

At this abstraction level, the derivation stands for different things. Read in one way, it is about combining proofs, but read in another, it is a recipe for turning any winning strategy of verifier in an evaluation game for $A \wedge (B \vee C)$ into one for the game $(A \wedge B) \vee C$. Insights about strategies work across logic games. ■

Teasing out a general calculus of strategies covering paradigmatic cases of reasoning about games seems to be a major open problem, but Chapter 4 has given some concrete examples of why and how it might be attractive.

18.3 Game equivalence

Thinking about logic games in general also raises the issue of the right structural level for understanding them. In Parts I and III, we saw a major choice between the more detailed action level of moves and strategies, and the rougher level of players' powers, calling two games equivalent when players have the same powers for forcing sets of outcomes. Even though logic games provide a lot of detail, and logic itself can be used to analyze the finer action level of games, it seems clear that logic games are used at the level of power equivalence. We saw this in Chapter 14, where logically equivalent formulas give players the same powers over every model where their evaluation game is played. This power equivalence will drive the general game algebra studied in Chapters 19 and 24. This is one of the intriguing spirals discussed in our Introduction when moving from game logics to logic games. While a logical language may record local details of extensive games, its own evaluation game may be power-oriented.

The theme of game equivalence occurs more generally. It is an interesting exercise to look at logic texts and ask when authors call two versions of a game obviously the same. Normally, one will find power level identifications. An illustration mentioned in Chapter 15 was telling in this respect, and we repeat it here.

EXAMPLE 18.2 Modified comparison games
Barwise & van Benthem (1999) have model comparison games starting with a finite partial isomorphism between two models. Starting from a partial isomorphism F, in each round, duplicator selects a set F^+ of partial isomorphisms such that, for every object a in one model, there is an object b in the other with $f \cup \{(a, b)\} \subseteq F^+$, and vice versa. Spoiler then chooses a pair in F^+, and so on. This is power equivalent to the standard game, but the scheduling and amount of local effort is different. ■

There is an obvious general transformation at work here, reversing players' roles in a way that is reminiscent of the Thompson transformations discussed in Chapter 3. Similar transformations have been used in computational logic to turn one player into a relatively passive chooser (see the appendix to this chapter).

Despite this leaning toward players' powers, there is no official account of equivalence and invariance for logic games that we are aware of. Getting more precise on this seems an interesting issue in the foundations of logic. However, our games suggest an interesting twist to the way the issues were phrased in Chapter 1.

Generic powers and game equivalence Recall the analysis of dialogue games in Chapter 17. Different ways of setting up the procedural rules led to different associated sets of validities, in particular, for intuitionistic versus classical logic, and Chapter 20 will add variations that validate linear logic. These sets of validities may be seen as powers for proponent, since they indicate what kind of game the distinguished player can win. How can we relate this view to that of Chapter 11, where powers of players applied only to specific games? What we are getting now is a general description of the forms of games that allow for a winning strategy of some distinguished player. We might call this an analysis in terms of "generic powers," and this time, different powers emerge by setting the level of control for the distinguished player. This seems to be a new way of thinking about game equivalence, and it will be elaborated in greater detail in the systems of Part V that merge ideas from game logics and logic games.

18.4 Connections between logic games

Through the various games presented in this part, the use of game-theoretic notions unifies themes across logic. For instance, in Chapter 16, strategies for one player in a construction game were proofs, while for the other, they were models.[204] Another source of unity is this. Logic games are not isolated activities: they can be related by transformations. The following nice illustration relates two basic logic games.

Linking evaluation games and comparison games The games presented in Chapters 14 and 15 are similar. When analyzed more in depth, the success lemma for model comparison games gives an explicit link between strategies across the two types of game, taken in their finite versions on models $\boldsymbol{M}$ and $\boldsymbol{N}$. We can do away with the formulas as an intermediary. Here is how. First-order differences φ of quantifier depth k between $\boldsymbol{M}$ and $\boldsymbol{N}$ corresponded to winning strategies for spoiler in the k-round comparison game between $\boldsymbol{M}$ and $\boldsymbol{N}$. Now, suppose that $\boldsymbol{M} \models \varphi$ while $\boldsymbol{N} \models \neg\varphi$. As we will see, this induces a winning strategy for verifier in the evaluation game $\boldsymbol{game}(\varphi, \boldsymbol{M})$ plus one for falsifier in $\boldsymbol{game}(\neg\varphi, \boldsymbol{M})$. Moreover, one can correlate all this directly, in a result that has been called the E-H Theorem (with a nod to Ehrenfeucht and Hintikka).

THEOREM 18.1 There is an effective correspondence between

(a) Winning strategies for spoiler in the n-round comparison game.

(b) Pairs of winning strategies for verifier and falsifier in some n-round evaluation game, played in opposite models.

Proof From (b) to (a), we argue as follows. Let an "H-pair" of depth n consist of a formula φ of quantifier depth n plus a winning strategy σ for verifier in the φ-game in one of the models, together with a winning strategy τ for falsifier in the φ-game in the other model. We sketch a way of merging σ and τ. Spoiler looks at the two evaluation games. Suppose that verifier wins φ in $\boldsymbol{M}$, and falsifier wins φ in $\boldsymbol{N}$. Formulas can be taken to be constructed from atoms with negations, disjunctions, and existential quantifiers only. If φ is a negation $\neg\psi$, spoiler switches to the obvious

204 More precisely, these logical notions are often not exactly the winning strategies themselves, but "invariants" to be used in wielding these strategies (cf. Chapter 4).

strategies for falsifier and verifier with respect to ψ.[205] If φ is a disjunction $\psi \vee \xi$, then spoiler uses the verifier-strategy in the one model to choose a disjunct. Spoiler's falsifier-strategy in the other model also wins against that disjunct. Proceeding in this way, the formula is broken down until an existential subformula $\exists x\psi$ is reached. Spoiler then uses the verifier-strategy σ in the model where it lives, say $\boldsymbol{M}$, to pick a witness object d. This $\boldsymbol{M}$ and d are spoiler's opening move in the first round of the $\boldsymbol{E}$-game.

What remains for spoiler is a winning strategy σ^- for ψ in $\boldsymbol{M}$ after this first move. Now, let duplicator respond with any object e in the other model $\boldsymbol{N}$. This can be seen as a move by verifier in the evaluation game for $\exists x\psi$ in $\boldsymbol{N}$. Now falsifier still has a winning strategy τ^- for ψ in $\boldsymbol{N}$ after this first move. Thus, by induction, we have an H-pair of depth $n-1$, which can be merged into a follow-up winning strategy for spoiler in the $(n-1)$-round comparison game for $\boldsymbol{M}$ and $\boldsymbol{N}$. In all, this is an n-round strategy for spoiler.

This inductive argument easily yields an algorithm for spoiler's computation.

The direction from (a) to (b) seems harder, as we must split spoiler's winning strategy into two separate strategies to form an H-pair. But we can follow our earlier construction in Chapter 14 of a first-order difference formula φ of depth n from a winning strategy for spoiler. This formula φ effectively induces two evaluation strategies as follows. Consider any winning strategy for spoiler in the n-round comparison game between two models $\boldsymbol{M}$ and $\boldsymbol{N}$. In the first move, spoiler chooses, say, model $\boldsymbol{M}$ and object d. Our desired formula then starts with an existential quantifier, and verifier has the winning strategy in $\boldsymbol{M}$. Let duplicator make any response e in the model $\boldsymbol{N}$. Spoiler still has an $(n-1)$-round winning strategy left in the two expanded models $(\boldsymbol{M}, d)$ and $(\boldsymbol{N}, e)$. Inductively, we find H-pairs of depth $n-1$ for each choice e made by duplicator. By the Finiteness Lemma, only finitely many non-equivalent formulas are involved in these pairs. Then, one existential quantification over a suitable conjunction of formulas of depth $n-1$ defines our desired H-pair of depth n. In particular, if it is verifier who has the winning strategy of a relevant H-pair φ in the model $\boldsymbol{M}$, put φ itself in the conjunction; otherwise, put its negation. ■

---

205 This is internal computation: the opponent in the comparison game sees no effect yet. Such silent interludes are typical of many arguments about logic games, and they may have an interesting calculus of their own.

Other links: The case of modal bisimulation There are many further connections between prima facie different logic games. Consider the basic notion of bisimulation from Chapter 1. We elaborate a bit on its fixed definition-point as discussed in Chapter 15. Recall that a bisimulation Exy on a finite vocabulary of unary predicate letters P and binary relations R can be defined syntactically as follows, where we introduce special unary proposition letters M and N for the domains of the relevant models:

$$BIS(M,N,E)(x,y) \quad \forall xy : (Mx \wedge Ny \wedge Exy) \to \bigwedge_P (Px \leftrightarrow Py) \wedge \bigwedge_R \forall z((Mz \wedge Rxz) \to \exists u(Nu \wedge Ryu \wedge Ezu)) \wedge \forall u((Nu \wedge Ryu) \to \exists z(Mz \wedge Rxz \wedge Ezu))$$

In this formula, all occurrences of E in the formula to the right-hand side are positive, and it follows as in Chapters 1 and 2 that there is a greatest fixed point definition in the first-order fixed point logic LFP(FO).

FACT 18.1 Two points s and t in models $\boldsymbol{M}$ and $\boldsymbol{N}$ stand in some bisimulation relation iff they belong to the greatest fixed point defined by the LFP(FO) formula $\nu E \bullet bis(M,N,E)(x,y)$ on the disjoint sum of $\boldsymbol{M}$ and $\boldsymbol{N}$.

Thus, we can match two kinds of logic games, from Chapters 14 and 15, when we spell out how greatest fixed points are computed in approximation stages.

FACT 18.2 The modal model comparison game of Section 15.7 matches an LFP(FO) evaluation game in the sense of Chapter 14 for $\nu E \bullet bis(M,N,E)(x,y)$.

Recent work on computational logic also contains an opposite perspective, where evaluation games for formulas are reduced to model comparison games between (some structure associated with) that formula and the given model. See Venema (2007) and Zanasi (2012) for a history, and an explanation of how this works.

These are just a few natural connections between our earlier logic games. A general theory linking all kinds that we have studied is an obvious further goal.

18.5 Operations on logic games

Our results in the preceding section and earlier chapters suggest that logic games support general operations. In the evaluation games of Chapter 14, Boolean operations were the ubiquitous game operations of *choice* and *role switch*. After that, things became delicate, since the usual first-order quantifiers are not game operations, but special actions, while the real operation at work was an implicit one, namely, *sequential composition*, which glues evaluation games together. These operations have an algebraic theory of their own that will be studied in Chapters 19 and 24. But other operations occurred as well. The tableau construction games of Chapter 16 suggested operations of *iteration* yielding repeated games, and the same notion emerged in the dialogue games of Chapter 17. The latter games also suggested a new game operation of *parallel merge* for actions.

Interestingly, by our earlier result, model comparison games combine evaluation games using generalized notions of parallel play. This has two features of general interest. In these games *communication* takes place between two playgrounds, by the definition of the back-and-forth moves, while the output of the game consists of partial functions, typical results of *collective action*. Most intuitions about parallel games so far have come from computation, but logic games might be another, and underexplored, source of examples and results.

Further constructions on logic games make sense, but the repertoire listed here seems significant, suggesting a general algebraic perspective on modes of combination that will be pursued in several chapters to follow. In particular, operations of iteration and parallel merge will be found in Chapters 19, 20, and 21.

18.6 Universal formats: Graph games

We have said that logic games are a small corner of the realm of all games. But how special are they really? Perhaps logic games provide a normal form or universal format of greater reach. One illustration is found in Chapter 24, where evaluation games for first-order logic are shown to be complete for the general game algebra of sequential operations. But there can be more concrete general formats. Consider the following simple notion, discussed briefly in Chapter 11.

DEFINITION 18.1 Graph game
Consider a graph $\boldsymbol{G}$ with two binary relations R, S, and fix a node s. In the *graph game* associated with $(\boldsymbol{G}, s)$, player $\boldsymbol{E}$ starts at s and picks an R-successor t. Player $\boldsymbol{A}$ then chooses an S-successor u of t, and so on. $\boldsymbol{E}$ wins all runs where $\boldsymbol{A}$ gets stuck at some finite stage, plus those that go on forever.[206] ■

206 Many variations exist, viewing the graph as an external game board, or also as a record of game-internal structure. For instance, one can give just one *move* relation, but then record players' turns at graph nodes. Another source of variation is the winning conditions

EXAMPLE 18.3
Here is a process graph with one action:

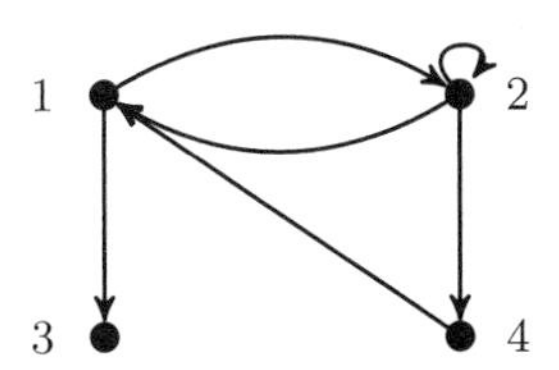

Player ***E*** can win the graph game starting from state 2. ***E*** has to move to state 4, then ***A*** must go to 1, and then ***E*** can move to state 3, immobilizing player ***A***. This power of ***E*** is expressed by a modal formula that is true at state 2: $\Diamond\Box\Diamond\bot$. ■

The following result makes a connection with the notions of Chapter 14.

FACT 18.3 Player ***E*** has a winning strategy in the graph game at $(\boldsymbol{G}, s)$ for a graph $\boldsymbol{G}$ of size n iff $\boldsymbol{G}, s \models (\langle R\rangle[S])^n\top$.

By the Gale-Stewart Theorem (Chapter 5) or by a direct proof, graph games are determined. Moreover, on finite graphs, they have the following important computational property.

FACT 18.4 Graph games are solvable in *Ptime* in the graph size.

This is a first sort of universality. Through this observation it is possible to determine computational complexity uniformly for many logical tasks, such as model checking modal formulas or testing for bisimulation.[207]

---

stipulating who wins on infinite histories. Interesting examples occur in Venema (2007), suggesting new game operations that dualize winning conditions, but not turns.

207 The following result by Martin Otto (personal communication) covers many cases. Consider any finite game tree with the properties that (a) maximal branch length is polynomial in some input parameter n, (b) maximal branching width is also polynomial in n, (c) each node can be encoded in space polynomial in n.Then solving its graph game will take only polynomial space in n. An application is model checking for first-order formulas, since trees for the evaluation games of Chapter 14 meet all conditions, and other examples are the construction games of Chapter 16 for modal logics, or suitable varieties of dialogue games in Chapter 17. Still, graph games also have their limitations, and further analysis is needed to deal with the often *Pspace*-complete tasks that are ubiquitous in more general games (cf. Papadimitriou 1994).

A second universal aspect of graph games is a more semantic one. Any extensive game form (cf. Chapter 1) can be viewed as a graph whose relations are the total *move* relations for both players. If we now provide winning end states for a player with a move for that player, but not the losing end states, then the graph game winning condition of never getting stuck simply encodes the original winning conditions, and we get an equivalent game. Of course, this is just brute redescription, and to make sense in general, the trick would need extension to games with preferences or imperfect information.

Graph games suggest connections with work on games in computer science that will be reviewed briefly in the appendix to this chapter. They will also return in Chapter 25 on merging logic games with game logics.

18.7 Conclusion

Many people see logic games as just useful didactical tools, which they certainly are. This more ambitious chapter may then come as a surprise. The above illustrations show the possibility of thinking about these techniques in general terms of strategies, transformations, game operations, and processes of gamification. Viewed at this level of generality, logic games are a particularly interesting subclass of games in general that can serve as a clean laboratory for broader notions and issues. Some issues raised here will be picked up in Parts V and VI of this book.

18.8 Literature

Beyond what was mentioned in the literature sections of Chapters 14 through 17, there are no canonical sources for the themes discussed in this chapter. For specific topics, the reader may consult the references occurring in the text.

18.9 Further directions

We list a few further perspectives, linking to earlier parts of this book.

Variations, richer game structure, and Theory of Play Once we have a logic game for some standard task, we see also how to change that game. For instance, evaluation games for first-order logic suggest moves that add or remove objects from a given model, and hence new notions arise of evaluation across changing models. Such notions will return in Chapter 23 on sabotage games, where we will see that logics over changing models make sense in several ways. But the same theme emerged in Chapter 17 on dynamic-epistemic logics of information change that transform the current epistemic, doxastic, or even preference structure of given models. Matching logic games will look ahead, or around, in universes of relevant models. As an illustration, van Benthem et al. (2006a) define model comparison games for the paradigmatic public announcement logic PAL of Chapter 7.

Another line of generalization is logic games that drop the competitive zero-sum character in all of the preceding in favor of notions of cooperation between players, a theme that will return in Chapter 21 on logic games with imperfect information, another notion from game theory that can be added generally to the games in our preceding chapters, as we did in Chapter 3. Further logic games in Chapter 22 add preferences that players might have in evaluation or update settings.

Finally, in line with Part II of this book, it makes sense to look at logic games from the standpoint of a Theory of Play, making players visible, and subject to variations. For instance, the pebble games of Chapter 14 endowed agents with modifiable memory resources, and this seems to be a significant way of introducing imperfect information into games in general. Also, diversity might occur inside one game, witness the speculation in Chapter 10 about logical systems as used by different kinds of agent, from finite automata to agents with perfect memory.

Logic games and game logics We end with a few points about the main themes of this book, logic *of* games and logic *as* games. The first is this. Logic *as* games is usually associated with just one kind of logic game, that is, evaluation games, and the reason is this. Much of the literature thinks of a "game semantics" as providing the meaning for logical formulas, and then, it seems most obvious to think of games about truth and evaluation, as in Chapter 14. But logic games are much more general than games of evaluation, so one might doubt this single-minded emphasis as the right way of thinking about what logic is.

More importantly, as will become clear in Part VI of this book, our main interest is not so much in this opposition, as in the duality of these two viewpoints. One can look at logic games as games, and hence in terms of game logics. For instance, modal languages of internal game structure can analyze what happens in logic games (see Chapter 25 on a strategy calculus for logic games). But in the opposite direction, one can use logic games to analyze game logics. We saw examples in Chapters 1 and 15, when noting how model comparison games can be used to probe the fine structure of degrees of bisimulation, the characteristic invariance for a modal logic.

18.10 Appendix on games in computational logic

In this chapter, we have surveyed some major themes running through logic games, finding points of contact with the game logics in Parts I and II of this book. But in order to develop a more systematic theory, an additional focus may be needed.

One such focus is computational systems that interact with their environment, where logic and games meet with automata in a hierarchy from finite state machines to Turing machines, and hence automata theory (cf. Hopcroft et al. 2001). What is crucial here is that *infinite streams* of input and output can be dealt with just as well as finite ones. This is a field with a fast-growing literature that falls outside the scope of this book (cf. Grädel et al. 2002, Venema 2006, 2007). In this appendix, we merely provide a small window onto this adjacent territory, by highlighting a few typical results, although experts in the area might well chose other lines.[208]

Processes, games, automata, and strategies Automata are a natural model of computation, while at the same time, they have intimate connections with logic and games. This shows in the graph games discussed in this chapter and earlier ones. For a start, a process graph as in Chapter 1 with moves for players can be viewed as an automaton with possible transitions. Play of a graph game unwinds this automaton into possible histories of actual process executions, where suitable winning conditions are added for the players. We have seen that, if winning conditions are simple enough (cf. Zermelo's Theorem in Chapter 1, or the Gale-Stewart Theorem in Chapter 5), the games will be determined. But the computational structure of the winning strategies themselves is of interest, too. Often it is given by simple devices, and once more, automata fit the bill. A classical result is the Buechi-Landweber Theorem, referring to simple ω-regular winning conditions for whose definition we refer to Thomas (2002).

THEOREM 18.2 Graph games over finite graphs with ω-regular winning conditions are determined with winning strategies that are defined by finite automata.

Many extensions exist of this way of analyzing the structure of strategies, which ties in with our theme of logical definability for strategies in Chapter 4.

---

208 I thank Erich Grädel and Yde Venema for helpful conversations about this appendix, although I do not claim their endorsement for anything that is said here.

Automata, tree logics, and games Another important interface with automata arises with logics describing the typical branching temporal structure that underlies computation and games, of which we have seen several instances already in Chapters 5, 7, and 9.

One well-known logic whose expressive power forms a natural limit for many of our formalisms is monadic second-order logic MSOL over trees with action relations and quantification over sets (monadic properties) only. The following landmark result by Rabin inaugurated this area.

THEOREM 18.3 MSOL over trees is decidable.

The key to the proof in Rabin (1968) is the introduction of a third partner next to formulas and models. This is an association of "tree automata" A_φ to formulas φ of MSOL in such a way that φ is true in a given tree model $\boldsymbol{M}$ iff the automaton A_φ *recognizes* $\boldsymbol{M}$ in some suitable way through its infinite runs.

Here we can assume that our models are trees with labeled transition relations plus some finite vocabulary P of unary predicates, defining a finite set of "colors" for nodes listing the presence or absence of each property. Rabin's proof was extended significantly from trees with a fixed bound on branching to trees with arbitrary finite branchings in Janin & Walukiewicz (1995), where tree automata of the following form were introduced:

> The automaton (A, δ, Ω) has a finite set of states A, a parity map Ω assigning each state a natural number, and a transition function δ given by a first-order formula $\delta(a, c)$ (a stands for a state, c for a color) in a monadic first-order logic with identity over the alphabet $P \cup A$, viewing states as unary predicate letters.

Automata and acceptance games Again, our notions get intertwined.[209]

Acceptance of a model by the above kind of automaton is defined by an alternating "acceptance game" with two players $\boldsymbol{E}$ and $\boldsymbol{A}$. Each round starts with a position (a, s) with $a \in A$ and $s \in \boldsymbol{M}$.[210] Now a new state must be reached subject to a condition linked to this current state:

---

209 The ideas to follow in this subsection go back to Gurevich & Harrington (1982) and Muller & Schupp (1995).

210 The reader may compare such states to the states (α, s) consisting of a formula α and a game board position s that were used in our logic games of Chapters 14 and 19.

(a) $\boldsymbol{E}$ picks a valuation for all $a \in A$ (viewed as atomic propositions) on the set $\sigma(s)$ of all immediate successors of s in the tree $\boldsymbol{M}$ in such a way that $\delta(a, c)$ holds for the submodel of $\boldsymbol{M}$ with domain $\sigma(s)$, where a is the state in the current pair (a, s) and c is the P-color of the current point s.

(b) $\boldsymbol{A}$ picks a position (t, b) with t satisfying b in the model chosen by $\boldsymbol{E}$.

The game is won or lost as follows. If $\boldsymbol{E}$ gets stuck at some finite stage (being unable to choose a valuation satisfying the condition $\delta(a, c)$), then $\boldsymbol{E}$ loses. For infinite histories, the convention is as for the fixed point logics in Chapter 14. Look at the Ω-values of all states a occurring infinitely often in the above rounds (a, s): $\boldsymbol{E}$ wins if the lowest occurring value is even, $\boldsymbol{A}$ wins if the lowest value is odd.

From alternating to nondeterministic automata Acceptance games correspond to the action of alternating automata going through alternate conjunctive and disjunctive states. However, simpler normal forms exist for the above games, where the first-order $\delta(a, c)$ gets simplified to a special functional syntax without essential conjunctions, corresponding to a game where $\boldsymbol{A}$'s role is restricted to choosing a successor, without options in $\boldsymbol{A}$'s choice of a state.[211] The following Simulation Theorem is from Muller & Schupp (1995).

THEOREM 18.4 Alternating tree automata can be faithfully simulated by nondeterministic automata.

For perspicuous proofs, see Venema (2006) and Zanasi (2012), where the role of the restricted transition syntax is made clear. Vardi & Wolper (1986) and Vardi (1995) provide background from the perspective of temporal logic and automata theory. (For another take, relating to topology, cf. Arnold & Niwinski 2007.)

These results and their proofs are interesting more generally, for instance, in light of Chapters 19 and 20. In showing that the nondetermined version suffices, one actually has to code up parallel play of many runs of the original game in one sequential play over suitably complicated states. Moreover, the non-trivial proofs in this literature have interesting general features whose formalization can serve as a challenge to our strategy logics of Chapters 4 and 25. One example is the heavy

---

211 This gets closer to nondeterminism in standard process-theoretic parlance, referring to one agent making a process work through a sequence of choices.

use of a device of "shadow matches" that do not show up in the official course of a game, but are essential to understanding players' strategies.[212]

Parity games and positional strategies Behind the above results, further insights have been found that make sense for games in general. One striking instance is "positional determinacy" (Emerson & Jutla 1991, Mostowski 1991) for parity games computing their winning conditions with an Ω-function on game states as in tree acceptance games. Positional strategies give a next move merely on the basis of the current game state, not its finite history. Such memory-free strategies will be important in Chapter 20.

THEOREM 18.5 Parity games are determined using only positional strategies.

This result covers a wide class of cases, including the evaluation games for fixed point logics such as the μ-calculus and LFP(FO) in Chapter 14. What the Positional Determinacy Theorem tells us exactly from a logical perspective is still a matter of some debate, but its technical convenience is beyond dispute. We will return to its significance in Chapter 25 when discussing two-level views of games.

The μ-calculus as a logical laboratory The modal μ-calculus that we have seen in Chapters 1, 13, 14, and others exemplifies many of the above features. For instance, its evaluation games are determined in positional strategies, but there is much more. In recent years, this system has become a testing ground for games and computation. The characteristic automata for this system (Janin & Walukiewicz 1995) are a simple modification of the earlier format for MSOL to one without identity atoms in defining formulas for the transition function. By now, there are constructive automata-theoretic proofs for many properties of the logic, both familiar and surprising ones, such as strong forms of interpolation. A well-known result is from Janin & Walukiewicz (1996).

212 As we noted in Chapter 5, these and other sophisticated arguments about games in the current literature use heavy imagery of players' *knowing* other parts of a game, strategies of other players, or their own strategies in other games, and then using this knowledge to play to their best advantage on the current history. While this might be considered harmless rhetoric in understanding proofs of pure existence statements, one could also take this knowledge more seriously. It actually ties in very well with our considerations in Parts I and II on epistemic aspects of games, and an epistemized version of the process theories discussed in this appendix with an added explicit knowledge component might be a natural meeting point with standard game theory.

THEOREM 18.6 The modal μ-calculus is the bisimulation-invariant fragment of monadic second-order logic MSOL.

Venema (2012) has an elegant short proof. Also highly relevant to our book is the explanation in Venema (2007) of two different views of μ-calculus automata: either as devices for evaluation, or as model comparison games in the style of Chapter 15 matching the structure of a formula (or its associated automaton) against that of a given model. What this suggests is once more a deeper unity across the logic games in this book.

Current perspectives Recent research has connected the above theoretical realm to real computation and game theory (cf. Apt & Grädel 2011). A more mathematical line has looked for the essence of many of the above results in the general framework of co-algebra: see Kupke & Venema (2005), Venema (2012). For further research lines, we refer to the work from the Aachen School, of which there are many good examples in Thomas (2002), Flum et al. (2007), Grädel et al. (2002), Berwanger (2005), and Kreutzer (2002). There are also active French and Polish communities, for which the recent dissertations Gheerbrant (2010), Fontaine (2010), Facchini (2011) provide extensive references.[213] Some of this work also links up with games in descriptive set theory, mentioned in Chapter 5.

Logic games after all It will be clear that many aspects of computation relate to the main line of our book, and especially, the logic games of Part IV. Verifying programs involves model-checking logical formulas that express intended specifications. Designing programs is more like solving satisfiability problems, and there are also natural scenarios mixing the two. Model comparison games, too, make sense for computation given their links with process invariances (cf. Chapter 1). Often, formalisms for computation are fixed point logics as in Parts I, II, and III, since these analyze definability and reasoning that involve the crucial phenomena of iteration and recursion.

The computational literature also contains themes with a general thrust for our understanding of logic itself. An example is the intriguing tradeoff between automata-theoretic methods and model comparison games in the composition method of Thomas (1997). More generally, the earlier automata provide a rich view

---

213 See also the informative website (`http://www.games.rwth-aachen.de/`) of the ESF project GAMES (Games in Design and Verification).

of "syntax as procedure" that might have far-reaching repercussions for received views of logical semantics.

Other uses of games in computation Uses of games in computation are more diverse than the particular approach to reactive systems, program verification, and program synthesis that we have seen here. Another major strand is proof-theoretic and category-theoretic (cf. Abramsky (1995)), and basic themes from the resulting game semantics for programming languages will occur in Chapter 20. Another major habitat of games in computer science is the field of agency (cf. Shoham & Leyton-Brown 2008), that connected in many ways with Parts I and II of this book. Games have also entered in the practical area of gaming (`http://en.wikipedia.org/wiki/Gamification`), far beyond the gamifications studied in this book so far (cf. also Chapter 23).

We are not aware of any standard reference work on games in computation, but it would be a welcome companion to this book on games and logic.

Conclusion to Part IV

In this fourth part of our book, we have introduced the major varieties of logic games: for evaluating formulas in models, comparing models, constructing models, and for engaging in proof dialogues. In each case, we defined the games, studied the structure of their strategies, and developed some basic theory about how the game functions, stressing general game-theoretic properties. For us, these games were not just tools for business as usual, even though it is quite true that game talk is a powerful metaphor for standard logic, packaging complex intuitions in a helpful concrete manner. But going beyond that, we have also presented the games as a novel way of thinking about logic as a family of dynamic multi-agent activities. Thus, to us, logic as games is a multi-faceted enterprise, going beyond what is sometimes called game semantics for logical languages. Driven by this dynamic interest, we also discussed in each case what new viewpoints in logic we get by taking the games seriously. In Chapter 18, the final chapter of Part IV, we brought together a number of these, emphasizing major open problems such as general strategy calculus, game equivalence, and operational repertoire, while opening a window to relevant work in the foundations of computation.

Having identified such topics of a broader nature, it is time to move on. Part V of this book is a study of general logics of game-forming operations that are inspired partly by logic games and partly by general logics for games and computation, showing how the two perspectives merge very naturally. Part VI will discuss further interfaces between logic games and game logics, arriving eventually at our best current understanding of the connections between these two perspectives.

V Operations on Games

Introduction to Part V

Parts I, II, and III of this book introduced game logics for understanding interactive behavior in terms of logical systems. Part IV then presented their counterpart of logic games that analyze logical tasks in terms of games. The next two parts of this book are about interfaces between these two perspectives. One focus for this contact is the game-theoretic view of the very logical constants. As we have seen, in logic games, these often become game-forming operations such as choice, switch, or composition. But clearly, this is at the same time a topic for the general logic of games: what are natural constructions forming new games out of old? In the two chapters of Part V, we will look at two major systems describing operations on games, one inspired by dynamic logics of programs, generalized to neighborhood models for strategic powers of players, the other arising from game semantics for linear logic and related systems that embody a proof-theoretic paradigm for interactive computation.

The very term semantics again highlights the duality of viewpoints at work here. One can think of games as providing meaning to some independently given logical system, which gives the game structures an auxiliary role. One can also think of logical systems as serving the goal of analyzing a fundamental independently given realm of games, adapting the logics to the needs of that structure. Such complementary viewpoints occur in many areas of logic; witness, say, the topological semantics for modal logics in Chapter 11 versus recent studies of modal logics of space. The two are closely related, and yet the different stances often lead to quite different questions to pursue. The same is true for games.

Still, a methodological distinction does not rule out productive coexistence, and the logical systems to be developed in the coming part, and also in Part VI, will have flavors from both sides.

19 Dynamic Logic of Sequential Game Operations

This chapter is about a system that describes powers of players in complex games on the analogy of propositional dynamic logic for programs, by substituting the forcing relations of Chapter 11 for the transition relations between states. This 'dynamic game logic' was proposed in Parikh (1985). There is much of interest to how this approach to logic and games works, and we tell the story slowly, with motivations and follow-up. Much of what follows is our own take, adding results on game algebra, bisimulation, and other logical themes.

19.1 Internal and external views of games

Logical description of games can proceed at various levels. The game logics of Parts I, II, and III were game-internal, providing statements that may hold at stages of play. But the logic games of Part IV, and especially the evaluation games studied in Chapter 14, took a game-external point of view, with formulas standing for whole games, and logical operations becoming game-forming operations of choice, composition, or role switch. The external viewpoint has its attractions, especially in describing game equivalences in an algebraic manner.[214] Internal and external viewpoints can be combined, as we will see in this part of our book. To do so, we step up the abstraction level for studying games. This chapter presents a dynamic logic for external game combination, at the level of forcing and outcomes. It encodes

214 Similar options occur with computation. Modal logics provide process-internal views, while process algebra (Bergstra et al. 2001) studies global process combination. The two views are related by bisimulation techniques, but they proceed differently.

an algebra of basic game-forming operations, while providing a new scope for modal techniques such as bisimulation.

19.2 From logic games to game logics

Instead of designing logic games for specific tasks, one can also study the combinatory structure of arbitrary games. Ideas developed for logic games then acquire a more general significance. In particular, we find natural operations that structure games. For instance, choice and role switch occurred across logic games, but they are much more general than that, as basic constructions that turn given games into new ones. Next, given such operations, we want to learn their general algebraic laws. For instance, it is important to see that the propositional distributive law

$$p \wedge (q \vee r) \leftrightarrow (p \wedge q) \vee (p \wedge r)$$

is valid in complete generality. It does not hinge on playing special test games for verifier and falsifier about atomic propositions at the end, or on claims made by proponent or opponent in a dialogue. The expressions p, q, r in this equivalence might stand for any games, of arbitrary complexity, plugged in at the final stages.

As we noted in Chapter 14, in the literature on logic games, logical formulas often do double duty: as static propositions, and as games (i.e., dynamic activities). These roles must now be pulled apart, following our distinction between games themselves and assertions one can make about their behavior. The system in this chapter is inspired by the dynamic logic of programs, a system that has occurred in this book ever since Chapter 1, and the evaluation games of Chapter 14 are a good analogy to keep in mind.[215]

19.3 Forcing models, games, and game boards

We start with a semiformal tour through the style of game analysis to be presented in this chapter, touching on all the major themes that arise.

---

215 The alternative system to be presented in Chapter 20 extends linear logic, with a link to the dialogue games of Chapter 17. Together, Chapters 19 and 20 span a space merging ideas from logic games and game logics.

Dynamic game logic: A quick preview We now consider a language like that of propositional dynamic logic, PDL, in Chapter 1, but this time with expressions for two-player games (with players $\boldsymbol{E}$ and $\boldsymbol{A}$) plus propositions.

DEFINITION 19.1 Language of dynamic game logic
The language of dynamic game logic is defined by these inductive syntax rules:

Formulas F	$p \mid \neg F \mid F \vee F \mid \{G\}F$
Game expressions G	$g \mid G \cup G \mid G^d \mid G\,;\,G \mid ?F$

A point of notation In this chapter, we will use variables G with primes when thinking primarily of games, but also variables $x, y, ...$ when thinking of the pure algebraic form of laws about games. Players will be denoted by $\boldsymbol{A}$ and $\boldsymbol{E}$, thought of generically as an opposite pair. ■

The game operations in this language are choice for player $\boldsymbol{E}$, dual ^{d} for role switch, and sequential composition ;. It is useful to also define a game conjunction $G \cap G'$ of choice for player $\boldsymbol{A}$ as $(G^d \cup G'^d)^d$. Parikh's system also has a game iteration G^* like the Kleene star of PDL that we will discuss separately later. Note the recursion in the syntax. Tests $?\varphi$ turn formulas φ into game expressions, while in the opposite direction, modalities $\{\ \}$ take game expressions to formulas.

The intuitive idea of the modality is as in Chapter 11, based on forcing relations (we will define the precise models a bit later on).

DEFINITION 19.2 Dynamic forcing modality
The *forcing modality* $\{G\}\varphi$ says, at any state, that player $\boldsymbol{E}$ has a strategy for playing game G guaranteeing, against any play by the opponent, a set of outcome states all of which satisfy the formula φ. ■

To make this work precisely, we need to define the following two notions:

$\rho_G\, s, X$	player $\boldsymbol{E}$ has a strategy for playing G starting in state s that is guaranteed to end up inside the set of states X
$s \models \{G\}\varphi$	for some $X : \rho_G s, X$ and $\forall x \in X : x \models \varphi$

The key relations ρ run not from states to states, as for programs, but from states to sets of states, and we recognize the forcing relations of Chapter 11. In particular, players need not be able to force unique outcomes of games, whence the set output.

DGL = neighborhood logic + game algebra Unlike modal operators $\Diamond$ and $\Box$, the modality $\{\ \}$ satisfies no distribution for disjunction or conjunction. As is easy to see in simple games, the following formulas are invalid:

$$\{G\}(\varphi \vee \psi) \leftrightarrow \{G\}\varphi \vee \{G\}\varphi, \qquad \{G\}(\varphi \wedge \psi) \leftrightarrow \{G\}\varphi \wedge \{G\}\psi$$

What remains valid is upward monotonicity:

$\{G\}\varphi$ implies $\{G\}\psi$ for any weaker proposition ψ implied by φ.

Going beyond Chapter 11, through its valid laws on top of the neighborhood logic, the logic encodes information about the basic game operations.[216] One simple example is the validity of commutativity of choice:

$$\{G \cup G'\}\varphi \leftrightarrow \{G' \cup G\}\varphi$$

Indeed, most of Boolean algebra holds for $\{\cup, ^d\}$. As for the further game operations, the forcing relations generalize standard relational algebra for relations of type $S \times S$, taking it to an algebra of relations of type $S \times P(S)$, with basic laws such as associativity:

$$\{G_1 \,;\, (G_2 \,;\, G_3)\}\varphi \leftrightarrow \{(G_1 \,;\, G_2) \,;\, G_3)\}\varphi$$

We will see much more of the mechanics of this system in Section 19.4.

Now let us become more precise about the semantic setting.

Forcing models, games, and game boards Models for this language are tuples

$$\boldsymbol{M} = (S, \{\rho_g \mid g \text{ atomic}\}, V)$$

with S a set of states, V a valuation assigning truth values to atomic propositions in states, and with atomic relations $\rho_g s, X$ assigned to basic game expressions g. These forcing relations ρ_g do not indicate players explicitly. One can take player $\boldsymbol{E}$ in mind, using determinacy of the game to find the powers of $\boldsymbol{A}$, in a way to be explained below, although we will also consider player-indexed forcing later. Forcing relations are closed under supersets, given their intended forcing interpretation:

if $\rho_g s, X$ and $X \subseteq Y$, then $\rho_g s, Y$.

216 In what follows, in line with the original system, we start by thinking of determined games, but this restriction will soon be lifted.

In terms of the earlier discussions in Chapters 1, 11, and 18, these models $\boldsymbol{M}$ are *game boards* or playgrounds, with states that model the external content of internal game states. One board can support many games with different turns and winning positions. For instance, with evaluation games, board states are variable assignments that change when a player makes a move $x := d$. But these states do not encode the internal changes with choices between disjuncts or conjuncts.

In line with this, atomic forcing relations ρ_g do not reflect structured games g, although the latter may arise in applications of the logic. This differs from the view in Chapter 11, where forcing relations for games used internal nodes and moves.

Computing complex forcing relations To get some generality that will be useful later when dropping the assumption of determinacy, we change the above models a bit, computing forcing relations for both players in one simultaneous recursion.

DEFINITION 19.3 Forcing relations for composite games
The following recursive clauses govern forcing relations:

$$
\begin{array}{lll}
\rho^{\boldsymbol{E}}_{G \cup G'} x, Y & \text{iff} & \rho^{\boldsymbol{E}}_{G} x, Y \text{ or } \rho^{\boldsymbol{E}}_{G'} x, Y \\
\rho^{\boldsymbol{E}}_{G \cup G'} x, Y & \text{iff} & \rho^{\boldsymbol{A}}_{G} x, Y \text{ and } \rho^{\boldsymbol{A}}_{G'} x, Y \\
\rho^{\boldsymbol{E}}_{G^d} x, Y & \text{iff} & \rho^{\boldsymbol{A}}_{G} x, Y \\
\rho^{\boldsymbol{A}}_{G^d} x, Y & \text{iff} & \rho^{\boldsymbol{E}}_{G} x, Y \\
\rho^{\boldsymbol{E}}_{G\,;\,G'} x, Y & \text{iff} & \exists Z : \rho^{\boldsymbol{E}}_{G} x, Z \ \& \ \forall z \in Z : \ \rho^{\boldsymbol{E}}_{G'} z, Y \\
\rho^{\boldsymbol{A}}_{G\,;\,G'} x, Y & \text{iff} & \exists Z : \rho^{\boldsymbol{A}}_{G} x, Z \ \& \ \forall z \in Z : \ \rho^{\boldsymbol{A}}_{G'} z, Y
\end{array}
$$

These clauses are probably self-evident, but they will return in the soundness for the axioms of dynamic game logic in Section 19.4. ■

Henceforth, we assume that atomic forcing relations satisfy the earlier-mentioned superset closure (monotonicity) as well as consistency (cf. Chapter 12):

$$\text{if } \rho^{\boldsymbol{E}}_{G} s, Y \text{ and } \rho^{\boldsymbol{A}}_{G} s, Z, \text{ then the sets } Y \text{ and } Z \text{ overlap.}$$

One can easily show the following lifting result by induction.

FACT 19.1 If the atomic relations on a game board satisfy monotonicity and consistency, then so do all complex forcing relations as defined here.

REMARK From determinacy to nondeterminacy
Parikh (1985) assumed that all games were determined, in the sense of Chapter 11:

$$\text{for each set } Y, \text{ either } \boldsymbol{E} \text{ can force } Y, \text{ or } \boldsymbol{A} \text{ can force } S - Y.$$

Then we just define forcing for player $\boldsymbol{E}$, simplifying the case of dual to $\rho^{\boldsymbol{E}}_{G^d}x, Y$ iff not $\rho^{\boldsymbol{E}}_{G}x, S - Y$. The relations for $\boldsymbol{A}$ are induced by $\rho^{\boldsymbol{A}}_{G^d}x, Y$ iff not $\rho^{\boldsymbol{E}}_{G}x, S - Y$.[217]

Test games Finally, we must define forcing relations for atomic test games. Parikh's original stipulation was as follows (see Pauly (2001) for more explanation):

$$\rho^{\boldsymbol{E}}_{?P}x, Y \quad \text{iff} \quad x \in Y \text{ and } P \text{ holds of } y$$

This has some debatable effects, such as the equivalence $\rho^{\boldsymbol{A}}_{?P}x, Y$ iff, if $x \in Y$, then P holds of y. Hence some authors have sidestepped the issue, dropping tests altogether, and having just a special "idle game" ι with:

$$\rho^{\boldsymbol{E}}_{\iota}x, Y \text{ iff } x \in Y, \qquad\qquad \rho^{\boldsymbol{A}}_{\iota}x, Y \text{ iff } x \in Y$$

In such a game, both players have the same power, that is, the power of being there. Information about propositions P holding at the current state x is then external, and not part of the game. Another line is to have two forcing modalities. The one so far relates only to players' powers of making sure the game ends in some set of positions. Since a test does not change the current state, its correct axiom is

$$\{?P\}\varphi \leftrightarrow \varphi$$

But another view of the forcing modality $\{G, \boldsymbol{E}\}\varphi$ brings in a game-internal property: player $\boldsymbol{E}$ has a strategy for playing G guaranteeing a set of outcome states that satisfy φ and are winning for $\boldsymbol{E}$. This will validate the following two equivalences:

$$\{?P, \boldsymbol{E}\}\varphi \leftrightarrow P \wedge \varphi \qquad\qquad \{?P, \boldsymbol{A}\}\varphi \leftrightarrow \neg P \wedge \varphi$$

The second reading satisfies the same laws for game operations as the first.[218] In what follows, we merely assume that some satisfactory interpretation is in place.

Things become clearer by looking at some basic motivating examples. The above system can be used in analyzing specific games, seeing what is general, and what are peculiarities of special states and moves.

---

217 To be consistent here, one has to check that the forcing relations for compound games remain determined, when starting from determined games at the bottom level.

218 In particular, note how the winning condition reverses when we move from $\boldsymbol{E}$ to $\boldsymbol{A}$.

EXAMPLE 19.1 First-order logic

We start with the first-order evaluation games of Chapter 14 for formulas φ on models $\boldsymbol{M}$. Their internal states were pairs

$$\langle \textit{current subformula, current variable assignment} \rangle.$$

The external board states are the variable assignments. Next, we split general formula games into atomic games plus general constructions:

role switch ^{d}, two choices $\cup, \cap$, plus composition ;

The atomic games were of two kinds:

(a) *atomic tests* $?P$ checking whether P holds under the current assignment.

(b) *object picking* for quantifiers $\exists x$, changing the current assignment s to a new assignment $s[x := d]$ by setting x equal to some object d.

Atomic tests can be seen as games for $\{\boldsymbol{E}, \boldsymbol{A}\}$ in various ways, making sure at states where P holds that the win is for $\boldsymbol{E}$, and elsewhere for $\boldsymbol{A}$. They have special properties, not shared with games in general. For example, it does not matter in which order we perform atomic tests, and performing the same test twice does not change the outcomes:

$$?P\,;\,?Q\; = ?Q\,;\,?P, \qquad\qquad ?P\,;\,?P\; = ?P$$

Atomic quantifier games are special, too, in that player $\boldsymbol{E}$ has complete control over outcomes: $\boldsymbol{E}$ can make sure that any next state occurs. Formally, their forcing relations are "distributive," satisfying a splitting condition:

$$\text{if } \rho^{\boldsymbol{E}}_{G^d} s, \bigcup_{i \in I} Y_i, \text{ then for some } i \in I,\ \rho^{\boldsymbol{E}}_{G^d} s, Y_i$$

By contrast, for atomic quantifier games $\exists x$, the passive player $\boldsymbol{A}$ has just one power, being the total set of all x-variants of the current assignment s.

Given these stipulations for atomic games, our earlier recursive clauses compute forcing relations for any first-order evaluation game $\boldsymbol{game}(\varphi, \boldsymbol{M})$. It is also easy to see that the result squares with the success lemma of Chapter 14. Starting the game from assignment s, $\boldsymbol{E}$ has a winning strategy if and only if $\boldsymbol{M}, s \models \varphi$. ■

Logical laws deconstructed All this is not mere reanalysis of what is known. As we have seen in Chapter 14, the game view decomposes the laws of first-order logic into several layers. Some are general game validities, having nothing to do with specific games of fact testing or object picking. This is true for our running example of Boolean distribution, or idempotence of choice: $\varphi \vee \varphi \leftrightarrow \varphi$. Other laws are special effects of the atomic repertoire, such as idempotence of composition for tests or quantifier games. Its game form $G\,;\, G = G$ fails as a general law. Still other laws may be called intermediate, such as distribution for existential quantifiers:

$$\exists x(\varphi \vee \psi) \leftrightarrow \exists x\varphi \vee \exists x\psi$$

In terms of games, its general form is

$$G\,;\,(G' \cup G'') = (G\,;\, G') \cup (G\,;\, G'')$$

This is not a valid law. We can see this quickly by substituting another game for $\boldsymbol{A}$: for example, a universal quantifier game. We then get the invalid $\forall x(\varphi \vee \psi) \leftrightarrow \forall x\varphi \vee \forall x\psi$. The special reason for the validity is the above distributive character of the game $\exists x$. Since distributivity is a ubiquitous property, its game law may be considered intermediate in force.

Thus, from a game perspective, the usual predicate-logical validities are a mixed bag. General repercussions of this observation will be found in Chapter 24.

Example 19.2 Modal logic

Other samples of this style of analysis are modal and dynamic logic. This time, modalities are atomic games. Existential modalities $\langle a \rangle$ make player $\boldsymbol{E}$ choose some R_a-successor of the current state, universal $[a]$ are duals for player $\boldsymbol{A}$.[219] This time, however, $\boldsymbol{E}$ can lose, when there is no such successor. Thus, we have games where players may have to move, but cannot, and there is an issue of defining the proper forcing relations again. If we disregard winning conditions, we get

$$\rho^{\boldsymbol{E}}_{\langle a \rangle} x, Y \quad \text{iff} \quad \exists y(R_a xy \wedge y \in Y)$$
$$\rho^{\boldsymbol{A}}_{\langle a \rangle} x, Y \quad \text{iff} \quad R_a[x] \subseteq Y$$

Both the first-order and the modal perspective will return in our analysis of the game algebra of sequential operations in Chapter 24.

219 This follows from the “standard translation” sending $\langle a \rangle q$ to $\exists y(R_a xy \wedge Qy)$.

19.4 Dynamic game logic

Basic axiom system Here is the basic proof system for dynamic game logic, DGL. It reasons about games by exploiting analogies with dynamic logic, PDL. We consider only the operations of ***E***'s choice ($\cup$), composition (;), and dual (^{d}), with $G_1 \cap G_2$ defined as above. For simplicity, we start with a version of the system for determined games.

DEFINITION 19.4 Dynamic logic for determined games
The *minimal dynamic game logic* (also called DGL) for determined games has the following principles:

(a) All valid principles of propositional logic: both axioms and rules.
(a) Monotonicity: if $\varphi \to \psi$ is provable, then so is $\{G\}\varphi \to \{G\}\psi$.
(a) Reduction laws for existence of strategies in compound games:

$$\begin{aligned}
\{G\,;\,G'\}\varphi &\leftrightarrow \{G\}\{G'\}\varphi \\
\{G \cup G'\}\varphi &\leftrightarrow \{G\}\varphi \vee \{G'\}\varphi \\
\{?P\}\psi &\leftrightarrow P \wedge \psi \\
\{G^d\}\varphi &\leftrightarrow \neg\{G\}\neg\varphi
\end{aligned}$$

A nondetermined version of this system will be presented below. ■

How it works Dynamic game logic encodes a good deal of "game algebra."

DEFINITION 19.5 Validity of game-algebraic identities
Two game expressions G and G' are *equal* in the sense of DGL if the following formula is valid: $\{G\}q \leftrightarrow \{G'\}q$, for some new proposition letter q. ■

This says that player ***E*** has the same powers for forcing outcomes in both games. In determined DGL, this implies that ***A*** also has the same powers.[220]

220 In nondetermined games, ***E***-equivalence by itself does not imply equivalent powers for both players: we will see a nice illustration in Chapter 21.

if $\{G\}q \leftrightarrow \{G'\}q$	is provable, then so is
$\{G\}\neg q \leftrightarrow \{G'\}\neg q$	by the substitution rule, whence
$\neg\{G\}\neg q \leftrightarrow \neg\{G'\}\neg q$	using propositional logic, and
$\{G^d\}q \leftrightarrow \{G'^d\}q$	by the axiom for dual.

To convey a sense of the mechanics of dynamic game logic, here are a few formal derivations for principles of general game algebra.

Game conjunction Intuitively, if $\boldsymbol{E}$ is to have a strategy guaranteeing φ, then one is needed in both games: $\{x\}\varphi \wedge \{y\}\varphi$. Using the definition of $\boldsymbol{A}$'s choice game $x \cap y$, we prove this:

$$\{x \cap y\}q \leftrightarrow \{(x^d \cup y^d)^d\}q \leftrightarrow \neg\{x^d \cup y^d\}\neg q \leftrightarrow \neg(\{x^d\}\neg q \vee \{y^d\}\neg q)$$
$$\leftrightarrow \neg\{x^d\}\neg q \wedge \neg\{y^d\}\neg q \leftrightarrow \neg\neg\{x\}\neg\neg q \wedge \neg\neg\{y\}\neg\neg q \leftrightarrow \{x\}q \wedge \{x\}q$$

Boolean distribution Next, consider distribution $(x \vee y) \wedge z = (x \wedge z) \vee (y \wedge z)$, that was analyzed informally earlier in this book using the definition of forcing:

$$\{(x \cup y) \cap z\}q \leftrightarrow \{x \cup y\}q \wedge \{z\}q \leftrightarrow (\{x\}q \vee \{y\}q) \wedge \{z\}q$$
$$\leftrightarrow (\{x\}q \wedge \{z\}q) \vee (\{y\}q \wedge \{z\}q) \leftrightarrow \{x \cap z\}q \vee \{y \cap z\}q$$
$$\leftrightarrow \{(x \cap z) \cup (y \cap z)\}q$$

Disjunction over conjunction (i.e., my choice over your choice) is proved similarly.

Composition and choice With distribution of sequential composition over choice, things are different. Here is a formal derivation for one version:

$$\{(x \cup y)\,;\, z\}q \leftrightarrow \{x \cup y\}\{z\}q \leftrightarrow \{x\}\{z\}q \vee \{y\}\{z\}q \leftrightarrow \{x\,;\, z\}q \vee \{y\,;\, z\}q$$
$$\leftrightarrow \{(x\,;\, z) \cup (y\,;\, z)\}q$$

And here is a failed attempt for the other version (already discarded earlier on):

$$\{x\,;\, (y \cup z)\}q \leftrightarrow \{x\}\{y \cup z\}q$$
$$\leftrightarrow \{x\}(\{y\}q \vee \{z\}q) \leftrightarrow \text{(not in DGL: only in PDL) } \{x\}\{y\}q \vee \{x\}\{z\}q$$
$$\leftrightarrow \{x\,;\, y\}q \vee \{x\,;\, z\}q \leftrightarrow \{(x\,;\, y) \cup (x\,;\, z)\}q$$ [221]

221 Dropping the distribution law $\{G\}(\varphi \vee \psi) \leftrightarrow \{G\}\varphi \vee \{G\}\psi$ while retaining the other $\{G \cup G'\}\varphi \leftrightarrow \{G\}\varphi \vee \{G'\}\varphi$, is reminiscent of process algebra (cf. Milner 1989).

Dual of composition Finally, here is a valid derivation for a key law of game dual:

$$\{(x\,;\,y)^d\}q \leftrightarrow \neg\{x\,;\,y\}\neg q \leftrightarrow \neg\{x\}\{y\}\neg q \leftrightarrow \neg\{x\}\neg\neg\{y\}\neg q \leftrightarrow \neg\{x\}\neg\{y^d\}q$$
$$\leftrightarrow \{x^d\}\{y^d\}q \leftrightarrow \{x^d\,;\,y^d\}q$$

Meta-theorems The key results about DGL are in Parikh (1985) and Pauly (2001).

THEOREM 19.1 Universal validity in dynamic game logic over forcing models with iteration, but without dual, is effectively axiomatized by the above DGL laws plus the iteration axioms of dynamic logic.

The relevant iteration axioms for the DGL modality will be stated below.

THEOREM 19.2 Dynamic game logic is decidable.

The proof of the first result uses a combination of standard neighborhood methods for completeness plus ideas from the completeness proof for PDL (cf. Harel et al. 2000), and we will give an impression in the final section of this chapter, using a nondetermined two-player version. Whether this also works for the full language including game dual is a longstanding open problem.

The second result follows by an effective reduction of DGL validity to PDL validity, in a way suggestive of the reasoning in Chapter 24 (cf. also Goranko 2003). The precise complexity of the SAT problem for this game logic is still unknown.

The general version Dropping determinacy gives an elegant version of DGL with separate modalities $\{G, \boldsymbol{E}\}\varphi$ and $\{G, \boldsymbol{A}\}\varphi$. Here are the basic axioms:

$$\begin{aligned}
\{G \cup G', \boldsymbol{E}\}\varphi &\leftrightarrow \{G, \boldsymbol{E}\}\varphi \vee \{G', \boldsymbol{E}\}\varphi \\
\{G \cup G', \boldsymbol{A}\}\varphi &\leftrightarrow \{G, \boldsymbol{A}\}\varphi \wedge \{G', \boldsymbol{A}\}\varphi \\
\{G^d, \boldsymbol{E}\}\varphi &\leftrightarrow \{G, \boldsymbol{A}\}\varphi \\
\{G^d, \boldsymbol{A}\}\varphi &\leftrightarrow \{G, \boldsymbol{E}\}\varphi \\
\{G\,;\,G', \boldsymbol{E}\}\varphi &\leftrightarrow \{G, \boldsymbol{E}\}\{G', \boldsymbol{E}\}\varphi \\
\{G\,;\,G', \boldsymbol{A}\}\varphi &\leftrightarrow \{G, \boldsymbol{A}\}\{G', \boldsymbol{A}\}\varphi
\end{aligned}$$

Of the domestic rules for forcing relations as such, we retain

$\{G, \boldsymbol{E}\}\varphi \rightarrow \{G, \boldsymbol{E}\}(\varphi \vee \psi)$	upward monotonicity
$\{G, \boldsymbol{A}\}\varphi \rightarrow \{G, \boldsymbol{A}\}(\varphi \vee \psi)$	upward monotonicity
$\{G, \boldsymbol{E}\}\varphi \rightarrow \neg\{G, \boldsymbol{A}\}\neg\varphi$	consistency

Practical uses and meta-properties of this version are similar to those above.

Iterated games DGL has a further operation of finite iteration G^*, satisfying the two standard PDL axioms of Chapter 1. In our notation, these are

$$\{G^*\}\varphi \leftrightarrow \varphi \wedge \{G\}\{G^*\}\varphi \qquad \text{fixed point axiom}$$
$$(\varphi \wedge \{G^*\}(\varphi \rightarrow \{G\}\varphi)) \rightarrow \{G^*\}\varphi \qquad \text{induction axiom}$$

The iteration game lets player ***E*** play some finite number of runs of game G where ***E*** need not say in advance how many, while infinite runs are blamed on ***A***.[222]

This concludes our presentation of the basics of dynamic game logic. Now we add some topics, linking up with earlier themes in Parts I and IV of this book.

19.5 Basic game algebra

As we have seen, the forcing semantics validates a basic algebra of game construction. Take a language of game expressions with variables for games and operations $\cup, {}^d, ;$. In addition, we add a name ι for the idle game, staying at the same state.

DEFINITION 19.6 Algebraic validity
An identity between two game expressions $G = G'$ is *valid* if interpreting these expressions in any DGL model gives both players the same forcing relations. The useful auxiliary relation $G \leq G'$ denotes the analogous valid inclusion.[223] ■

We now explore some basic validities, following van Benthem (1999).

FACT 19.2 The following principles are valid in game algebra:

(a) "De Morgan algebra" for disjunction, conjunction, and negation, whose laws are defined separately below.

---

222 Other notions of iteration occur in this book: in the evolutionary games of our Introduction, infinite evaluation games for fixed point logics in Chapter 14, or games of model comparison in Chapter 15, where it is ***E*** who wants to keep the game going forever. It is an interesting problem how to incorporate these into DGL.

223 One natural variation that does not matter to the algebraic laws would define validity by requiring forcing bisimulation, introduced below, between the games denoted by two terms in different models.

(b) $G\,;(G'\,;G'') = (G\,;G')\,;G''$ — associativity

$(G \cup G')\,;G'' = (G\,;G'') \cup (G'\,;G'')$ — left-distribution

$(G\,;G')^d = G^d\,;G'^d$ — dualization

(c) *if* $G \leq G'$, *then* $H\,;G \leq H\,;G'$ — right-monotonicity

(d) $G\,;\boldsymbol{\iota} = G = \boldsymbol{\iota}\,;G$ — unit game

Definition 19.7 De Morgan algebra
The system of De Morgan algebra has the standard axioms for a distributive lattice, plus an idempotent negation:

$$
\begin{array}{ll}
x \cup x = x & x \cap x = x \\
x \cup y = y \cup x & x \cap y = y \cap x \\
x \cup (y \cup z) = (x \cup y) \cup z & x \cap (y \cap z) = (x \cap y) \cap z \\
x \cup (y \cap z) = (x \cup y) \cap (x \cup z) & x \cap (y \cup z) = (x \cap y) \cup (x \cap z) \\
(x^d)^d = x & \\
(x \cup y)^d = x^d \cap y^d & (x \cap y)^d = x^d \cup y^d
\end{array}
$$

Non-valid in De Morgan algebra are excluded middle and non-contradiction. This makes sense. For instance, as we saw in the Introduction, for games in general, $x \vee \neg x$ is no longer valid, since the game x need not be determined. ■

In this axiomatization, the crucial second set of laws is reminiscent of relational algebra (with ; behaving like composition of binary relations),[224] but the behavior of the game dual is sui generis. Further valid identities are derivable from these principles by algebraic manipulation.[225] Soundness follows by direct inspection, or via a more technical route sketched in the next remark.

Remark Soundness via translation
A link with classical logic clarifies the above results. Starting from relation symbols $\rho_a^{\boldsymbol{E}} xY, \rho_a^{\boldsymbol{A}} xY$ for basic game expressions, write the power relations for both players in complex games using the earlier recursive clauses. These expressions are logically equivalent for both game terms in the given algebraic laws. For example, propositional distribution follows by Boolean propositional distribution applied to power relation formulas. Similarly, left distribution follows by mere predicate logic.

---

224 Game algebra lacks the right distribution: $G\,;(G' \cup G'') = (G\,;G') \cup (G\,;G'')$ of relation algebra. The reason for this failure has already been explained.

225 The reader may want to try the case of $(G \cap G')\,;G'' = (G\,;G'') \cap (G'\,;G'')$.

EXAMPLE 19.3 Non-valid principles
The translation is illustrated by two earlier-mentioned non-valid principles.

(a) $G \cap (G' \cup \neg G') = G$. On the left, one has $\rho^{\boldsymbol{E}}_{G}xY \wedge (\rho^{\boldsymbol{E}}_{G'}xY \vee \rho^{\boldsymbol{A}}_{G'}xY)$ for $\boldsymbol{E}$'s powers. This is not equivalent to the formula $\rho^{\boldsymbol{E}}_{G}xY$ on the right: $\rho^{\boldsymbol{E}}_{G'}xY \vee \rho^{\boldsymbol{A}}_{G'}xY$ is not a game tautology.

(b) $G\,;\,(G' \cup G'') = (G\,;\,G') \cup (G\,;\,G'')$. On the left, player $\boldsymbol{E}$ gets $\exists Z : \rho^{\boldsymbol{E}}_{G}xZ \wedge \forall z \in Z : (\rho^{\boldsymbol{E}}_{G'}zY \vee \rho^{\boldsymbol{E}}_{G''}zY)$. This is not equivalent to $\boldsymbol{E}$'s powers on the right-hand side, given by $\exists Z(\rho^{\boldsymbol{E}}_{G}xZ \wedge \forall z \in Z \rho^{\boldsymbol{E}}_{G'}zY) \vee \exists Z(\rho^{\boldsymbol{E}}_{G}xZ \wedge \forall z \in Z : \rho^{\boldsymbol{E}}_{G'}zY)$. ■

The following result shows that our analysis is on the mark.

THEOREM 19.3 Basic game algebra is complete for algebraic validity.

Proof One uses a reduction of game algebra to modal logic, reducing forcing modalities { } to modal combinations $\Diamond\Box$. See Goranko (2003) for this technique, going back to Parikh (1985), that will also be used in Chapter 24.[226] ■

Open problems remain. For instance, how does the algebra change when we assume determinacy for all games? In Chapters 20 and 21, we will study laws for a richer repertoire, including parallel operations on games.

19.6 Bisimulation, invariance, and safety

In Chapters 1 and 11, we raised the issue of when two games are the same in a natural sense, finding answers in versions of bisimulation. The language of dynamic game logic does not describe games directly, but we can ask when two game boards are the same for DGL.

DEFINITION 19.8 Forcing bisimulation
A *forcing bisimulation* between two models $\boldsymbol{M}$ and $\boldsymbol{N}$ is any binary relation E between their states that satisfies the following conditions:

(a) *Atomic harmony* If xEy, then x and y verify the same proposition letters.

(b) *Back-and-forth* For each player i and each atomic forcing relation:

(i) If xEy and $\rho^{\boldsymbol{M},i}_{g}x, U$, then there is a V with $\rho^{\boldsymbol{N},i}_{g}y, V$ & $\forall v \in V \exists u \in U : uEv$.

226 Venema (2003) presents a more purely algebraic proof.

(ii) If xEy and $\rho_g^{\boldsymbol{N},i}y, V$, then there is a U with $\rho_g^{\boldsymbol{M},i}x, U$ & $\forall u \in U \exists v \in V : uEv$

This is essentially the notion of power bisimulation from Chapter 11. ■

Forcing bisimulation fits well in the context of this chapter.

FACT 19.3 (a) The DGL language is invariant for forcing bisimulations. (b) If two finite models $\boldsymbol{M}, s$ and $\boldsymbol{N}, t$ satisfy the same DGL formulas, then there exists a forcing bisimulation E between them with sEt.

Proof The inductive proof of (a) follows the invariance facts in Chapter 1. The proof of (b) follows the analogous fact about power bisimulation in Chapter 11. ■

As with PDL programs in Chapter 1, the proof of (a) involves a notion of safety with forcing relations for complex game terms, that we will now explain.

Safety for bisimulation In proving the invariance proposition, we need a special step. A forcing bisimulation E only guarantees the correct back-and-forth behavior for atomic relations ρ_g. But in proving that arbitrary formulas $\{G\}\varphi$ are invariant, we need back-and-forth behavior of E for all relations $\rho_G^{\boldsymbol{M}}$ for complex game expressions G in our models. The latter were constructed out of the former by our recursive rules. At this point, we introduce a notion that is known from dynamic logic (van Benthem 1996).

DEFINITION 19.9 Safety for forcing bisimulation

A game operation $G\#G'$ is *safe for forcing bisimulation* if for any forcing bisimulation E between models that satisfies the above back-and-forth conditions for the power relations $\rho_G, \rho_{G'}$, E also satisfies these conditions for $\rho_{G\#G'}^{\boldsymbol{M}}$. ■

We state the following results for forcing in determined games, but our arguments also work for the nondetermined two-player format.

FACT 19.4 All operations of DGL are safe for forcing bisimulation.

Proof Suppose that a relation E bisimulates between models $\boldsymbol{M}$ and $\boldsymbol{N}$ with respect to the two forcing relations ρ_G and $\rho_{G'}$ ($). We show that it also bisimulates for the defined relations $\rho_{G\cup G'}$, $\rho_{G\,;\,G'}$, and ρ_{G^d}, and that for both players.

Case (a1) Let xEz, and $\rho_{G\cup G'}^{\boldsymbol{M},\boldsymbol{E}}x, Y$. Then either $\rho_G^{\boldsymbol{M},\boldsymbol{E}}x, Y$ or $\rho_{G'}^{\boldsymbol{M},\boldsymbol{E}}x, Y$: say, the former. By ($) there is a set V such that $\rho_G^{\boldsymbol{N},\boldsymbol{E}}x, V$ and every point $v \in V$ has an E-related point $y \in Y$. But then a fortiori also $\rho_{G\cup G'}^{\boldsymbol{N},\boldsymbol{E}}x, V$, while every $v \in V$ still has an E-related point $y \in Y$. The other direction goes analogously. *Case (a2)* We need to show the same invariance for player $\boldsymbol{A}$. In this case, the given set Y has both $\rho_G^{\boldsymbol{M},\boldsymbol{A}}x, Y$ and $\rho_{G'}^{\boldsymbol{M},\boldsymbol{A}}x, Y$. By ($), we can find suitably matching sets V and W in $\boldsymbol{N}$, and the union of these is the required total match for Y, being a power for player $\boldsymbol{A}$ in both games.

Case (b) Let xEz, while $\rho_{G\,;\,G'}^{\boldsymbol{M},\boldsymbol{E}}x, Y$. By the definition of forcing relations, there is a set U with $\rho_G^{\boldsymbol{M},\boldsymbol{E}}x, U$ and $\forall u \in U : \rho_{G'}^{\boldsymbol{M},\boldsymbol{E}}u, Y$. Using ($) once more, there is a set V with $\rho_G^{\boldsymbol{N},\boldsymbol{E}}z, V$ where every $v \in V$ has an E-related $u \in U$. Again by ($), this time applied to the links uEv, there exist sets W_v with $\rho_{G'}^{\boldsymbol{N},\boldsymbol{E}}v, W_v$ while every $w \in W_V$ has an E-related $y \in Y$. Now, the union of all of these sets W_v is a set W that serves as a counterpart for the initial Y in $\boldsymbol{M}$, as is easy to check. The opposite direction, and also the cases for player $\boldsymbol{A}$, are similar.

Case (c) Finally, let xEz, while also $\rho_{G^d}^{\boldsymbol{M},\boldsymbol{E}}x, Y$. By the definition of determined forcing relations, $\neg\rho_G^{\boldsymbol{M},\boldsymbol{E}}x, (M - Y)$, where M is the universe of $\boldsymbol{M}$. To find a match to Y, consider the set $V = E[Y]$, where $E[Y]$ consists of all points z in $\boldsymbol{N}$ that are E-related to some $y \in Y$:

Claim $\rho_{G^d}^{\boldsymbol{N},\boldsymbol{E}}z, V$ and $\forall v \in V \exists y \in Y : yEv$.

The second conjunct is clear from the definition. To prove the first, recall its meaning: $\neg\rho_G^{\boldsymbol{N},\boldsymbol{E}}z, (N - V)$ with N the universe of $\boldsymbol{N}$. Suppose that $\rho_G^{\boldsymbol{N},\boldsymbol{E}}z, (N - V)$. By assumption (#), there must be a set U in $\boldsymbol{M}$ such that $\rho_G^{\boldsymbol{M},\boldsymbol{E}}x, U$ while every $u \in U$ has an E-related point $v \in (N - V)$. In particular, this means that the set U is disjoint from the initial Y. This is so because any point in U is E-related to some point v in $(N - E[Y])$, where by definition, the latter has no E-relations with Y-points. By monotonicity then, we would have that $\rho_G^{\boldsymbol{M},\boldsymbol{E}}z, (M - Y)$. But this contradicts the initial assumption. ■

Not all operations are safe for forcing bisimulation, and we give an illustration.

Example 19.4 Unsafe game operations
A counterexample is *thirteen*(G), letting players force just those sets of states that contain at least thirteen elements. It is easy to find a concrete counterexample, using the fact that forcing bisimulations cannot count numbers of states.

Natural operations and bisimulation Our analysis illustrates a general desideratum for a process theory: the choice of an operational repertoire must fit the relevant notion of structural equivalence. For instance, van Benthem (1996) studied safety for complex programs viewed as binary state-to-state relations, finding the following theorem on expressive completeness of the PDL repertoire.

THEOREM 19.4 The first-order definable operations on programs that are safe for bisimulation are precisely those that can be defined by arbitrary applications of union, composition, test negation, and atomic tests.

Pauly (2001) analyzes syntactic formats for safety in dynamic game logics, showing how forcing bisimulation lives in harmony with the operations of DGL.

Coda: Two levels again Bisimulation and invariance arose by thinking about internal structure of single games. But here we have seen how they also make sense for game combination. Perhaps the internal and external perspectives distinguished at the beginning of this chapter are not so different after all.

19.7 Conclusion

The main points We have explored a dynamic logic of sequential game operations merging ideas from logic games and game logics at a well-chosen abstraction level. It encodes some of the basic reasoning about strategies behind many of the systems of Part IV. The theory still looks like a propositional dynamic logic of programs, but now for complex games over neighborhood models with forcing bisimulation, and we have developed its basic model-theoretic and axiomatic properties. In addition, we found interesting new features, such as a decidable algebra of game operations.

Open problems Many questions arise now, connecting to earlier topics in this book. First, it makes sense to add imperfect information, preferences, and other features of real games, to see whether the elegant compositional design of our logic survives such an extension. Next, there is the issue of a natural operational repertoire for games, and in particular, dealing with operations for which PDL has not been so suitable historically as a process theory, including parallel game combinations. Also, much remains to be understood concerning the connection of DGL to infinite games and temporal or modal fixed point logics, a topic that will return in Chapters 20 and 25. Finally, as to game equivalences, DGL is clearly power-oriented in the sense of Chapter 11. Could there be similar logics for games at finer action levels of detail, such as the bisimulations in Chapter 1?

19.8 Literature

The presentation in this chapter is from the lecture notes of van Benthem (1999). An extension of the framework to parallel products of games is given in van Benthem et al. (2008).

Key sources on DGL are Parikh (1985) and Pauly (2001). Also relevant is the survey article of van der Hoek & Pauly (2006). Sources for general neighborhood semantics were given in Chapter 11.

19.9 Further directions

Here are two further threads following up on basic dynamic game logic. We have put them in this separate final section to avoid making the chapter top-heavy.

DGL with parallel operations Dealing with parallel composition of programs and distributed computation has not been PDL's strongest suit, and systems of process algebra took over around 1980 (cf. Milner 1999, Bergstra et al. 2001). Even so, a system of "concurrent PDL" exists using sets of local states as the output of computations (cf. Goldblatt 1992, van Benthem et al. 1994). For games, too, parallel products make sense, as we will see in Chapters 20 and 21. Indeed, standard strategic games already allowed for simultaneous moves (cf. Chapter 12).

Concurrent PDL can be merged with DGL into a natural system that performs one more set lifting (van Benthem et al. 2008). Here is a sketch. Models for concurrent PDL have accessibility relations Rs, X of type $S \times \wp(S)$ with the output read conjunctively: from state s, the program can produce all of the set X together.

DEFINITION 19.10 Game boards for concurrent *DGL*
Models for *concurrent* DGL are like models for DGL, but now with forcing relations of type $S \times \wp(\wp(S))$ running from states to families of sets of states. The interpretation of the inner set layer is as in concurrent PDL, but the outer set layer is disjunctive: the game can end in one of the sets of states collected there.[227] ■

227 This allows for two notions of monotonicity: increasing outer level sets is a form of weakening like in DGL, but increasing inner level sets means stronger output.

Our earlier forcing definitions for sequential game operations lift to this setting in a straightforward manner. But more interesting is the emergence of new operations.

DEFINITION 19.11 Forcing for parallel game product
Here is a natural *parallel product* $G \times G'$ of games, collecting outputs for subgames:

$$\rho^i_{G \times G'} sX \quad \text{iff} \quad \exists Y, Z : \rho^i_G s, Y \mathbin{\&} \rho^i_{G'} s, Z \mathbin{\&} X = \{y \cup z \mid y \in Y \mathbin{\&} z \in Z\}$$

This operation fits natural games that allow for simultaneous moves, such as those of Chapters 12 and 13. The modality of the system is now interpreted as follows

$$\boldsymbol{M}, s \models \{G, i\}\varphi \quad \text{iff} \quad \exists X : \rho^{\boldsymbol{M},i}_G s, X \mathbin{\&} \text{for all } x \in \bigcup X : M, x \models \varphi$$

Further details are as in our earlier modal logics of forcing. ■

The following theorem can be shown by a standard completeness argument.

THEOREM 19.5 The logic of the sequential game operations plus the above parallel product is axiomatizable and decidable.

The crucial DGL style decomposition axiom for the product is

$$\{G \times G', i\}\varphi \leftrightarrow \{G, i\}\varphi \wedge \{G', i\}\varphi$$

This encodes a game algebra, which has not yet been axiomatized equationally.

REMARK Collective action
The language and interpretation chosen here looks only at local properties of states. But simultaneous action is often at the same time collective action. It is also shown in van Benthem et al. (2008) how to lift the indices of evaluation to "collective set states" X instead of single points s, in a format

$$\boldsymbol{M}, X \models \{G, i\}\varphi,$$

where we can now also have a richer language with modalities referring to the natural inclusion structure of these new collective states. This modified logic of collective action for games has not yet been explored in a game setting.

Concurrent DGL seems a good fit with the IF logic of Chapter 21, although the only current study is Galliani (2012b) on connections with the dependence logics of Väänänen (2007). But quite different approaches make sense as well. Our next chapter will pursue a line on parallel games based on proof theory and linear logic.

Our second theme is logics for neighborhood models that may give the reader a better feeling for the mechanics of what has been proposed in this chapter.

PDL in neighborhood semantics The system DGL without dual is much like a propositional dynamic logic on neighborhood models with monotonic accessibility relations as in Chapter 11. It is of interest to see how its completeness works, since it ties together many notions introduced in this chapter.

THEOREM 19.6 PDL on forcing relations in neighborhood models is completely axiomatized by the modal base logic of forcing plus the standard recursive axioms of PDL for complex programs.

What follows is a compact summary of the main steps leading up to this result.

Semantics Accessibility relations for complex programs with choice and composition are defined as before, but with simplified notation, since we can now ignore player markings. The crucial further clause is for the iteration operator. It can be stated in a relational fixed point format such as that of Chapters 1, 2, and 8, now for point-to-set relations:

$$R_{\pi^*} = \mu S, x, X \bullet (x \in X \lor \exists Y(xR_\pi Y \land \forall y \in Y : ySX)).$$ [228]

Next, for formulas, we define the usual forcing modality:

$$\boldsymbol{M}, s \models \langle \pi \rangle \varphi \quad \text{iff} \quad \exists X : sR_\pi X \land \forall x \in X : \boldsymbol{M}, x \models \varphi.$$ [229]

Deduction in this system involves familiar principles from earlier in this book.

Axiomatics The logic validates all axioms of PDL minus modal distribution, now replaced by monotonicity. We still have left-distribution by the recursion axiom for program union. For iteration, we adopt the following inference principles:

$$\varphi \lor \langle \pi \rangle \langle \pi^* \rangle \varphi \to \langle \pi \rangle \varphi$$

$$\text{if} \vdash (\varphi \lor \langle \pi \rangle \alpha) \to \alpha, \text{ then} \vdash \langle \pi^* \rangle \varphi \to \alpha$$

Now we come to the proof of our main result.

---

228 While this looks like the usual notion of transitive closure, there is no guarantee that the approximation sequence stops at stage ω: finite approximations need not suffice.

229 Given the monotonicity of the relations, this is equivalent to $sR_\pi [\![\varphi]\!]^{\boldsymbol{M}}$, where the double brackets stand for the usual denotation in the model of all worlds satisfying φ.

Completeness Fix any valid formula φ, and stick to the finite sublanguage induced by it in the Fisher-Ladner closure $FL(\varphi)$ of φ (see Blackburn et al. 2001 for this standard device). Each atom s has a canonical finite description $\#s$ by enumeration, and each (finite) set of atoms X has a finite description $\#X$ by disjunction of descriptions for its members. We define the following relation for each program π:

$$sS_\pi X \text{ iff } \#s \wedge \langle\pi\rangle\#X \text{ is consistent.}$$

Taking this explanation for atomic programs only, and then proceeding inductively by the truth definition gives us the standard semantic relations R_π. The following connection between these two relations is crucial.

Inclusion Lemma $\quad S_\pi \subseteq R_\pi$

Proof Case (a). Atomic programs a satisfy the inclusion by definition. *Case (b).* For program union, first use the PDL inference from $\langle\pi_1 \cup \pi_2\rangle\varphi$ to $\langle\pi_1\rangle\varphi \vee \langle\pi_2\rangle\varphi$ to show that, if $\#s \wedge \langle\pi_1 \cup \pi_2\rangle\#X$ is consistent, then so is $\#s \wedge \langle\pi_1\rangle\#X$ or $\#s \wedge \langle\pi_2\rangle\#X$. The rest follows by the inductive hypothesis. *Case (c).* Program composition requires a slightly more complex argument. Using one direction of the PDL composition axiom, if $\#s \wedge \langle\pi_1 ; \pi_2\rangle\#X$ is consistent, then so is $\#s \wedge \langle\pi_1\rangle\langle\pi_2\rangle\#X$. All atoms t whose $\#t$ are consistent with $\langle\pi_2\rangle\#X$ form a set Y. The following implication states how these notions behave.

Claim $\quad \langle\pi_2\rangle\#X$ provably implies $\#Y$.[230]

The claim implies that we also have $\#s \wedge \langle\pi_1\rangle\#Y$ consistent, while, by the definition of Y, $\#t \wedge \langle\pi_2\rangle\#X$ is consistent for all $t \in Y$. But then, by the inductive hypothesis for $\pi = \pi_1 and \pi_2$, the desired inclusion $S_\pi \subseteq R_\pi$ follows. *Case (d).* The inclusion for test programs is straightforward by the inductive hypothesis. *Case (e).* Finally, dealing with program iteration crucially uses the smallest fixed point induction rule. Let $\#s \wedge \langle\pi^*\rangle\#X$ be consistent for some set of atoms X. We show that s belongs to the smallest fixed point of the following procedure. Start with X and keep applying the map

$$F(Y) = X \cup \{s \mid sS_\pi Y\}$$

---

230 To see this, suppose that the implication is not derivable. Then there is a maximally consistent set Σ containing $\langle\pi_2\rangle\#X$ and $\neg\#Y$. Restricting Σ to the $FL(\varphi)$ sublanguage gives an atom t in the set Y: but all of these atoms are ruled out by the presence of $\neg\#Y$. This is a contradiction, and so the claim is proved.

If we can show this fact about s, the rest of the proof follows in a standard manner, relying on the inductive hypothesis. Now, in our finite model, the given procedure stops at some finite stage in a finite set A where

$$A = X \cup \{s \mid sS_\pi A\} \qquad (\$)$$

The following claim relates this observation to a fact about provability in PDL.

CLAIM The implication $(\#X \vee \langle\pi\rangle\#A) \rightarrow \#A$ is provable.[231]

Given the implication of the claim, by the smallest fixed point rule, $\langle\pi^*\rangle\#X \rightarrow \#A$ is provable. But then, since $\#s \wedge \langle\pi^*\rangle\#X$ was consistent, $\#s \wedge \#A$ is consistent, too, and this must mean that $s \in A$. ■

The final piece of the completeness argument is the usual Truth Lemma.

TRUTH LEMMA The equivalence $\varphi \in s$ iff $s \models \varphi$ holds for all states and formulas.

Proof The proof is by induction on formulas, with a subinduction on programs for $\langle\pi\rangle\varphi$. The latter's direction from left to right uses the inclusion lemma. From right to left, the inductive steps use half of the PDL program axioms, appealing only to the monotonicity of the logic. We examine three cases. *Case (a).* Consider an atomic program a with $s \models \langle a\rangle\varphi$. By the truth definition, we have a set X with sR_aX, i.e., $\#s \wedge \langle\pi\rangle\#X$ is consistent in our logic, and for all $x \in X : x \models \varphi$. Then by the inductive hypothesis, $\varphi \in x$. Now we observe that all this implies $\vdash \#X \rightarrow \varphi$, by a standard argument about maximally consistent sets and their $FL(\varphi)$ restrictions. But then by monotonicity, $\#s \wedge \langle a\rangle\varphi$ is also consistent, and hence $\langle a\rangle\varphi \in s$. *Case (b).* Consider a program composition with $s \models \langle\pi_1 ; \pi_2\rangle\varphi$. By the truth definition, we have a set X with $sR_{\pi_1 ; \pi_2}X$ and for all $x \in X : x \models \varphi$. Thus, there is a set Y with $sR_{\pi_1}Y$ and for all $y \in Y : yR_{\pi_2}X$ while for all $x \in X : x \models \varphi$. In other words, for all $y \in Y : y \models \langle\pi_2\rangle\varphi$. Then by the inductive hypothesis, $\langle\pi_1\rangle\langle\pi_2\rangle\varphi \in s$, and using one half of the PDL axiom for composition, also $\langle\pi_1 ; \pi_2\rangle\varphi \in s$. *Case (c).* Consider a program iteration with $s \models \langle\pi^*\rangle\varphi$. Again, using the truth definition, we either φ have true at s, and hence by the inductive hypothesis in s, where it implies the presence of $\langle\pi^*\rangle\varphi$ by half of our fixed point axiom. Or, we have sets X, Y with

231 To see this, note first that $\#X \rightarrow \#A$ is provable, as $X \subseteq A$. If $\langle\pi\rangle\#A \rightarrow \#A$ is not provable, then there is a maximally consistent Σ including $\langle\pi\rangle\#A, \neg\#A$. Restricting Σ to the $FL(\varphi)$ sublanguage, we get an atom t with $tS_\pi A$ that is not in A because of the presence of $\neg\#A$ in Σ. This contradicts ($).

$sR_\pi Y \wedge \forall y \in Y : yR_{\pi^*}X$ and all $x \in X$ satisfy φ. But then all $y \in Y$ satisfy $\langle\pi^*\rangle\varphi$, and so s satisfies $\langle\pi\rangle\langle\pi^*\rangle\varphi$. Thus, by the inductive hypothesis, $\langle\pi\rangle\langle\pi^*\rangle\varphi$ is in s. By the remaining part of the fixed point axiom, we also get $\langle\pi^*\rangle\varphi$ in s. ■

The earlier open problem concerning completeness for DGL with game dual added emerges in this setting when we try to modify the above proof to deal with player-dependent accessibility relations.

20
Linear Logic of Parallel Game Operations

Alternative logics of game construction come from another computational tradition: infinite communication and interaction games for program semantics, with historical links to logic games for dialogue. In this chapter, we define operations for playing several games in parallel. Instead of modal logics, these lead to connections with linear logic, showing how game operations are related to intuitions about resources. The inspiration for the latter approach is proof-theoretic, representing a logical paradigm with a vast literature of its own. This chapter will only make a brief connection with the model-theoretic mainstream of this book, although we do provide a window with further references.

20.1 From logic games to game logics, once more

Logic games for argumentation and dialogue were introduced in Chapter 17. These had three main parameters: (a) rules for attack and defense of complex propositions, (b) a format of what constitutes a two-person dialogue, and (c) procedural conventions, including winning conditions and constraints on admissible repetitions.

The settings chosen by Lorenzen linked his games to intuitionistic logic, with intuitionistic proofs matching winning strategies for the proponent $\boldsymbol{P}$ of the initial thesis.[232] These games involve contrived features such as asymmetries in attack and defense rights for the proponent and the counterplayer opponent $\boldsymbol{O}$. This led to important amendments in Blass (1992), with a further modern development in the game semantics of Abramsky (1995).

232 One could get classical logic systems by fiddling with rights and duties under (c).

Again, this involves a move from logic games to game logics. In Chapter 17, atomic propositions were still black boxes: winning was maneuvering your opponent into a contradiction without any fact checking. At this point, however, we make a conceptual change, and think of atomic games as variables for real games. Say, $p \vee q$ may become a compound game, Chess or Cricket. When reaching an atom, we open the box and play the corresponding game. This simple move has many repercussions. For instance, refuting excluded middle will no longer involve tinkering with procedural rules. It suffices to plug in a *nondetermined game* p in the dialogue game for $p \vee \neg p$. Further interesting features of the resulting abstract game logic will unfold in this chapter.

REMARK Infinite games and nondeterminacy
Nondetermined games of perfect information must be infinite, by the Gale-Stewart Theorem of Chapter 5. Such games had non-open winning conditions, and their existence involved an appeal to the Axiom of Choice.[233] Infinite games are natural for independent reasons, as we have seen in Chapter 4 and later in this book. The focus in this chapter is on infinite games, although finite ones are included.

20.2 Parallel operations

From sequential to parallel Our first new idea is parallel game operations. Different atoms in a dialogue game stand for different subgames that may become active, just as different atoms were different tasks in the model construction games of Chapter 16. There is no compulsory sequential schedule triggered by one leading formula, as with evaluation games, so we need to go beyond the operations of choice, switch, and sequential composition of the preceding chapter. To do so, the compositional structure of infinite games must also include parallel operations for interaction. In the dialogue games of Chapter 17, switching had to do with repeating attacks or defenses on the same atom. The systems in this chapter re-encode their procedural conventions in a more elegant game format.

Parallel conjunction of games Our pilot example is a new parallel conjunction $A \times B$ stipulating carefully who gets to switch between two games A and B (taking

233 Thus, constructivizing game logic involves games obtained by nonconstructive means. We will see a possible alternative later, in terms of games with imperfect information.

the initiative) and what are the wins. Think of Chess $\times$ Cricket, where we make some moves on a board, hit some balls, return to the board, and so on. We give an informal description here, but the details will come later.

The game $A \times B$ gives the initiative for switching to opponent $\boldsymbol{O}$. We let $\boldsymbol{O}$ win the infinite runs that have at least one infinite projection to moves in the subgame A or B that is a win for $\boldsymbol{O}$. (Thus, $\boldsymbol{P}$ wins a completed run iff both projections are either wins for $\boldsymbol{P}$, or if the run is finite.) A *parallel disjunction game* $A + B$ is like $A \times B$, but with the roles of the players $\boldsymbol{O}$ and $\boldsymbol{P}$ reversed throughout.

Copy-cat strategy A paradigmatic example for these games is simultaneous Chess. If you want to hold your own against any opponent, propose the game

$$Chess + Chess^d$$

Here, the dual game $Chess^d$ is *Chess* with the order for Black and White reversed. In the parallel sum game, you are $\boldsymbol{P}$, and you can play essentially the same Chess game twice, once as $\boldsymbol{P}$ and once as $\boldsymbol{O}$. A simple method works as follows.

DEFINITION 20.1 Copy-cat strategy
Copy-cat works like this. Let the other player open in one game, and copy that opening move across to the other game. Wait for the other player's response there, and play that in the original game, and so on. ■

This copying strategy produces two identical runs on both sides:

	Chess	$Chess^d$
$\boldsymbol{O}$	m_1	
$\boldsymbol{P}$		m_1
$\boldsymbol{O}$		m_2
$\boldsymbol{P}$	m_2	
	...	

By the winning convention, one of the projections is a win for $\boldsymbol{P}$, or both are draws, so $\boldsymbol{P}$ never loses the whole game. Note that *communication* between subgames is of the essence here, which makes these games a paradigm for parallel computation. In games without draws, copy-cat is even a winning strategy for $\boldsymbol{P}$.[234]

234 Chess is a finite game, so $\boldsymbol{P}$ could choose the sub game with a Zermelo non-losing strategy directly. But copy-cat is much less complex, and it also works for infinite games.

New operations The above setting supports new operations without classical counterparts. An example is the finite repetition $R(G)$ of Blass (1992), which resembles the DGL iteration G^* of Chapter 19. It lets one player open finitely many copies of G, exploiting what goes on in all of them to win at least one.

20.3 The games defined

The games We now define the structures we will work with in what follows.

DEFINITION 20.2 Infinite games
There are two players $\boldsymbol{P}$ and $\boldsymbol{O}$. *Linear games* (or just games, for brevity) are tuples $G = (M, \delta, Q, W)$ with M a set of moves, δ the turn function assigning a player to each move, Q the set of admissible positions (the legitimate finite sequences of moves), and W the set of infinite (only!) runs all of whose finite initial segments are in Q that we designate as wins for player $\boldsymbol{P}$. ■

With a few twists, the reader will recognize the infinite models of Chapters 4 and 5. This format is very flexible, by varying admissible positions. Many games are scheduled like the dialogues of Chapter 18: $(\boldsymbol{OP})^*$, but other orders are possible. Also, winning can depend on many types of condition. In addition to their uses in program semantics (see our earlier references), games such as this also occur in the French school of ludics (cf. Danos et al. 1996, Girard 1998a, and Girard 1998b).

Game constructions We now go over the basic operations, some reminiscent of earlier chapters, explained more precisely as constructions on games.

DEFINITION 20.3 Dual, choice, and parallel conjunction

(a) The *dual game* $G^d = (M, \delta^-, Q, M^\infty - W)$ has a turn function δ^- switching the roles of the two players, as well as the winning convention for infinite runs.

(b) *Choice games* $A \vee B$ have an initial choice by $\boldsymbol{P}$, just as in Chapter 19, putting two disjoint copies of the component games side by side under a common root:

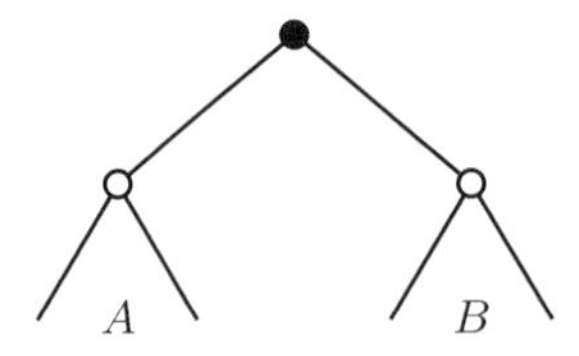

(c) Finally, *parallel game conjunction* $A \times B$ is defined as follows. Take the disjoint union of moves in both games, with their original turn functions. Acceptable histories are those whose projections to both A and B are acceptable in those games. That is, one plays in each component always resuming where one left off. Moreover, only $\boldsymbol{O}$ is allowed to switch games. The winning convention is that $\boldsymbol{O}$ wins a run r if at least one projection $r|G$ ($G \in \{A, B\}$) is a win for $\boldsymbol{O}$ in subgame G.

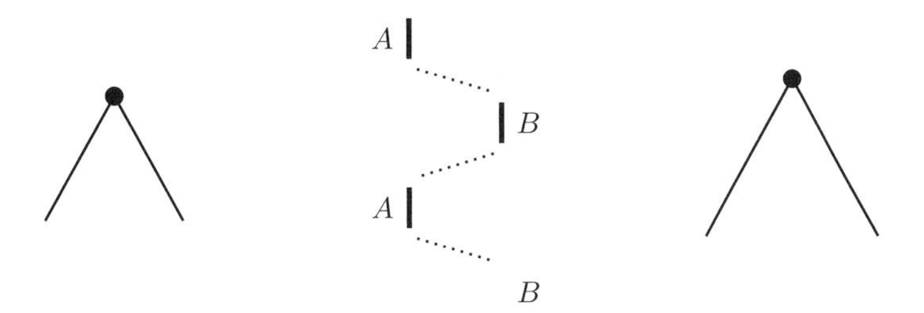

Tree operations like this occurred in Chapter 14, and will return in Chapter 25. ■

One motivation for the rules in the parallel game is this: $\boldsymbol{O}$ attacks the conjunction, and is allowed to choose all the time where $\boldsymbol{P}$ has to defend. But the switching can be viewed just as well as cooperation, initiated by requests from $\boldsymbol{O}$.[235]

Derived game operations We now provide three useful derived game operations:

$\boldsymbol{O}$'s choice	$A \wedge B \ = (A^d \vee B^d)^d$
parallel sum	$A + B \ = (A^d \times B^d)^d$
implication	$A \rightarrow B = A^d + B$

Parallel sum is the dual of product. Implication has a related flavor: player $\boldsymbol{P}$ can use information obtained in the subgame A to play successfully in game B. This control is reflected in basic logical inference patterns.

EXAMPLE 20.1 Modus ponens
Proponent $\boldsymbol{P}$ has a winning strategy for a modus ponens game:

$$(A \times (A \rightarrow B)) \rightarrow B$$

The reader should work out the rights and duties for both players here. ■

---

235 The scheduling $(\boldsymbol{OP})^*$ in the game $A \times B$ is like a finite-state machine (Abramsky 1995) with transitions between states $\langle$*player to move in* G_1, *player to move in* $G_2\rangle$.

We will concentrate on the logic of the new parallel operations, since the simpler $\vee$ and $\wedge$ were already studied in Chapter 19.

20.4 Logical validity of game expressions

The new operations show some classical behavior. For instance, parallel product and sum are commutative and associative. But the differences are more striking. Even classical idempotence does not hold. G and choice product $G \wedge G$ amount to the same game, but in general, G is not equivalent to the parallel game $G \times G$.[236] Having one game to play, or various copies of it simultaneously, is not at all the same thing. An example of this phenomenon with parallel sum was the game

$$G + G^d$$

where $\boldsymbol{P}$ had a winning strategy, even when G and G^d are nondetermined. Thus, our earlier discussions about determinacy in the Introduction and Chapter 14 of this book become substantially more sophisticated. Whether excluded middle is valid really depends on the game disjunction used.

DEFINITION 20.4 Validity of game expressions
An expression G is *valid* if there is a winning strategy for player $\boldsymbol{P}$ in each actual game arising from taking concrete games for the atomic expressions in G. ■

The system of game validities looks like propositional logic, but with some super-constructive differences that are even more striking than intuitionistic logic. As in Chapter 18, generic powers of players show nonclassical behavior, but they obey the laws of a natural nonclassical logic proposed independently in the 1980s.

20.5 Linear logic and resources

The non-equivalence of winning powers in games $G, G \times G, G + G$ is reminiscent of linear logic, a system of resource-conscious reasoning from the 1980s Girard 1993 as a fine structure analysis of structural rules in classical proof theory.

---

236 Compare this with the nondetermined free ultrafilter game G of Chapter 5 and the associated strategy stealing argument. In product play $G \times G$, $\boldsymbol{O}$ can always win, using $\boldsymbol{P}$'s moves in one game one step later against $\boldsymbol{P}$ in the other copy of G.

Occurrences and resources The main intuition is a computational view of logical formulas, say the premises of an inference, as resources that can be used only once. Validity is having exactly the right resources to reach the conclusion.

EXAMPLE 20.2 Modus ponens and its relatives
An implication $A \to B$ is a function that wants exactly one argument A to produce a value B. This is exemplified perfectly in a modus ponens inference $A, A \to B \Rightarrow B$. More than one copy of A leaves an unused resource. Therefore, in linear logic,

$$\text{a premise sequence } A, A, A \to B \text{ implies: not } B, \text{ but } A \times B,$$

with the product $\times$ storing the still available resources. On the other hand, an implication $A \to (A \to B)$ needs two premises A to get the B, not one or three:

$$A, A \to (A \to B) \text{ implies } A \to B, \text{ but not } B \text{ by itself.}$$

Classical logic ignores resources: many copies of a formula amount to just one. ■

Motivations for the present delicate view of inference come from grammars for natural language (van Benthem 1991), where occurrences of formulas stand for occurrences of linguistic categories. Game semantics (Abramsky 1995) stresses links with communication and interaction, where resources are like calls, moving from a game logic perspective to a wider view of interactive computation. (For further motivations and details for linear logic, cf. Girard 1993.)

Sequent calculus Proof systems for linear logic differ from those for classical logic in two main respects. Let us first fix a widely used proof format, Gentzen sequents $\varphi_1, \ldots, \varphi_n$ (cf. Troelstra & Schwichtenberg 2000). Classical logic reads commas in these sequents as disjunctions, and the sequent expresses that the disjunction of the given formulas is valid. Linear logic reads the commas as parallel sums $+$.

DEFINITION 20.5 Sequent transformation
The classical sequent version of modus ponens $A, A \to B \Rightarrow B$ translates into $\Rightarrow (A \times (A \to B)) \to B$ that becomes the linear sequent

$$A^d, A \times B^d, B$$

after unpacking the definition of the linear implication. ■

A standard sequent calculus for classical logic contains three central ingredients. (a) There are axiomatic sequents $\Sigma, \varphi, \neg\varphi$, with Σ any finite sequence of formulas. (b) Logical rules take valid sequents to new ones, introducing one new logical operator at a time.[237] (c) Structural rules manipulate sequents as sets of formulas: permuting formulas, contracting more occurrences of a formula into one, or expanding one formula into several.

In linear logic, sequents stand for so-called multi-sets or bags of formulas where every occurrence counts. As a consequence, the familiar structural rules of classical logic disappear, such as contraction of identical formulas, monotonicity conjoining formulas to antecedents, or disjunctive weakening of consequents. Only the classical principles of permutation of formulas and the cut rule remain. There is also an initial sequent axiom $A, \neg A$, looking like the classical excluded middle, but as we shall soon see, it says something different. Its validity reflects the above observation about $\boldsymbol{P}$'s having a winning strategy in all games of the form $A + A^d$. In particular, no other formulas may be put alongside A, A^d, as the relevant copy-cat strategy would no longer work.

Weaker base, richer vocabulary Typically, the new paradigm shows a feature found with many weak logics. Moving to a weaker proof-theoretic basis supports a variety of non-equivalent operations beyond standard logic. In particular, classical connectives split: there are multiplicative and additive versions of conjunction and disjunction. Intuitively, multiplicative $A \times B$ is a combination of an A-type and a B-type resource, while additive $A \wedge B$ is both an A-type and a B-type resource. The difference shows clearly in the respective introduction rules, which are as follows.

EXAMPLE 20.3 Introduction rules for additive and multiplicative conjunction
The reader may want to ponder the difference between the following two rules:

$$\frac{X \Rightarrow A \qquad Y \Rightarrow B}{X, Y \Rightarrow A \times B} \qquad\qquad \frac{X \Rightarrow A \qquad X \Rightarrow B}{X \Rightarrow A \wedge B}$$

With the classical structural rules, $\times$ and $\wedge$ are interderivable. ■

In the rest of this chapter, we concentrate on the multiplicatives $\times, +$. For convenience, we push negations inside to atoms via the duality laws for $\times$ and $+$ as in the earlier definitions.

237 Examples of such rules are: from Σ, φ to $\Sigma, \varphi \vee \psi$, or from Σ, φ and Σ, ψ to $\Sigma, \varphi \wedge \psi$.

20.6 An axiom system

We now present the proof principles for resource-conscious reasoning.

DEFINITION 20.6 Sequent calculus for linear logic
The *basic proof system* for linear logic has the following principles:

linking axiom	$A, \neg A$
product rule	$\dfrac{\Sigma, A \qquad \Delta, B}{\Sigma, \Delta, A \times B}$
permutation	from Σ, derive any of its permutations
product rule	$\dfrac{\Sigma \qquad \Delta}{\Sigma, \Delta}$

Rules for the defined multiplicative $+$ follow from those for $\times$ and dual ^{d}. ■

Variants of the calculus add rules. For instance, "affine linear logic" has a weakening rule concluding from Σ to Σ, A. If we add in all structural rules of classical logic, the whole repertoire collapses into the standard Boolean operations.

Some simple linear derivations show how the austere basic proof system works.

EXAMPLE 20.4 Derivations in linear logic
(a) As we saw, modus ponens $A \times (A \to B) \Rightarrow B$ becomes $\neg(A \times (\neg A + B)) + B$, which works out to a linear sequent $\neg A, A \times \neg B, B$ that can be proved as follows:

$$\frac{\neg A, A \qquad \neg B, B}{\neg A, A \times \neg B, B} \quad \text{(product introduction)}$$

(b) Under the same translation, transitivity $(A \to B) \to ((B \to C) \to (A \to C))$ comes out as $(A \times \neg B), (B \times \neg C), \neg A, C$, and this can be proved as follows:

$$\frac{\neg A, A \quad \dfrac{\neg B, B \quad \neg C, C}{\neg B, B \times \neg C, C}\ \text{(product introduction)}}{A \times \neg B, B \times \neg C, \neg A, C} \quad \text{(product introduction + permutation)}$$

With the many links that exist between $+$, $\times$ and $\to$ via the dual operation ^{d}, one can often read the same sequent in various equivalent ways. ■

As with the dynamic logic of Chapter 19, there is much more to linear logic than we can discuss here, including a crucial role for cut elimination theorems, and the use of sophisticated representation methods such as proof nets modeling interaction. We refer the reader to textbooks such as Troelstra (1993), while there are also many modern resources.

20.7 Soundness and completeness

The main theoretical results of relevance to our game logic are as follows.

Soundness Unlike the usual soundness arguments for logical calculi, the one for linear logic on its game interpretation is non-trivial and illuminating.

THEOREM 20.1 Every provable sequent of linear logic is valid in our game sense.

Proof All proof rules express significant facts about combining strategies for $\boldsymbol{P}$ in games for the premises into one for the conclusion game. We give two cases.

Product rule Suppose that player $\boldsymbol{P}$ has strategies σ and τ for winning the games Σ, A and Δ, B, respectively. Then $\boldsymbol{P}$ can also win the sequent game for $\Sigma, \Delta, A \times B$ as follows. If a move is made in Σ, let $\boldsymbol{P}$ respond with σ, and if it is in Δ, let $\boldsymbol{P}$ respond with τ. If $\boldsymbol{O}$ chooses a conjunct A or B, then respond with the move prescribed by σ for A, and τ for B. This procedure is a winning strategy, as can be seen by inspecting the resulting histories. Either $\boldsymbol{P}$ obtains an infinite subhistory that is a win in Σ or in Δ, and either is enough to win the parallel disjunction game, or if not, $\boldsymbol{P}$ has winning subhistories in both A and B, making $\boldsymbol{P}$ win the total game for $\Sigma, \Delta, A \times B$ via $A \times B$.

Cut rule Suppose $\boldsymbol{P}$ has strategies σ and τ for winning Σ, A and $\Delta, \neg A$, respectively. The following describes a winning strategy for $\boldsymbol{P}$ in Σ, Δ: play σ in Σ, and play τ in Δ. However, these strategies may prescribe moves that go into the subgames A or $\neg A$. If this happens, $\boldsymbol{P}$ plays a "virtual match," acting as $\boldsymbol{P}$ in A using σ, and also as $\boldsymbol{O}$ using τ from $\neg A$.[238] Now crucially, such a virtual episode cannot go on forever. If it did, some tail of the resulting infinite history would project to a loss for $\boldsymbol{P}$, contradicting the fact that σ and τ were winning strategies for $\boldsymbol{P}$ in Σ, A

238 This is reminiscent of the shadow matches in the strategic reasoning in Chapter 18.

and $\Delta, \neg A$.[239] Therefore, some exit move is produced in the episode, surfacing in one of Σ or Δ again, and this will be $\boldsymbol{P}$'s official response in the game for Σ, Δ. Again, checking some cases for the resulting histories of the game will show that the procedure described is a winning strategy. ∎

There are some nasty bookkeeping details when such arguments get fully spelled out. This is a sort of curse of game semantics, at least, to outsiders.

Completeness The next major result is the converse implication.

THEOREM 20.2 Every game-valid sequent is provable in multiplicative linear logic.

A proof is beyond the scope of this book, and we refer to Abramsky & Jagadeesan (1994). Moreover, these authors obtain an even stronger result that relates to a persistent theme in this book, namely, the role of strategies as fundamental logical objects in their own right (cf. Chapters 4, 5, and 17). They improve the preceding standard completeness theorem, curing the $\exists$-sickness of Chapter 18, to obtain so-called "full completeness."

THEOREM 20.3 There is an effective correspondence between uniform winning strategies for a game sequent and linear proofs for that same sequent.

The proof of this stronger result involves categories of games whose morphisms are strategies, with the crucial cut rule of the system reflecting a natural associative composition of strategies.[240]

An interesting feature of this analysis is also that *history-free* strategies turn out to suffice for establishing completeness of linear logic. Like copy-cat, these only use the last move played, not the full history of the game so far. History-free strategies are like the positional strategies for graph games in Chapter 18, but we leave a comparison between the two settings to the reader.

20.8 From proof theory to program semantics

While our treatment has tied the games in this chapter to logic, this may be misleading. The intuitions behind proof structures in linear logic are linked closely to

---

239 For motivation, recall the strategy stealing argument analyzed in Chapter 5.

240 The match is between strategies and proof nets. The completeness proof uses a well-chosen category analogous to a logical term model, in which morphisms are proof nets.

multi-agent interaction, giving a computational thrust to the topics discussed here. The game semantics in this chapter is also a model of interactive computation in its own right, with strategies as algorithmic styles of system behavior, and deep uses of proof-theoretic and matching category-theoretic methods. This paradigm has been applied widely in the semantics of programming languages, with connections to denotational semantics (Scott & Strachey 1971) and domain theory (Scott 1976, Abramsky & Jung 2001). The resulting game-based area of interactive system behavior studied by logical techniques is beyond the scope of this book. For some highlights, see Abramsky & McCusker (1999) and Curien (2005, 2006). Abramsky (2012) discusses where this paradigm is leading in terms of natural levels for modeling the realities of modern computation as levels of system behavior.

20.9 Conclusion

The main points Once more, we have followed a natural path from logic games to game logics. The game semantics in this chapter treats logical formulas as game terms, and as in earlier chapters, logical operations then become general game constructions. We have seen how new parallel operations arise in this way, whose origins came from resource-based proof theory, and program semantics for interactive computation. We gave the basic definitions of game semantics, discussed its connections with linear logic, and stated soundness and completeness results showing how this works. But we can also see the system in this chapter as a continuation of the dynamic game logic of Chapter 19, providing a high-level analysis of natural operations on games.

Open problems The logic of parallel games presented here raises quite a few further issues. In particular, one can study many of the topics that came up in Chapter 19, starting with how the two approaches to logic of game constructions compare. A few such lines will be found in Section 20.11 on further directions.

20.10 Literature

We have mentioned some classic sources for linear logic and game semantics in the text, including Blass (1992) and Girard (1993). For the ensuing French School on games in this setting, see Danos et al. (1996), Girard (1998a), Girard (1998b), and the tutorial Curien (2005). For game semantics of programming languages, various papers by Abramsky are recommended, starting from the classic Abramsky & Jagadeesan (1994). A recent tutorial aimed at logicians is Abramsky (2008b). Also relevant is the discussion of information and computation in Abramsky (2008a). An alternative view of logic, games, and computation is found in Japaridze (1997).

20.11 Further directions

As usual, this chapter raises many further issues, of which we only mention a few.

Varieties of parallelism Interaction games have natural choice-points for defining parallel conjunction and disjunction. Here are at least two variants.

One might make things less uniform by letting players switch whose turn it is in both games. Then $\boldsymbol{P}$ can be a switcher after all in $A \times B$, say, when A and B both start with a move for $\boldsymbol{P}$. A more radical approach would introduce switching policies as an additional argument of the operation: $\#(A, B, \pi)$.[241] Netchitailov (2001) is a study of parallel operations with controlled switching in finite games of imperfect information.

Also, our parallel operations interleaved moves. What about games with simultaneous moves, as in Chapter 12? Natural models here would have pair states (s, t) that allow for either componentwise or simultaneous transitions. On such games, many new types of winning conditions make sense beyond the simple Boolean combinations of winning and losing in separate games used in this chapter. For instance, they might now refer to the results of *collective action*, such as saying that the sum of the yields of s and t exceeds some threshold.

Thus, the realm of natural parallel game combinations has by no means been exhausted in this chapter.

Comparisons with dynamic game logic The dynamic game logic of Chapter 19 was about sequential operations, a subset of those studied here. But it described arbitrary outcomes, with a richer language than linear sequents. It is therefore of interest to compare the expressive resources of the two logics. The above linear

241 In a similar vein, process algebra has worked with labeled parallel process operators $A||_c B$ where c is a communication method (cf. Bergstra et al. 2001).

sequents are pure game expressions,[242] and so their logic resembles the earlier game algebra. We might rework linear logic into an equational calculus for game equivalence, and then merge approaches, taking the notion of forcing to linear games, and adding modalities $\{G\}\varphi$ with G a linear game term and φ a statement about infinite runs resulting from it, perhaps in a branching temporal logic.

Adding temporal logics But then, a broader issue merges. The games discussed in this chapter are also extensive games in the sense of Parts I and II of this book. Many of the logics introduced there make sense for linear games, and they form natural combinations with the syntax of this chapter. For instance, Chapter 5 contained a few cases of strategic reasoning when a temporal logic is added that describes the histories of games explicitly. This suggests combining temporal logics and logics with explicit game terms describing the current game. We will discuss this topic briefly in Chapter 25. Such richer description languages also link to our perennial issue of game equivalence, since merging global and local views combines game equivalences at different levels of detail.

Strategy calculus In particular, merging logics allow us to resume a theme from Chapters 4 and 5: making the reasoning about strategies explicit that is supposed to be so central, but often stays inside the metalanguage. Crucial strategies for $\boldsymbol{P}$ in this chapter are simply definable, with copy-cat as a prime example. We considered definability of strategies in Chapter 4, and a joint logic with the above linear formulas as game terms, and our earlier logics for more game-internal structure would be a good candidate. A good benchmark for such a richer logic would be to formalize the earlier non-trivial soundness proof for linear logic. In a sense, the proof of the full completeness theorem delivers this, but one would also want an analysis in a formalism closer to the mainstream of this book.

More proof theory and category theory While the main thrust of this book is clearly model-theoretic, Chapters 17 and 20 have shown how basic notions and techniques from the area of proof theory, too, can enter into a profitable symbiosis with games. In particular, proof terms witnessing propositions in type theories are very much like strategies in games. This view came up in the concrete samples of strategy calculus in Chapter 4, and it suggests a whole range of analogies between

242 They also stood proxy for a statement about these games: viz. that player $\boldsymbol{P}$ has a winning strategy. We have taken care to distinguish these two perspectives in earlier chapters, and one should also do so here.

proof theory and game analysis that is beyond the scope of our study. Our earlier references on game semantics show how these contacts can be studied in great generality in a category-theoretic framework.

Connections with real game theory The infinite games of this chapter are reminiscent of infinite evolutionary games in game theory. Likewise, identity strategies such as copy-cat play a basic role in the latter area, under names such as strategy stealing, or Tit for Tat. These connections seem obvious, and yet they have not been explored. Likewise, it makes sense to add structure that we have studied in Parts I and II. For example, with imperfect information present, even finite games can be nondetermined. What would imperfect information versions of dynamic and linear game logics be? There may be difficult issues in bringing these worlds together, since in game theory, the interesting games are typically thought of as being noncompositional, whereas compositional analysis is the core methodology in the foundations of computation. But clearly, knowledge and imperfect information make sense for the games of this chapter, and there may well be an interesting interface with game theory here.

Conclusion to Part V

We have shown how ideas from logic games and game logics come together in two major logics of compositional game structure. The dynamic game logic of Chapter 19 analyzes players' powers, generalizes modal and dynamic logic to neighborhood models, and provides an interesting abstraction level from which to view sequential structure in general. Moreover, in dealing with both game expressions and assertions about local states, it represents an intriguing combination of external and internal perspectives on games. We have explained the basics of the system, including techniques for analyzing definability and axiomatic completeness. Next, linear game logic arose from a very different motivation, adding proof-theoretic intuitions of resources in interaction. Chapter 20 explained the basic ideas, and pointed at broader links to game logics for infinite games, even though the original motivations for the framework come from category-theoretic semantics rather than model-theoretic semantics.

Both systems admitted algebraic viewpoints that form a natural extension of the game logics studied in Parts I, II, and III. Indeed, they may be viewed as formalizing part of an abstract meta-theory of games. But also, as we have seen in several passages, both systems make for interesting connections to the modern foundations of computation.

At this stage, many questions arise. Perhaps a bit inward looking, how can logicians reconcile the model-theoretic and proof-theoretic aspects of the game logics presented here? And more outward looking, can these logics accommodate significant structure from real game theory such as imperfect information or preferences? Also, in that encounter, will the logician's and computer scientist's methodology of compositionality hit its limits, given the recalcitrant entangled structure of real games? Finally, and perhaps more urgent from the perspective of this book, how do the systems of Part V relate to the earlier logic games in Part IV and game logics in Parts I, II, and III? We have no definite answers to all this, but several chapters of Part VI will give a view of their own on how all of these different approaches and mentalities relate.

VI Comparisons and Merges

Introduction to Part VI

This final part is our last chance of harmonizing the two main strands of this book, logic *of* games and logic *as* games. Readers expecting a grand synthesis or even a mutual reduction will be disappointed, however. No such grand unity seems on the horizon. What we hope to have shown already in Parts I through V is how the two faces of the contact between logic and games can live together, perhaps even belong together. What we will do in Part VI is expand this view in a number of explorations.

One immediate, but by no means futile test of compatibility is the creation of healthy hybrids, and we will do so by adding game-theoretic structure to quantifier evaluation games with imperfect information in Chapter 21. Another creative contact is the emergence of new games that naturally combine different flavors from this book. Chapter 22 is about new knowledge games played with group information that are reminiscent of the dynamic-epistemic logics of information flow in Part II, and Chapter 23 introduces sabotage games that make standard computational tasks interactive. These chapters show how both game logic and logic game perspectives arise naturally in concrete scenarios. At the same time, they also raise several issues of independent interest, such as game modeling of information flow in observation and communication, and the status of games as a natural continuation of computation. Next, a third, more theoretical line looks for technical connections and reductions between the main themes of this book. Chapter 24 shows exactly how evaluation games for first-order or modal logic are expressively complete for a general algebra of game operations, and Chapter 25 discusses merged logical systems combining features of both logic games and game logics, providing a layered view of games.

Our considered view on the DNA of this book is deferred to the Conclusion.

21
Logic Games with Imperfect Information

A concrete way of relating logic games with general game logic is by fusion of ideas. One striking illustration is an innovative incorporation of the game-theoretic notion of imperfect information into the logic games of Part IV, that hitherto involved perfect information only. This chapter is mainly a brief discussion of this game-theoretic semantics and its accompanying IF logic (Hintikka & Sandu 1997). This framework has foundational aims of its own concerning the notions of dependence and independence in logic that may make its games more of an auxiliary device. But we will discuss it as an example of the many surprising themes that arise by fusing ideas from two traditions. Many topics in this book appear in a new light.[243]

21.1 IF games and imperfect information

Recall the following example of a first-order evaluation game from Chapter 14.

EXAMPLE 21.1 Evaluating $\forall x \exists y \, x \neq y$
Here is an extensive game with perfect information that evaluates the first-order formula $\forall x \exists y \, x \neq y$ on a two-element domain $\{1, 2\}$:

---

243 Of course, imperfect information is just one interface, still close to the informational aspect of logic, and it would also be of interest to see what happens when we bring in evaluative notions such as preferences related to the purposes of logical activity.

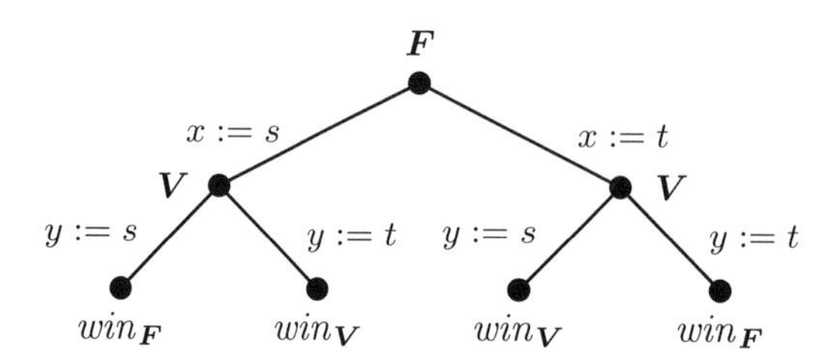

Verifier $\boldsymbol{V}$ had a winning strategy, reflecting the obvious truth of the assertion. ■

But now, consider the following variant in an extension of first-order logic:

$$\forall x \exists y/\boldsymbol{x}\; x \neq y$$

where the slash / indicates that verifier does not have access to the object chosen by falsifier. As in the imperfect information games of Chapter 3, this may be because falsifier's move is hidden, or verifier did not notice, or forgot. The matching evaluation game is called an *IF game*, where IF stands for independence-friendly, for reasons having to do with quantifier scope that will be explained below.

EXAMPLE 21.2 Evaluating $\forall x \exists y/\boldsymbol{x}\; x \neq y$
This modified slash formula has the following game tree, with falsifier's uncertainty link marked by a dotted line:

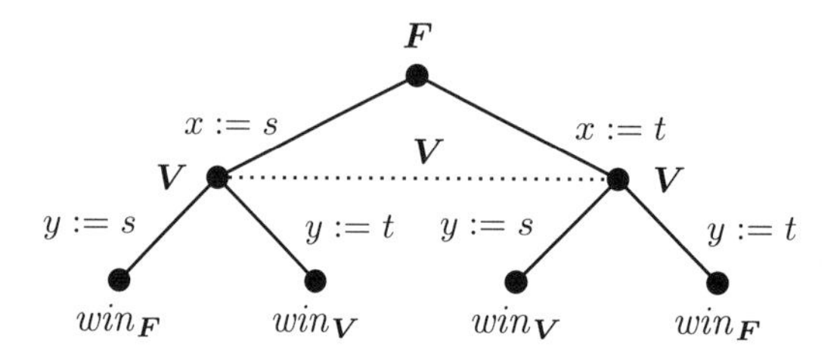

At the outcome level of Chapter 11, where this chapter will operate, this game is nondetermined. Falsifier had no winning strategy before, and still does not have one, as verifier may get lucky by accident. But verifier loses two powers, including the former winning strategy, as the only usable strategies for verifier are the two uniform ones *left; left* and *right; right*, neither of them winning. ■

A more detailed view The preceding game can be analyzed equally well at the action level of Chapter 3 with knowledge and ignorance of players en route.

In particular, verifier's plight at the intermediate nodes was described by the modal-epistemic formulas $K_{\mathbf{V}}(\langle y := s\rangle win_{\mathbf{V}} \vee \langle y := t\rangle win_{\mathbf{V}})$ and $\neg K_{\mathbf{V}}\langle y := s\rangle win_{\mathbf{V}} \wedge \neg K_{\mathbf{V}}\langle y := t\rangle win_{\mathbf{V}}$. While we will pursue players' powers only, it would be of interest to see whether a matching theory could be developed at this finer level.

21.2 The new perspective motivated

Branching quantifiers The original motivation for IF games came from the semantics of natural language.

EXAMPLE 21.3 The villager-townsman sentence
Consider the sentence, "Some relative of each villager and some relative of each townsman hate each other." Its intuitive reading should have the branching pattern:

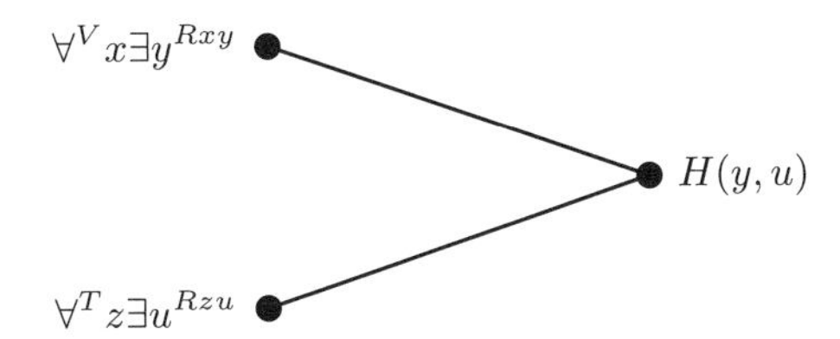

Any linear syntactic form for this natural language sentence coming from standard logical languages would enforce too many scope dependencies.[244] ■

The intended information flow can be described by an IF game that is like the standard evaluation game for the linear first-order formula:

$$\forall x \exists y \forall z \exists u\, R(x, y, z, u)$$

but now with the dependency relaxed: by the time $\exists u$ is reached, verifier no longer has access to the value of x mentioned at the start of the game by falsifier:

$$\forall x \exists y \forall z \exists u / \boldsymbol{x}\; R(x, y, z, u)$$

---

244 Whether branching occurs in natural language remains a matter of debate: see Keenan & Westerståhl (1997) on quantifiers in linguistics. The dissertation by Paperno (2011) studies coordination phenomena in natural language quantification via parallel game products as proposed in van Benthem (2003), using the game semantics of Chapter 20.

IF games have also been proposed in the foundations of mathematics, parallel computation, and in modeling uncertainty and indeterminacy in quantum physics.

Skolem functions Next, consider the strategic aspect of evaluation. Any first-order assertion of the form $\forall x \exists y R(x, y)$ is equivalent to the second-order formula

$$\exists f \forall x\, R(x, f(x))$$

where the "Skolem function" f is a choice function picking witnesses from the domain, dependent on the choice of the object y. One can think of this function as a winning strategy for verifier in the evaluation games of Chapter 14. With longer quantifier prefixes, one gets a sequence of Skolem functions. An illustration is the equivalence between

$$\forall x \exists y \forall z \exists u\, R(x, y, z, u) \quad \text{and} \quad \exists f \exists g \forall x \forall z\, R(x, f(x), z, g(x, z))$$

Now, the slash versions suppress dependency of these functions on some arguments. Thus, the intended interpretations become as follows:

$\forall x \exists y/\boldsymbol{x}\ x \neq y$	$\exists f \forall x\, R(x, f)$ with f a 0-place function,
$\forall x \exists y \forall z \exists u/\boldsymbol{x}\ R(x, y, z, u)$	$\exists f \exists g \forall x \forall z\, R(x, f(x), z, g(z))$ where f and g are both 1-place functions

This liberates logical syntax from accidents of the standard linear notation.[245]

21.3 The IF language and its games

Slash syntax Start with first-order logic, but allow additional quantifiers of forms

$$\exists x/\boldsymbol{W}, \forall x/\boldsymbol{W} \qquad \text{where } \boldsymbol{W} \text{ is a set of variables.}$$

The corresponding game instruction will be to choose a value for x *independently from*, or without knowing the values of the variables in the set W. Thus, the above branching reading would correspond to

$$\forall x \exists y \forall z \exists u/\boldsymbol{x}\ R(x, y, z, u)$$

245 Skolem functions do not capture verifier's strategies entirely, as evaluation also requires propositional choices for disjunctions. But this is a quibble that is easily remedied.

This extension leads to allowing connectives $\vee/\boldsymbol{W}, \wedge/\boldsymbol{W}$, where the choices by verifier or falsifier must be made independently of W values (Sandu & Väänänen 1992). We can even make choices at occurrences of connectives independent from earlier occurrences of connectives.

Signaling This move comes with delicate problems. Suppose that in the propositional formula $(p \wedge (q \vee (r \wedge s))$, at later stages, falsifier no longer knows what choice was made at the first conjunction. Then falsifier would not know the game being played now: that for p or for $(q \vee (r \wedge s))$. However, the fact of having to make a choice between r and s right now will tell falsifier which game is being played. This is called *signaling*: some relevant information, even if officially unavailable, may be derived by a chain of signals from the past. There are many triggers for signaling in IF games, and discussions persist (Hodges 1997, Janssen & Dechesne 2006).

How to play? Moving from syntax to semantics, a surprising issue arises. How should one play IF games? As we have discussed already in Parts I and II, imperfect information games are a record of uncertainty, but they do not tell us which specific scenario produced these. Informal explanations for IF games have ranged from incomplete observation to memory loss. Some such scenarios might have to be somewhat strange, witness the formula

$$\forall x \exists y \forall z \exists u/\boldsymbol{x} \forall v \exists s \, Rxyzuvs$$

for players with intermittent memory lapses. Its Skolem form is

$$\exists f \exists g \forall x \forall z \, R(x, f(x), z, g(z))$$

but this does not help in telling us which activity took place, while it also seems biased toward the perspective of verifier only (more on this will follow).

It has been suggested that IF games have *groups* or coalitions of players, which explains independence patterns by restrictions on internal group communication.

Agency and information flow One more theme from Part II seems relevant here. Some ignorance is part of the official design of a game: think of dealing hands to card players. These games can be played by players with ideal capacities for reasoning and observation. A quite different source of ignorance is *players' limitations*: they may not pay attention, have bounded memory, cheat, and so on. These things are often run together in the story of IF games, making them a delightful challenge for the style of analysis in this book. We will only explore this lightly in this chapter, but we hand the reader all the strands needed for a more extensive investigation.

21.4 IF games: Extended examples

Let us now look at a few more elaborate examples to see what can happen.

Concrete examples In Chapter 14, we defined extensive games $\boldsymbol{game}(\varphi, \boldsymbol{M}, s)$ for any first-order formula φ, model $\boldsymbol{M}$, and variable assignment s. What about IF syntax? We already had $\forall x \exists y/\boldsymbol{x}\ x \neq y$ on a two-object domain, and its form should be clear for arbitrary models $\boldsymbol{M}$. Next, consider two more complex signaling examples, where we switch from inequalities to equalities for simplicity. The first is a beautiful scenario from Hodges (1997).

EXAMPLE 21.4 An IF game with signaling
Consider a modification of $\forall x \exists y/\boldsymbol{x}\ x \neq y$, inserting one vacuous quantifier $\exists z$:

$$\forall x \exists z \exists y/\boldsymbol{x}\ x \neq y$$

In standard logics, a vacuous quantifier makes no difference. But things are different in this IF game, whose behavior is unlike our earlier example. This time, verifier does have a uniform winning strategy:

Use your z-move to copy falsifier's first move, then copy that as your own y-move.

Turning toward its game tree on $\{s, t\}$, it may not be immediately clear how to draw dotted lines for verifier. A recipe is implicit in the following picture:

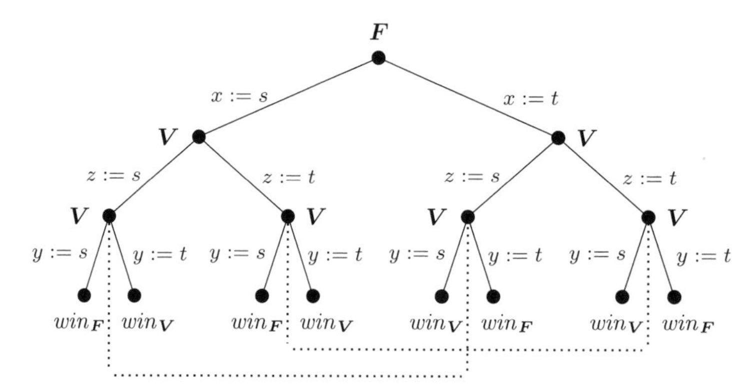

The dotted lines represent verifier's uncertainty in the third round about falsifier's first move. At the same time, in that round, the lines show that verifier knows the move that was made in the second round. We indicate a uniform winning strategy for verifier with bold-face arrows in the following diagram:

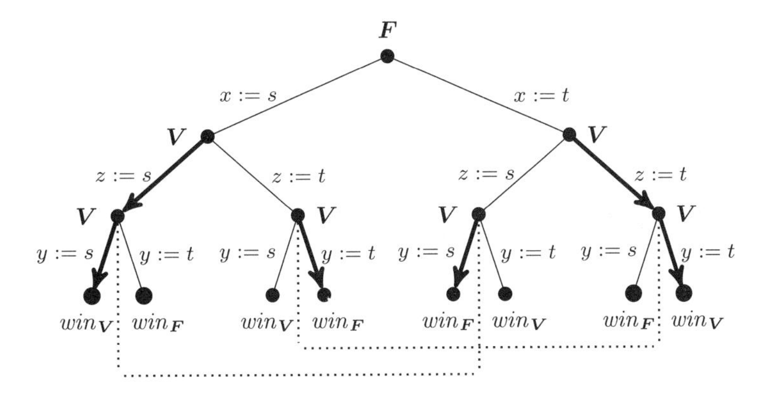

Verifier's strategy assigns the same move to indistinguishable states. ■

There are subtleties in this example that illustrate key points from Chapters 3 and 4. In the third round as depicted here, verifier will not know that the current strategy is winning, as verifier considers it possible that falsifier played another move, in which case the move of the strategy would result in a loss. But if we also assume that verifier is aware of playing the stated strategy, our game tree is no longer a correct representation, since we would have to drop the innermost four branches. In that case, no uncertainty is left at the third stage.

Here, another point from Chapters 3 and 4 returns. A player can have a uniform winning strategy without knowing at each stage that playing the rest of that strategy is in fact winning. Think of following a guide through a bog, having forgotten why we trusted the guide in the first place. In game-theoretic terms, the point is that the game as pictured lacks perfect recall for verifier. In the third round, verifier forgot the information known in the second round.

However, if we also factor in that players know their own strategy throughout, the game simplifies as stated, although it may no longer be one that corresponds to an IF formula. Moreover, representing this additional information systematically may well take us to the complexities of modeling discussed in Chapter 6.[246]

Now we ask what happens when verifier gets even less information.

246 Much more can be said about this example, but the present discussion will do to show the game-theoretic interest of even simple IF scenarios.

Example 21.5 A game for $\forall x \exists z \exists y/\{\boldsymbol{x}, \boldsymbol{z}\}\ x \neq y$
The preceding game may be usefully contrasted with one for $\forall x \exists z \exists y/\{\boldsymbol{x}, \boldsymbol{z}\}\ x \neq y$, where only the dot pattern for the uncertainty of verifier changes:

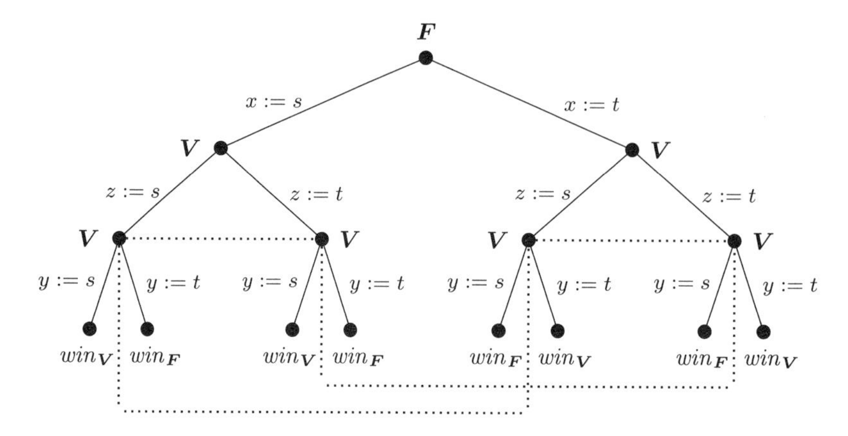

As we see here, simple changes can have dramatic consequences: verifier does not at all know the present location. With the uncertainty lines drawn in this game, verifier has no strategy known to work. ■

A general algorithm determining game trees for IF formulas in arbitrary models can be found in Mann et al. (2011).

21.5 IF games, algebra, and logic of imperfect information games

Now it is time to look at things from our general perspective on imperfect information games in this book. Recall the notions of uniform strategy and power equivalence from Chapters 3 and 11. IF games are often called three-valued since they are nondetermined, but of course, in such games, what we are really after is the power structure for *both players*. We analyze the games through the basic viewpoint of when they are the same.

Uniform power equivalence We now proceed with a simple illustration.

Example 21.6 Powers in IF games
In our first IF game, falsifier kept the original powers from the perfect information version, but verifier lost two former powers, as follows:

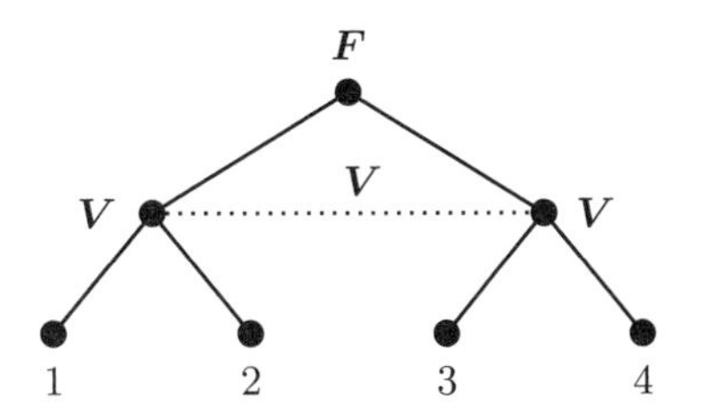

powers of $\boldsymbol{F}$: $\{1,2\}, \{3,4\}$, powers of $\boldsymbol{V}$: $\{1,3\}, \{2,4\}$

This seems poorer than the perfect information version, but it is really a more subtle form of power sharing, making the two players more equal.[247] ■

Game equivalence may again be analyzed in terms of powers in this new sense, as we saw in Chapters 3 and 11. In particular, we are now in a position to answer an obvious question about our first example: What is a correct game equivalent for $\forall x \exists y/\boldsymbol{x}\, x \neq y$? The first-order answer $\exists y \forall x\, x \neq y$ given in the original IF literature is a nonstarter: the slash game is nondetermined, while the game for a first-order formula is determined. The mistake is that this equivalence works for verifier only. To deal with both players, we need a much more pleasing symmetric formula.

FACT 21.1 The formula $\forall x \exists y/\boldsymbol{x}\, x \neq y$ is equivalent to $\exists y \forall x/\boldsymbol{y}\, x \neq y$.

Proof The IF game for the proposed new formula corresponds to the earlier game for $\forall x \exists y/\boldsymbol{x}\, x \neq y$, but now with its turns and outcomes interchanged:

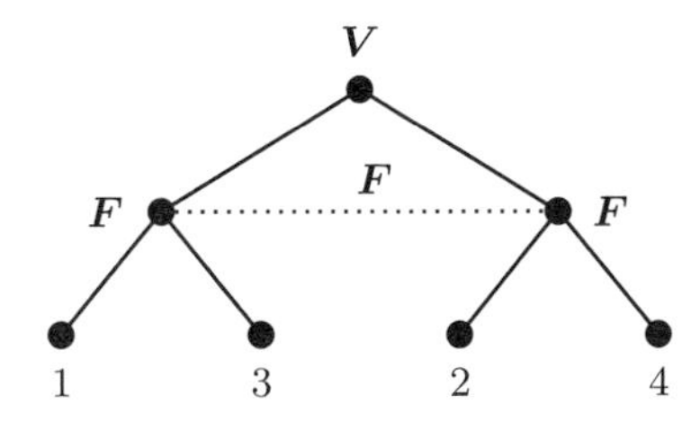

The players' uniform powers are exactly the same here as in the above game. This equivalence reflects one of the basic Thompson transformations' of game theory discussed in Chapter 3. ■

247 As we observed in Chapter 3, imperfect information is crucial to designing social scenarios with a delicate distribution of tasks and powers among agents.

This style of power analysis works more broadly. For instance, it can be shown that Hodges' formula $\forall x \exists z \exists y/\boldsymbol{x}\ x \neq y$ is power equivalent to a slash-free determined formula $\forall x \exists y\, x \neq y$.

IF logic as game calculus The preceding examples raise some interesting logical issues. We can see the equivalential part of IF logic as a calculus for game equivalence, just as first-order logic encodes an algebraic calculus for perfect information games (cf. Chapter 24). For instance, the new valid quantifier equivalence

$$\forall x \exists y/\boldsymbol{x}\ x \neq y \leftrightarrow \exists y \forall x/\boldsymbol{y}\ x \neq y$$

that we found above is a beautiful new distribution law not dreamt of in classical logic. Following this lead, does IF logic, like the first-order logic it extends, contain a decidable core logic of perspicuous operator equivalences that axiomatize uniform power equivalence over games? This would then be the basic mechanics of IF-style games, while the remaining complexity reflects special features of first-order assignment and uncertainty.

Parallel games There is also a different road to logic. Abramsky (2000) embeds IF games in game semantics (cf. Chapter 20), using parallel product to model imperfect information. Another view (van Benthem 2002b) goes back to the earlier branching quantifier prefixes that make choices for variables independently, bringing them together at the end to test a matrix atom. Such games involve the imperfect information of matrix games (Chapter 12) with ignorance of others' moves played at the same time.

Technically, the relevant game operation is one that was defined at the end of Chapter 19, namely, a product $G \times H$ whose runs are pairs of separate runs for G and H with the product of their end states as the total end state. We recall the powers of players in such a game.

FACT 21.2 Players' powers in a product game satisfy the equivalence

$$\rho^i_{G\times H}(s,t), X \text{ iff } \exists UV : \rho^i_G s, U \ \&\ \rho^i_H t, V : U \times V \subseteq X$$

where $U \times V$ stands for the Cartesian product of the sets U and V.

This product operation validates new principles beyond the dynamic logic DGL for sequential game operations, which were discussed in Chapter 19. For present purposes, we couch the discussion in terms of algebraic identities.

FACT 21.3 The following identities of game algebra hold for product games:

$$\begin{aligned} A \times (B \cup C) &= (A \times B) \cup (A \times C) \\ (A \cup B) \times C &= (A \times C) \cup (B \times C) \\ (A \times B)^d &= A^d \times B^d \\ A \times B &= B \times A \end{aligned}$$

This may be proved by straightforward analysis of players' powers.[248]

IF logic with product What does it mean to play a product game $\varphi \times \psi$? The slash formula $\forall x \exists y \forall z / \{\boldsymbol{x}, \boldsymbol{y}\} \exists u / \{\boldsymbol{x}, \boldsymbol{y}\}\ Rxyzu$ for our branching quantifier in Example 21.3 may now be written as follows, with a test game at the end:

$$((\forall x\,;\, \exists y) \times (\forall z\,;\, \exists u))\,;\, ?Rxyzu$$

Game-algebraic laws then have IF instances that manipulate quantifier prefixes.

FACT 21.4 The following identity is valid in IF game algebra:

$$(\forall x\,;\, \exists y) \times ((\forall z\,;\, \forall u) \cup (\exists v\,;\, \exists u)) = ((\forall x\,;\, \exists y) \times (\forall z\,;\, \forall u)) \cup ((\forall x\,;\, \exists y) \times (\exists v\,;\, \exists u))$$

The converse is also true, valid principles of IF logic show up as algebraic validities. For instance, the earlier equivalence $\forall x \exists y / \boldsymbol{x}\ \varphi \leftrightarrow \exists y \forall x / \boldsymbol{y}\ \varphi$ says the following in game-algebraic terms.

FACT 21.5 The following identity is valid in IF game algebra:

$$(G \times H)\,;\, K = (H \times G)\,;\, K$$

This follows easily from the above algebraic identities. At the same time, IF logic can also witness invalid principles of core game algebra.

EXAMPLE 21.7 An algebraic invalidity
The following formula fails in the algebra of IF games:

$$(A \times B)\,;\, C = (A\,;\, C) \times (B\,;\, C)$$

An IF counterexample is the slash formula $\exists y \forall x / \boldsymbol{y}\ x \neq y$, whose evaluation game is not equivalent to that for the simultaneous game for $\exists x Rxy$ and $\forall y Rxy$. ■

---

248 The fourth identity takes component order in product states (s, t) to be immaterial.

Could IF logic be a complete format for an abstract algebra of parallel combination for imperfect information games?

21.6 The underlying epistemic dynamics

We conclude with a central theme in this book: the information dynamics underneath games viewed as traces of behavior. Let us leave the power level, and look at what happens in IF games as played by agents, the untold story behind the paradigm. General information dynamics was studied in Part II of this book in dynamic-epistemic logics of information-driven action. For a start, as we saw in Chapter 3, imperfect information games support standard epistemic languages, and it would be natural to use these to describe what happens inside IF games, putting the two systems together. In epistemic program logics, we can then even define players' uniform strategies, replacing Skolem functions by the knowledge programs of Chapter 4. There have been a few attempts in this direction (cf. van Benthem 2006), but nothing systematic exists.[249]

This is the major challenge that we see in this chapter, but we leave the interface of IF logic and information dynamics as a future task.[250]

21.7 Conclusion

We have shown how significant game-theoretic features can be added to logic games. Our case study was IF logic that incorporates imperfect information into logic games of evaluation. We have shown how these fit well with the general game-theoretic perspectives in this book, using notions such as power equivalence, game algebra, uniform strategies, and epistemic logics.

In doing so, IF logic turned out to be far more than juxtaposition of existing ideas: new phenomena came to light of interest to both logic and game theory. In particular, making sense of IF games highlights subtleties for games in general.

---

249 Galliani (2012b) has a few systems that are relevant here, merging dependence, belief, and public announcement logic. Rebuschi (2006) even investigates a carousel EL(IF(EL(IF$\cdots$ of game logics, alternating epistemizing and gamifying.

250 Given the links between IF logic and dependence logic (Väänänen 2007), this would also address the issue of how to combine two basic notions of information in logic today, based on semantic range and on dependence (van Benthem & Martínez 2008).

Also, we found a new core game algebra extending that in Chapter 19 inside the more complex full system.

Our treatment left many lines unexplored. We saw how truly understanding playable IF games requires an underlying Theory of Play, as studied in Parts I and II of this book. Likewise, there must be a more general, possibly decidable, core logic of games behind this chapter that needs to be teased out. And of course, interfacing logic games and game theory has many aspects beyond imperfect information, with adding preferences as the most obvious desideratum.

21.8 Literature

This chapter is based on van Benthem (1999), van Benthem (2004b), and van Benthem (2006a).

Many key papers are mentioned in the text, but Sandu (1993), Hintikka & Sandu (1997) are classic introductions. An excellent state of the art source is Mann et al. (2011). A related framework with a fast recent development is the dependence logic of Väänänen (2007). Various merged systems in between IF logic, dependence logic, and DEL logics are studied in Galliani (2012b).

21.9 Further directions

As usual, our treatment suggests many further lines of investigation. We summarize a few from our main text, and add some more.

Logical aspects The literature on IF logic emphasizes its higher-order complexity. It will be clear to discerning readers that we are a bit suspicious of this feature. It comes from quantifying over Skolem functions: in our view, a poor projection of an intuitive semantics for IF games. More generally, the claim that a base logic of dependence and independence should have very high complexity may result from a confusion between core calculus and accidental details of the first-order setting. The jury still seems to be out on complexity and axiomatizability.[251]

---

251 Recall our distinction in Chapter 3 between the complexity of a logic and of the activity it describes. That statements about IF games are complex does not preclude that these games might be easier to play than perfect information games, since players might be simpler, including automata (Chapter 18) or memory-bounded agents (Chapter 7).

Next, here are a few other logical themes arising in this chapter. First, one important component of a Theory of Play would be an explicit general strategy calculus (cf. Chapters 4 and 18) for IF logic, moving away from specifics of concrete first-order models. But also at a coarser power level, the external game algebra of IF logic is unknown. And even if we cannot find that yet, we might be able to extend the representation theorem for perfect information games via evaluation games in Chapter 24 to IF logic.[252] Finally, one would like to understand better what the system in this chapter really means. Could there be a general transformation of "IF-ing" that produces imperfect information versions for any logic game, from proof to model comparison? Introducing limited memory and knowledge of players makes equal sense in all of them, but can they be transformed uniformly?[253]

Game-theoretic aspects IF games introduce standard game-theoretic structure in logic games. Thus, classical solution methods for games apply. In particular, we can take a strategic form as in Chapter 12, and apply iterated removal of strictly (or weakly) dominated strategies to prune the game. The spirit is conveyed by the following example from van Benthem (2004b).

EXAMPLE 21.8 Branching quantifiers revisited
Recall the earlier IF formula $\forall x \exists y \forall z \exists u/\boldsymbol{x}\ H(y, u)$ for the villager/townsman sentence. For a start, here is a model where this formula is neither true nor false:

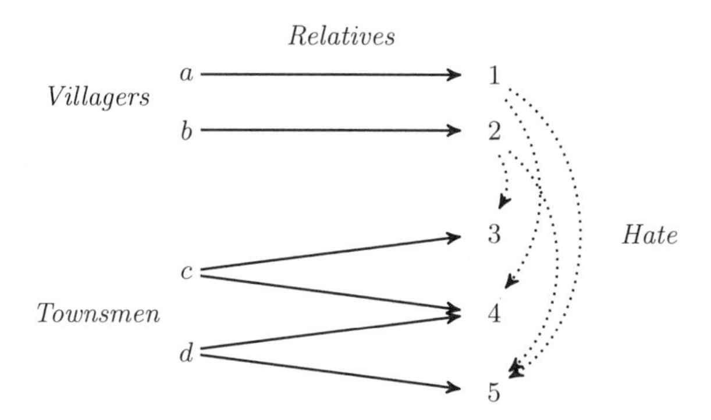

252 Galliani (2012b) claims some first results in this direction, but see also Abramsky (2006) on IF logics and the game semantics of Chapter 20.

253 Other options exist. The pebble games of Chapter 14 were an interesting general device for giving players explicit resources or memory.

$\forall x \exists y \forall z \exists u/\boldsymbol{x}\ H(y,u)$ does not hold with u independent from x. However, it is true without the slash, and so it is not false in IF semantics over this model either. ■

On these models, we can now apply completely standard game-theoretic methods, such as the solution algorithms of Chapters 12 and 13.

EXAMPLE 21.9 Applying game solution algorithms to IF games

Looking at best actions for players, it is clear that verifier must pick 1 and 2 for a and b, respectively, so that relevant strategies are essentially just the choices at c and d. The strategies for falsifier are just the choices for the variables x and z. Tabulating outcomes as usual in game theory, we get:

F \ ***V***	34	35	44	45
ac	−	−	+	+
ad	+	+	+	+
bc	+	+	−	−
bd	−	+	−	+

Here, + stands for a win for verifier and − stands for a loss. Iterated removal of weakly dominated strategies removes columns or lines as follows:

F \ ***V***	34	35	44	45
ac	−	−	+	+
bc	+	+	−	−
bd	−	+	−	+

F \ ***V***	35	45
ac	−	+
bc	+	−
bd	+	+

F \ ***V***	35	45
ac	−	+
bc	+	−

Thus, looking at the last table, in the end falsifier should play either a or b, but always c, while verifier should play 5 for d, and either 3 or 4 for c. ■

The output of this procedure needs interpretation, since it has no matching logical syntax. But this very fact suggests that combining IF logic with standard game theory may yield intriguing new views of roles of players, dependency structure of imperfect information games, and links with the shape of the playground model.

Adding probability A basic insight in game theory mentioned earlier is that strategic equilibria may only exist when we move from pure to mixed strategies (Osborne & Rubinstein 1994) where basic actions are played with certain probabilities. Such solutions also arise in IF games (again we follow van Benthem 2004b).

EXAMPLE 21.10 Probabilistic structure in IF games
Recall the pilot game discussed in Example 21.2:

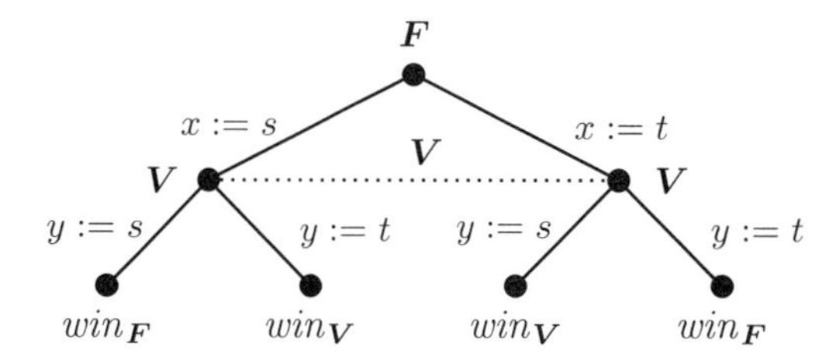

This is really just the matching pennies game. It has an optimal value $(1/2, 1/2)$, achieved by players using their uniform strategies with probability $1/2$. ■

While these numerical values may seem to be an artifact of a game, mixed Skolem functions might in fact correspond to viable styles of play. (For recent results on probabilistic solutions to IF games, see Galliani 2009b, Mann et al. 2011.)[254]

Dependence logic Independence as raised in IF logic is a crucial notion in many areas of reasoning. Dependence logics (Väänänen 2007) take sets of assignments as the basic indices of evaluation, whose interpretation is that of "teams." IF games describe procedural dependencies in making or evaluating assertions, but teams might also model objective dependencies in nature (Dretske 1981). Galliani (2012b) links dependence logics to the dynamic-epistemic logics of Chapter 7, and the game algebra of Chapter 19. Like IF logic, dependence logics have a high second-order complexity that can be questioned.[255] By contrast, the dependence logic of "generalized assignment models" for first-order logic (Németi 1995, van Benthem 1996) does the opposite, becoming decidable by dropping the usual feature of first-order

254 Further topics in van Benthem (2004b) included connections with other results linking logic and probability such as the zero-one laws, and the issue of whether the true habitat of IF logic is in probabilized models, where not just ways of playing strategies, but also the base objects themselves have become probabilistic mixtures.

255 Here, Henkin-style generalized models (Väänänen 2007) might alleviate complexity on models with restricted ranges of Skolem functions.

semantics that values for variables can always be modified independently of those for other variables. Do the logics in this chapter incur their high second-order complexity by the nature of dependence, or as an accident of modeling?

From information to control There is an alternative to the line of thought in this chapter, moving from dependence to the basic notion of control (cf. Chapter 12). Harrenstein (2004) defines Boolean games about control of values for variables by agents in shared activities.

Preference in logic games After imperfect information, the next aspect of real games that may make sense in logic games are finer preferences as in Parts I, II, and III of this book. Natural preferences in logic games might have to do with using few resources, or they could model the value of a claim or a proof.[256]

This is not a routine addition. In a setting with preferences, the laws of logic, interpreted as laws for games in the style of Part IV, may change. We repeat some earlier illustrations from Chapter 2.

EXAMPLE 21.11 Propositional logic with preferences
Call two formulas equivalent if their strategic equilibria are the same for each concrete version with players' preferences. Propositional distribution may then fail:

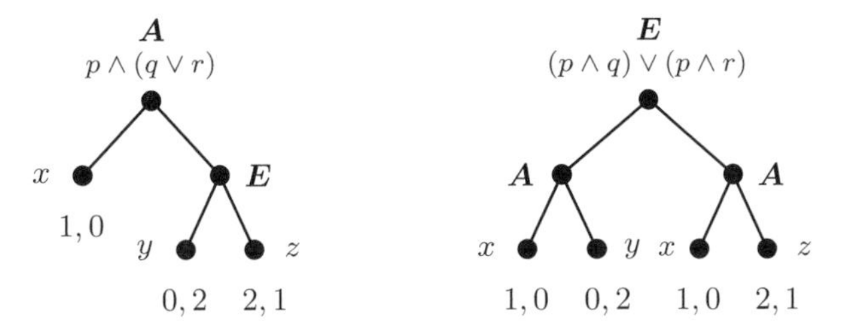

Backward Induction computes the following pairs (***A***-value, ***E***-value):

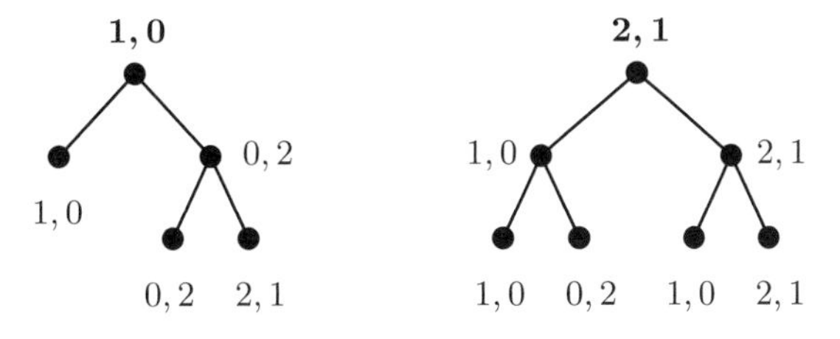

256 Further motivation may come from many-valued logic; see Fermüller & Majer (2013).

These trees represent different outcomes for joint behavior of players. Accordingly, we predict outcome x on the left, but z on the right. ■

Still, some standard laws of logic are valid with preference.

FACT 21.6 Absorption $A = A \cap (A \cup B)$ holds when preferences are zero-sum.

The complete propositional logic of finite games with preference is unknown. Its set of validities might also offer clues as to an extended notion of preference-based game equivalence, a desideratum already mentioned in Chapter 2.

More agents, and groups While logic games tend to have two players, nothing prevents us from having more. At least on the proof side, it seems significant that argumentation is often not a two- but a three-role game, involving a proponent, an opponent, and a referee (cf. Chapter 23). Moving from individuals to groups of players made sense at various places in this book. As we said earlier, IF players might be teams whose informational dependencies derive from limitations on internal communication, and likewise, dependence logic has a robust team interpretation. This view of social action in logic systems deserves a serious hearing, including the introduction of notions of group knowledge from Parts I and II.

22
Knowledge Games

Our paradigm for the dynamics of information-driven agency in Part II were systems of dynamic-epistemic logic describing stepwise changes on current models. These single update steps congregated in longer-term informational procedures, where the total playground is a set of branching histories over time, as in the epistemic-temporal models of Chapters 5, 6, and 9. Such models embody an informational protocol in their selection of admissible histories (Fagin et al. 1995, van Benthem et al. 2009a), and games are obvious concretizations of this huge space. In this chapter we very briefly discuss an intriguing recent development, the knowledge games of Ågotnes & van Ditmarsch (2011) that can be played over epistemic models, showing how they merge ideas from logic and game theory.

22.1 Group communication on epistemic models

An epistemic model represents the information state of a group of agents. One dynamic perspective on groups is that they engage in shared activities. To see how this can work on an epistemic model, let us start with a cooperative scenario.[257]

Tell all: Maximal communication Consider two epistemic agents in an information model $\boldsymbol{M}$ at an actual world s. They can tell each other things they know, cutting down the model. What is the best correct information they can give?

---

257 This section is largely from van Benthem (2011e). For simplicity, we will mainly use finite models with two epistemic equivalence relations.

EXAMPLE 22.1 The best agents can do by internal communication
What is the best that can be achieved in the following model?

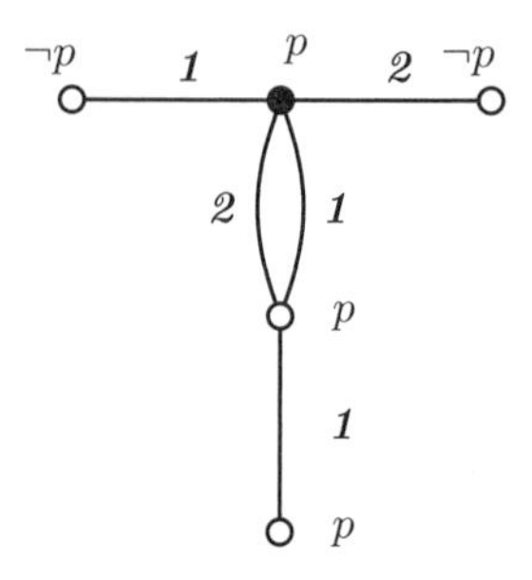

Geometrical intuition suggests that the outcome must be the following model:

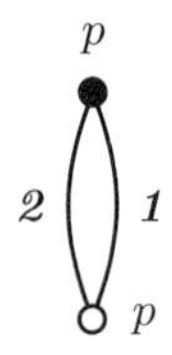

This is correct. For instance, *1* might say "I don't know if p is the case," ruling out the right-most world, and then *2* might say "I don't know either," ruling out the left-most and bottom world. They could also say these things simultaneously and get the same effect. ■

Any sequence of updates where agents say all they know must terminate in some submodel that can no longer be reduced. This limit is reached when everything each agent knows is true in every world, and hence has become common knowledge. The resulting "core model" is studied in van Benthem (2011e), bringing to light some technical subtleties.[258] For our purpose in this chapter, the following suffices. Recall the notion of "distributed knowledge" in Chapter 3, being what a group knows implicitly in the intersection of all individual epistemic accessibility relations

258 For instance, on finite bisimulation-contracted models, the core can be reached by one simultaneous announcement of what each agent knows to be true. On such models with two agents, the core is also reached by two consecutive single-agent announcements plus one more contraction. But already with three agents, successive statement of all that the agents know need not reach the core, and outcomes become order-dependent.

(Fagin et al. 1995). Ignoring complex epistemic assertions, with a free choice of announcements that agents can make about what they know, the following is a common property of many informational protocols: iterated communication turns distributed knowledge into common knowledge.[259]

Whether common knowledge is in fact reached is scenario-dependent, and information flow can be subtle. Recall the puzzle of the Muddy Children of Chapter 7. The children are highly constrained in what they are allowed to say; but even when stating only their current knowledge or ignorance of their mud status, common knowledge of the real physical facts resulted. Scenarios of maximal communication that iterate one fixed assertion φ also occurred in Chapters 8, 9, and 13 when we analyzed game solution procedures by means of repeated announcements of players' rationality. But in the scenarios to follow, assertions may vary from stage to stage.

22.2 Games on epistemic models

Consider games where players have competing interests. Real conversation is after all a mixture of cooperation and competition.

The first observation to make is that conversation scenarios are delicate.

EXAMPLE 22.2 Be the first to know

In the model drawn, at the actual world x, *1* knows that p is the case, but does not know about q. Agent *2* knows that q is the case, but does not know about p:

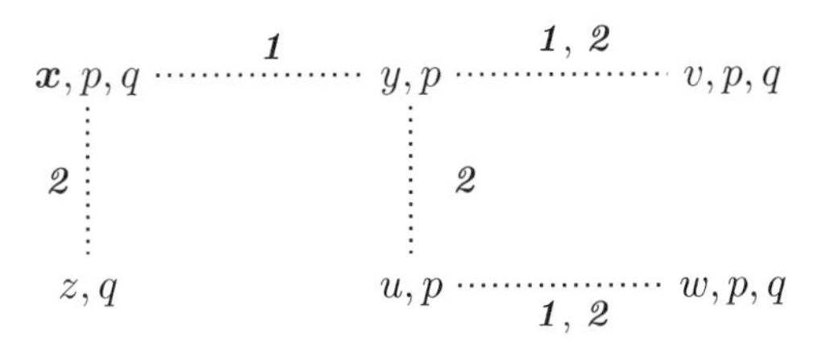

Let us assume that agents have to say something they know to be true, and moreover, that it has to be non-trivial. Agents speak in turn, and *1* starts. The goal of both players is to be the first to know the actual world, while the other does not. Whoever has won must claim victory. What happens in this game? Let us

259 There are some technical issues with a best account of distributed knowledge, and some authors use the dynamic property stated here as its definition.

describe assertions as subsets of the model, to avoid writing complex epistemic formulas. The knowledge condition means that the set mentioned is closed under the accessibility relation for the speaker.

Now *1* does not claim victory, which rules out world z. *1* might start by saying $\{x, y, v\}$, this being the strongest assertion that *1* knows. Then *2* learns what the actual world is, and wins the game. Before we look at *1*'s other options, there is already a subtlety here. While this assertion makes *2* win in the actual world, *1* does not know that it will make *2* win, since in world y, it would not. Next, *1* has two other options saying less, $\{x, y, v, z\}$ and $\{x, y, v, u, w\}$. The former makes no difference, since z was already ruled out. The latter still informs *2* that the actual world is x.

Let us also look at the case where y is the actual world. In that case, *2*'s response to $\{x, y, v, u, w\}$ has to be $\{y, v, u, w\}$, after which *1* has to say $\{y, v\}$. Then *2* loses, being unable to claim victory, and having nothing non-trivial left to say.

Other interesting scenarios arise by changing the winning conditions. ■

It may be clear, even from this crude example, that epistemic models support interesting scenarios. In the same way, any epistemic model used in this book can be mined for games.

EXAMPLE 22.3 Card games

Recall the Three Cards game of our Introduction. Does any player have a strategy for being the first to know in the following model?

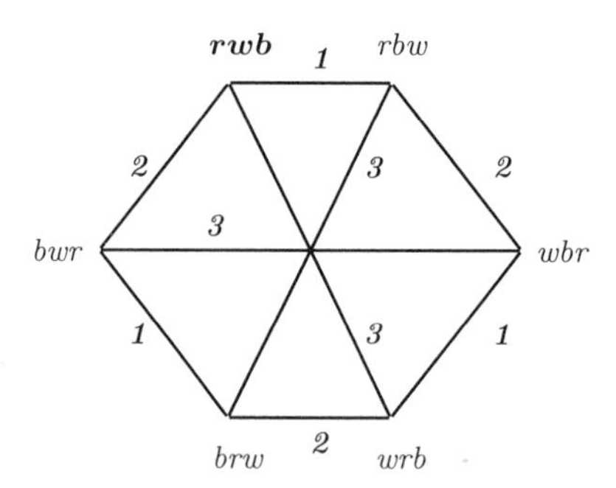

The answer is that no player has such a pure strategy. Games such as this will return below, where they suggest probabilistic equilibrium solutions. ■

These games are somewhat group-internal, in that people merely convey information about some fixed world. In reality, games may also involve new fact-finding, but for the purposes of this chapter, we will ignore such outdoor scenarios.

22.3 Announcement games

A nice setting for games over epistemic models was proposed in Ågotnes & van Ditmarsch (2011), arriving at more definite results by judiciously constraining scenarios. This time, players speak only once, and they do that simultaneously, as in the simplest version of our cooperative scenario. But now their goals may be different. Here we define a goal as an epistemic formula φ that can be true or false of a pointed model $(\boldsymbol{M}, s)$. The earlier game of being the first to know can be defined in this way, and so can the original cooperative scenario, in terms of shared goals of achieving common knowledge about some assertions.

DEFINITION 22.1 Announcement games
Given any epistemic model $\boldsymbol{M} = (S, \{\sim_i\}_{i \in I}, V)$, we define a strategic *announcement game* as follows. There are two players $\boldsymbol{1}$ and $\boldsymbol{2}$. A move is an announcement $!\varphi$ of some epistemic assertion φ. For convenience, again, we think of these as subsets of the domain S. Players all have the same assertions available, but we make our old proviso: they can only say things that they know to be true. The goals of the players are given as two formulas γ_1 and γ_2. Now we define utilities for joint assertions made in world s of the model $\boldsymbol{M}$:

$$U_i^s(!\varphi, !\psi) = 1 \text{ if } \boldsymbol{M}, s \models \langle !(\varphi \wedge \psi) \rangle \gamma_1, \text{ and } 0 \text{ otherwise.}^{260}$$

The resulting strategic game allows all uniform strategies for players, that is, the assertion made is the same between worlds that they cannot distinguish epistemically. Then the total utility for a strategy profile (σ, τ) is defined as an average

260 The condition evaluates whether the pair of the two given assertions would achieve the goal of player i. In Ågotnes & van Ditmarsch (2011), it is not the statements themselves that are used in this clause, but players' true announcements that they know, or do not know them to be true. However, the present simpler format can still define this by changing the formulas, and placing any relevant requirements on utterance in a protocol of admissible assertions in the game.

utility over specific worlds, assuming a uniform probability distribution:

$$U_i(\sigma, \tau) = \Sigma_{s \in S} u_i^s(\sigma(s), \tau(s))/|S|$$

This is a one-shot game, longer sequences are not included. ■

The results of the cited paper clarify our earlier examples, in terms of an appropriate game-theoretic notion of equilibrium (Osborne & Rubinstein 1994).

FACT 22.1 Announcement games need not have Nash equilibria in pure uniform strategies. However, their equilibria in mixed strategies are the same as those of an associated Bayesian game in the standard sense of game theory.

Perhaps most interesting from a logical point of view is further fine structure, in the form of a connection between winning and the logical form of goals. Suppose that the goal statements are both "universal," constructed from literals applying only conjunction, disjunction, knowledge operators, and dynamic modalities with universal announcements. Such special formulas are preserved under submodels of any current model where they hold (see Chapter 7).

FACT 22.2 Announcement games with universally definable goals possess Nash equilibria in pure strategies.

Indeed, both players making the strongest assertion about what they know weakly dominates any other strategy that they might play.

The use of goals connects with other parts of this book. As noted in Chapter 9, evaluating outcomes for games in terms of goals rather than preferences is attractive, and it fits with priority-based logics of preference (cf. Liu 2011).

More game theory As with the imperfect information games of Chapter 21, many further topics make sense now, including standard game solution methods such as SD^∞ or WD^∞ (cf. Chapter 13). What does their output tell us about the initial epistemic model? And more generally, is there a natural logical interpretation of the probabilistic equilibrium values obtained in general announcement games?

22.4 Question games

Real conversation does not involve statements only: its dynamics is driven by questions and answers. This, too, leads to epistemic games that can be dealt with in the above style (cf. Ågotnes et al. 2011). We only present a sketch.

DEFINITION 22.2 Question games
A *question game* uses the same structure as the earlier announcement games. We only state what needs to change. This time, players ask one question "φ?" each, simultaneously. Answers by the other player are compulsory: "Yes" (in case you know that φ is true), "No" (if you know φ to be false), or else "Don't know," and this is automatically followed by the next round. This updates the model as before with the corresponding statements. Utilities are computed over the answers, where the relevant answer can be read off from the local world s. ■

The main results in the paper are formally similar to those for announcement games. Still, the dynamics is interestingly different, since players can now force others to respond. The cited paper shows that assertion games and question games may have very different outcomes on the same epistemic model.[261]

Dynamics of questions In the above games, questions only serve as a switch: one player determines what the other has to say, modulo uncertainty about the answer. Therefore, the games do not represent explicitly what a question does beyond allowing us to interpret the answer. Real question-answer scenarios tend to be longer, leading to extensive versions of the above games that have not been investigated yet. In such extensive games, questions themselves might be represented as moves, using the dynamics of questions in van Benthem & Minică (2012) and Minică (2011). An outline of the relevant acts of issue management has been given in Chapter 7, and knowledge games might then be analyzed in a combined dynamic logic of knowledge and issues.

22.5 Epistemic games, imperfect information, and protocols

Let us now step back and look at the above in the light of our earlier chapters. Eventually, we want to consider extensive games, since that is where the flavor of conversation or inquiry unfolds best.

It will be clear that all scenarios considered so far are games of imperfect information in the sense of Chapter 3. In fact, they can be viewed in two ways. The

261 This is only a surface comparison, and underneath there might be translations between the two sorts of games.

assertions made are public moves that can be identified with sets of worlds.[262] There is then a game tree of possible moves where players know at each stage which assertions have been made. The imperfect information resides, so to speak, in the epistemic models associated with nodes of this game tree, as projections in the sense of Chapters 6 and 25.

These imperfect information games fit into the general framework proposed in Parts I and II of this book, where admissible assertions starting from some initial epistemic model can be constrained by means of temporal "protocols."

Epistemic protocol models Knowledge games take place in epistemic forests in the sense of Chapters 5 and 6, where we start with an initial epistemic model, and histories are produced by successive announcements. Constraints can be imposed just as in the protocol models of Chapters 7 and 8, now specialized to public announcement logic, PAL (see Hoshi 2009 for the general logic of such scenarios). Each world in the initial model comes with a protocol of admissible assertions, and since what can be said may differ across worlds (think of only saying something that one knows), the protocol is "local." At least two conditions are at work here. Statements come with a precondition, say, that the speaker knows them to be true, which puts closure conditions on the sets of worlds. But also, assertions must be made in the submodel arising from the last assertion.[263] There can also be procedural constraints, such as not repeating the same assertion more than once, something that might refer to the actual syntax of the sequences in the protocol.

This is a big jump in generality. An epistemic forest is not yet a game, but we can create games over the protocol by adding turns and scheduling for players, sequential or parallel. Still, as imperfect information games, all this is simple. There is public observation of moves, so uncertainties are only those that are propagated from the epistemic accessibility links in the initial model. This suggests a restatement of the representation theorems discussed in Chapter 9.

FACT 22.3 Every imperfect information game with public observation can be represented as an extensive-form knowledge game.

---

262 How these sets of worlds are described by players in language may vary with the scenario, and the current model.

263 This requires a slight technical generalization of our earlier PAL protocols, or at least, some care in defining the admissible sequences of assertions in the initial model.

Proof Using perfect recall and no miracles for the players, the idea is to chase all uncertainties downward to the root, where the remaining structure is the initial model. Moves can then be replaced by announcing their preconditions. ■

Thus, the knowledge games of this chapter may look special, but they model a much broader class of epistemic scenarios.

Game-theoretic aspects once more Imperfect information games work with uniform strategies, while their Nash equilibria can be complex. We will not pursue these issues here, but the results in the well-chosen special scenario of Section 22.3 may be harder to obtain in our general setting.

22.6 Further logical lines

We conclude with a few further directions leading from where we are now.

Languages, expressive power, and axiomatization Knowledge games support languages with forcing modalities with uniform powers as in Chapter 11, and it would be of interest to determine what these conditions can say about the initial epistemic model. Depending on the winning conditions, special epistemic features may arise that were discussed in Chapters 3 and 4, such as knowing that you have won if you have. One would also like to have complete epistemic forcing logics for knowledge games (Minică 2011 has a first exploration), perhaps related to the protocol version of public announcement logic in Chapter 7.

Structural invariance through games Knowledge games are a nice way of exploring the structure of some given epistemic model, beyond the immediate truth of epistemic assertions. Intuitively, we then get a new dynamic notion of model equivalence as the sameness of the associated exploration games. Does this make sense? At least this invariance seems to go beyond the basic epistemic language, since distributed knowledge has to do with intersections. Sadzik (2009) studies an analogous issue in game theory about games having the same correlated equilibria whose explanation requires internal communication.

Connections with other knowledge games In the setting of this book, an interesting connection would be links between knowledge games as defined here and the information-oriented IF games of Chapter 21. Here current work on connections with dynamic-epistemic logic may be relevant (Galliani 2009a, 2012b). Another obvious comparison is with current logics of questions and inquiry in philosophy and linguistics. Hamami (2010) links dynamic-epistemic logics of issue management to Hintikka's game paradigm for inquiry, and the special issue Hamami & Roelofsen (2013) presents an overview of the state of the art in logics of questions.

22.7 Conclusion

Knowledge games seem to be an interesting playground showing standard epistemic models in a new light. They may be viewed in different ways. Are they models for real conversation, or a new sort of logic game for epistemic logic? The answer is, "both." Thus, again, we see that natural games may have aspects of the two major strands entangled in this book.

Of course, as we said, these games are internal to a given model. Hence, they have all evidence on the table, a bit like in a legal procedure (cf. Chapter 17). Taking the design further to include learning new facts in scenarios of inquiry (cf. Hamami 2010) might be the next step down the current road.

22.8 Literature

This chapter is largely based on Ågotnes & van Ditmarsch (2011). A few ideas come from Ågotnes et al. (2011), while van Benthem & Minică (2012) and Minică (2011) are also relevant. Further links between information games and game theory include van Rooij (2003a) and van Rooij (2003b) on questions and decision problems in natural language, Clark (2011) on signaling games that establish meanings, and Feinberg (2007) on strategic conversation. Another congenial approach is recent uses of knowledge games of inquiry for defining the very notion of knowledge itself (Baltag et al. 2012) in a tradition going back to Hintikka (1973).

23
Sabotage Games and Computation

Logic games gamified basic logical tasks of semantic evaluation or reasoning. But gamification applies just as well to other phenomena, and in fact, games are fast becoming a general model for interactive computation. In this chapter, we look at one particular game that arose in the study of the computational core task of search. We will find a logical background to the game, as well as further repercussions. Thus, logic games and game logic meet in the compass of one concrete scenario.

23.1 From algorithms to games: The sabotage game

Standard algorithmic tasks are single-agent jobs. But sometimes, it makes sense to turn a basic computational task into a social game, prying things apart with roles for different agents. To see how this works, recall the key search problem of graph reachability under adverse circumstances from the Introduction.

EXAMPLE 23.1 Disrupting a travel network
The following picture gives a travel network between two European centers of logic and computation, Amsterdam and Saarbrücken:

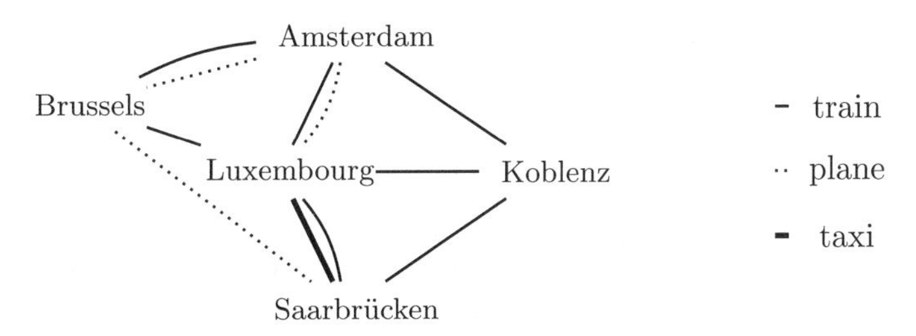

It is easy to plan trips either way. But now transportation breaks down, and a demon starts canceling connections, anywhere in the network. At every stage of our trip, let the demon first take out one connection, while the traveler then follows a remaining link. This turns a one-agent planning problem into a two-player sabotage game, and the question is who can win where. Simple Zermelo-style reasoning shows that, from Saarbrücken, a German traveler still has a winning strategy, while in Amsterdam, the demon has the winning strategy against the Dutch traveler. In particular, the symmetry of the original search problem is broken. ■

This game is graph reachability "sabotaged," and it gives a fair impression of the reality some years ago when the Dutch Railways were constantly being disrupted, and coming home from work each evening turned into a game of high complexity. More generally, any algorithmic task over graphs turns into a two-player game with one player, "runner," trying to do the original job, and another player, "blocker," taking out edges at each stage. Of course, different schedulings and winning conditions are still possible in such a scenario.

Example 23.2 Sabotaging traveling salesman
An undirected graph is given, and runner must complete a circuit. This time, let runner start each round, after which blocker takes out a link. Players must move as long as they can: the game stops the first time a player cannot move. Runner wins if the end situation contains a completed circuit; otherwise, blocker wins.

Which player has a winning strategy in the following game, which starts with runner at the black dot?

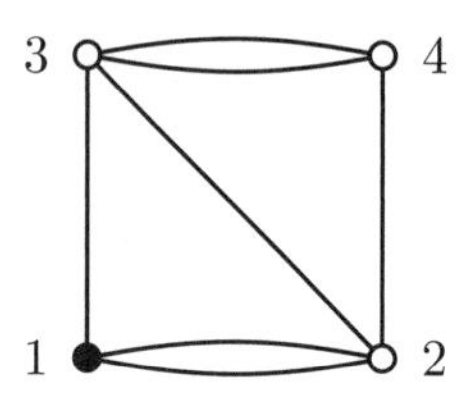

It is blocker who has the winning strategy, by first cutting one upper link. ■

Fact 23.1 Sabotaged graph games are determined.

Proof This follows by Zermelo's Theorem (cf. Chapter 1). It is easy to draw all possible positions in the extensive game tree for a sabotage game by moving the black dot around in an ever-decreasing graph. The game must stop at the latest when all links are removed, and it always ends in winning or losing. ■

23.2 Logical and computational features

The game may be more interesting than the original single-agent task, but how does the computational complexity change when we solve its game version? Graph reachability is in *Ptime*, but what happens now? There are two opposing forces here: the sabotage game creates a multiplication of graphs as positions, but these graphs also get simpler and simpler.

An upper bound on the complexity comes from a logical perspective. Solving the sabotage game amounts to model checking with an appropriate first-order formula. For convenience, think of the given graph $\boldsymbol{G}$ as a set of edges.

FACT 23.2 Runner has a winning strategy in the graph $\boldsymbol{G}$ iff the following first-order formula quantifying over edges is true in $\boldsymbol{G}$: $\exists x_1 \forall y_1 \exists x_2 : x_2 \neq y_1 \ \forall y_2 \exists x_3 : x_3 \neq y_1, x_3 \neq y_2 \cdots \alpha(x_1, \ldots, x_k)$, where α is the standard condition of reachability adapted to the remaining edges.[264]

The length of the defining first-order formula is linear in the size of the graph, but we have quantifier interchanges that generate complexity (the original reachability problem had only existential quantifiers). Thus, an upper bound for the complexity is that of uniform model checking of first-order formulas over finite models, which takes polynomial space in the size of the model plus that of the formula. Hence, given the dependence of the formula on the graph, solving the sabotage game takes at most *Pspace* in the size of the graph.

But what about a lower bound? One can give a heuristic argument for a *Pspace* lower bound in terms of connections with quantified Boolean formulas (QBF, cf. Papadimitriou 1994). For sabotaged graph reachability, the real solution is in Rohde (2005), which shows that this complexity jumps up from *Ptime*.

THEOREM 23.1 The solution complexity of the sabotage game is *Pspace*-complete.

Proof This uses a nice combinatorial argument with a technique of "encoding gadgets" for representing QBF problems as sabotage tasks. ■

What this means is that sabotage games take a polynomial amount of memory space, like Go or Chess. Many variants have been analyzed by now, using similar methods, of which Kurzen (2011) is an up-to-date treatment.

---

264 We can also write this as a first-order property of points in the graph.

23.3 Other interpretations: Learning

Link cutting games such as the above have also picked up interesting further interpretations in terms of learning. Our Introduction mentioned a variant, where a teacher tries to prevent a student from reaching a position of escape, while versions where the teacher tries to force the student to reach a certain point make sense as well. This learning interpretation of sabotage has turned out to be more serious than it may appear at first sight, and it was developed further in Gierasimczuk et al. (2009), while Gierasimczuk (2010) connects it to general perspectives from formal learning theory.[265] In such practical settings, the *Pspace* complexity shows in the absence of a simple invariant that could guarantee winning.[266]

23.4 Computational and game-theoretic aspects of gamification

Computational complexity Sabotage is just one way of turning algorithmic tasks into interactive games. The upward jump in complexity from *Ptime* to *Pspace* that we noted above need not hold in all gamifications. For instance, here is another scenario for graph reachability, where someone tries to stop a player en route. This catching game bears a vague resemblance to the parlor game of Scotland Yard.

DEFINITION 23.1 Catch me if you can
Starting from an initial position $(\boldsymbol{G}, x, y)$ with player *1* at x and player *2* at y, *1* moves first, then *2*, and so on. *1* wins by reaching the goal in a fixed finite number of moves without meeting *2*. Otherwise, *2* wins. ■

Thus, player *2* wins by catching player *1* before the goal region, and also if *1* gets stuck, or if the game continues indefinitely. Other ways of casting these winning conditions allow draws.

---

265 The game has also been used in psychological experiments and education (see the projects listed at `http://www.talentenkracht.nl/`). Board versions with travel maps have been developed to play it with seven-year-olds as a means of studying the emergence of strategic thinking in children.

266 The original paper, van Benthem (2005a), also discusses games for social skills such as avoiding obnoxious people, or seeking out celebrities at receptions. Such scenarios may involve a third player who influences the chances of meeting other people.

The difference with the sabotage scenario is that the graph remains fixed during the game. And indeed, the computational complexity stays lower.

FACT 23.3 The catching game is solvable in *Ptime*.

Proof The game can be recast over a new graph with positions $(\boldsymbol{G}, x, y)$ with players' moves as before while *2* gets free moves when *1* is caught or gets stuck. Player *2* then wins if *2* can keep moving forever. As noted in Chapter 18, graph games such as this can be solved in *Ptime* in the size of the graph. ■

For details of the preceding argument, and also for other games analyzed at the interface of logic and complexity theory, see Sevenster (2006). Still, it seems fair to say that we lack a general understanding of what determines the complexity of gamified algorithms compared with that of their original tasks.[267]

Sabotaging games themselves One can also disturb any game itself. Consider a tree, with one player for convenience, and at the start of each round, let a blocker prune one move from the tree. This local disturbance is not dramatic.

FACT 23.4 Sabotaged games are still solvable by Backward Induction.

Proof One can still compute winning positions inductively in the original game, using Zermelo's algorithm. At any node, runner has a winning strategy against blocker iff runner has winning strategies in the subtrees for at least two of the available moves. From right to left, blocker can only affect at most one of those subtrees in the first round, leaving runner free to go to the other. From left to right, if there were at most one such subtree, then blocker cuts the move to that, forcing runner to choose a move to a losing position. ■

Our sabotaged graph tasks were not like this. By removing only one edge from the graph, the demon cut travel options for traveler at many stages in the associated game. This prunes a whole set of moves simultaneously from the corresponding game tree. The effects of such global changes are not easy to compute inductively (Jones 1978), as reflected in the greater complexity of our original sabotage game.

---

267 There are also other unresolved issues. Ron van der Meijden (personal communication) has pointed out that our analysis gamifies a given task, but not the original algorithm solving that task. What transformation takes a standard search algorithm and turns it into a Zermelo strategy for a sabotage game?

Powers and coalitions Gamified algorithms suggest various links with standard game theory. Sabotaging makes n-person games into $(n+1)$-person games, linking up with coalitional powers of groups of players, including blocker.

EXAMPLE 23.3 Sabotage games with coalitions
Here is an extremely simple example showing what can happen:

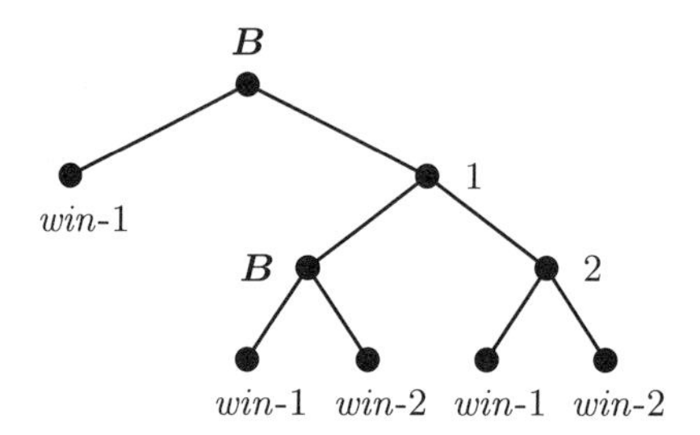

Working upward, inductively, one computes winning coalitions as in Chapter 11. One winning coalition of this mixed sort at the root is the set $\{\boldsymbol{B}, 1\}$. ■

Epistemic versions and imperfect information Epistemic versions of sabotage with observational limitations lead to imperfect information games as in Chapter 3. Perhaps we cannot observe exactly what changes are made by the demon, as happens in war and many parlor games. Nash equilibria for such modified games may involve mixed strategies, and our study of gamification would have to include probabilistic considerations. This would also be natural in real network problems, where demon or blocker may be a random opponent, not necessarily doing the very best against us. If we turn algorithms into games of imperfect information, solution complexity may jump up again (once more, see the classic Jones 1978).[268]

23.5 Logical aspects of sabotage

One can also view sabotage games as logic games such as those of Chapter 14, although of an unusual sort: the model itself changes in the process of evaluation.[269]

268 Sevenster (2006) presents an extensive study of epistemized gamified algorithms, linking up with the IF logic of Chapter 21.

269 Thus, sabotage logic seems related to the reactive Kripke models of Gabbay (2008).

A standard reachability problem amounts to evaluating modal formulas $\Diamond \cdots \Diamond p$ where the proposition p holds in the goal states. Uniform model checking for modal formulas on finite models has complexity *Ptime* in the size of the model and the formula. This is lower than the *Pspace*-complexity for first-order formulas. However, sabotaged modal formulas involve changing the model as we evaluate.

DEFINITION 23.2 Sabotage modality for link deletion
A cross-model *sabotage modality* refers to submodels where arrows were removed:

$$\boldsymbol{M}, s \models \langle - \rangle \varphi \quad \text{iff} \quad \text{there is a link } (s,t) \text{ in } R \text{ with } \boldsymbol{M}[R := R - \{(s,t)\}], s \models \varphi$$

This is a new type of modality, but semantic evaluation is entirely classical. ■

Sabotage modal logic The language of sabotage modal logic has the old standard internal modality $\Diamond$ and the new external modality $\langle - \rangle$, and the two combine. For instance, the well-known fact that syntactically universal formulas are preserved under submodels with fewer links shows in validities of this new modal logic.

FACT 23.5 The formula $\Box p \rightarrow [-]\Box p$ is valid for factual atomic statements p.

Sabotage modal logic lacks some standard modal features. For instance, the preceding validity does not hold for arbitrary modal formulas φ instead of p, since their truth value may change when going to the new model. A concrete counterexample is easily found, but we leave this task to the reader.

More spectacularly, the new language is not invariant for bisimulation, unlike the basic modal logics used in this book.

FACT 23.6 The sabotage modal formula $\langle - \rangle \Box \bot$ is not invariant for bisimulation.

Proof The formula $\langle - \rangle \Box \bot$ holds in an irreflexive 2-cycle, but it fails in the bisimilar model of two points with a universal accessibility relation. ■

Nevertheless, we do retain the following standard feature of modal languages.

FACT 23.7 All formulas of sabotage modal logic are translatable into the first-order language of graphs.

Proof We just give an illustration: $\langle - \rangle \Box \bot$ translates into the formula

$$\exists y \exists z (Ryz \land \neg \exists u (Rxu \land \neg (x = y \land z = u))),$$

and the general translation method should be clear from this specific case. ■

The blow up in length of this first-order translation is only polynomial, so model checking is at most *Pspace*. Rohde (2005) adds a lower bound, showing that sabotage modal logic sides with first-order logic in general.

THEOREM 23.2 Model-checking for sabotage modal logic is *Pspace*-complete.

The translation leaves the complexity of decidability open, but Rohde also proves that the logical effects of model change during evaluation can be drastic.

THEOREM 23.3 Sabotage modal logic is undecidable.

Sabotage also suggests modifying our logic games in Part IV. For instance, as mentioned in Chapters 14 and 18, what happens to first-order evaluation games as we change models by adding or removing objects, or change facts during atomic tests? Motivations for such variations are easy to find in the scenarios of game change considered in Chapter 9.

General logics of model change Something general is going on here. There is a growing family of logics whose modalities involve evaluating at some model different from the initial one. Examples are the public announcement modalities $[!\varphi]\psi$ of the system PAL in Chapter 7, that shifted evaluation to a definable submodel, while more general dynamic-epistemic logics provided further instances with product models. Another important illustration is the modal "bisimulation quantifier" $\langle p\rangle\varphi$ of Hollenberg (1998) that requires truth of φ in some bisimulation variant for the current model of evaluation modulo truth values for p.[270] Our final example is moving to arbitrary submodels, not throwing away arrows one by one: as in the logic APAL of Balbiani et al. (2009).

The general theory of model-changing logics remains to be understood. This is a major foundational problem raised by the logical dynamics program in this book.

23.6 Conclusion

The sabotage game is natural, and it has led to interesting follow-up in practice and theory. What it shows is how computational tasks can easily become interactive,

270 Bisimulation quantifiers provide an alternative syntax for the modal μ-calculus.

and as such, it offers one general method for gamifying single-agent scenarios.[271] In the context of this book, we end with a question. Is the sabotage scenario a logic game, or a game logic? The point is that the distinction does not matter: sabotage has features of both. This shows how the two major strands in this book can be naturally entangled.

23.7 Literature

This chapter is based on the original sabotage scenarios in van Benthem (2005a).

We have mentioned important follow-up literature in the dissertations of Rohde (2005), Gierasimczuk (2010), and Kurzen (2011). Recent computational variations on sabotage games can be found in Klein et al. (2009).

271 More general notions of gamification and epistemization as moves toward a social paradigm of computation are discussed in van Benthem (2007b).

24
Logic Games Can Represent Game Logics

This chapter presents one technical case study of what logic games can do for general game logics. Logic games are special-purpose activities, and as such, just a tiny corner of the world of all games. But sometimes, these special games provide a kind of normal form for arbitrary games. This chapter establishes such a link by proving completeness of the evaluation games of Chapter 14 for the game algebra of Chapter 19. This may be seen as a paradigm for further representation theorems linking logic games and game logics. We present the material in telegram style, mainly weaving together threads in this book that have been explained elsewhere.

24.1 Forcing relations and game operations

Forcing relations and players' powers were defined and studied in Chapters 11 and 19 over abstract game models (game boards)

$$\boldsymbol{M} = (S, \{\rho_a \mid a \in A\}, V)$$

where the ρ_a are atomic forcing relations on the state set S, given at the outset. Basic game algebra was about the game constructions of disjunction (choice), dual, and composition. These were defined in the following manner.

Definition 24.1 Forcing relations for complex games
We consider just two player games, and exploit the standard upward monotonicity of powers to simplify clauses:

$$
\begin{array}{lll}
\rho^{\boldsymbol{E}}_{G \cup G'} x, Y & \text{iff} & \rho^{\boldsymbol{E}}_{G} x, Y \text{ or } \rho^{\boldsymbol{E}}_{G'} x, Y \\
\rho^{\boldsymbol{A}}_{G \cup G'} x, Y & \text{iff} & \rho^{\boldsymbol{A}}_{G} x, Y \text{ and } \rho^{\boldsymbol{A}}_{G'} x, Y \\
\rho^{\boldsymbol{E}}_{G^d} x, Y & \text{iff} & \rho^{\boldsymbol{A}}_{G} x, Y \\
\rho^{\boldsymbol{A}}_{G^d} x, Y & \text{iff} & \rho^{\boldsymbol{E}}_{G} x, Y \\
\rho^{\boldsymbol{E}}_{G\,;\,G'} x, Y & \text{iff} & \exists Z : \rho^{\boldsymbol{E}}_{G} x, Z \;\&\; \forall z \in Z : \rho^{\boldsymbol{E}}_{G'} z, Y \\
\rho^{\boldsymbol{A}}_{G\,;\,G'} x, Y & \text{iff} & \exists Z : \rho^{\boldsymbol{A}}_{G} x, Z \;\&\; \forall z \in Z : \rho^{\boldsymbol{A}}_{G'} z, Y
\end{array}
$$

These clauses generalize the operations of relational algebra to relations of type object to set-of-objects, while adding one new notion (role switch) with no relation-algebraic counterpart, as it reflects the extra player structure. ■

Henceforth, in this chapter, we assume that all games G are determined in the strong sense of Chapters 1 and 19: for each set of outcomes Y, either player $\boldsymbol{E}$ can force G to end in Y, or $\boldsymbol{A}$ can force the game to end in $S - Y$. Then, we can just define forcing relations for player $\boldsymbol{E}$, replacing the above clause for dual by

$$\rho^{\boldsymbol{E}}_{G^d} x, Y \quad \text{iff} \quad \rho^{\boldsymbol{A}}_{G} x, (S - Y) \qquad (\$)$$

In what follows, this is what we mean by the forcing relation of a game.

24.2 Forcing bisimulations

Another basic notion from Chapter 11 was that of a forcing bisimulation between two game boards $\boldsymbol{M}$ and $\boldsymbol{N}$. We recall some facts from Chapter 19. These bisimulations preserved all statements in the language of dynamic game logic, DGL, whose crucial feature were the strategic forcing modalities $\{G, i\}\varphi$ of Chapter 11 with games made explicit.

FACT 24.1 All formulas of DGL are invariant for forcing bisimulations.

The inductive proof with complex terms G involved a new notion of safety.

DEFINITION 24.2 Safety for bisimulation
A game operation $\#$ is *safe for bisimulation* if, whenever E is a forcing bisimulation for forcing relations ρ and ρ', the same E is a forcing bisimulation for $\#(\rho, \rho')$. ■

24.3 Basic game algebra

Basic game algebra calls two game terms equivalent when their interpretations always have the same powers for both players. Thus,

> $R = S$ is valid if the forcing relations of R and S for each player are the same on any game board $\boldsymbol{M}$.

Some typical valid laws of the resulting game algebra were:

for $\cap, \cup,^d$	all Boolean laws, except for those involving constants $\boldsymbol{t}, \boldsymbol{f}$
for ;	associativity, left-distribution with respect to disjunction, dualization of composition: $(G\,;\,H)^d = G^d\,;\,H^d$

Typically, though, right-distribution of composition over choice was not valid:

$$G\,;\,(G_1 \cup G_2) = (G\,;\,G_1) \cup (G\,;\,G_2)$$

These matters have been explained extensively in Chapter 19, which showed in particular that basic game algebra was axiomatizable and decidable.

24.4 First-order evaluation games and general game laws

Evaluation games for predicate logic were compounds of atomic games of fact testing (for atoms Px) and object picking (for quantifiers $\exists x$). These were set up in such a way that a success lemma holds.

FACT 24.2 A first-order formula φ is true in a model $\boldsymbol{M}$ iff verifier has a winning strategy in $\boldsymbol{game}(\varphi, \boldsymbol{M})$.

Concrete forcing relations for games $\boldsymbol{game}(\varphi, \boldsymbol{M})$ follow the above inductive clauses for players' strategic powers, now starting from the atomic stipulations

$$\begin{array}{lll} \rho^V_{Px}s, Y & \text{iff} & P^{\boldsymbol{M}}(s(x)) \text{ and } s \in Y \\ \rho^V_{\exists x}s, Y & \text{iff} & \text{for some } d \text{ in } \boldsymbol{M}, s[x := d] \in Y \end{array}$$[272]

272 It follows that $\rho^V_{\forall x}s, Y$ iff all x-variants of the assignment s belong to the set Y.

Now, as discussed in Chapter 14, the usual predicate-logical validities will fall into levels. The most general are formulas satisfying the following requirement.

DEFINITION 24.3 Schematic validity
A predicate-logical formula is *schematically valid* if it remains valid under any replacement of atomic subgames by other games. ■

Examples are all predicate-logical instances of the general laws of game algebra, such as Boolean distribution or associativity of composition. A counterexample was the distribution of the existential quantifier over disjunction

$$\exists x(Ax \vee Bx) \leftrightarrow \exists xAx \vee \exists xBx$$

that exemplified the invalid right-distribution law for games. It only holds in first-order logic because of a special splitting property for player $\boldsymbol{E}$ in existential quantifier games. Its non-validity could be demonstrated concretely, as we saw in Chapter 19, by plugging in a universal quantifier (game) $\forall x$ in the $\exists$ position. This is significant, as we shall see.

24.5 The main result

The method of first-order counterexamples for general game laws is quite powerful. The main point of this chapter is a representation result: first-order evaluation games are complete for game algebra. The precise sense in which this is true will become clear from the following proof leading up to a statement of the theorem.

We sketch a sequence of stages, turning abstract counterexamples for game identities $R = S$ into concrete ones involving just first-order evaluation games.

Step 1. Take an abstract counterexample to $R = S$ on some game board $\boldsymbol{M}$. Take some set of outcomes in which the forcing relations of $R^{\boldsymbol{M}}, S^{\boldsymbol{M}}$ for player $\boldsymbol{E}$ differ, and call it p, where p is some proposition letter.

Step 2. Next, define an auxiliary two-sorted first-order model $\boldsymbol{N}$ as follows:

(a) State-objects are as in $\boldsymbol{M}$, sets of states are added as new objects.

(b) First-order relations R_a are defined between state objects and state-set-objects matching the atomic forcing relations ρ_a in $\boldsymbol{M}$, while we add membership E:

$$R_a sA \text{ iff } \rho_a sA, \qquad EsA \text{ iff } s \in A$$

The predicate-logical language for the model is clear from this description. If desired, add a unary predicate letter P to represent the above proposition letter p.

Step 3. The next model $\boldsymbol{K}$ is a concrete game board:

> States are assignments of appropriate objects in $\boldsymbol{N}$ to two variables x and y (intuitively, states and sets).

Now we define abstract forcing relations over $\boldsymbol{K}$, using an auxiliary notion:

> The "value map" f sends any assignment s to the object $s(x)$.

Next, define relations between states and sets of states on $\boldsymbol{K}$ using f-images:

> $R_a s\Sigma$ if there exists some set A such that $\rho_a s(x) A$ and all assignments of the form $\{(x, d), (y, A)\}$ with $d \in A$ belong to Σ.

Finally, we define the valuation for atomic predicates:

$$s \models p \quad \text{iff} \quad f(s) \models p$$

The important thing to note is the following feature.

FACT 24.3 The map f induces a forcing bisimulation between the abstract model $\boldsymbol{K}$ and the initial state model $\boldsymbol{M}$.

The main check for the back-and-forth clauses is straightforward by the definition of the relations $R_a s\Sigma$. By the invariance of dynamic game logic for bisimulations, then, there is a counterexample to the original identity $R = S$ on $\boldsymbol{K}$.

Now we show that $\boldsymbol{K}$ is the scene of matching first-order evaluation games.

Step 4. Here is the key idea. The basic forcing relations in $\boldsymbol{K}$ are definable as the standard forcing relations of suitable evaluation games over the above model $\boldsymbol{N}$:

$$\exists y.R_a xy\,;\,\forall x.Exy$$

FACT 24.4 The computed forcing relation for this evaluation game on $\boldsymbol{N}$ equals the above relation $R_a s\Sigma$ on $\boldsymbol{K}$.

This requires an elementary verification. Unpacking the definition of forcing for the above formula, it matches the definition of the relations $R_a s\Sigma$. ■

This concludes our construction. We now state what we have achieved.

THEOREM 24.1 First-order evaluation games are complete for basic game algebra.

This justifies our choice of evaluation games as a key paradigm in this book.[273] At least for sequential game operations, they say it all.

24.6 Discussion

Normalizing to standard formulas The formulas produced by the construction are not yet standard first-order formulas. Two more things are needed. First, formulas for atomic games are quantifier prefixes lacking a matrix. Here, one can just add some assertion "True" that succeeds for any player.

Next, with complex game expressions, the operations $\neg$ and $\vee$ have direct counterparts in first-order formulas; but composition ; does not. Here is how to solve this. Have some proposition letter p mark the difference between the forcing relations for the games R and S. This induces an atomic predicate Px that can be added at the end of our prefixes as a matrix assertion. These prefixes, for complex games, are built by our atomic two-quantifier routine, plus dual, choice, and composition. To clinch things, the following syntactic rules turn any combination of the form "complex quantifier prefix P plus matrix formula φ" into a standard predicate-logical formula $P|\varphi$ having the same associated game.

DEFINITION 24.4 Reduction rules for quantifier prefixes
Here are the decomposition rules for complex quantifier prefixes:

$$\begin{array}{lcl} \text{atomic quantifier prefix } P|\varphi & = & P\varphi \\ (P_1 \vee P_2)|\varphi & = & (P_1|\varphi) \vee (P_2|\varphi) \\ (P_1 \,;\, P_2)|\varphi & = & P_1|(P_2|\varphi) \end{array}$$

One can check how this construction works concretely by looking at non-valid game identities such as $a\,;\,b = b\,;\,a$ or $a\,;\,(b \cup c) = a\,;\,b \,\cup\, a\,;\,c$.[274] ■

Bounded versus unbounded quantifiers The formulas substituted for atomic games in our proof use bounded quantifiers. Can we also make do with just the absolute quantifiers of standard first-order logic? For a start, the superficially attractive

---

273 In fact, first-order logic with bounded quantifiers suffices for the result.

274 Often, there are simpler refutations than those of this general recipe.

substitution formula $\exists y \,;\, R_g xy \,;\, \forall x \,;\, Exy$ works out wrong.[275] And other tricks fare no better. In fact, the usual definability of bounded quantifiers in terms of absolute ones may break down in the game version of first-order logic.

To end on a more positive note, keeping the bounded quantifiers just as they are, the proof of our main theorem also shows how modal evaluation games suffice for capturing game algebra.

24.7 Extensions and desiderata

Our completeness result is stronger than stated. The above construction easily extends to the complete dynamic language of games defined in Chapter 19.

THEOREM 24.2 Every statement of dynamic game logic that is falsifiable on abstract games can also be falsified in first-order evaluation games.

Proof This is obvious from the proof for our main theorem, using the fact that the forcing bisimulation that we defined leaves all such statements invariant. ■

Further issues abound here. How can we extend our results to create further links between game logics and logic games? For instance, all games in this chapter are determined. Do our results extend to the algebra and dynamic logic of abstract nondetermined games, introduced in Chapters 11 and 19? In particular, are the IF games of Chapter 21 a complete class of counterexamples? An answer might involve adding new game operations to the above vocabulary, such as the parallel products of Chapters 19 and 20.[276]

But even with sequential game operations, further questions arise. Bisimulation invariance extended to modal languages with fixed point operators. Can we show completeness of the evaluation games for first-order fixed point logic LFP(FO) or the modal μ-calculus (cf. Chapters 14 and 18) with respect to stronger fixed point versions of the above game algebra, or of dynamic game logic?

---

275 Verifier's forcing relation for $\forall x \,;\, Exy$ yields either the whole state set (if the formula is true), or there is no output set at all. But we want all assignments where Rxy holds.

276 Galliani (2012a) has some relevant results, although not yet a representation theorem for the algebra, let alone the epistemic DGL logic, of imperfect games. His results use the model transformations in the above proof to recast the dependence logic of Väänänen (2007) in terms of a dynamic logic of powers based on transitions between teams.

24.8 Literature

This chapter is the core content of van Benthem (2003). Follow-up questions about versions with parallel operations are found in van Benthem et al. (2008).

25
Merging Logic Games with Game Logics

The view of this book is that logic games and game logics form a natural connection. Many chapters have shown contacts between these two stances, and putting them together invariably suggested interesting new issues. In this final chapter, we will briefly discuss a few further threads, drawing on earlier themes. This will not result in a grand theory, but in a better view of the entanglement of our main strands.

25.1 Logical systems with game terms

One place where ideas from both realms met naturally was Part V of this book on logics for game operations. There we presented systems of dynamic game logic and linear game logic that combined features of interpreting logic as games and of using logic to understand the compositional structure of arbitrary games.

Something needs to be clarified here. First, the logic game aspect was not primarily about all the games found in Part IV, but intuitions seemed to revolve around *evaluation games*. This has to do with the game semantics idea that the meaning of a logical language consists of assigning games to its formulas. One does not normally think of that semantic meaning as also comprising the other kinds of game that we considered, say, those for proof in Chapter 17, although a proponent of "proof-theoretic semantics" might want to do just that. Even the model comparison games of Chapter 15 analyze expressive power in a very natural way. In fact, developments in computational logic discussed at the end of Chapter 18 (cf. Venema 2007) suggested that they, too, hold keys to understanding a logic.

It is important to keep in mind that formulas of logical systems then get a double aspect. Interpreted in game semantics, a formula is a complex algebraic term for a game. Interpreted in a standard manner, it is an assertion about some situation, perhaps a game. Thus two very different functions of logical syntax come together.

A second point is that the logic game perspective does not have exclusive rights. There are other ways of thinking about the systems in Part V. Starting just from the viewpoint of the game logics in Parts I and II of this book, they make a very natural step, namely, bringing games themselves into the language by adding a suitable syntax for game terms. While the modal and temporal logics of those earlier parts worked on a game by game basis, one now puts a reflection of their space of different models into the formalism.[277] One can see this as a momentous change in methodology, where we now study games from the outside, understanding their internal structure through their modes of combination. But even on the game combination view, logic games remain relevant. Our representation theorem in Chapter 24 showed how first-order evaluation games provide a normal form for the abstract algebra of sequential game operations. This result arose originally from thinking about what makes the dynamic game logic of Chapter 19 tick, and this again suggested open problems of finding similar results for game logics with parallel operations (Chapter 20) and imperfect information (Chapter 21).

In the rest of this chapter, we offer a few themes showing how the perspectives of internal structure and external game description form a natural combination. We will look at three topics: multi-level views of games, strategy calculus and merged logics, and uses of logic games to elucidate fine structure of game logics.

25.2 Tracking a game at different levels

A first general topic that makes different views coexist is this. One game can be looked at from different perspectives, finer and coarser. This allows for tracking one level by another, in a way that relates to the simulations of Chapter 1.

Tracking and simulation The following pattern lies behind several notions and results in Chapters 6, 14, 18, and 19.

DEFINITION 25.1 Simulation pairs
A simulation pair consists of a game G and a coarser process graph $\boldsymbol{M}$ (the "tracking model"), related by a map F sending game states to process states:

277 Compare how process algebra (Bergstra et al. 2001) moves from local modal logics inside specific processes to a general algebra of process terms.

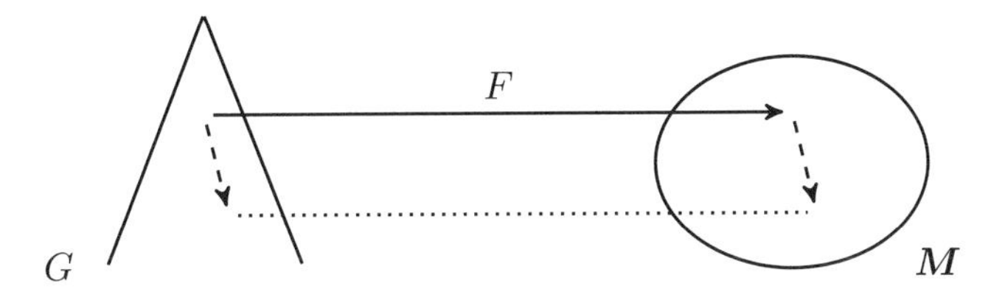

The connection F correlates internal game predicates of winning and losing with externally visible properties.[278] For final outcome states s in G:

$G, s \models win_{\boldsymbol{E}}$ iff $\boldsymbol{M}, F(s) \models \omega_{\boldsymbol{E}}$, for some formula $\omega_{\boldsymbol{E}}$ in the language of $\boldsymbol{M}$.

$G, s \models p$ iff $\boldsymbol{M}, F(s) \models p$, for atomic predicates p in the language of $\boldsymbol{M}$.

F also tracks the dynamics of the game in the tracking $\boldsymbol{M}$, correlating internal game moves with external ones. Assume that players have their own moves in the game G. The model $\boldsymbol{M}$ has just two binary relations R_E and R_A.

The relations in $\boldsymbol{M}$ satisfy back-and-forth clauses via F for moves in G.

Thus, F works roughly like a bisimulation (cf. Chapter 1). ∎

Now, essentially, the modal invariance analysis of Chapter 1 applies. Consider any formula φ in a modal game language (with fixed point operators). There is an obvious translation $t(\varphi)$ of the more detailed formula φ into the modal language of the process graph, replacing *win*-predicates by their ω-versions, and modalities for game moves by modalities in terms of the R-relations.

FACT 25.1 $G, s \models \varphi$ iff $\boldsymbol{M}, F(s) \models \mathrm{t}(\varphi)$

Instead of the obvious proof, we give some concrete instances.

Evaluation games The success lemma for evaluation games in Chapter 14 had the following format, with $\boldsymbol{E}$ for verifier:

$$\boldsymbol{game}(\varphi, \boldsymbol{M}), \langle s, \varphi \rangle \models \{\boldsymbol{E}\} win_{\boldsymbol{E}} \text{ iff } \boldsymbol{M}, s \models \varphi$$

The map F deletes the formula component of a game state, leaving the variable assignment. Players' moves are similar on both sides, and obey an obvious back-and-forth condition. Finally, winning is an external property, since the game state

278 Winning a game often just depends on some external property of the game board.

$\langle s, ?P\rangle$ is a win for $\boldsymbol{E}$ iff $\boldsymbol{M}, s \models P$. Therefore, the preceding analysis applies. The forcing statement $\{\boldsymbol{E}\}win_{\boldsymbol{E}}$ can be defined by a modal fixed point formula matching the definition of forcing in Chapter 1. Its F-translation is a modal fixed point formula that is true in the game board $\boldsymbol{M}$.

This still does not explain the simple form on the right in the success lemma, with just one first-order formula. The reason for that simplification is the inductive structure of the game determined by the formula φ, and the fact that the compositional forcing clauses of Chapters 11 and 19 match up precisely with the standard truth conditions for the operators inside φ.

Graph games and invariants The graph games of Chapter 18 provide another illustration. Here the above map F removes the internal turn information from a game form G, and an external reduction results of the following form:

> $G, \langle s, \boldsymbol{E}\rangle \models WIN_{\boldsymbol{E}}$ iff the playground $graph(G), s$ satisfies some suitable matching fixed point assertion.

What is meant by suitable depends on how we define winning conditions in the graph game (van Benthem 1999). The translated formulas act as external invariants that can be maintained during play, as suggested in Chapter 4.

Further examples of two-level tracking occur with the distinction between games and game boards that was central in Chapter 19 (see also Section 25.3). Another example is the invariants used in logic games, such as the model comparison games of Chapter 15. Finally, external state descriptions also occur in game theory, witness the invariants for game states discussed in Binmore (1992).

Here is one concrete source of invariants: coarse representation in memory.

Memory as tracking Think of a finite-state machine performing a simulation between a game and memory transitions.

EXAMPLE 25.1 Copy-cat as a finite-state automaton
Our key strategy in Chapter 20 was of this sort. Tracking the whole game projects things to the following machine (assuming just two moves a and b for both players):

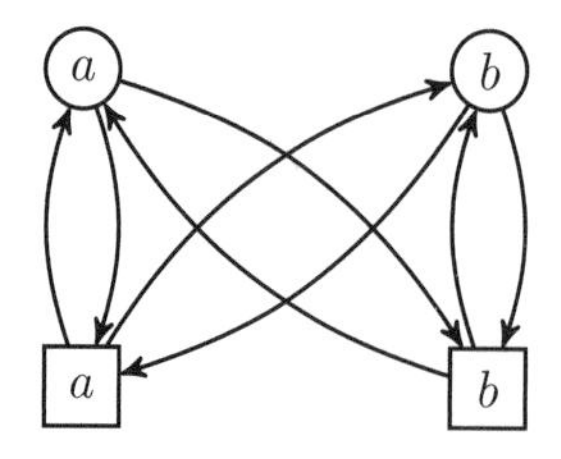

last move made by $\boldsymbol{A}$

last move made by $\boldsymbol{E}$

The copy-cat strategy is a definable subrelation of this graph. ■

This example illustrates how useful simple strategies in games can live at some coarser level of tracking. A discussion of finite automata-based strategies in game theory can be found in Osborne & Rubinstein (1994). More sophisticated strategies occur in computational logic, witness the appendix to Chapter 18.[279]

Here is one more example of this style of thinking. The Positional Determinacy Theorem of Chapter 18 said that graph games with a parity condition for winning are determined, with a winning strategy that is positional. That is, the next move to be played depends only on the graph position in the current game state, not on the whole game history leading up to that state. What this basically says is that the property of winnability at states in these games is correlated exactly, via the natural projection map F, with a set of points in the graph. But then, it makes sense to ask for an explanation of positional determinacy in terms of logical invariants. Could it be that parity games support a systematic *translation* $\boldsymbol{t}$ in the background sending forcing statements φ in the game (see Chapters 5 and 11) to equivalent modal statements $\boldsymbol{t}(\varphi)$, perhaps in the μ-calculus of the graph?

25.3 Making games explicit

Our next scenario for coexistence of different levels in logic and games comes from Chapter 19. The game boards used there did not reveal internal structure of games being played, only their external forcing effects. Games remained a sort of hidden variable. We can make the games explicit by means of a representation argument, this time, reconstructing a finer level behind the coarser one.

DEFINITION 25.2 Games behind game boards
We first enrich our semantics, starting with games $game(g, \boldsymbol{M}, s)$ for each atomic game expression g given a model $\boldsymbol{M}$ and state s. We then induct upward:

(a) The game for $G_1 \vee G_2$ is the natural disjoint sum of the games, put under one new root indicating a choice for $\boldsymbol{E}$ with two arrows leading to the roots of $game(G_1, \boldsymbol{M}, s)$ and $game(G_2, \boldsymbol{M}, s)$.

(b) The game for $G_1 \,;\, G_2$ arises by tree substitution at G_1-leaves.

(c) The game for G^d arises by reversing turn predicates at all nodes.

---

279 Information tracking in computational logic is studied in Berwanger & Kaiser (2010).

Next, let an atomic assignment $game(g, \boldsymbol{M}, s)$ come with a "decoration" δ_g mapping nodes in the extensive game tree to states in $\boldsymbol{M}$ (the root goes to s). Now define

$\rho_g sX$ if, for some U contained in the terminal nodes of $game(g, \boldsymbol{M}, s)$ that can be forced by some strategy for player $\boldsymbol{E}$, we have $\delta_g[U] \subseteq X$.

It is easy to see that the general relation

$$\Delta_g s, X := \delta_g[U] \subseteq X$$

between states s and sets X in $\boldsymbol{M}$ can be lifted to all games. For example, for a disjoint sum, the lifted Δ extends the relations for the component games, mapping the new root and its immediate successors to the same initial state s, reflecting the fact that choices are game-internal, without visible external effects.[280] ■

A simple induction then shows the following harmony.

FACT 25.2 For all game expressions G, $\rho_G = \Delta_G$.

Proof The maps δ are forcing bisimulations in the sense of Chapters 11 and 19, and one can then use the safety of our game operations. ■

Now recall the representation results of Chapter 11. Each forcing relation satisfying upward monotonicity, consistency, and completeness, can be represented faithfully as a two-player two-move game over the state space of the DGL game board to produce real games with the given powers.

Putting everything together, we have proved this result (cf. Pauly 2001).[281]

THEOREM 25.1 Every game board can be represented as a two-level structure with real underlying games providing the forcing relations.

25.4 Making strategies explicit

Another meeting point for different description levels of games arose by juxtaposing Chapters 4 and 5 on game logics for strategies with Chapter 18 on running themes in logic games. In the latter realm, strategies encoded formulas, models, proofs,

280 The map Δ and its properties are reminiscent of the representation in Chapter 24.

281 This analysis should be liftable to nondetermined games with imperfect information.

simulations, and so on. Still, there were many shared features. We would like to have a general calculus of reasoning about strategies that contains the basic arguments shared between many different games. We merely found a promising format so far, with operations of

storing strategies for a player who is not to move	$\langle\, , \rangle$
using a strategy from a list	$(\ )_i$
executing the first action of a strategy	*head*( )
executing the remaining strategy	*tail*( )
choosing depending on a test	IF THEN ELSE

In Chapter 18, we noted how these abstract operations have useful interpretations, and satisfy obvious laws. Without elaborating a calculus here, we note another area where this dual viewpoint emerges naturally.

Strategizing DGL The dynamic game logic of Chapter 19 still suffered from the $\exists$-sickness discussed at several places in this book: strategies were in the background, but they remained implicit. For instance, the soundness of the crucial DGL axioms required a basic argument on combining strategies in games that remained informal. Making such reasoning explicit is an obvious benchmark for a logical system. We still need to add strategy terms to our logic, say, in the following style.[282]

DEFINITION 25.3 Explicit DGL modalities
The *explicit DGL modality* $\{\sigma, G\}\varphi$ says that the internal strategy σ forces G into some set of end nodes that are mapped by the above tracking map δ into states of the matching game board $\boldsymbol{M}$ satisfying φ. ■

Upon inspection, the proof for soundness of DGL in Chapter 19 then involves strategic reasoning with explicit principles such as the following:

(a) $\sigma : G \Rightarrow (L, \sigma) : G \cup G'$

(b) $\sigma : G \Rightarrow (R, \sigma) : G' \cup G$

(c) $\sigma : G \cup G' \Rightarrow \mathit{tail}(\sigma) : \mathit{IF}\ \mathit{head}(\sigma) = L\ \mathit{THEN}\ G\ \mathit{ELSE}\ G'$

282 An analogous move has led to a flourishing theory in modal provability logic enriched with explicit proof terms, now developed into a broader paradigm of justification logics (cf. Artemov 1994, 2008).

where $L = left, R = right, head(\sigma)$ picks one of the games G and G', and $tail(\sigma)$ is what remains of the strategy σ inside the chosen game.

Other authors have further discussion of explicit strategy versions of DGL (cf. Ghosh 2008). Also, Chapters 5 and 6 provided several concrete illustrations of explicit strategy calculus applied to classical results about games and strategies.[283]

Strategizing linear game logic What is true for dynamic game logic is just as true for the linear game logic of Chapter 20. Looking at the minimum needed to formalize natural game-theoretic arguments, the soundness of linear logic in its game semantics involved a very basic piece of strategic reasoning about the strategy copy-cat that occurs far beyond logic games.[284]

Again, we would like to see an explicit strategy calculus for this, beyond what is recorded by the principles of linear logic at a higher level of game algebra. Such a calculus needs to access the internal structure of the games. The infinite games of Chapter 20 are really branching time models of the sort studied in Chapters 5 and 6, and in Part II, yielding playgrounds that can be described using simple temporal logics. The temporal logic can also describe computational properties such as safety and liveness that are more like the winning conditions for infinite model comparison games in Chapter 15. An example of this was the compact notation $\{\boldsymbol{E}\}\varphi \vee \{\boldsymbol{A}\}\mathcal{G}\neg\{\boldsymbol{E}\}\varphi$ in Chapter 5 for weak determinacy in a temporal forcing language. Here the forcing modalities express global features of the game, while the temporal modality $\mathcal{G}$ looks at all future events on the current history.[285]

Even more structure arises when we add game terms to the forcing modalities, in a format $\{\boldsymbol{E}, G\}\varphi$. One would like to have a joint system with linear logic taking care of game combinations, and temporal logic of the internal run descriptions.[286] Such an abstract game logic might also throw new light on connections between two modes of computation that were kept separate in Part V of this book. For instance, parallel play can sometimes be coded sequentially, as in the automata

283 We also saw that keeping the original existential game modality $\{G\}\varphi$ makes sense, as its negation is needed to say that an agent lacks a strategy for achieving goal φ.

284 The argument also had a "shadow match" whose logic may need additional resources such as operators for hiding information.

285 This language could deal with basic game-theoretic arguments, such as the proof of the Gale-Stewart Theorem, but also with the strategy stealing proof behind the classical example of an infinite nondetermined game.

286 Abramsky (2000) proposes a relevant calculus of "information structures."

constructions of Muller & Schupp (1995). A general logic of strategies should get clear on the power and range of such transformations.

Merged languages, many levels, and mixed equivalences The combined languages discussed here are a natural counterpart to the many-level view of games in the preceding sections. They also offer a coexistence of different game equivalences (cf. Chapters 1, 3, and 11). Equivalence at the level of game terms, whether for DGL or in linear logic, seems largely power-based, but the internal description level of a modal or temporal language is much closer to bisimulation-like comparisons of extensive games. In other words, while we have emphasized traveling between zoom levels in this book, maybe the better stance is to keep more of them around.

25.5 Logic games as fine structure for game logics

Our third encounter between logic games and game logics is of a different kind. The style of analysis in logic games is a natural addition to topics in game logics.

Playing games about games In Chapter 15, we studied Ehrenfeucht-Fraïssé games that test models for similarity over some specified number k of rounds. These games captured definability of similarities or differences between the models up to quantifier depth k in concrete first-order or modal formulas. In this finer perspective, following a bisimulation was a winning strategy for establishing similarity in a comparison game with an infinite number of rounds. Playing the associated game over finite stretches investigates the actual dynamics of tracking along the simulation, while showing definable invariants or differences on the way. Thus, logic games provide a way of refining our earlier basic question in Chapter 1 of "When are two games the same?" to a more sophisticated one,

"How similar are two given games?"

This is just one instance of how logic games can throw new light on game logics.

25.6 Logic games and game logics entangled

We end with three thoughts on where we find ourselves with our twists and turns.

The first point is one of logical methodology. As we have suggested in our Introduction, the duality of perspectives in logic *of* games and logic *as* games can be iterated. For instance, our suggested use of logic games to provide fine structure for invariance notions between games is such a perspective. We can apply logic-as-is to games, but also logic in the guise of logic games, asking, say, how the evaluation game for the description language of a game relates to that game itself. Conversely, we can apply games to analyze logical tasks, but also the logical description languages for these games, asking, say, whether first-order logic is in harmony with the modal forcing language of first-order evaluation games. Of course, not every such iteration may make sense. Still, we should think of our duality not as a closed cycle, but as a helix-type structure for the DNA of the logic-games interface.

Next, consider a hard-nosed mathematical perspective on this book. Presumably, there are general transformations at work behind many examples in this book, taking us from games to matching logics, and from logics to matching games. But then it makes sense to study mathematical reductions between the two realms, and algebraic laws for these transformations. While we doubt that there is one reduction tying the whole field together, it seems of great interest to see what can be achieved along these lines.

Finally, the stances in this book may influence our philosophical understanding of logic. If we take the view of logic as games as seriously as we have done, the question arises of what our fundamental understanding of logical constants and logical reasoning should be. Positions in the field seem to diverge. On what may be called the Strong Thesis, games *embody* the meaning of the basic logical notions, with other accounts as derivative or at least less fundamental. Few logicians seem to be willing to go this far, although Lorenzen's original view of grounding logic in dialogue practice comes close, and perhaps also that of some game semanticists. I myself subscribe to the Weak Thesis that games are a natural and useful way of understanding logic, on a par with others, as has been the line in the Introduction to this book. The Weak Thesis is clearly more reasonable, and therefore, the Strong Thesis is more exciting.

25.7 Conclusion

Drawing together threads from preceding chapters, we have seen how merging ideas from logic games and game logics makes sense and generates new questions. This chapter just drew a few lines, trying to show that the issues are legitimate. One major concern was combining different structure levels and languages for games.

But there are many further forms of cross-traffic. Logic games have long been a conceptual laboratory for interesting devices that may well have general game-theoretic import, such as the use of pebbles for memory or resources for the players, whose general scope seems to extend far into game logics. Also, at the end of Part IV, we mentioned the large body of theory concerning logic games and automata in a computational setting. The proofs of major results there provide a wealth of general ideas about strategic reasoning in games that may have a much broader potential (cf. Berwanger 2008).

Going in the opposite direction, it might be of interest to focus the concerns of our Theory of Play in Part II on logic games. As we discussed briefly in Chapter 10, many unusual issues arise. What happens when we drop the uniformity assumptions that we have all been taught implicitly, and assume a variety of agents in logic, say, from finite automata to Turing machines that compete or cooperate in deduction? This book has put us at a threshold of thinking about these issues, but it remains for the reader to cross it.

Conclusion to Part VI

Game logics and logic games meet in many ways. This part has explored a number of these, although it fell short of presenting a systematic theory. We started with natural combinations of logic games and ideas from standard game theory. The IF games of evaluation under imperfect information in Chapter 21 exemplify major themes in this book, at both power and action levels, while also illuminating the logical core topic of quantifier dependence. Next, Chapter 22 presented knowledge games played over epistemic models that combine features of logic and general games in new ways. Chapter 23 was about a new genre of sabotage games that continue computation by other means, while also raising new issues in logics of model change, an important theme in the dynamics of Part II. Chapter 24 took a more theoretical line, and compared the two realms by showing how logic games can act as normal forms for general games, in the algebra of sequential game constructions. Finally, Chapter 25 lightly drew together a number of lines in the book, highlighting encounters between logic games and game logics. In particular, it was shown how ideas from both sides can merge into useful mixed systems with different description levels coexisting.

Thus, logic of games and logic as games mix well, even if we did not find one crystalline mathematical duality between the two perspectives. Some further thoughts on where this leaves us will be found in the Conclusion to this book.

VII Conclusion

Conclusion

This book is an investigation of encounters between logic and games. There were two main directions to this interface, that we called logic *of* games and logic *as* games, and some readers might view the resulting text as two separate books.

Content In Parts I, II, and III, we used logical systems to analyze the structure of games. Part I chose the established road of standard modal, dynamic, epistemic, and temporal logics for analyzing game structure, with the notion of rationality as a guiding theme. In doing so, we brought major themes from logic to bear on games, such as invariance, definability, and the use of small models that stay close to social reasoning practices. Part II then continued with the perspective of logical dynamics of intelligent agency, looking at games not as static tree structures, but as a family of events and activities whose logic can be studied as such. This merging of ideas from game theory and dynamic logics led to a view that what we are after is really a richer agenda of a Theory of Play combining aspects of both fields. The Theory of Play is still largely a programmatic proposal, although we have tested its viability in Part III on games in strategic form with simultaneous action.

In Part IV of the book, we then made a turn to the field of logic games that show how many core notions of logic rest on activities of evaluation, comparison, or proof, that can themselves be cast as games. While such games are often viewed as just intuitive metaphors for teaching logic, we took them much more seriously, as embodying a conception of logic as activity. Moreover, as we saw when looking at them in their generality, they offer a clean laboratory for studying general structural issues about games. Moreover, in their various spillovers into computer science, logic games have already generated a sophisticated mathematical theory of interactive behavior that may well come to impact game theory.

In our view, the perspectives of logic games and game logics are natural duals that are entangled in various ways. First of all, casting logic in the form of games is obviously congenial to the program of logical dynamics, and it is an even more radically dynamic view of logic than game semantics where formulas denote just one particular kind of game. But also, Parts V and VI of the book show that there is a large number of merges between the two perspectives. Part V discussed dynamic and linear logics for game operations connected to logical constants, and showed how the resulting basic issues cannot easily be classified as either logic of games or logic as games. These chapters also showed a contrast between model-theoretic and proof-theoretic views of games, a not unimportant design choice for logicians that we have downplayed in this book to avoid getting into too many discussions at once. Part VI then presented a number of systems combining ideas from logic games and game theory, as well as some formal perspectives on how they relate. Here, combination was the main theme rather than reduction. This may be the most convincing link anyway, since producing viable joint offspring is usually considered the best test of compatibility between life forms.

Challenges This book contains a lot of material, both standard topics and results based on the author's own publications, as mentioned in the literature sections for the separate chapters. Still, it does not present a finished job by any means. There are major open problems inside each of its two main strands, and at their interface. Theory of Play is a program, not a reality, requiring a deeper understanding of how logical dynamics interfaces with logic as games. For instance, looking at logical systems from the perspective of player diversity might lead to major changes in our understanding of the field, that have not yet been thought through. In all, this book offers many techniques and results, but for each answer provided, it asks several new questions.

Also, our open problems are not all confined to games. One of the most perplexing issues in understanding logic is the general duality between implicit and explicit approaches to phenomena (van Benthem 2010c). Intuitionistic and linear logic take knowledge seriously by making the very logical constants information-based, but, for instance, epistemic logic studies knowledge via explicit operators added to a classical base. Likewise, game logics, even in their dynamic guise, add explicit game descriptions to a classical base logic, while game semantics makes games the stuff that interprets the logical constants themselves, often resulting in deviant base logics. How these two perspectives are related is an issue in the philosophy of logic that goes well beyond this book. But it seems highly relevant, for its own sake, and also for a better understanding of the perennial issue what logicality really is.

Omissions In addition to grand open problems, more down-to-earth things have been neglected in this book. Clearly, infinite games, simultaneous action, coalitions and collective action, complexity theory of games, evolutionary game theory, dynamical systems, and probabilistic approaches deserve much more attention than they have received here. Consider this paragraph an honest apology to all of them.

Further disciplines While this book has maintained a technical focus on logic and games, with occasional excursions toward game theory and computer science, there are also broader interdisciplinary aspects. We have already mentioned some philosophical issues in the understanding of logic, but one might also think of the growing role of games in epistemology and philosophy of science, continuing the agenda of dynamic-epistemic information-driven agency in van Benthem (2011e). There are also many further links with the foundations of computation, that have only been addressed very lightly in this book (see van Benthem 2007b on the thesis that computation is conversation). And the ideas presented here at the logic and games interface may also come to affect more practical areas such as informal logic and argumentation theory (see van Benthem 2010a and van Benthem 2012c).

Finally, we mention connections between this book and more empirical matters. Clearly, this book invites a junction with uses of games in natural language (van Benthem 2008, Clark 2011) and in cognitive science (van Benthem 1990, 2007a) where games are a natural paradigm for intelligent behavior, far richer than machine models. While of great interest to the present author, more linguistic or cognitive themes have not been included in this book, to avoid its getting top heavy.

So? We have presented the interface of logic and games as one can see it today, with a number of rich and sometimes perplexing perspectives. Whether we have pulled off the feat of making this look like a coherent enterprise must be left to the reader. In the worst case, you can read Parts I through III and Parts IV through VI as separate books, and not be bothered by the rest. Sometimes, life deals us two books for the price of one.

Bibliography

Abramsky, S. (1995). Semantics of interaction. Lecture notes. Department of Computer Science, University of Edinburgh, UK.

Abramsky, S. (2000). Concurrent interaction games. In A. W. R. J. Davies and J. Woodcock (Eds.), *Millennial Perspectives in Computer Science*, pp. 1–12. Basingstoke, UK: Palgrave.

Abramsky, S. (2006). Socially responsive, environmentally friendly logic. In T. Aho and A.-V. Pietarinen (Eds.), *Truth and Games: Essays in Honour of Gabriel Sandu*, pp. 17–45. Helsinki: Acta Philosophica Fennica.

Abramsky, S. (2008a). Information, processes and games. In P. Adriaans and J. van Benthem (Eds.), *Handbook of the Philosophy of Information*, pp. 483–549. Amsterdam: Elsevier Science.

Abramsky, S. (2008b). Tutorial on game semantics. Computing Laboratory, Oxford University, UK.

Abramsky, S. (2012). Foundations of interactive computation. Lecture, Logic Colloquium/Alan Turing Centennial Year, Manchester.

Abramsky, S. and R. Jagadeesan (1994). Games and full completeness for multiplicative linear logic. *Journal of Symbolic Logic 59*(2), 543–574.

Abramsky, S. and A. Jung (2001). Domain theory. In S. Abramsky, D. Gabbay, and T. Maibaum (Eds.), *Handbook of Logic in Computer Science*, Volume 3: Semantic Structures, pp. 1–168. Oxford, UK: Oxford University Press.

Abramsky, S. and G. McCusker (1999). Game semantics. In H. Schwichtenberg and U. Berger (Eds.), *Computational Logic: Proceedings of the 1997 Marktoberdorf Summer School*, pp. 1–56. Berlin: Springer-Verlag.

Ågotnes, T., V. Goranko, and W. Jamroga (2007). Alternating-time temporal logics with irrevocable strategies. In D. Samet (Ed.), *TARK'07*, pp. 15–24. New York: ACM.

Ågotnes, T., J. van Benthem, H. van Ditmarsch, and S. Minica (2011). Question-answer games. *Journal of Applied Non-Classical Logics 21*(3–4), 265–288.

Ågotnes, T. and H. P. van Ditmarsch (2011). What will they say? Public announcement games. *Synthese (KRA) 179* (Suppl.1), 57–85.

Aiello, M., I. Pratt-Hartmann, and J. van Benthem (Eds.) (2007). *Handbook of Spatial Logics*. Dordrecht, The Netherlands: Springer Science Publishers.

Aiello, M. and J. van Benthem (2002). A modal walk through space. *Journal of Applied Non-Classical Logics 12*(3–4), 319–363.

Alur, R., T. A. Henzinger, and O. Kupferman (2002). Alternating-time temporal logic. *Journal of the ACM 49*(5), 672–713.

Anand, P., P. Pattanaik, and C. Puppe (Eds.) (2009). *Handbook of Rational and Social Choice*. Oxford, UK: Oxford University Press.

Andersen, M. B., T. Bolander, and M. H. Jensen (2012). Don't plan for the unexpected: Planning based on plausibility models. Department of Informatics and Mathematical Modelling (IMM), Technical University of Denmark (DTU).

Andréka, H., M. Ryan, and P.-Y. Schobbens (2002). Operators and laws for combining preference relations. *Journal of Logic and Computation 12*(1), 13–53.

Andréka, H., J. van Benthem, and I. Németi (1998). Modal logics and bounded fragments of predicate logic. *Journal of Philosophical Logic 27*(3), 217–274.

Apt, K. (2007). The many faces of rationalizability. *Topics in Theoretical Economics 7*(1), 1–39.

Apt, K. (Ed.) (2011). *Proceedings of the 13th Conference on Theoretical Aspects of Rationality and Knowledge (TARK-2011)*, New York: ACM.

Apt, K. and E. Grädel (Eds.) (2011). *Lectures in Game Theory for Computer Scientists*. Cambridge, UK: Cambridge University Press.

Areces, C. and B. ten Cate (2006). Hybrid logic. In Blackburn et al. (2006), pp. 821–868.

Arnold, A. and D. Niwinski (2007). Continuous separation of game languages. *Fundamenta Informaticae 81*(1–3), 19–28.

Artemov, S. (1994). Logic of proofs. *Annals of Pure and Applied Logic 67*, 29–59.

Artemov, S. (2008). The logic of justification. *The Review of Symbolic Logic 1*, 477–513.

Aumann, R. J. (1976). Agreeing to disagree. *The Annals of Statistics 4*(6), 1236–1239.

Aumann, R. J. (1987). Correlated equilibrium as an expression of Bayesian rationality. *Econometrica 55*(1), 1–18.

Aumann, R. J. (1990). Nash equilibria are not self-enforcing. In J. J. Gabszewicz, J.-F. Richard, and L. A. Wolsey (Eds.), *Economic Decision-Making: Games, Econometrics and Optimisation*, pp. 201–206. Amsterdam: Elsevier Science.

Aumann, R. J. (1995). Backward induction and common knowledge of rationality. *Games and Economic Behavior 8*(1), 6–19.

Aumann, R. J. (1999). Interactive epistemology I: Knowledge. *International Journal of Game Theory 28*(3), 263–300.

Axelrod, R. (1984). *The Evolution of Cooperation.* New York: Basic Books.

Balbiani, P., A. Baltag, H. van Ditmarsch, A. Herzig, T. Hoshi, and T. de Lima (2009). "Knowable" as "known after an announcement." *Review of Symbolic Logic 1*(3), 305–334.

Baltag, A. (2002). A logic for suspicious players: Epistemic actions and belief update in games. *Bulletin of Economic Research 54*(1), 1–46.

Baltag, A., V. Fiutek, and S. Smets (2012). Playing for "knowledge." Dynamics Group, Institute for Logic, Language and Computation ILLC, University of Amsterdam.

Baltag, A., N. Gierasimczuk, and S. Smets (2011). Belief revision as a truth-tracking process. In Apt (2011), pp. 187–190.

Baltag, A., D. Grossi, A. Marcoci, B. Rodenhäuser, and S. Smets (Eds.) (2012). *Dynamics Yearbook 2011.* Amsterdam: ILLC.

Baltag, A., L. Moss, and S. Solecki (1998). The logic of public announcements and common knowledge and private suspicions. In Gilboa (1998), pp. 43–56.

Baltag, A. and S. Smets (2008). A qualitative theory of dynamic interactive belief revision. In Bonanno et al. (2008), pp. 13–60.

Baltag, A. and S. Smets (2009). Group belief dynamics under iterated revision: Fixed points and cycles of joint upgrades. In Heifetz (2009), pp. 41–50.

Baltag, A., S. Smets, and J. A. Zvesper (2009). Keep "hoping" for rationality: A solution to the backward induction paradox. *Synthese 169*(2), 301–333.

Baltag, A., J. van Benthem, and S. Smets (2014). *The Music of Knowledge. A Dynamic-Logical Approach to Epistemology.* Book to appear. University of Amsterdam.

Barendregt, H. (2001). Lambda calculi with types. In S. Abramsky, D. Gabbay, and T. Maibaum (Eds.), *Handbook of Logic in Computer Science*, Volume 2: Background. Computational Structures, pp. 117–309. Oxford, UK: Oxford University Press.

Barringer, H., D. Gabbay, and J. Woods (2012). Temporal, numerical and meta-level dynamics in argumentation networks. *Argument and Computation 3*(2–3), 143–202.

Barth, E. M. and E. C. W. Krabbe (1982). *From Axiom to Dialogue: A Philosophical Study of Logics and Argumentation.* Berlin: Walter de Gruyter.

Barwise, J. and J. Etchemendy (1999). *Language, Proof and Logic.* New York & Stanford: Seven Bridges Press & CSLI Publications.

Barwise, J. and J. van Benthem (1999). Interpolation, preservation, and pebble games. *Journal of Symbolic Logic 64*(2), 881–903.

Battigalli, P. and G. Bonanno (1999a). Recent results on belief, knowledge and the epistemic foundations of game theory. *Research in Economics 53*(2), 149–225.

Battigalli, P. and G. Bonanno (1999b). Synchronic information, knowledge and common knowledge in extensive games. *Research in Economics 53*(1), 77–99.

Battigalli, P. and M. Siniscalchi (1999). Hierarchies of conditional beliefs and interactive epistemology in dynamic games. *Journal of Economic Theory 88*(1), 188–230.

Belnap, N. D., M. Perloff, and M. Xu (2001). *Facing the Future.* Oxford, UK: Oxford University Press.

Benz, A., G. Jäger, and R. van Rooij (Eds.) (2005). *Game Theory and Pragmatics.* New York: Palgrave McMillan.

Bergstra, J. A., A. Ponse, and S. A. Smolka (Eds.) (2001). *Handbook of Process Algebra.* Amsterdam: North-Holland.

Berwanger, D. (2005). *Games and logical expressiveness.* Ph. D. thesis, Department of Computer Science, RWTH Aachen, Germany.

Berwanger, D. (2008). Infinite coordination games. In G. Bonanno, B. Löwe, and W. van der Hoek (Eds.), *LOFT*, Volume 6006 of *Lecture Notes in Computer Science*, pp. 1–19. Berlin: Springer-Verlag.

Berwanger, D. and L. Kaiser (2010). Information tracking in games on graphs. *Journal of Logic, Language and Information 19*(4), 395–412.

Beth, E. W. (1955). Semantic entailment and formal derivability. *Koninklijke Nederlandse Akademie van Wentenschappen, Proceedings of the Section of Sciences 18*, 309–342.

Bicchieri, C. (1988). Common knowledge and backward induction: A solution to the paradox. In Vardi (1988), pp. 381–393.

Bicchieri, C., R. Jeffrey, and B. Skyrms (Eds.) (1999). *The Logic of Strategy.* Oxford: Oxford University Press.

Binmore, K. (1992). *Fun and Games – A Text on Game Theory.* Lexington, MA: D. C. Heath & Co.

Blackburn, P., M. de Rijke, and Y. Venema (2001). *Modal Logic*. Number 53 in Cambridge Tracts in Theoretical Computer Science. Cambridge, UK: Cambridge University Press.

Blackburn, P., J. van Benthem, and F. Wolter (Eds.) (2006). *Handbook of Modal Logic*, Volume 3 of *Studies in Logic and Practical Reasoning*. Amsterdam: Elsevier Science.

Blass, A. (1992). A game semantics for linear logic. *Annals of Pure and Applied Logic 56*(1–3), 183–220.

Board, O. (1998). Belief revision and rationalizability. In Gilboa (1998), pp. 201–213.

Bod, R., R. Scha, and K. Sima'an (2003). *Data-Oriented Parsing*. Stanford, CA: CSLI Publications.

Bolander, T. and M. B. Andersen (2011). Epistemic planning for single and multi-agent systems. *Journal of Applied Non-Classical Logics 21*(1), 9–34.

Bonanno, G. (1992a). Players' information in extensive games. *Mathematical Social Sciences 24*(1), 35–48.

Bonanno, G. (1992b). Set-theoretic equivalence of extensive-form games. *International Journal of Game Theory 20*(4), 429–447.

Bonanno, G. (1993). The logical representation of extensive games. *International Journal of Game Theory 22*(2), 153–169.

Bonanno, G. (2001). Branching time, perfect information games, and backward induction. *Games and Economic Behavior 36*(1), 57–73.

Bonanno, G. (2004a). A characterization of Von Neumann games in terms of memory. *Synthese 139*(2), 281–295.

Bonanno, G. (2004b). Memory and perfect recall in extensive games. *Games and Economic Behavior 47*(2), 237–256.

Bonanno, G. (2007). Axiomatic characterization of the AGM theory of belief revision in a temporal logic. *Artificial Intelligence 171*(2–3), 144–160.

Bonanno, G. (2012). Reasoning about strategies and rational play in dynamic games. Lecture at Lorentz Workshop on Modeling Reasoning about Strategies, Leiden 2012.

Bonanno, G. (2013). A dynamic epistemic characterization of backward induction without counterfactuals. *Games and Economic Behavior 78*(C), 31–43.

Bonanno, G., W. van der Hoek, and M. Wooldridge (Eds.) (2008). *Logic and the Foundations of Game and Decision Theory (LOFT-7)*, Volume 3 of *Texts in Logic and Games*. Amsterdam: Amsterdam University Press.

Boutilier, C. (1994). Conditional logics of normality: A modal approach. *Artificial Intelligence 68*(1), 87–154.

Bradfield, J. and C. Stirling (2006). Modal μ-calculi. In Blackburn et al. (2006), pp. 721–756.

Brandenburger, A. (2007). Forward induction. Stern School of Business, NYU New York.

Brandenburger, A., P. Battigalli, A. Friedenberg, and M. Siniscalchi (2014). *Game Theory, An Epistemic Approach.* Singapore: World Scientific Publishing.

Brandenburger, A. and A. Friedenberg (2008). Intrinsic correlation in games. *Journal of Economic Theory 141*(1), 28–67.

Brandenburger, A. and H. J. Keisler (2006). An impossibility theorem on beliefs in games. *Studia Logica 84*(2), 211–240.

Broersen, J. (2009). A STIT logic for extensive form group strategies. In *Web Intelligence-IAT Workshops*, pp. 484–487. Los Alamitos, CA: IEEE.

Broersen, J. (2011). Deontic epistemic STIT logic distinguishing modes of mens rea. *Journal of Applied Logic 9*(2), 137–152.

Broersen, J., A. Herzig, and N. Troquard (2006). Embedding alternating-time temporal logic in strategic STIT logic of agency. *Journal of Logic and Computation 16*(5), 559–578.

Burgess, J. P. (1981). Quick completeness proofs for some logics of conditionals. *Notre Dame Journal of Formal Logic 22*(1), 76–84.

Chellas, B. (1980). *Modal Logic: An Introduction.* Cambridge, UK: Cambridge University Press.

Cho, I.-K. and D. M. Kreps (1987). Signaling games and stable equilibria. *The Quarterly Journal of Economics 102*(2), 179–221.

Ciuni, R. and J. Horty (2013). STIT logics, games, knowledge, and freedom. To appear in A. Baltag & S. Smets (Eds.). *Johan van Benthem on Logical Dynamics.* Dordrecht: Springer Science Publishers.

Clark, R. (2011). *Meaningful Games.* Cambridge, MA: MIT Press.

Cui, J. (2012). Dynamic epistemic characterizations for IERS. Institute of Logic and Cognition, Sun Yat-sen University, Guangzhou, China.

Curien, P.-L. (2005). Introduction to linear logic and ludics. *Advances in Mathematics 24*(5), 513–544.

Curien, P.-L. (2006). Notes on game semantics. PPS, Université Paris VII.

D'Agostino, G. and M. Hollenberg (2000). Logical questions concerning the μ-calculus: Interpolation, Lyndon and Los-Tarski. *Journal of Symbolic Logic 65*(1), 310–332.

Danos, V., H. Herbelin, and L. Regnier (1996). Game semantics and abstract machines. In *LICS*, pp. 394–405. Los Alamitos, CA: IEEE Computer Society.

Dawar, A., E. Grädel, and S. Kreutzer (2004). Inflationary fixed points in modal logic. *ACM Transactions of Computational Logic 5*(2), 282–315.

de Bruin, B. (2004). *Explaining games: On the logic of game-theoretic explanations.* Ph. D. thesis, Institute for Logic, Language and Computation, University of Amsterdam (UvA), Amsterdam, The Netherlands. ILLC Dissertation series DS-2004-03.

de Bruin, B. (2010). *Explaining Games: The Epistemic Programme in Game Theory*, Volume 346 of *Synthese Library Series.* Dordrecht, The Netherlands: Springer Science Publishers.

Dechesne, F. (2005). *Game, set, maths: Formal investigations into logic with imperfect information.* Ph. D. thesis, Department of Philosophy, University of Tilburg, Tilburg, The Netherlands.

Dechesne, F., J. van Eijck, W. Teepe, and Y. Wang (2009). Dynamic epistemic logic for protocol analysis. In J. van Eijck and R. Verbrugge (Eds.), *Discourses on Social Software*, Volume 5 of *Texts in Logic and Games*, pp. 147–161. Amsterdam, The Netherlands: Amsterdam University Press.

Dégremont, C. (2010). *The temporal mind: Observations on belief change in temporal systems.* Ph. D. thesis, Institute for Logic, Language and Computation, University of Amsterdam (UvA), Amsterdam, The Netherlands. ILLC Dissertation series DS-2010-03.

Dégremont, C., B. Löwe, and A. Witzel (2011). The synchronicity of dynamic epistemic logic. In Apt (2011), pp. 145–152.

Dégremont, C. and O. Roy (2009). Agreement theorems in dynamic-epistemic logic. In Heifetz (2009), pp. 91–98.

Doets, K. (1996). *Basic Model Theory.* Stanford, CA: CSLI Publications.

Doets, K. (1999). Evaluation games for first-order fixed-point logic. Institute for Logic, Language and Computation, University of Amsterdam.

Dretske, F. (1981). *Knowledge and the Flow of Information.* Cambridge, MA: MIT Press.

Dung, P. M. (1995). On the acceptability of arguments and its fundamental role in nonmonotonic reasoning, logic programming and n-person games. *Artificial Intelligence 77*(2), 321–357.

Dutilh-Novaes, C. (2007). *Formalizing Medieval Logical Theories. Suppositio, Consequentiae and Obligationes.* Berlin: Springer-Verlag.

Ebbinghaus, H.-D. and J. Flum (1999). *Finite Model Theory* (2nd ed.). Berlin: Springer-Verlag.

Ehrenfeucht, A. (1961). An application of games to the completeness problem for formalized theories. *Fundamenta Mathematicae 49*, 129–141.

Emerson, E. A. and C. S. Jutla (1991). Tree automata, μ-calculus and determinacy (extended abstract). In *FOCS*, pp. 368–377. Washington, DC: IEEE Computer Society.

Escardo, M. and P. Oliva (2010). Selection functions, bar recursion, and backward induction. *Mathematical Structures in Computer Science 20*(2), 127–168.

Euwe, M. (1929). Mengentheoretische betrachtungen über das schachspiel. *Handelingen KNAW, Afdeling Natuurkunde 32*(5), 633–642.

Facchini, A. (2011). *A study on the expressive power of some fragments of the modal μ-calculus.* Ph. D. thesis, University of Lausanne & University of Bordeaux.

Fagin, R. and J. Halpern (1988). Belief, awareness, and limited reasoning. *Artificial Intelligence 34*(1), 39–76.

Fagin, R., J. Halpern, Y. Moses, and M. Vardi (1995). *Reasoning About Knowledge.* Cambridge, MA: MIT Press.

Feinberg, Y. (2007). Meaningful talk. In van Benthem et al. (2007), pp. 41–54.

Felscher, W. (2001). Dialogues as a foundation for intuitionistic logic. In D. Gabbay and F. Günthner (Eds.), *Handbook of Philosophical Logic*, Volume III, pp. 341–372. Dordrecht, The Netherlands: Kluwer Academic Publishers.

Fermüller, C. and O. Majer (2013). *Games and Many-Valued Logics.* Berlin: Springer-Verlag.

Fitting, M. (2011). Reasoning about games. *Studia Logica 99*(1–3), 143–169.

Flum, J., E. Grädel, and T. Wilke (Eds.) (2007), *Logic and Automata.* Volume 3 of *Texts in Logic and Games.* Amsterdam, The Netherlands: Amsterdam University Press.

Fontaine, G. (2010). *Modal fixpoint logic: Some model-theoretic questions.* Ph. D. thesis, Institute for Logic, Language and Computation, University of Amsterdam (UvA), Amsterdam, The Netherlands. ILLC Dissertation series DS-2010-09.

Fraïssé, R. (1954). Sur quelques classifications des systèmes de relations. *Publications des Sciences de l'Université d' Algérie, Série A 1*, 35–182.

Gabbay, D. (2008). Introducing reactive Kripke semantics and arc accessibility. In A. Avron, N. Dershowitz, and A. Rabinovich (Eds.), *Pillars of Computer Science*, Volume 4800 of *Lecture Notes in Computer Science*, pp. 292–341. Berlin: Springer-Verlag.

Gabbay, D. and J. Woods (Eds.) (2002). *Handbook of the Logic of Argument and Inference: The Turn Towards the Practical.* Amsterdam: Elsevier Science.

Galliani, P. (2009a). Dependence logic, coalitions and announcements. Working paper, ILLC, University of Amsterdam.

Galliani, P. (2009b). Probabilistic dependence logics. Second GASICS meeting, Aachen, Germany. Working paper, ILLC, University of Amsterdam.

Galliani, P. (2012a). Dynamic logics of imperfect information and transition semantics. Working paper, ILLC, University of Amsterdam.

Galliani, P. (2012b). *The dynamics of imperfect information.* Ph. D. thesis, Institute for Logic, Language and Computation, University of Amsterdam (UvA), Amsterdam, The Netherlands. ILLC Dissertation series DS-2012-07.

Gärdenfors, P. (1988). *Knowledge in Flux: Modeling the Dynamics of Epistemic States.* Cambridge, MA: MIT Press.

Gärdenfors, P. and H. Rott (1994). Belief revision. In *Handbook of Logic in Artificial Intelligence and Logic Programming*, Volume 4: Epistemic and Temporal Logics, pp. 35–132. Oxford, UK: Oxford University Press.

Geanakoplos, J. (1992). Common knowledge. *Journal of Economic Perspectives 6*(4), 53–82.

Geanakoplos, J. D. and H. M. Polemarchakis (1982). We can't disagree forever. *Journal of Economic Theory 28*(1), 192–200.

Gheerbrant, A. (2010). *Fixed-point logics on trees.* Ph. D. thesis, Institute for Logic, Language and Computation, University of Amsterdam (UvA), Amsterdam, The Netherlands. ILLC Dissertation series DS-2010-08.

Ghosh, S. (2008). Strategies made explicit in dynamic game logic. In J. van Benthem and E. Pacuit (Eds.), *Proceedings of the Workshop on Logic and Intelligent Interaction*, organized as part of the European Summer School on Logic, Language and Information (ESSLLI) 2008, pp. 74–81. Hamburg, Germany.

Ghosh, S. and R. Ramanujam (2011). Strategies in games: A logic-automata study. In N. Bezhanishvili and V. Goranko (Eds.), *ESSLLI*, Volume 7388 of *Lecture Notes in Computer Science*, pp. 110–159. Berlin: Springer-Verlag.

Gierasimczuk, N. (2010). *Knowing one's limits. Logical analysis of inductive inference.* Ph. D. thesis, Institute for Logic, Language and Computation, University of Amsterdam (UvA), Amsterdam, The Netherlands. ILLC Dissertation series DS-2010-11.

Gierasimczuk, N., L. Kurzen, and F. R. Velázquez-Quesada (2009). Learning and teaching as a game: A sabotage approach. In He et al. (2009), pp. 119–132.

Gierasimczuk, N. and J. Szymanik (2011). A note on a generalization of the Muddy Children puzzle. In Apt (2011), pp. 257–264.

Gigerenzer, G., P. M. Todd, and the ABC Research Group (1999). *Simple Heuristics That Make Us Smart.* Oxford, UK: Oxford University Press.

Gilboa, I. (Ed.) (1998). *Proceedings of the 7th Conference on Theoretical Aspects of Rationality and Knowledge (TARK-98)*, San Mateo, CA: Morgan Kaufmann.

Gintis, H. (2000). *Game Theory Evolving.* Princeton, NJ: Princeton University Press.

Girard, J.-Y. (1993). Linear logic: A survey. In L. F. Bauer, W. Brauer, and H. Schwichtenberg (Eds.), *Proceedings of the International Summer School of Marktoberdorf*, NATO Advanced Science Institutes, Series F94, pp. 63–112. Berlin: Springer-Verlag. Also in P. De Groote (Ed.), *The Curry-Howard Isomorphism*, pp. 193–255, Département de Philosophie, Université Catholique de Louvain, Cahiers du Centre de Logique 8, Academia Press.

Girard, J.-Y. (1998a). On the meaning of logical rules, I: Syntax vs. semantics. Unpublished manuscript, University of Marseille, Luminy.

Girard, J.-Y. (1998b). On the meaning of logical rules, II: Multiplicatives and additives. Unpublished manuscript, University of Marseille, Luminy.

Girard, P. (2008). *Modal logic for belief and preference change.* Ph. D. thesis, Department of Philosophy, Stanford University, Stanford, CA. ILLC Dissertation Series DS-2008-04, University of Amsterdam.

Girard, P., F. Liu, and J. Seligman (2012). General dynamic dynamic logic. In T. Bolander, T. Bräuner, S. Ghilardi, and L. Moss (Eds.), *Proceedings of the 9th International Conference on Advances in Modal Logic (AiML'12)*, pp. 239–260. London: College Publications.

Goldblatt, R. (1992). Parallel action: Concurrent dynamic logic with independent modalities. *Studia Logica 51*(3–4), 551–578.

Goranko, V. (2003). The basic algebra of game equivalences. *Studia Logica 75*(2), 221–238.

Goranko, V., W. Jamroga, and P. Turrini (2013). Strategic games and truly playable effectivity functions. *Autonomous Agents and Multi-Agent Systems 26*(2), 288–314.

Goranko, V. and P. Turrini (2012). Non-cooperative games with preplay negotiations. *CoRR abs/1208.1718.*

Grädel, E., W. Thomas, and T. Wilke (Eds.) (2002). *Automata, Logics, and Infinite Games: A Guide to Current Research*, [outcome of a Dagstuhl seminar, February 2001], Volume 2500 of *Lecture Notes in Computer Science*, Berlin: Springer-Verlag.

Greenberg, J. (1990). *The Theory of Social Situations.* Cambridge, UK: Cambridge University Press.

Grossi, D. (2010). On the logic of argumentation theory. In W. van der Hoek, G. A. Kaminka, Y. Lespérance, M. Luck, and S. Sen (Eds.), *AAMAS'2010*, pp. 409–416. Richland, SC: IFAAMAS.

Grossi, D. (2012). Introduction to abstract argumentation theory. Lecture notes, ESSLLI 2012 Summer School, Opole.

Grossi, D. and P. Turrini (2012). Short sight in extensive games. In W. van der Hoek, L. Padgham, V. Conitzer, and M. Winikoff (Eds.), *AAMAS*, pp. 805–812. Richland, SC: IFAAMAS. Report ULCS-11-005, University of Luxemburg.

Guo, M. and J. Seligman (2012). Making choices in social situations. In Baltag et al. (2012), pp. 176–202.

Gurevich, Y. and L. Harrington (1982). Trees, automata, and games. In H. Lewis, B. Simons, W. Burkhard, and L. Landweber (Eds.), *STOC*, pp. 60–65. New York: ACM.

Gurevich, Y. and S. Shelah (1986). Fixed-point extensions of first-order logic. *Annals of Pure and Applied Logic 32*, 265–280.

Halpern, J. (2001). Substantive rationality and backward induction. *Games and Economic Behavior 37*(2), 425–435.

Halpern, J. (2003a). A computer scientist looks at game theory. *Games and Economic Behavior 45*(1), 114–131.

Halpern, J. (2003b). *Reasoning about Uncertainty.* Cambridge, MA: MIT Press.

Halpern, J. and R. Pass (2012). Iterated regret minimization: A new solution concept. *Games and Economic Behavior 74*(1), 184–207.

Halpern, J. and L. Rêgo (2006). Extensive games with possibly unaware players. In H. Nakashima, M. Wellman, G. Weiss, and P. Stone (Eds.), *AAMAS*, pp. 744–751. New York: ACM.

Halpern, J. and M. Vardi (1989). The complexity of reasoning about knowledge and time, I. Lower bounds. *Journal of Computer and System Sciences 38*(1), 195–237.

Hamami, Y. (2010). *The interrogative model of inquiry meets dynamic epistemic logics.* Master's thesis, Institute for Logic, Language and Computation, University of Amsterdam (UvA), Amsterdam, The Netherlands. ILLC Master of Logic Thesis Series MoL-2010-04.

Hamami, Y. and F. Roelofsen (2013). Logics of questions. Forthcoming special issue of *Synthese.*

Hamblin, C. L. (1970). *Fallacies.* London: Methuen.

Hansen, H., C. Kupke, and E. Pacuit (2009). Neighbourhood structures: Bisimilarity and basic model theory. *Logical Methods in Computer Science 5*(2), 1–38.

Hanson, S. O. (2001). Preference logic. In D. Gabbay and F. Günthner (Eds.), *Handbook of Philosophical Logic*, Volume IV, pp. 319–393. Dordrecht, The Netherlands: Kluwer Academic Publishers.

Harel, D. (1985). Recurring dominoes: Making the highly undecidable highly understandable. *Annals of Discrete Mathematics 24*, 51–72.

Harel, D., D. Kozen, and J. Tiuryn (2000). *Dynamic Logic.* Cambridge, MA: MIT Press.

Harrenstein, P. (2004). *Logic in conflict.* Ph. D. thesis, Institute of Computer Science, University of Utrecht.

He, X., J. F. Horty, and E. Pacuit (Eds.) (2009). *Proceedings Second International Workshop on Logic, Rationality, and Interaction (LORI-2009)*, Volume 5834 of *Lecture Notes in Computer Science*, Berlin: Springer-Verlag.

Heifetz, A. (Ed.) (2009). *Proceedings of the 12th Conference on Theoretical Aspects of Rationality and Knowledge (TARK-2009)*, New York: ACM.

Herzig, A. and E. Lorini (2010). A dynamic logic of agency, I: STIT, capabilities and powers. *Journal of Logic, Language and Information 19*(1), 89–121.

Hesse, H. (1943). *Das Glassperlenspiel.* Zürich: Fretz und Wasmuth.

Hintikka, J. (1973). *Logic, Language Games and Information. Kantian Themes in the Philosophy of Logic.* Oxford, UK: Clarendon Press.

Hintikka, J. and G. Sandu (1997). Game-theoretical semantics. In van Benthem & ter Meulen (1997), pp. 361–410.

Hirsch, R. and I. Hodkinson (2002). *Relation Algebras by Games.* Amsterdam: Elsevier.

Hodges, W. (1997). Compositional semantics for a language of imperfect information. *Logic Journal of the IGPL 5*(4), 539–563.

Hodges, W. (2001). Logic and games. Stanford Encyclopedia of Philosophy, `http://plato.stanford.edu/entries/logic-games/`.

Hodges, W. (2006). *Building Models by Games.* Mineola, NY: Dover Publications.

Hodkinson, I. and M. Reynolds (2006). Temporal logic. In Blackburn et al. (2006), pp. 655–720.

Hofbauer, J. and K. Sigmund (1998). *Evolutionary Games and Population Dynamics.* Cambridge, UK: Cambridge University Press.

Hollenberg, M. (1998). *Logic and bisimulation.* Ph. D. thesis, Philosophical Institute, Utrecht University, Utrecht, The Netherlands.

Holliday, W. (2012). Epistemic logic, relevant alternatives, and the dynamics of context. Department of Philosophy, Stanford University.

Holliday, W., T. Hoshi, and T. Icard (2011). Schematic validity in dynamic epistemic logic: Decidability. In van Ditmarsch et al. (2011), pp. 87–96.

Holliday, W., T. Hoshi, and T. Icard (2012). A uniform logic of information dynamics. In T. Bolander et al. (Eds.), *Proceedings of the 9th International Conference on Advances in Modal Logic (AiML'12)*, pp. 348–367. London: College Publications.

Holliday, W. and T. Icard (2010). Moorean phenomena in epistemic logic. In L. Beklemishev, V. Goranko, and V. Shehtman (Eds.), *Advances in Modal Logic*, pp. 178–199. College Publications.

Hopcroft, J., R. Motwani, and J. Ullman (2001). *Introduction to Automata Theory, Languages, and Computation.* Reading, MA: Addison-Wesley.

Horty, J. and N. Belnap (1995). The deliberative STIT: A study of action, omission, ability, and obligation. *Journal of Philosophical Logic 24*(6), 583–644.

Horty, J. F. (2001). *Agency and Deontic Logic.* Oxford, UK: Oxford University Press.

Hoshi, T. (2009). *Epistemic dynamics and protocol information.* Ph. D. thesis, Department of Philosophy, Stanford University, Stanford, CA. ILLC Dissertation Series DS-2009-08.

Hu, T. and M. Kaneko (2012). Critical comparisons between the Nash noncooperative theory and rationalizability. Department of Social Systems and Management, University of Tsukuba.

Huizinga, J. (1938). *Homo Ludens.* Haarlem, The Netherlands: Tjeenk Willink.

Hutegger, S. and B. Skyrms (2012). Emergence of a signaling network with "probe and adjust." In B. Calcott, R. Joyce, and K. Sterelny (Eds.), *Signaling, Commitment and Emotion.* Cambridge, MA: MIT Press.

Huth, M. and M. Ryan (2004). *Logic in Computer Science: Modelling and Reasoning about Systems.* Cambridge, UK: Cambridge University Press.

Icard, T. (2013). *Inference and active reasoning.* Ph. D. thesis, Department of Philosophy, Stanford University, Stanford, CA.

Immerman, N. and D. Kozen (1989). Definability with bounded number of bound variables. *Information and Computation 83*(2), 121–139.

Isaac, A. and T. Hoshi (2011). Taking mistakes seriously: Equivalence notions for game scenarios with off-equilibrium play. In van Ditmarsch et al. (2011), pp. 111–124.

Jacobs, B. (1999). *Categorical Logic and Type Theory.* Amsterdam: Elsevier Science.

Janin, D. and I. Walukiewicz (1995). Automata for the modal μ-calculus and related results. In J. Wiedermann and P. Hájek (Eds.), *MFCS*, Volume 969 of *Lecture Notes in Computer Science*, pp. 552–562. Berlin: Springer-Verlag.

Janin, D. and I. Walukiewicz (1996). On the expressive completeness of the propositional μ-calculus with respect to monadic second order logic. In U. Montanari and V. Sassone (Eds.), *CONCUR*, Volume 1119 of *Lecture Notes in Computer Science*, pp. 263–277. Berlin: Springer-Verlag.

Janssen, T. and F. Dechesne (2006). Signalling: A tricky business. In J. van Benthem, G. Heinzmann, and H. Visser (Eds.), *The Age of Alternative Logics: Assessing the Philosophy of Logic and Mathematics Today*, pp. 223–242. Dordrecht, The Netherlands: Kluwer.

Japaridze, G. (1997). A constructive game semantics for the language of linear logic. *Annals of Pure and Applied Logic 85*(2), 87–156.

Jiang, H. (2012). Modeling lifted preference formation. Department of Philosophy, Tsinghua University, Beijing.

Johansen, L. (1982). On the status of the Nash-type of noncooperative equilibrium in economic theory. *Scandinavian Journal of Economics 84*, 421–441.

Jones, N. D. (1978). Blindfold games are harder than games with perfect information. *Bulletin EATCS 6*, 4–7.

Kamlah, W. and P. Lorenzen (1967). *Logische Propädeutik.* Mannheim, Germany: Bibliographisches Institut.

Kaneko, M. (2002). Epistemic logics and their game-theoretic applications. *Economic Theory 19*, 7–62.

Kaneko, M. and N. Suzuki (2003). Epistemic models of shallow depths and decision making in games: Horticulture. *Journal of Symbolic Logic 68*(1), 163–186.

Kechris, A. (1994). *Classical Descriptive Set Theory.* Berlin: Springer-Verlag.

Keenan, E. L. and D. Westerståhl (1997). Generalized quantifiers in linguistics and logic. In van Benthem & ter Meulen (1997), pp. 837–893. Updated in the second revised and expanded edition 2010.

Kelly, K. T. (1996). *The Logic of Reliable Inquiry.* Oxford, UK: Oxford University Press.

Klein, D., F. G. Radmacher, and W. Thomas (2009). The complexity of reachability in randomized sabotage games. In F. Arbab and M. Sirjani (Eds.), *FSEN*, Volume 5961 of *Lecture Notes in Computer Science*, pp. 162–177. Berlin: Springer-Verlag.

Kohlberg, E. and J.-F. Mertens (1986). On the strategic stability of equilibria. *Econometrica 54*(5), 1003–1037.

Kolaitis, P. (2001). Combinatorial games in finite model theory. Lecture notes. ESSLLI 2001 Summer School, Helsinki.

Kooistra, S. (2012). *Logic in classical and evolutionary games.* Master's thesis, Institute for Logic, Language and Computation, University of Amsterdam (UvA), Amsterdam, The Netherlands. ILLC Master of Logic Thesis Series MoL 2012-18.

Kremer, P. and G. Mints (2007). Dynamic topological logic. In Aiello et al. (2007), pp. 565–606.

Kreutzer, S. (2002). *Pure and applied fixed-point logics.* Ph. D. thesis, Department of Computer Science, RWTH Aachen.

Kreutzer, S. (2004). Expressive equivalence of least and inflationary fixed-point logic. *Annals of Pure and Applied Logic 130*(1–3), 61–78.

Kupke, C. and Y. Venema (2005). Closure properties of coalgebra automata. In *LICS*, pp. 199–208. Washington, DC: IEEE Computer Society.

Kurtonina, N. and M. de Rijke (1997). Bisimulations for temporal logic. *Journal of Logic, Language and Information 6*(4), 403–425.

Kurz, A. and A. Palmigiano (2012). Product update on an intuitionistic basis. ILLC, University of Amsterdam & Institute for Informatics, University of Munich.

Kurzen, L. (2011). *Complexity in interaction.* Ph. D. thesis, Institute for Logic, Language and Computation, University of Amsterdam (UvA), Amsterdam, The Netherlands. ILLC Dissertation series DS-2011-10.

Lagerlund, H., S. Lindström, and R. Sliwinski (Eds.) (2006). *Modality Matters.* Number 53 in Uppsala Philosophical Studies. Uppsala: University of Uppsala.

Lang, J. and L. van der Torre (2008). From belief change to preference change. In M. Ghallab, C. D. Spyropoulos, N. Fakotakis, and N. M. Avouris (Eds.), *ECAI*, Volume 178 of *Frontiers in Artificial Intelligence and Applications*, pp. 351–355. Amsterdam: IOS Press.

Leitgeb, H. (2007). Beliefs in conditionals vs. conditional beliefs. *Topoi 26*(1), 115–132.

Lewis, D. (1969). *Convention: A Philosophical Study.* Cambridge, MA: Harvard University Press.

Lewis, D. (1973). *Counterfactuals.* Oxford, UK: Blackwell.

Leyton-Brown, K. and Y. Shoham (2008). *Essentials of Game Theory: A Concise Multidisciplinary Introduction.* Cambridge, UK: Cambridge University Press.

Libkin, L. (2004). *Elements of Finite Model Theory.* Texts in Theoretical Computer Science. Berlin: Springer-Verlag.

Liu, C. (2012). Logic for priority-based games with short sight. Department of Computer Science, Peking University.

Liu, F. (2008). *Changing for the better. Preference dynamics and agent diversity.* Ph. D. thesis, Institute for Logic, Language and Computation, University of Amsterdam (UvA), Amsterdam, The Netherlands. ILLC Dissertation series DS-2008-02.

Liu, F. (2009). Diversity of agents and their interaction. *Journal of Logic, Language and Information 18*(1), 23–53.

Liu, F. (2011). *Reasoning about Preference Dynamics*, Volume 354 of *Synthese Library.* Dordrecht: Springer Science Publishers.

Liu, F. and Y. Wang (2013). Reasoning about agent types and the hardest logic puzzle ever. *Minds and Machines 23*(1), 123–161.

Lorenzen, P. (1955). *Einführung in die Operative Logik und Mathematik.* Berlin: Springer-Verlag.

Lorini, E., F. Schwarzentruber, and A. Herzig (2009). Epistemic games in modal logic: Joint actions, knowledge and preferences all together. In He et al. (2009), pp. 212–226.

Löwe, B. (2003). Determinacy for infinite games with more than two players with preferences. ILLC Publication Series PP-2003-19, University of Amsterdam. To appear in *Journal of Applied Logic.*

Luce, R. D. and H. Raiffa (1957). *Games and Decisions.* New York: John Wiley.

Mann, A., G. Sandu, and M. Sevenster (2011). *Independence-Friendly Logic: A Game-Theoretic Approach.* Cambridge, UK: Cambridge University Press.

Marx, M. (2006). Complexity of modal logic. In Blackburn et al. (2006), pp. 139–179.

Maynard-Smith, J. (1982). *Evolution and the Theory of Games.* Cambridge, UK: Cambridge University Press.

McClure, S. (2011). Decision making. Lecture slides SS100, Introduction to Cognitive Science and Information Science, Stanford University.

Miller, J. S. and L. S. Moss (2005). The undecidability of iterated modal relativization. *Studia Logica 79*(3), 373–407.

Milner, R. (1989). *Communication and Concurrency.* Englewood Cliffs, NJ: Prentice Hall.

Milner, R. (1999). *Communicating and Mobile Systems: The Pi Calculus.* Cambridge, MA: MIT Press.

Minică, S. (2011). *Dynamic logic of questions.* Ph. D. thesis, Institute for Logic, Language and Computation, University of Amsterdam (UvA), Amsterdam, The Netherlands. ILLC Dissertation series DS-2011-08.

Mittelstaedt, P. (1978). *Quantum Logic.* Dordrecht, The Netherlands: Reidel.

Moller, F. and G. Birtwistle (Eds.) (1996). *Logics for Concurrency – Structure versus Automata. Proceedings 8th Banff Higher Order Workshop, August 27 – September 3, 1995,* Volume 1043 of *Lecture Notes in Computer Science,* Berlin: Springer-Verlag.

Moore, R. C. (1985). A formal theory of knowledge and action. In J. R. Hobbs and R. C. Moore (Eds.), *Formal Theories of the Commonsense World,* Volume 1 of *Ablex Series in Artificial Intelligence,* pp. 319–358. Westport, CT: Greenwood Publishing Group Inc.

Moschovakis, Y. N. (1974). *Elementary Induction on Abstract Structures.* Studies in Logic and the Foundations of Mathematics. Amsterdam: North-Holland.

Moschovakis, Y. N. (1980). *Descriptive Set Theory.* Amsterdam: North-Holland.

Mostowski, A. (1991). Games with forbidden positions. Technical Report 78, University of Danzig, Institute of Mathematics and Informatics.

Muller, D. E. and P. E. Schupp (1995). Simulating alternating tree automata by nondeterministic automata: New results and new proofs of the theorems of Rabin, McNaughton and Safra. *Theoretical Computer Science 141*(1–2), 69–107.

Németi, I. (1995). Decidable versions of first-order logic and cylindric-relativized set algebras. In L. Csirmaz, D. Gabbay, and M. de Rijke (Eds.), *Logic Colloquium '92, Veszprem, Hungary,* Studies in Logic, Language and Information, pp. 177–241. Stanford: CSLI Publications.

Netchitailov, I. (2001). *An extension of game logic with parallel operators.* Master's thesis, Institute for Logic, Language and Computation, University of Amsterdam (UvA), Amsterdam, The Netherlands. ILLC Master of Logic Thesis Series MoL-2001-02.

Niwinski, D. and I. Walukiewicz (1996). Games for the μ-calculus. *Theoretical Computer Science 163*(1–2), 99–116.

Nozick, R. (1981). *Philosophical Explanations.* Cambridge, MA: Harvard University Press.

Osborne, M. J. and A. Rubinstein (1994). *A Course in Game Theory.* Cambridge, MA: MIT Press.

Osherson, D. and S. Weinstein (2012). Preference based on reasons. *The Review of Symbolic Logic 5*(1), 122–147.

Pacuit, E. (2007). Some comments on history based structures. *Journal of Applied Logic 5*(4), 613–624.

Pacuit, E. and O. Roy (2011). A dynamic analysis of interactive rationality. In van Ditmarsch et al. (2011), pp. 244–257. New version in Baltag et al. 2012, 303–320.

Pacuit, E. and O. Roy (2013). *Interactive Rationality.* Forthcoming book. Departments of Philosophy, University of Groningen and University of Tilburg.

Pacuit, E. and S. Simon (2011). Reasoning with protocols under imperfect information. *The Review of Symbolic Logic 4*(3), 412–444.

Papadimitriou, C. H. (1996). Computational aspects of organization theory (extended abstract). In J. Díaz and M. J. Serna (Eds.), *ESA*, Volume 1136 of *Lecture Notes in Computer Science*, pp. 559–564. Berlin: Springer-Verlag.

Papadimitriou, C. M. (1994). *Computational Complexity.* Reading, MA: Addison-Wesley.

Paperno, D. (2011). *Non-standard coordination and quantifiers.* Ph. D. thesis, Department of Linguistics, University of California at Los Angeles, Los Angeles, CA.

Parikh, R. (1985). The logic of games. *Annals of Discrete Mathematics 24*, 111–140.

Parikh, R. and R. Ramanujam (2003). A knowledge based semantics of messages. *Journal of Logic, Language and Information 12*(4), 453–467.

Parikh, R., C. Tasdemir, and A. Witzel (2011). The power of knowledge in games. Working paper, CUNY Graduate Center & New York University.

Pauly, M. (2001). *Logic for social software.* Ph. D. thesis, Institute for Logic, Language and Computation, University of Amsterdam (UvA), Amsterdam, The Netherlands. ILLC Dissertation series DS-2001-10.

Perea, A. (2011). Belief in the opponents' future rationality. Working paper, Epicenter, Department of Quantitative Economics, University of Maastricht.

Perea, A. (2012). *Epistemic Game Theory: Reasoning and Choice.* Cambridge, UK: Cambridge University Press.

Prakken, H. (1997). *Logical Tools for Modelling Legal Argument. A Study of Defeasible Reasoning in Law.* Dordrecht, The Netherlands: Kluwer.

Rabin, M. (1968). Decidability of second-order theories and automata on infinite trees. *Bulletin American Mathematical Society 74*(5), 1025–1029.

Rahman, S. and H. Rückert (Eds.) (2001). *New Perspectives in Dialogical Logic.* Guest-edited volume of *Synthese* 127.

Ramanujam, R. (2008). Some automata theory for epistemic logic. Invited lecture at Workshop on Intelligent Interaction, ESSLLI Summer School, August 11–15, Hamburg.

Ramanujam, R. (2011). On growing types. Talk at Workshop on New Trends in Logical Dynamics, Department of Philosophy, Beijing University.

Ramanujam, R. and S. Simon (2008). A logical structure for strategies. In Bonanno et al. (2008), pp. 183–208.

Ramanujam, R. and S. Simon (2009). Dynamic logic of tree composition. In K. Lodaya, M. Mukund, and R. Ramanujam (Eds.), *Perspectives in Concurrency Theory*, pp. 408–430. Hyderabad: Universities Press; Boca Raton, FL: CRC Press.

Rebuschi, M. (2006). IF and epistemic action logic. In J. van Benthem, G. Heinzmann, and H. Visser (Eds.), *The Age of Alternative Logics*, pp. 261–281. Dordrecht, The Netherlands: Kluwer.

Reiter, R. (2001). *Knowledge in Action: Logical Foundations for Specifying and Implementing Dynamical Systems*. Cambridge, MA: MIT Press.

Restall, G. (2000). *An Introduction to Substructural Logics*. London: Routlege.

Rodenhäuser, B. (2001). *Updating epistemic uncertainty*. Master's thesis, Institute for Logic, Language and Computation, University of Amsterdam (UvA), Amsterdam, The Netherlands. ILLC Master of Logic Thesis Series MoL-2001-07.

Rohde, P. (2005). *On games and logics over dynamically changing structures*. Ph. D. thesis, Rheinisch-Westfälische Technische Hochschule, Aachen.

Rott, H. (2001). *Change, Choice and Inference: A Study of Belief Revision and Nonmonotonic Reasoning*. Number 42 in Oxford Logic Guides. Oxford Science Publications.

Rott, H. (2006). Shifting priorities: Simple representations for 27 iterated theory change operators. In Lagerlund et al. (2006), pp. 359–384.

Roy, O. (2008). *Thinking before acting: Intentions, logic, rational choice*. Ph. D. thesis, Institute for Logic, Language and Computation, University of Amsterdam (UvA), Amsterdam. ILLC Dissertation series DS-2008-03.

Roy, O. (2011). Qualitative deontic reasoning in social decisions. Working paper, Center for Mathematical Philosophy, Ludwig Maximilians University, Munich.

Roy, O., A. Anglberger, and N. Gratzl (2012). The logic of best actions from a deontic perspective. Center for Mathematical Philosophy, University of Munich.

Sadrzadeh, M., A. Palmigiano, and M. Ma (2011). Algebraic semantics and model completeness for intuitionistic public announcement logic. In van Ditmarsch et al. (2011), pp. 394–395.

Sadzik, T. (2006). Exploring the iterated update universe. Technical Report PP-2006-26, Institute for Logic, Language and Computation (ILLC), University of Amsterdam (UvA).

Sadzik, T. (2009). Beliefs revealed in Bayesian-Nash equilibrium. Department of Economics, New York University, New York.

Sandu, G. (1993). On the logic of informational independence and its applications. *Journal of Philosophical Logic 22*, 29–60.

Sandu, G. and J. Väänänen (1992). Partially ordered connectives. *Zeitschrift für Mathematische Logik und Grundlagen der Mathematik 38*, 361–372.

Savant, M. V. (2002). Ask Marilyn. *Parade Magazine*, March 31, *San Francisco Chronicle.*

Schaeffer, J. and H. J. van den Herik (2002). Games, computers, and artificial intelligence. *Artificial Intelligence 134*(1–2), 1–7.

Schelling, T. (1978). *Micromotives and Macrobehavior.* New York: Norton.

Scott, D. and C. Strachey (1971). Toward a mathematical semantics for computer languages. Technical Report PRG-6, Oxford, UK: Oxford University.

Scott, D. S. (1976). Data types as lattices. *SIAM Journal of Computation 5*(3), 522–587.

Segerberg, K. (1995). Belief revision from the point of view of doxastic logic. *Logic Journal of the IGPL 3*(4), 534–553.

Seligman, J. (2010). A hybrid logic for analyzing games. Lecture at the Workshop Door to Logic, Tsinghua University, Beijing.

Sevenster, M. (2006). *Branches of imperfect information: Logic, games, and computation.* Ph. D. thesis, Institute for Logic, Language and Computation, University of Amsterdam (UvA), Amsterdam. ILLC Dissertation series DS-2006-06.

Shoham, Y. and K. Leyton-Brown (2008). *Multiagent Systems: Algorithmic, Game-Theoretic, and Logical Foundations.* Cambridge, UK: Cambridge University Press.

Skyrms, B. (1996). *Evolution of the Social Contract.* Cambridge, UK: Cambridge University Press.

Skyrms, B. (2004). *The Stag Hunt and the Evolution of Social Structure.* Cambridge, UK: Cambridge University Press.

Stalnaker, R. (1999). Extensive and strategic form: Games and models for games. *Research in Economics 53*(2), 93–291.

Stirling, C. (1995). Modal and temporal logics for processes. In Moller & Birtwistle (1996), pp. 149–237.

Stirling, C. (1999). Bisimulation, modal logic and model checking games. *Logic Journal of the IGPL 7*(1), 103–124.

Szymanik, J. (2013). A note on the complexity of backward induction games. Department of Philosophy, University of Groningen & ILLC Amsterdam.

Tamminga, A. M. and B. P. Kooi (2008). Conditional obligations in strategic situations. In G. Boella, G. Pigozzi, M. P. Singh, and H. Verhagen (Eds.), *NORMAS*, pp. 188–200.

Tan, T. and S. Werlang (1988). A guide to knowledge and games. In Vardi (1988), pp. 163–177.

Thomas, W. (1992). Infinite trees and automaton-definable relations over omega-words. *Theoretical Computer Science 103*(1), 143–159.

Thomas, W. (1997). Ehrenfeucht games, the composition method, and the monadic theory of ordinal words. In J. Mycielski, G. Rozenberg, and A. Salomaa (Eds.), *Structures in Logic and Computer Science*, Volume 1261 of *Lecture Notes in Computer Science*, pp. 118–143. Berlin: Springer-Verlag.

Thomas, W. (2002). Infinite games and verification (extended abstract of a tutorial). In E. Brinksma and K. Larsen (Eds.), *CAV*, Volume 2404 of *Lecture Notes in Computer Science*, pp. 58–64. Berlin: Springer-Verlag.

Thompson, F. (1952). Equivalence of games in extensive form. Technical Report RM 759, The Rand Corporation.

Toulmin, S. E. (1958). *The Uses of Argument.* Cambridge, UK: Cambridge University Press.

Troelstra, A. (1993). *Lectures on Linear Logic.* CSLI Lecture Notes. Stanford, CA: CSLI Publications.

Troelstra, A. S. and H. Schwichtenberg (2000). *Basic Proof Theory* (2nd ed.). Number 43 in *Cambridge Tracts in Theoretical Computer Science.* Cambridge, UK: Cambridge University Press.

Troelstra, A. S. and D. van Dalen (1988). *Constructivism in Mathematics*, Volume 1–2. Amsterdam: North-Holland.

Uckelman, S. (2009). *Modalities in Medieval Logic.* Ph. D. thesis, Institute for Logic, Language and Computation, University of Amsterdam (UvA), Amsterdam. ILLC Dissertation series DS-2009-04.

Väänänen, J. (2007). *Dependence Logic.* Cambridge, UK: Cambridge University Press.

Väänänen, J. (2011). *Models and Games.* Cambridge, UK: Cambridge University Press.

van Benthem, J. (1990). Computation versus play as a paradigm for cognition. *Acta Philosophica Fennica 49*, 236–251.

van Benthem, J. (1991). *Language in Action: Categories, Lambdas and Dynamic Logic.* Amsterdam: North Holland. Paperback also published with the MIT Press, 1995.

van Benthem, J. (1996). *Exploring Logical Dynamics.* Stanford, CA: CSLI Publications.

van Benthem, J. (1997). Modal foundations for predicate logic. *Logic Journal of the IGPL 5*(2), 259–286.

van Benthem, J. (1999). Logic in games. Lecture notes, ILLC, University of Amsterdam & Department of Philosophy, Stanford University, Stanford, CA.

van Benthem, J. (2001a). Action and procedure in reasoning. *Cardozo Law Review 22*, 1575–1593.

van Benthem, J. (2001b). Games in dynamic-epistemic logic. *Bulletin of Economic Research 53*(4), 219–248.

van Benthem, J. (2002a). Extensive games as process models. *Journal of Logic, Language, and Information 11*(3), 289–313.

van Benthem, J. (2002b). Invariance and definability: Two faces of logical constants. In W. Sieg, R. Sommer, and C. Talcott (Eds.), *Reflections on the Foundations of Mathematics. Essays in Honor of Sol Feferman*, Number 15 in *ASL Lecture Notes in Logic*, pp. 426–446. ASL & Cambridge University Press.

van Benthem, J. (2003). Logic games are complete for game logics. *Studia Logica 75*, 183–203.

van Benthem, J. (2004a). De kunst van het vergaderen (The art of conducting meetings). In W. van der Hoek (Ed.), *Liber Amicorum John-Jules Charles Meijer 50*, pp. 5–7. Onderzoeksschool SIKS, Utrecht.

van Benthem, J. (2004b). Probabilistic features in logic games. In D. Kolak and J. Symons (Eds.), *Quantifiers, Questions, and Quantum Physics*, pp. 189–194. Berlin: Springer-Verlag.

van Benthem, J. (2004c). Update and revision in games. Lecture notes. ILLC University of Amsterdam & Philosophy Stanford University.

van Benthem, J. (2005a). An essay on sabotage and obstruction. In D. Hutter and W. Stephan (Eds.), *Mechanizing Mathematical Reasoning*, Volume 2605 of *Lecture Notes in Computer Science*, pp. 268–276. Berlin: Springer-Verlag.

van Benthem, J. (2005b). Open problems in logic and games. In S. Artëmov, H. Barringer, A. d'Avila Garcez, L. C. Lamb, and J. Woods (Eds.), *We Will Show Them!*, Volume 1, pp. 229–264. London: College Publications.

van Benthem, J. (2006a). The epistemic logic of IF games. In R. Auxier and L. Hahn (Eds.), *The Philosophy of Jaakko Hintikka*, Schilpp Series, pp. 481–513. Chicago: Open Court Publishers.

van Benthem, J. (2006b). Logical construction games. In T. Aho and A.-V. Pietarinen (Eds.), *Truth and Games: Essays in Honour of Gabriel Sandu*, pp. 123–138. Helsinki: Acta Philosophica Fennica 78.

van Benthem, J. (2007a). Cognition as interaction. In G. Bouma, I. Krämer, and J. Zwarts (Eds.), *Cognitive Foundations of Interpretation*, pp. 27–38. Amsterdam: KNAW.

van Benthem, J. (2007b). Computation as conversation. In B. Cooper, B. Löwe, and A. Sorbi (Eds.), *New Computational Paradigms: Changing Conceptions of What Is Computable*, pp. 35–58. Berlin: Springer-Verlag.

van Benthem, J. (2007c). Dynamic logic for belief revision. *Journal of Applied Non-Classical Logics 17*(2), 129–155.

van Benthem, J. (2007d). Rational dynamics. *International Game Theory Review 9*(1), 13–45. Erratum reprint: *9*(2), 377–409.

van Benthem, J. (2007e). Rationalizations and promises in games. In *Philosophical Trends*, pp. 1–6. Beijing: Chinese Academy of Social Sciences.

van Benthem, J. (2008). Games that make sense: Logic, language and multi-agent interaction. In K. Apt and R. van Rooij (Eds.), *New Perspectives on Games and Interaction*, Volume 4 of *Texts in Logic and Games*, pp. 197–209. Amsterdam: Amsterdam University Press.

van Benthem, J. (2009). The information in intuitionistic logic. *Synthese 167*(2), 251–270.

van Benthem, J. (2010a). A logician looks at argumentation theory. *Cogency 1*(2). Santiago de Chili: Universidad Diego Portales.

van Benthem, J. (2010b). *Modal Logic for Open Minds.* CSLI Lecture Notes. Stanford, CA: CSLI Publications.

van Benthem, J. (2010c). Two stances: Implicit versus explicit in logical modeling. Invited Lecture, Hans Kamp 70 Conference, University of Stuttgart. Appeared as ILLC Preprint PP-2013-02, University of Amsterdam.

van Benthem, J. (2011a). Belief update as social choice. In P. Girard, O. Roy, and M. Marion (Eds.), *Dynamic Formal Epistemology*, Volume 351 of *Synthese Library*, pp. 151–160. Dordrecht, The Netherlands: Springer Science Publishers.

van Benthem, J. (2011b). Exploring a theory of play. In Apt (2011), pp. 12–16.

van Benthem, J. (2011c). Logic games: From tools to models of interaction. In J. van Benthem, A. Gupta, and R. Parikh (Eds.), *Proof, Computation and Agency*, Volume 352 of *Synthese Library*, pp. 183–216. Dordrecht: Springer Science Publishers.

van Benthem, J. (2011d). Logic in a social setting. *Episteme 8*(3), 227–247.

van Benthem, J. (2011e). *Logical Dynamics of Information and Interaction.* Cambridge, UK: Cambridge University Press.

van Benthem, J. (2011f). On keeping things simple. Talk at Workshop on New Trends in Logical Dynamics, Department of Philosophy, Beijing University.

van Benthem, J. (2012a). In praise of strategies. In J. van Eijck and R. Verbrugge (Eds.), *Games, Actions, and Social Software*, Volume 7010 of *Lecture Notes in Computer Science*, pp. 96–116. Berlin: Springer-Verlag.

van Benthem, J. (2012b). Modeling reasoning in a social setting. ILLC Tech report PP-2012-21, University of Amsterdam. To appear in *Studia Logica*, special issue on Logic and Games, Fall 2013.

van Benthem, J. (2012c). The nets of reason. *Argumentation and Computation 3*(2–3), 83–86.

van Benthem, J. (2012d). Problems concerning qualitative probabilistic update. Institute for Logic, Language and Computation, University of Amsterdam.

van Benthem, J. (2012e). Some thoughts on the logic of strategies. Lecture at Lorentz Workshop on Modeling Reasoning about Strategies, Leiden 2012.

van Benthem, J. (2012f). Two logical faces of belief revision. In R. Trypuz (Ed.), *Segerberg Volume*, Outstanding Logicians Series. Dordrecht, The Netherlands: Springer Science Publishers. To appear, Fall 2013.

van Benthem, J. (2013). Reasoning about strategies. In B. Coecke, L. Ong, and P. Panangaden (Eds.), *Computation, Logic, Games, and Quantum Foundations*, Volume 7860 of *Lecture Notes in Computer Science*, pp. 336–347. Berlin: Springer-Verlag.

van Benthem, J. and G. Bezhanishvili (2007). Modal logics of space. In Aiello et al. (2007), pp. 217–298.

van Benthem, J., N. Bezhanishvili, and I. M. Hodkinson (2012). Sahlqvist correspondence for modal μ-calculus. *Studia Logica 100*(1–2), 31–60.

van Benthem, J. and C. Dégremont (2008). Bridges between dynamic doxastic and doxastic temporal logics. In G. Bonanno, B. Löwe, and W. van der Hoek (Eds.), *LOFT*, Volume 6006 of *Lecture Notes in Computer Science*, pp. 151–173. Berlin: Springer-Verlag.

van Benthem, J., J. Gerbrandy, T. Hoshi, and E. Pacuit (2009a). Merging frameworks for interaction. *Journal of Philosophical Logic 38*(5), 491–526.

van Benthem, J., J. Gerbrandy, and B. Kooi (2009b). Dynamic update with probabilities. *Studia Logica 93*(1), 67–96.

van Benthem, J. and A. Gheerbrant (2010). Game solution, epistemic dynamics and fixed-point logics. *Fundamenta Informaticae 100*(1–4), 19–41.

van Benthem, J., S. Ghosh, and F. Liu (2008). Modelling simultaneous games in dynamic logic. *Synthese 165*(2), 247–268.

van Benthem, J., S. Ghosh, and R. Verbrugge (Eds.) (2013). *Modeling Reasoning about Strategies*. Berlin: Springer-Verlag. To appear in *Springer Lecture Notes in Computer Science*, FoLLI Series.

van Benthem, J., P. Girard, and O. Roy (2009c). Everything else being equal: A modal logic for ceteris paribus preferences. *Journal of Philosophical Logic 38*(1), 83–125. Also appeared in The Philosopher's Annual 2009.

van Benthem, J., S. Ju, and F. Veltman (Eds.) (2007). *Proceedings First International Workshop on Logic, Rationality, and Interaction (LORI-1)*, London: College Publications.

van Benthem, J. and F. Liu (1994). Diversity of logical agents in games. *Philosophia Scientiae 8*(2), 163–178.

van Benthem, J. and F. Liu (2007). Dynamic logic of preference upgrade. *Journal of Applied Non-Classical Logics 17*(2), 157–182.

van Benthem, J. and M. Martínez (2008). The stories of logic and information. In P. Adriaans and J. van Benthem (Eds.), *Philosophy of Information*, Handbook of the Philosophy of Science, pp. 217–280. Amsterdam: North-Holland.

van Benthem, J. and S. Minică (2012). Toward a dynamic logic of questions. *Journal of Philosophical Logic 41*(4), 633–669.

van Benthem, J. and E. Pacuit (2006). The tree of knowledge in action: Towards a common perspective. In G. Governatori, I. Hodkinson, and Y. Venema (Eds.), *Advances in Modal Logic, 2006*, pp. 87–106. London: College Publications.

van Benthem, J. and E. Pacuit (2011). Dynamic logics of evidence-based beliefs. *Studia Logica 99*(1), 61–92.

van Benthem, J. and E. Pacuit (2012). Connecting logics of choice and change. To appear in T. Mueller (Ed.), *Belnap Volume*, Outstanding Logicians Series, Fall 2013, Dordrecht, The Netherlands: Springer Science Publishers.

van Benthem, J., E. Pacuit, and O. Roy (2011). Toward a theory of play: A logical perspective on games and interaction. *Games 2*(1), 52–86.

van Benthem, J. and A. ter Meulen (Eds.) (1997). *Handbook of Logic and Language*. Amsterdam: Elsevier Science. Expanded electronic second edition appeared with Elsevier Insights, 2011.

van Benthem, J., J. van Eijck, and B. Kooi (2006a). Logics of communication and change. *Information and Computation 204*(11), 1620–1662.

van Benthem, J., J. van Eijck, and V. Stebletsova (1994). Modal logic, transition systems and processes. *Journal of Logic and Computation 4*(5), 811–855.

van Benthem, J., S. van Otterloo, and O. Roy (2006b). Preference logic, conditionals and solution concepts in games. In Lagerlund et al. (2006), pp. 61–76.

van Benthem, J. and F. R. Velázquez-Quesada (2010). The dynamics of awareness. *Synthese (Knowledge, Rationality and Action) 177* (Supplement 1), 5–27.

van den Herik, H. J., H. Iida, and A. Plaat (Eds.) (2011). *Computers and Games – 7th International Conference, CG 2010, Kanazawa, Japan, September 24–26, 2010, Revised Selected Papers*, Volume 6515 of *Lecture Notes in Computer Science*, Berlin: Springer-Verlag.

van der Hoek, W. and M. Pauly (2006). Modal logic for games and information. In Blackburn et al. (2006), pp. 1077–1148.

van der Hoek, W. and M. Wooldridge (2003). Cooperation, knowledge, and time: Alternating-time temporal epistemic logic and its applications. *Studia Logica 75*(1), 125–157.

van der Meyden, R. (1996). The dynamic logic of permission. *Journal of Logic and Computation 6*(3), 465–479.

van Ditmarsch, H., W. van der Hoek, and B. Kooi (2007). *Dynamic Epistemic Logic*, Volume 337 of *Synthese Library Series*. Dordrecht, The Netherlands: Springer Science Publishers.

van Ditmarsch, H. P., J. Lang, and S. Ju (Eds.) (2011). *Proceedings Third International Workshop on Logic, Rationality, and Interaction (LORI-2011)*, Volume 6953 of *Lecture Notes in Computer Science*, Berlin: Springer-Verlag.

van Eijck, J. (2008). Yet more modal logics of preference change and belief revision. In K. Apt and R. van Rooij (Eds.), *New Perspectives on Games and Interaction*, Volume 4 of *Texts in Logic and Games*, pp. 81–104. Amsterdam: Amsterdam University Press.

van Eijck, J. (2012). PDL as a multi-agent strategy logic. CWI and ILLC Amsterdam. Poster presentation at TARK Chennai 2013.

van Emde Boas, P. (2011). Pregame communication in dynamic-epistemic logic. ILLC Crazy Talk, University of Amsterdam.

van Otterloo, S. (2005). *A strategic analysis of multi-agent protocols.* Ph. D. thesis, Department of Computer Science, University of Liverpool, Liverpool, UK. ILLC Dissertation series DS-2005-05.

van Otterloo, S., W. van der Hoek, and M. Wooldridge (2004). Knowledge as strategic ability. *Electronic Notes in Theoretical Computer Science 85*(2), 152–175.

van Rooij, R. (2003a). Quality and quantity of information exchange. *Journal of Logic, Language and Information 12*(4), 423–451.

van Rooij, R. (2003b). Questioning to resolve decision problems. *Linguistics and Philosophy 26*, 727–763.

van Rooij, R. (2004). Signalling games select Horn strategies. *Linguistics and Philosophy 27*, 493–527.

van Ulsen, P. (2000). *E.W. Beth als logicus, A scientific biography.* Ph. D. thesis, Institute for Logic, Language and Computation (ILLC), University of Amsterdam (UvA), Amsterdam. ILLC Dissertation series DS-2000-04.

Vardi, M. (Ed.) (1988). *Proceedings of the Second Conference on Theoretical Aspects of Reasoning about Knowledge (TARK-88)*, San Mateo, CA: Morgan Kaufmann.

Vardi, M. (1995). An automata-theoretic approach to linear temporal logic. In Moller & Birtwistle (1996), pp. 238–266.

Vardi, M. and P. Wolper (1986). An automata-theoretic approach to automatic program verification (preliminary report). In *LICS*, pp. 332–344. Los Alamitos, CA: IEEE Computer Society.

Velázquez-Quesada, F. R. (2011). *Small steps in dynamics of information.* Ph. D. thesis, Institute for Logic, Language and Computation, University of Amsterdam (UvA), Amsterdam. ILLC Dissertation series DS-2011-02.

Veltman, F. (1996). Defaults in update semantics. *Journal of Philosophical Logic 25*, 221–261. Also appeared in *The Philosopher's Annual*, 1997.

Venema, Y. (2003). Representation of game algebras. *Studia Logica 75*(2), 239–256.

Venema, Y. (2006). Algebras and co-algebras. In Blackburn et al. (2006), pp. 331–426.

Venema, Y. (2007). Lectures on the modal μ-calculus. Institute for Logic, Language and Computation, University of Amsterdam.

Venema, Y. (2012). Expressiveness modulo bisimilarity: A co-algebraic perspective. ILLC, University of Amsterdam.

Vervoort, M. (2000). *Games, walks, and grammars.* Ph. D. thesis, Institute for Logic, Language and Computation, University of Amsterdam (UvA), Amsterdam. ILLC Dissertation series DS-2000-03.

Vreeswijk, G. (2000). Representation of formal dispute with a standing order. *Artificial Intelligence and Law 8*(2–3), 205–231.

Walton, D. N. and E. C. W. Krabbe (1995). *Commitment in Dialogue: Basic Concepts of Interpersonal Reasoning.* Albany, NY: State University of New York Press.

Walukiewicz, I. (2002). Monadic second-order logic on tree-like structures. *Theoretical Computer Science 275*(1–2), 311–346.

Wang, Y. (2010). *Epistemic modelling and protocol dynamics.* Ph. D. thesis, Centrum voor Wiskunde & Informatica (CWI), Amsterdam. ILLC Dissertation series DS-2010-06.

Wang, Y. (2011). On axiomatizations of PAL. In van Ditmarsch et al. (2011), pp. 314–327.

Xu, M. (2010). Combinations of STIT and actions. *Journal of Logic, Language and Information 19*(4), 485–503.

Zanasi, F. (2012). *Expressiveness of monadic second-order logics on infinite trees of arbitrary branching degree.* Master's thesis, Institute for Logic, Language and Computation, University of Amsterdam (UvA), Amsterdam. ILLC Master of Logic Thesis Series MoL-2012-08.

Zermelo, E. (1913). Uber eine anwendung der mengenlehre auf die theorie des schachspiels. In E. W. Hobson and A. E. H. Love (Eds.), *Proceedings of the Fifth International Congress of Mathematicians*, Volume II, pp. 501–504. Cambridge, UK: Cambridge University Press.

Zhang, J. and F. Liu (2007). Some thoughts on Mohist logic. In van Benthem et al. (2007), pp. 85–102.

Zvesper, J. A. (2010). *Playing with information.* Ph. D. thesis, Institute for Logic, Language and Computation, University of Amsterdam (UvA), Amsterdam. ILLC Dissertation series DS-2010-02.

Index